Fourier, Laplace, and the Tangled Love Affair with Transforms

Sofen Kumar Jena

Fourier, Laplace, and the Tangled Love Affair with Transforms

The Art of Signal Synthesis and Analysis

Springer

Sofen Kumar Jena
Department of Process & Energy,
Faculty of Mechanical Engineering
Technical University of Delft
Delft, Zuid-Holland, The Netherlands

ISBN 978-3-031-80167-9 ISBN 978-3-031-80165-5 (eBook)
https://doi.org/10.1007/978-3-031-80165-5

Cover Artist: Trisuk Moharana

This Springer imprint is published by the registered company Springer Nature Switzerland AG
The registered company address is: Gewerbestrasse 11, 6330 Cham, Switzerland

i

"To the imaginary realms of my life, where dreams blur with reality yet are deeply felt, as real as the love that lingers in every whispered thought and untold story."

Joseph Fourier

Siméon Poisson

Pierre-Simon Laplace

Leonhard Euler

Daniel Bernoulli

d'Alembert

Oliver Heaviside

Thomas Bromwich

Paul Dirac

Laurent Schwartz

James Cooley

John Tukey

The Founding Architects of The Subject

Preface

One cannot understand ... the universality of laws of nature, the relationship of things, without an understanding of mathematics. There is no other way to do it.
—Richard P. Feynman

Welcome to "mathematical foundations for modern signal processing and systems analysis." This book is designed to provide a comprehensive introduction to fundamental mathematical concepts pivotal in understanding and solving problems in signal processing, systems analysis, and applied mathematics. Whether you are a student embarking on a journey into these fields, a professional seeking to refresh your knowledge, or a researcher looking for a solid foundation, this book aims to bridge theoretical concepts with practical applications. The term "synthesis" describes constructing something by combining its basic elements. In the context of the Fourier series and inverse Fourier transform, signals are formed by assembling fundamental sinusoidal components. Conversely, "analysis" refers to the mathematical breakdown of a complex entity to identify its essential building blocks, precisely the role Fourier analysis plays.

In addition to covering the essential mathematical tools, this book delves into the historical development of these subjects, providing context and understanding of how these concepts have evolved over time. By integrating historical perspectives with mathematical content, I hope to provide a solid foundation in these critical areas and an appreciation of the historical context that has shaped modern mathematical and engineering practices. By exploring the history behind these mathematical techniques, I hope the readers will gain insight into their origins, key contributors, and the milestones that have shaped modern practices.

This book has been thoughtfully curated to support a one-semester course at the senior undergraduate level for students in electrical engineering, electronics, communications, and information sciences. The material is tailored to introduce students to the fundamental concepts of signals and systems, providing them with the foundational tools necessary for deeper study in these fields. It covers essential topics that are critical for understanding how signals behave, are manipulated, and are analyzed in various real-world applications, such as communications systems, control theory, and digital signal processing.

Additionally, the content of this book is well-suited for postgraduate courses in applied mathematics. It can serve as a comprehensive text for an entire semester course at the master's level, where the focus may shift more toward the mathematical rigor behind signal processing, system theory, and their real-world applications. For mathematics students, the book is an essential precursor to more advanced topics, particularly in integral transforms and complex analysis, which are central to higher-level mathematical physics and engineering problems.

Moreover, for engineering students, the mathematical background provided in this book is invaluable preparation for advanced studies in areas like turbulence, hydrodynamic stability analysis, image processing, and data science. These disciplines require a deep understanding of differential equations, Fourier analysis, and complex systems, topics that are introduced and thoroughly explored in this text. Whether the student's goal is to work in cutting-edge research or practical, high-level engineering, this book lays the groundwork for further exploration of complex, dynamic systems and their behavior across a range of technical and scientific fields.

Here's a glimpse of what you will find:

Basics of Trigonometry: The basics of trigonometry are introduced, laying the foundation for the comprehension of periodic phenomena and oscillatory behaviors. The fundamentals of angular measurements and trigonometric functions are covered, along with an exploration of the historical development of trigonometry from ancient civilizations to its present-day applications. The properties, operations, and applications of complex numbers are also examined, as their importance in numerous mathematical models used in engineering and physics is emphasized. The historical context of their development is traced, highlighting their introduction by early mathematicians and their evolution into a vital tool in modern science.

Signals and Systems: A comprehensive understanding of signals and systems facilitates the analysis of how different signals interact with systems. Various signal types, system properties, and their interactions are explained, and historical perspectives on the development of signal theory and systems analysis are also provided. Significant contributions from key figures in the field are acknowledged.

Fourier Series and Fourier Analysis: Fourier series decompose complex periodic signals into simpler sinusoidal components. This section discusses the Fourier series and Fourier analysis principles and provides a historical overview of Jean-Baptiste Joseph Fourier's groundbreaking work and its impact on signal processing. The Fourier transform extends the concept of Fourier series to non-periodic functions. We explore the Fourier transform of generalized functions and trace the historical development of this mathematical tool, highlighting its application and significance in various scientific and engineering contexts.

Laplace Transform: The Laplace transform simplifies the analysis of linear time-invariant systems by transforming differential equations into algebraic equations. This section covers its properties and applications and provides historical insights into its development, including contributions from Pierre-Simon Laplace and its evolution into a central tool in control theory.

Discrete Fourier Transform (DFT) and Fast Fourier Transform (FFT): Discrete transforms are essential for processing discrete signals. We cover the DFT and the FFT, an efficient algorithm for computing the DFT, and a historical account of their development, including the contributions of Cooley and Tukey and their transformative impact on digital signal processing.

Convolution, Correlation, and Stochastic Analysis: Convolution and correlation describe how signals combine and interact. This section explores these operations, their mathematical properties, the historical development of these concepts, and their applications in signal processing.

Linear Systems: Linear systems are central to control theory and signal processing. We discuss their characteristics and how they can be modeled and analyzed, including a historical perspective on the development of linear systems theory and its key contributors.

Z Transform: The Z transform analyzes discrete-time signals and systems. This section covers its properties and applications, along with a historical look at its development and its role in the advancement of digital signal processing.

Multidimensional Fourier Transform: The multidimensional Fourier transform extends signal analysis to higher dimensions. We cover its principles and applications, providing historical context on its development and its significance in fields such as image processing and multidimensional data analysis.

Applications of Fourier Transform for Solution of PDEs and ODEs: Fourier transforms are crucial for solving partial differential equations (PDEs) and ordinary differential equations (ODEs). We discuss their applications and provide historical insights into how Fourier's techniques have been used to tackle complex problems in physics, engineering, and beyond.

This book aims to be both a reference and a guide, offering clear explanations, practical examples, and illustrative diagrams to support your learning journey. Each chapter builds upon the previous ones, ensuring a coherent and progressive understanding of the topics covered.

One challenge in the subject is the diversity of notation. Although various authors have used different notations, the core principles remain unchanged. In this text, the variable x is used to represent both time and space, a deliberate choice to facilitate the transition to higher-dimensional Fourier transforms. This uniformity simplifies the visualization of these transforms, making them appear analogous to their one-dimensional counterparts.

Happy reading, and may your exploration of these mathematical foundations be both enlightening and inspiring.

"**Perfection is reserved for almighty**"

Utmost effort has been taken to draft the manuscript accurately. Despite the best efforts to ensure accuracy, I believe some errors may have slipped through. I am confident that your feedback will help us improve the manuscript. Your feedback is invaluable in our pursuit of excellence. Feel free to email your comments and feedback (s.k.jena@tudelft.nl, sofenkumarjena@gmail.com).

Delft, The Netherlands Sofen Kumar Jena
October 2024

Acknowledgments

The search for truth is more precious than its possession.
—Albert Einstein

This book is dedicated to the mentors, teachers, and supporters who have guided me through an extraordinary journey of learning, discovery, and passion for applied mathematics.

I would like to express my deepest gratitude to Professor Sundar from IIT Madras, whose support was instrumental in my journey, beginning with the opportunity to attend the 2011 workshop on Computational Partial Differential Equations: Modelling and Simulations (CPDEMS). My sincerest thanks go to Professor Helmut Neunzert, a distinguished German industrial mathematics figure who awarded me a visiting student fellowship in Germany and recognized my performance at the workshop. My time at Fraunhofer ITWM (Fraunhofer Institute for Industrial Mathematics) was pivotal in developing my ability to tackle real-world problems through the lens of mathematics, equipping me with the tools to apply advanced mathematical concepts to practical challenges. The course "Special Functions in Mathematical Physics", taught by Professor Willi Freeden at TU Kaiserslautern, was life-changing. It gave me the confidence to see myself as an engineer within the world of applied mathematics. Professor Freeden's insights opened my eyes to the vast potential of applied mathematics, forging a connection between my mechanical engineering background and more profound studies in fluid mechanics and turbulence.

I owe an immense debt of gratitude to Professor Brad Osgood of Stanford University, whose lectures have a unique, almost musical quality for those with a love for mathematics. His online lectures on "The Fourier Transforms and Its Applications", which I avidly followed throughout my PhD, are not just mathematics but the arts. They did not merely enhance my understanding of the subject; they reshaped the very way I approach applied mathematics, instilling in me a deeper appreciation for the elegance and power of Fourier theory in solving complex problems. I am equally grateful to Professor Herbert Gross of Massachusetts Institute of Technology (MIT), whose exceptional online lecture series on complex analysis gave me more than just knowledge; it gave me clarity and a profound mastery of this essential field. Professor Gross's ability to demystify intricate concepts was invaluable, allowing me to approach the subtleties of complex analysis with newfound confidence and intellectual rigor.

Through their remarkable teaching, each professor has played a transformative role in my mathematical journey, making what once seemed like complex and abstract concepts not only comprehensible but deeply enriching. Their guidance has profoundly influenced my career as a fluid dynamicist. I've worked extensively on turbulence, regularly using Fourier transforms to analyze and characterize the chaotic structures known as eddies. Any aspiring fluid dynamicist who has encountered the seminal work "The Theory of Homogeneous Turbulence" by G.K. Batchelor will quickly recognize the importance of mastering the foundational concepts covered in this book, which serves as a critical stepping stone to such advanced studies.

During my tenure at Faurecia, I had the opportunity to lead courses on Fourier analysis tailored for practicing engineers. During this period, I became acutely aware of the confusion and challenges many engineers face when grappling with these ideas. I frequently encountered curious questions from colleagues, eager to understand the workings of the Fast Fourier Transform (FFT) and how this "miraculous" tool could extract meaningful information from seemingly chaotic, noisy data. Their curiosity and the need to demystify such powerful mathematical tools inspired me to compile and simplify the material I had been teaching. The seeds for this book were planted over eight years ago as I crafted those course materials to address real-world concerns.

This book represents the culmination of over a decade of work, research, and a deep-seated passion for analyzing turbulence signals. It is an extension of my continuous effort to make challenging mathematical methods, like Fourier analysis, accessible to both engineers and researchers, enabling them to unlock the mysteries hidden within complex, turbulent data. Whether you are a student, engineer, or researcher, I hope this book clarifies the intricacies of Fourier analysis and instils in you the same fascination and utility that has fueled my career.

I want to extend my deepest gratitude to my PhD advisors, Professor Amitava Sarkar and Professor Swarup Mahapatra, whose unwavering support and guidance have been instrumental in shaping my academic journey. Their mentorship refined my research approach and instilled in me the resilience and intellectual curiosity required to tackle complex problems in fluid dynamics. I am profoundly thankful for their wisdom and constant encouragement, which served as a steady source of motivation throughout my life.

I am equally indebted to all my previous employers, who provided me with the professional environment and opportunities to grow and apply my research in real-world settings. Their support allowed me to bridge the gap between theoretical knowledge and practical applications, a crucial step in my development as both a researcher and an engineer.

A special mention must go to Professor Rémi Manceau of the University of Pau, France, and Dr. Willem J. Havarkort of TU Delft, whose trust in my capabilities opened new avenues for my research. Their belief in me provided a pivotal platform in advancing my work. The opportunities they offered broadened my academic horizons and allowed me to collaborate with a network of brilliant minds, driving my research forward. Their trust, encouragement, and the invaluable experiences I gained under their mentorship have been integral to the progress of my research endeavors.

Ultimately, none of this achievement would have been possible without the unwavering love, encouragement, and support of my parents. I am profoundly grateful to my parents, Sabita Jena and Pramod Kumar Jena, who have been my steadfast guiding lights throughout this journey. I dedicate this work to them with all my heartfelt appreciation. I want to acknowledge the unwavering encouragement of my uncle, Late Sangram Keshari Nayak, who sadly left for his heavenly abode just a month before the publication of this book. His support has been a guiding light throughout this journey.

I extend my heartfelt gratitude to everyone who has been a cherished part of this beautiful journey. This book is a testament to all that I have learned, shaped by the myriad influences and inspirations I have encountered. Your invaluable contributions, whether through words of encouragement, thoughtful discussions, or shared insights, have woven a rich tapestry that has brought this work to fruition. It is not merely a reflection of my efforts; it embodies the collective spirit of those who have walked alongside me, illuminating the path and enriching my experience. Thank you for being an integral part of this endeavor; your presence has made all the difference.

October 2024 Sofen Kumar Jena

Contents

Chapter 1
Introduction

The profound study of nature is the most fertile source of mathematical discoveries. Not only does this study, by offering a definite goal to research, have the advantage of excluding vague questions and futile calculations, but it is also a sure means of molding analysis itself, and discerning those elements in it which it is still essential to know and which science ought to conserve. These fundamental elements are those which recur in all natural phenomena.

— Joseph Fourier

Abstract This chapter introduces the fundamentals of the Fourier series, beginning with a historical look at Joseph Fourier's work on heat distribution using trigonometric series, which paved the way for modern Fourier analysis. It covers essential mathematical concepts, including the properties of sinusoids, orthogonal functions, and the roles of odd and even functions in decomposing complex periodic signals. The chapter also discusses function series expansions, enabling the representation of complex functions as infinite sums of simpler terms. Key principles of complex analysis are introduced, with a focus on essential differentiation and integration formulas for complex variables. These mathematical foundations are established to support a deeper understanding of the Fourier series and its applications.

Keywords Sinusoids · Law of sines · Law of cosines · Periodic phenomena · Wavenumber · Frequency · Orthogonal functions · Odd and even functions · Series expansion · Complex variables · Cauchy-Riemann equation · Contour integration

1.1 Historical Development of the Subject

Jean-Baptiste Joseph Fourier (1768–1830) is recognized as one of the most outstanding French mathematical scientists. Fourier began his education at Pallais Elementary School, where he studied Latin and French. Initially interested in

S. K. Jena, *Fourier, Laplace, and the Tangled Love Affair with Transforms*,
https://doi.org/10.1007/978-3-031-80165-5_1

becoming a priest, he abandoned theology to pursue his remarkable aptitude for mathematics and mechanics. By the age of 13, he had thoroughly studied six volumes of Étienne Bézout's (1730–1783) "Cours de mathématique." In 1783, Fourier was awarded first prize for his study of Charles Bossut's (1730–1814) Mécanique. At a very young age, he presented a research paper on algebraic equations to the Académie Royale des Sciences. Subsequently, he enrolled at the École Royale Militaire to further his studies in science and mathematics. In 1790, Fourier returned to the Benedictine College, École Royale Militaire of Auxerre, as a mathematics teacher.

Fourier was arrested twice for his revolutionary activities during the tumultuous French Revolution in 1793–1794. Despite the profound pain and suffering he endured during his imprisonment, Fourier remained determined to fight for his life and freedom, yet he sought to avoid further political entanglements. In 1794, he was nominated to study at the École Normale in Paris, where he met leading French mathematicians of the time, including Joseph-Louis Lagrange (1736–1813), Pierre-Simon Laplace (1749–1827), and Gaspard Monge (1746–1818). By 1797, Fourier held the Chair of Analysis and Mechanics at the École Polytechnique. The following year, he joined Napoleon Bonaparte's Egyptian expedition as a scientific advisor, aiming to alleviate the severe social and political issues in Egypt. Napoleon founded the Institute d' Égypte in Cairo, appointing Monge as its president and Fourier as permanent secretary. During this period, Fourier compiled various archaeological, literary, and scientific findings in Egypt, resulting in the publication of "Description de l' Égypte", prefaced by Napoleon. Fourier returned to Paris in 1801 to resume his professorship at the École polytechnique. In 1802, he was appointed Prefect (Governor) of Isére in Grenoble, where he managed significant administrative duties. In recognition of his achievements, Napoleon awarded him the title of baron.

During his tenure in Grenoble, Fourier discovered the theory of heat conduction through his innovative mathematical ideas, experiments, and observations. He completed a seminal memoir titled "On the propagation of heat in solid bodies" and submitted it to the Academy of Sciences in Paris in 1807 for a research prize. A prestigious committee, including Lagrange, Laplace, Monge, Adrien-Marie Legendre (1752–1833), and Sylvestre François Lacroix (1765–1843), reviewed his work. While they acknowledged the importance and novelty of Fourier's research, they were divided over the quality and rigor of his mathematical treatment, particularly his use of trigonometric (Fourier) series expansions without sufficient justification. The controversy stemmed from Fourier's demonstration of the paradoxical property of equality in a finite interval between algebraic results of a totally different form. Additionally, the committee criticized his derivation of the heat equation in a continuous solid, which they felt was based on inadequate physical principles. Consequently, his memoir was rejected for the prize. Despite facing strong objections from some contemporaries, the academy recognized the significance of Fourier's work and encouraged him to resubmit it for a grand prize in 1812. Meanwhile, rivals such as Siméon Denis Poisson (1781–1840) and Jean-Baptiste Biot (1774–1862) expressed serious concerns and significant criticisms of Fourier's mathematical and physical approach to the theory of heat in solids.

Interestingly, Biot praised Poisson's less original work on the same topic. Prior to Fourier, Daniel Bernoulli (1700–1782) had proposed that a continuous function over the interval $(0, \pi)$ could be represented by an infinite series of sine functions in his solution to the problem of vibrating strings. This proposal was based on Bernoulli's physical intuition and faced severe criticism from contemporary mathematicians.

Fourier possessed remarkable talent and exceptional courage, defending his work in an unprecedented manner. His powerful rebuttal to Poisson and Biot's unfounded criticisms was outlined in his unpublished "Historical Précis" and in letters to Laplace around 1808–1809, which included a note on specific analytic expressions related to the equations of the theory of heat. Following this, all criticisms dissipated, leaving Fourier's reputation untarnished. He continued his research and resubmitted his revised memoir in 1811, adding new material on the cooling of infinite solids, radiant and terrestrial heat, and comparisons of his theory with experimental observations and equations of heat movement in fluids. Despite some lingering disagreements, Fourier was awarded the grand prize for his work in 1812, though his memoir was not recommended for publication in the memoirs of the academy. In 1822, Fourier became the permanent secretary of the mathematics and physics division of the Académie des Sciences. That same year, the académie published his seminal work, "La Théorie Analytique de la Chaleur (The Analytical Theory of Heat)", which was based on Isaac Newton's (1642–1727) law of cooling. According to this law, heat flow between two adjacent molecules is proportional to the very small temperature difference between them. Fourier's analytical theory of heat was highly mathematical, relying on a general equation of heat propagation with various initial and boundary conditions. Fourier became interested in the problem of heat diffusion because he wanted to determine the optimal depth for his cellar to maintain a perfect temperature for storing wine throughout the year. He aimed to understand how heat spreads across the surface and through the ground. An ideal cellar should maintain a constant temperature year-round. Since heat propagates through the Earth's crust like a wave, the hottest day on the surface wouldn't coincide with the hottest day inside the cellar (Cadeddu and Cauli, 2018). His fundamental heat conduction equation for temperature distribution in a three-dimensional Cartesian space (named after French mathematician René Descartes) is given by

$$\frac{\partial T}{\partial t} = \alpha \left(\frac{\partial^2 T}{\partial x^2} + \frac{\partial^2 T}{\partial y^2} + \frac{\partial^2 T}{\partial z^2} \right) \tag{1.1}$$

where α is the thermal diffusivity (a property of solid). This partial differential equation describing the conductive diffusion of heat has made significant contributions to both mathematics and physics. This equation is dimensionally homogeneous (dimensions of both the left and right sides of the equation are consistent). Fourier made important contributions to dimensional analysis. The above heat equation is first order in time derivatives and second order in space derivatives, unlike the dynamical equations of wave propagation for vibrating string, which is second order in both time and space derivatives. Further, Fourier claimed that any continuous or discontinuous function could be represented as a series of sines and cosines of multiples of the variable. Although this claim isn't entirely correct without further conditions, Fourier's insight that some discontinuous functions can be expressed as infinite series was revolutionary. The question of when a Fourier series converges has been crucial for centuries. Joseph-Louis Lagrange (1736–1813) identified specific cases of this idea of series expansion of sines and cosines, hinted at its general applicability, and did not explore it further. Johann Peter Gustav Lejeune Dirichlet (1805–1859) later provided a satisfactory proof with certain restrictive conditions. This work laid the groundwork for what is now known as the Fourier transform.

The Fourier series and transform will be explored in detail in the upcoming chapters. However, this chapter will first cover the essential mathematical concepts required for beginners in this subject.

1.2 The Sinusoids

Sinusoids are the fundamental components of Fourier analysis, with trigonometric sine and cosine functions being prime entities. The term trigonometry, which means the measurement of three angles, originated from studying the relationships between the sides and angles of triangles. Sumerian astronomers were pioneers in angle measurement, dividing circles into 360 degrees. Both Sumerians and Babylonians investigated the ratios of the sides of similar triangles and discovered properties of these ratios, though they did not develop a systematic method for calculating sides and angles. The ancient Nubians employed a similar method. In third century BC, Hellenistic mathematicians such as Euclid and Archimedes studied the properties of chords and inscribed angles in circles, proving theorems equivalent to modern trigonometric formulas, though their presentations were geometric rather than algebraic. In 140 BC, Hipparchus (190 BC–120 BC) of Nicaea created the first chord tables, akin to modern sine tables, and used them to solve problems in trigonometry and spherical trigonometry. In second century AD, the Greco-Egyptian astronomer Claudius Ptolemy (AD 100–AD 170) of Alexandria compiled detailed trigonometric tables known as Ptolemy's Table of Chords. The modern definition of the sine function first appeared in the *Surya Siddhanta*, with further elaboration by the Indian mathematician and astronomer Aryabhata (476 CE–550 CE) in fifth century AD. These Greek and Indian works were translated and expanded by medieval Islamic mathematicians. In AD 830, the Persian mathematician Habash al-Hasib al-Marwazi (AD 766–AD869) produced the first table of cotangents. By tenth century AD, Persian mathematician Abu al-Wafa al-Buzjani (940–998) used all six trigonometric functions. Nasir al-Din al-Tusi (1201–1274), a Persian polymath, is credited with establishing trigonometry as a distinct mathematical discipline. Knowledge of trigonometric functions and methods spread to Western Europe through Latin translations of Ptolemy's Almagest and the works of Persian and Arab astronomers like Al Battani (858–929) and Nasir al-Din al-Tusi. One of the earliest northern European works on trigonometry is De Triangulis by the

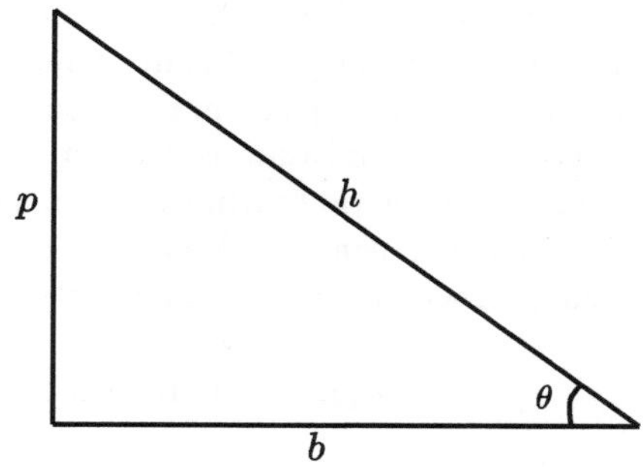

Fig. 1.1 Trigonometric relations in a right angle

fifteenth-century German mathematician Johannes Mûller von Kônigsberg (1436–1476) (also known as Regiomontanus). Around the same time, George of Trebizond (1395–1486) completed another translation of the *Almagest* from Greek into Latin. In the sixteenth century, trigonometry was still relatively unknown in Northern Europe, prompting Nicolaus Copernicus (1473–1543) to include two chapters on its basic concepts in his "De revolutionibus orbium coelestium". Driven by the need for navigation and accurate mapping, trigonometry became a major mathematics branch. Bartholomaeus Pitiscus (1561–1613) first used the term trigonometry in his work Trigonometria in 1595. Gemma Frisius (1508–1555) introduced the method of triangulation, which is still used in surveying today. Leonhard Euler (1707–1783) later fully incorporated complex numbers into trigonometry. Contributions of Scottish mathematicians James Gregory (1638–1675) in the seventeenth century and Colin Maclaurin (1698–1746) in the eighteenth century were significant in the development of trigonometric series.

In modern mathematics, trigonometric functions themselves have become central objects of study, a field known as analysis. Sines and cosine functions are used to analyze various periodic phenomena in science and engineering. The classical triangle approach is the most foundational way of introducing these functions. Figure 1.1 represents a right-angle triangle with the base b, perpendicular length p, and hypotenuse h. Physical aspects of these trigonometric functions are defined from the ratio of these lengths as follows:

$$\sin\theta = \frac{p}{h} \tag{1.2}$$

$$\cos\theta = \frac{b}{h} \tag{1.3}$$

$$\tan\theta = \frac{p}{b} \tag{1.4}$$

These relations measure an angle θ in terms of the sides of a right-angle triangle. Sine and cosine functions are dimensionless. Dimensions of the angle θ may be radians, gradians, or degrees. A sexagesimal system divides a circle into 360 basic units or degrees. The degree is divided into 60 minutes, which, in turn, is divided into 60 seconds. Degree ^{0} is the common measure of the angle. In the metrication system, each circle quadrant is divided into 100 grads and the whole circle into 400 gradians. Radial measure is much more useful for discussing analytical properties of the trigonometric functions. When a circle with unit radial length is unbent into a line element, then the circumference of the circle becomes 6.28319 ... or 2π. The circumference of a circle obtains 2π radians across the center.

$$\begin{aligned}
\text{Degrees to Radians: radians} &= \text{degrees} \times \frac{\pi}{180} \\
\text{Radians to Degrees: degrees} &= \text{radians} \times \frac{180}{\pi} \\
\text{Gradians to Radians: radians} &= \text{gradians} \times \frac{\pi}{200} \\
\text{Radians to Gradians: gradians} &= \text{radians} \times \frac{200}{\pi}
\end{aligned}$$

The Greek philosopher and mathematician Pythagoras (570 BC–500 BC) established the famous Pythagorean theorem, which describes the relationship between the sides of a right-angled triangle. The theorem states that in a right-angled triangle, the square of the length of the hypotenuse (the side opposite the right angle) is equal to the sum of the squares of the lengths of the other two sides. Mathematically, this is expressed as

$$h^2 = p^2 + b^2 \tag{1.5}$$

This theorem is fundamental in geometry and has numerous applications in various fields of science and engineering. Historically, the Pythagorean theorem was used in construction work in ancient civilizations such as Sumer, Babylonia, and Egypt. These early cultures applied the principles of the theorem to design and build structures, demonstrating their practical understanding of geometric relationships long before the formal proof attributed to Pythagoras. Pythagorean theorem in conjunction with Eqs. (1.2)–(1.4) give some important trigonometric relations.

$$\begin{aligned}
p^2 &= h^2 \sin^2\theta \\
b^2 &= h^2 \cos^2\theta
\end{aligned} \tag{1.6}$$

$$\sin^2\theta + \cos^2\theta = 1.0 \tag{1.7}$$

$$\tan\theta = \frac{\sin\theta}{\cos\theta} \tag{1.8}$$

Many other well-known results in trigonometry, such as the law of sines, the law of cosines, and formulas for the sine and cosine of sums of angles, can also be derived using straightforward geometric and algebraic techniques. These results extend the foundational relationships established by the Pythagorean theorem and provide essential tools for solving a wide range of problems in mathematics and its applications.

Law of Sines The law of sines relates the lengths of the sides of a triangle to the sines of its angles:

$$\frac{a}{\sin A} = \frac{b}{\sin B} = \frac{c}{\sin C} \tag{1.9}$$

where a, b, and c are lengths of sides opposite angles A, B, and C, respectively.

Law of Cosines The law of cosines generalizes the Pythagorean theorem to all triangles:

$$c^2 = a^2 + b^2 - 2ab\cos C \tag{1.10}$$

where a, b, and c are lengths of sides of the triangle and C is the angle opposite side c.

Sum of Angles Formulas The sine and cosine of the sum of two angles can be derived as follows:

$$\begin{aligned} \sin(A+B) &= \sin A\cos B + \cos A\sin B \\ \cos(A+B) &= \cos A\cos B - \sin A\sin B \end{aligned} \tag{1.11}$$

These formulas and laws are crucial in trigonometry, allowing for the analysis and solution of more complex geometric and trigonometric problems.

Angles greater than 90° do not make physical sense within the context of a triangle. However, it is useful to generalize sine and cosine functions to accommodate angles greater than 90°. In Fig. 1.2, a circle of radius R is divided into four sections or quadrants, labelled from I to IV in a counterclockwise direction. The line drawn at an angle θ from the origin (the center of the circle) to the circumference is known as the radius vector. The angle θ is measured from the horizontal axis to the radius vector and is considered positive when swept in a counterclockwise direction. To construct the right triangle, a line is dropped from the tip of the radius vector to the horizontal axis, forming the triangle's opposite side (p). The value or length of this side is measured from the horizontal axis to the tip of the radius vector. A line drawn from the origin to the point of intersection of the p side and the horizontal axis forms the adjacent side (b) and completes the construction of the triangle. The length of this side is measured from the origin and can be positive or negative depending on the angle θ. Generalized sine and cosine functions of the angle θ are defined as in Eqs. (1.2)–(1.4), but now p and b are allowed to take on negative values. Note that the length of the radius vector is always considered positive. Depending on the angle θ, the right triangle can be located in any of the

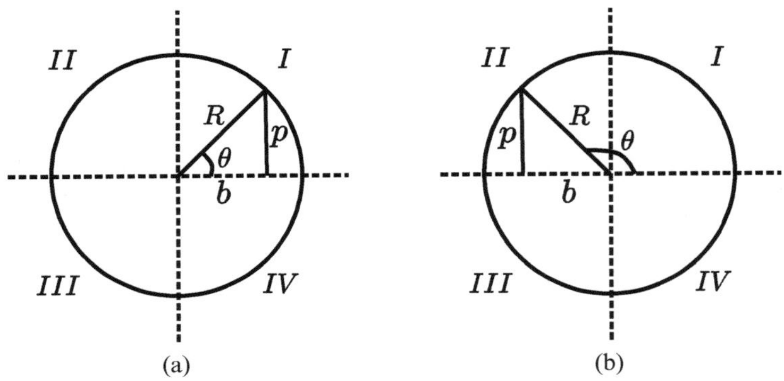

Fig. 1.2 Generalized trigonometric relations

four quadrants (I to IV). For example, Fig. 1.2b shows an angle θ that places the triangle in the second quadrant. It is important to note that for these generalized sine and cosine functions, the angle θ does not necessarily lie between the hypotenuse and the adjacent side of the triangle. Instead, it is always measured from the positive horizontal axis to the radius vector in a counterclockwise direction.

The triangle-solving equations remain applicable to triangles irrespective of their quadrant placement. For instance, in Fig. 1.2b, the angle between the hypotenuse and the adjacent side is $180° - \theta$. Solving this triangle and comparing the results with the generalized trigonometric functions gives

$$\begin{aligned} \sin(180° - \theta) &= \sin\theta \\ \cos(180° - \theta) &= -\cos\theta \end{aligned} \tag{1.12}$$

This construction can be extended one step further to illustrate the functional behavior of sine and cosine functions in Fig. 1.3. The length of the side p, plotted as $R\sin\theta$, is graphed against θ on the horizontal axis. This straightforward construction involves graphically projecting the side p onto the graph for different values of θ. This clearly demonstrates the functional behavior of the sine function. A similar construction is demonstrated for the cosine function on the vertical graph. The motion described by these functions is known as sinusoidal or simple harmonic motion. As the figure shows, the sine and cosine functions repeat their values every 2π radians or 360°, corresponding to one complete revolution of the radius vector. Mathematically, this is expressed as

$$\begin{aligned} \sin(\theta + 2\pi) &= \sin\theta \\ \cos(\theta + 2\pi) &= \cos\theta \end{aligned} \tag{1.13}$$

Figure 1.4 shows a circle with unit radius. The arc length $P - A - P' = 2\theta$. The length $P - P' = 2\sin\theta$ and $R - R' = 2\tan\theta$. From the above figure

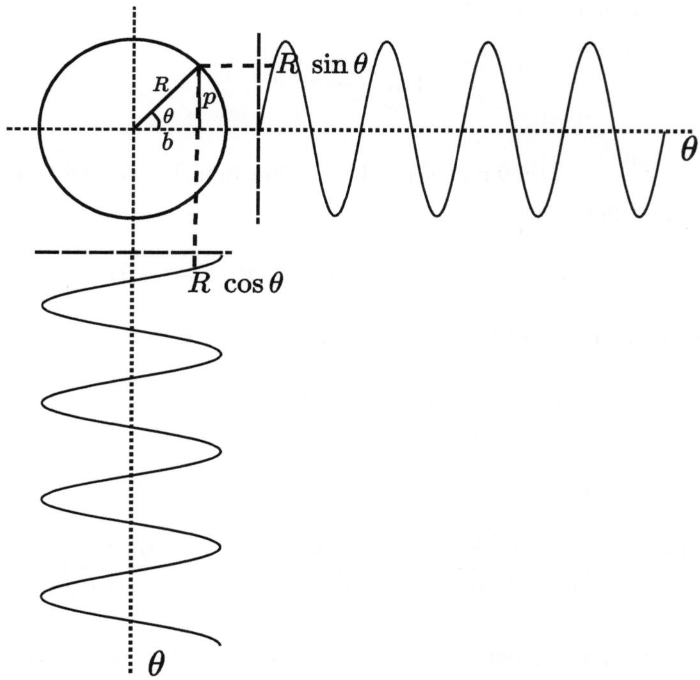

Fig. 1.3 Periodic behavior of trigonometric functions

Fig. 1.4 Similar triangles on unit circle

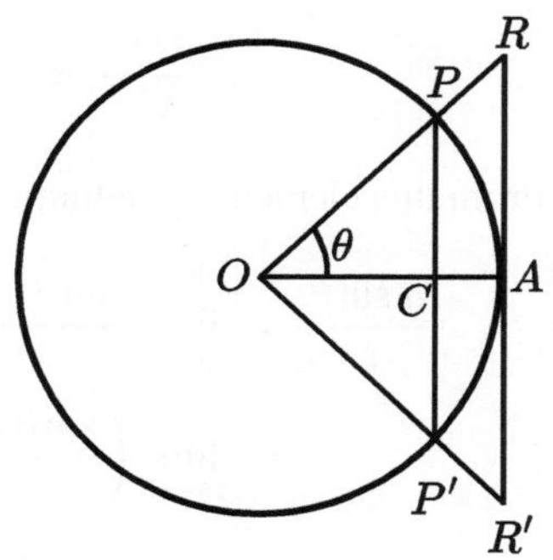

$$
\begin{aligned}
P - P' \le P - A - P' \le R - R' \\
\sin\theta \le \theta \le \tan\theta
\end{aligned}
\tag{1.14}
$$

Dividing $\sin\theta$ in both sides of Eq. (1.14) and taking the reciprocal results,

$$
\cos\theta \le \frac{\sin\theta}{\theta} \le 1 \tag{1.15}
$$

In the limiting condition $\theta \to 0,\ \cos\theta \to 1$.

$$\lim_{\theta\to 0} \frac{\sin\theta}{\theta} = 1 \tag{1.16}$$

The function $\frac{\sin\theta}{\theta} = sinc\theta$ represents the sinc function. The second required limit is obtained as follows:

$$\sin^2\theta = 1 - \cos^2\theta = (1+\cos\theta)(1-\cos\theta) \tag{1.17}$$

Now dividing both sides of Eq. (1.17) by $\theta(1+\cos\theta)$ results in

$$\frac{\sin^2\theta}{\theta(1+\cos\theta)} = \frac{1-\cos\theta}{\theta} \tag{1.18}$$

Thus,

$$\lim_{\theta\to 0} \frac{1-\cos\theta}{\theta} = \lim_{\theta\to 0} \frac{\sin^2\theta}{\theta(1+\cos\theta)} = \left(\lim_{\theta\to 0} \frac{\sin\theta}{\theta}\right)\left(\lim_{\theta\to 0} \frac{\sin\theta}{1+\cos\theta}\right) = (1)(0) = 0 \tag{1.19}$$

The above two limiting relationships are useful to calculate the derivative of the sine and cosine function. From elementary calculus, the definition of the derivative of a function f is

$$\frac{df(x)}{dx} = f'(x) = \lim_{\Delta x\to 0} \frac{f(x+\Delta x) - f(x)}{\Delta x} \tag{1.20}$$

From this elementary definition, the derivative of the sine function is

$$\begin{aligned}\frac{d\sin\theta}{d\theta} &= \lim_{\Delta\theta\to 0} \frac{\sin(\theta+\Delta\theta) - \sin\theta}{\Delta\theta}\\ &= \lim_{\Delta\theta\to 0} \left(\frac{\sin\theta\cos\Delta\theta + \sin\Delta\theta\cos\theta - \sin\theta}{\Delta\theta}\right)\\ &= \lim_{\Delta\theta\to 0} \sin\theta\left(\frac{\cos\Delta\theta - 1}{\Delta\theta}\right) + \lim_{\Delta\theta\to 0}\left(\frac{\sin\Delta\theta}{\Delta\theta}\right)\cos\theta\\ &= \cos\theta\end{aligned} \tag{1.21}$$

Analogous derivation from the elementary definition gives

$$\frac{d\cos\theta}{d\theta} = -\sin\theta \tag{1.22}$$

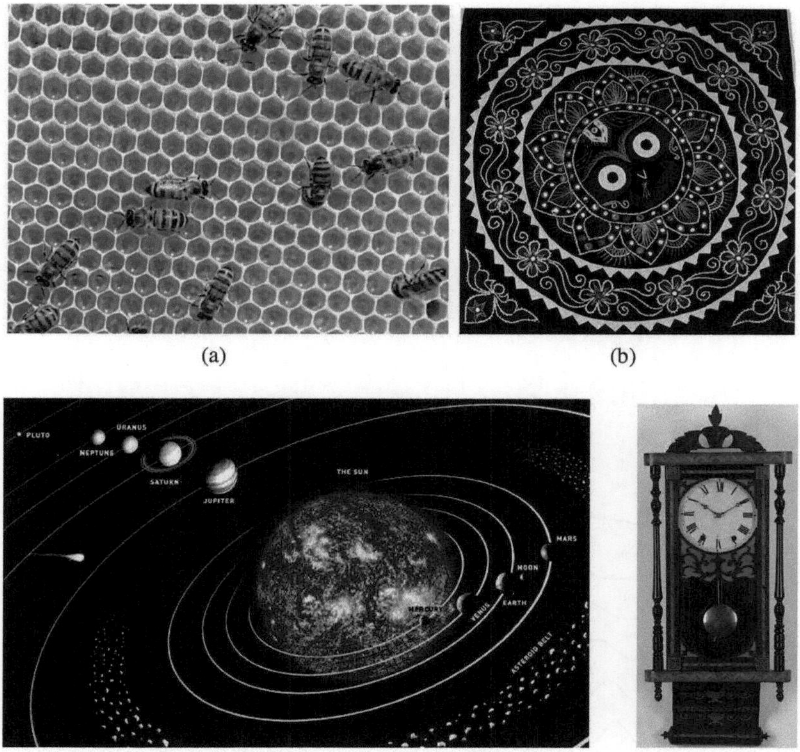

(a) (b) (c) (d)

Fig. 1.5 Some periodic phenomena in space and time: (**a**) Patterns in honeycomb, (**b**) Lord Jagannath in Pipili Chandua, (**c**) solar system, (**d**) pendulum watch. (**a**) Periodic in space through repetition. (**b**) Periodic in space through symmetry. (**c**) Periodic in time due to rotation. (**d**) Periodic in time due to oscillation

Given that the derivative of the sine function is the cosine function and the derivative of the cosine function is the negative sine function, we can observe that both sine and cosine functions have derivatives of all orders. In fact, their derivatives repeat in a cycle every four derivatives.

What is periodic? To get a suitable answer, we must watch the beautiful world around us, from the honeycomb to wall mounts in the drawing-room wall. Looking at the flower's petals, ACGT (adenine, cytosine, guanine, thymine) arrangement in DNA, and El Niño in the ocean, we have to appreciate the periodic things in the world. Periodic phenomena are of two types: periodic in space and periodic in time, periodic in space due to repetition of the pattern (Fig. 1.5a patterns in honeycomb that repeats after a certain distance) or due to symmetry in objects (Fig. 1.5b patterns in the Chandua). (The Chandua is used for wall decorative purposes and crafted in Pipli, a place close to lord Jagannath temple in Puri, India.) If an object repeats its motion with time, then it is periodic in time. The earth's and other planets' rotation

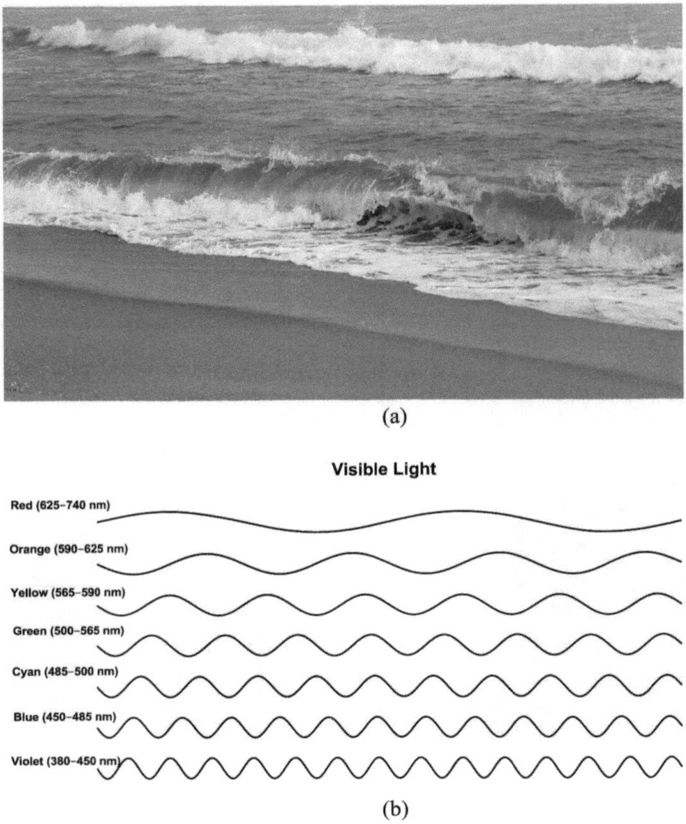

Fig. 1.6 Periodic phenomena in both space and time: (**a**) Water waves in the Bay of Bengal. (**b**) Visible light waves

around the sun is a periodic phenomenon in time. The oscillation of the pendulum in the wall clock is periodic in time. A few periodic phenomena exist in nature, such as wave motion, which is periodic in space and time. Standing on the sea beach, we can notice the temporal periodicity of the oceanic water waves. After an interval of time, these water waves wash over the feet. The height of these water waves is a periodic function of time, while water waves' recurring pattern is periodic in space. Visible light is composed of periodic waves of different colors (Fig. 1.6b). Sound is a longitudinal pressure wave that results from periodic compression and rarefaction of the air.

A real valued function $f(x)$ is periodic in an interval $[a, b]$, if it is defined for all x in the interval, except at some discrete points, and there exists some positive number P (period of the function) for which

$$f(x \pm P) = f(x) \tag{1.23}$$

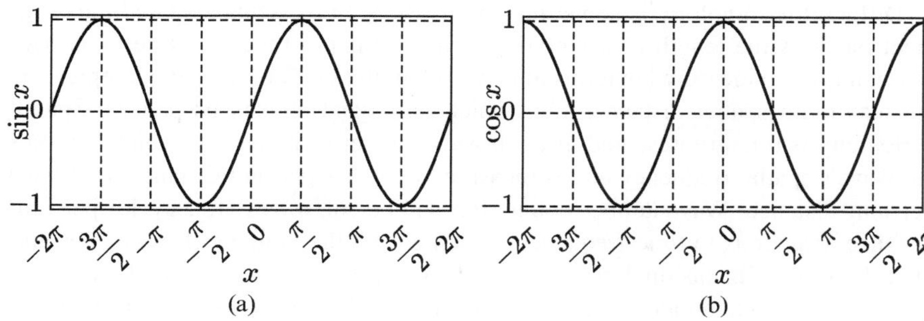

Fig. 1.7 Sinusoids are the simplest, smooth, and periodic functions

The smallest positive value of P is called the fundamental period of function f, and any integral multiple of the fundamental period is also a period.

$$f(x \pm nP) = f(x) \quad \text{for} \quad n = 0, 1, 2, \cdots \tag{1.24}$$

Sinusoids are the simplest mathematical periodic functions defined in the interval $[-\infty, \infty]$ having a period of 2π radians (Fig. 1.7). Both sine and cosine functions oscillate between $+1$ and -1. Both functions are continuous and infinitely differentiable in $[-\infty, \infty]$, which makes these functions unique in mathematical analysis. Let us now consider the functions $A\sin(Bx)$ and $A\cos(Bx)$, which are more general forms of sine and cosine functions. Here, the constant A is known as the amplitude, and B is referred to as the radial frequency. The amplitude A is a constant that scales the height of the sine and cosine waves, causing them to oscillate between $+A$ and $-A$. The radial frequency B measures how frequently the functions repeat. While $\sin(x)$ repeats every2π radians, $A\sin(Bx)$ repeats every $\frac{2\pi}{B}$ radians.

Since arguments of both sine and cosine functions are in radians, the radial frequency B has the units of radians per unit x. For instance, if x represents time in seconds, then B has units of radians per second. Whenever a function is periodic in time, its periodicity is measured by frequency (ν). Frequency is measured as the reciprocal of the time period having a unit of "Hertz".

$$\nu = \frac{1}{T} \tag{1.25}$$

Radians per second is a somewhat abstract unit, so frequency defines the number of cycles that a periodic function passes through a unit interval of time. Frequency multiplied by 2π is called circular or angular frequency ($\omega = 2\pi\nu$). The period T of $A\sin(2\pi\omega x)$ or $A\cos(2\pi\omega x)$ is defined as the number of x units required to complete one cycle or 2π radians.

Different terminology is sometimes used when x represents a spatial measurement (such as the length of a vibrating string). When a function is periodic over space, it is often measured in terms of its wavelength (λ). The wavelength represents the maximum spatial distance, after which the signal repeats its pattern. Spatial periodicity is often represented through wavenumber or repetency, which signifies the wave's spatial frequency and is measured in cycles per unit distance or radians per unit distance. Wavenumber counts the number of waves (full cycle) per unit distance analogous to time-frequency, which counts the number of cycles per unit time. In a multidimensional physical problem, the wave vector is called reciprocal space, and its magnitude is represented through the wavenumber. Wavenumber is an important factor in optics and spectroscopy studies. In the fluid instability study, wavenumber is frequently used. Spectroscopic wavenumber is defined as the reciprocal of the wavelength.

$$\tilde{\nu} = \frac{1}{\lambda} \tag{1.26}$$

Spectroscopic wavenumber multiplied by 2π gives angular wavenumber ($k = 2\pi\tilde{\nu}$), which measures the number of radians per unit distance and is more often used in instability and turbulence study. When a function is periodic both in space and time (regularly moving disturbance), it is defined in terms of wavelengths (or wavenumber) and frequency. The total distance travelled by a signal is the product of its velocity (u) times time.

$$\begin{aligned} \lambda &= u \times \frac{1}{\nu} \\ u &= \nu\lambda \end{aligned} \tag{1.27}$$

There exist periodic functions that repeat themselves after a time period of P as shown in Fig. 1.8, but these functions are not smooth in the entire domain.

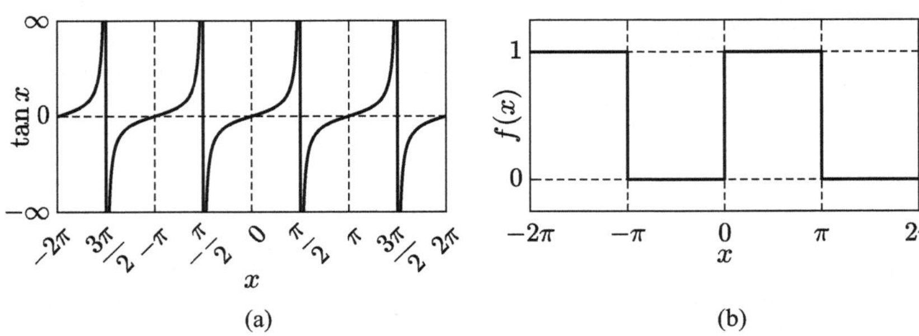

Fig. 1.8 Periodic but discontinuous functions. (**a**) $\tan x$. (**b**) Square wave

$\tan x$, $\cot x$ and square pulse are such examples. $\tan x$ has sharp discontinuity at the point $\pm\pi/2$, $\pm 3\pi/2$, ..., where its value is not defined. Such functions are known as continuous functions with finite discontinuity.

1.3 Orthogonal Functions

For mathematical analysis, functions are often treated similarly to geometric vectors. The inner product of two mathematical functions $f_1(x)$ and $f_2(x)$ in an open interval $[a, b]$ in $\mathbb{R}^3$ is expressed as

$$(f_1, f_2) = \int_a^b f_1(x) f_2(x) dx \quad (1.28)$$

Two geometric vectors in the Euclidean space $\mathscr{E}$ are considered orthogonal (perpendicular) when their inner product vanishes. In the functional space, zero inner product of two functions f_1 and f_2 in an interval $[a, b]$ results in orthogonality.

$$(f_1, f_2) = \int_a^b f_1(x) f_2(x) dx = 0 \quad (1.29)$$

Mathematically, the functional space possessing the real valued orthogonal functions is known as Hilbert or L^2 space. Hilbert space is named after German mathematician David Hilbert (1862–1943). Hilbert space extends the vector algebra and calculus methods from the finite Euclidean space to any higher dimensional finite or infinite functional space. Hilbert space is an abstract vector space possessing an inner product structure that allows length and angle to be measured.

A Hilbert space, named after the German mathematician David Hilbert (1862–1943), is a complete abstract vector space equipped with an inner product. This inner product allows for the measurement of length and angle, thereby extending vector algebra and calculus methods from finite-dimensional Euclidean spaces to higher-dimensional or even infinite-dimensional functional space. Hilbert spaces are significant in the context of real-valued orthogonal functions (L^2 space). The structure of a Hilbert space ensures that every Cauchy sequence converges within the space, making it a complete metric space under the distance function induced by the inner product.

Example 1.1 (Orthogonal Functions) Two functions $f_1(x) = x$ and $f_2(x) = x^2$ are said to be orthogonal in an interval $[-1, 1]$, since

$$(f_1, f_2) = \int_{-1}^{1} x.x^2 dx = \frac{1}{4} x^4 \Big|_{-1}^{1} = 0$$

In a functional space, orthogonal is the synonym of perpendicular, and they do not represent any geometric significance. A zero function is orthogonal to every function.

1.3.1 Set of Orthogonal Functions

A set of real values functions $\{f_1, f_2, f_3, ...\}$ in $\mathbb{R}^3$ are said to be orthogonal if

$$(f_m, f_n) = \int_a^b f_m(x) f_n(x) dx = 0, m \neq n \tag{1.30}$$

The norm represents the magnitude of vectors, and the dot product between two vectors represents the angle. Similarly, a function's square norm is obtained by taking the dot product of the function with itself in an interval. Squared norm of a function $f_n(x)$, in an interval $[a, b]$, is expressed as

$$\| f_n(x) \|^2 = \int_a^b f_n^2(x) dx \tag{1.31}$$

Norm or generalized length of a function in an interval $[a, b]$ is expressed as

$$\| f_n(x) \| = \sqrt{\int_a^b f_n^2(x) dx} \tag{1.32}$$

If $\{f_n(x)\}$ is an orthogonal set of function having $\| f_n(x) \| = 1$ in the interval $[a, b]$, then $\{f_n(x)\}$ is said to be orthonormal in $[a, b]$. Any set of orthogonal functions can be made orthonormal by dividing the functions with their respective norms, and the process is known as orthogonal function normalization.

Example 1.2 (Orthonormal Sets) Let us check the orthogonality of the functional set $\{1, \cos x, \cos 2x, ..., \cos nx\}$ in the interval $[-\pi, \pi]$ and find its orthonormal basis.

$$(f_m, f_n) = \int_{-\pi}^{\pi} \cos mx \cos nx dx$$

$$= \frac{1}{2}\int_{-\pi}^{\pi} [\cos (m+n)x \cos (m-n)x]dx$$

$$= \frac{1}{2}\left[\frac{\sin(m+n)x}{(m+n)} + \frac{\sin (m-n)x}{(m-n)}\right]_{-\pi}^{\pi}$$

$$= 0, m \neq n$$

This ensures that the function set $\{1, \cos x, \cos 2x, ..., \cos nx\}$ is orthogonal. Let us calculate the norm of functions

$$\| f_0 (x)\| = \sqrt{\int_{-\pi}^{\pi} 1dx} = \sqrt{2\pi}$$

For any $n > 0$

$$\| f_1 (x)\| = \sqrt{\int_{-\pi}^{\pi} \cos^2 xdx} = \sqrt{\frac{1}{2}\int_{-\pi}^{\pi} (1 + \cos 2x)dx} = \sqrt{\pi}$$

$$\| f_n (x)\| = \sqrt{\int_{-\pi}^{\pi} \cos^2 nxdx} = \sqrt{\frac{1}{2}\int_{-\pi}^{\pi} (1 + \cos 2nx)dx} = \sqrt{\pi}$$

The orthonormal basis of the functional set in the interval $[-\pi, \pi]$ becomes

$$\left\{\frac{1}{\sqrt{2\pi}}, \frac{\cos x}{\sqrt{\pi}}, \frac{\cos 2x}{\sqrt{\pi}}, ..., \frac{\cos nx}{\sqrt{\pi}}\right\}$$

Sinusoids in the interval $[-\pi, \pi]$ are orthogonal and forms an orthonormal basis

$$\int_{-\pi}^{\pi} \sin nx \sin mx dx = 0 \qquad (n \neq m)$$
$$\int_{-\pi}^{\pi} \cos nx \cos mx dx = 0 \qquad (n \neq m) \tag{1.33}$$
$$\int_{-\pi}^{\pi} \sin nx \cos mx dx = 0 \qquad (n \neq m \quad \text{or} \quad n = m)$$

1.4 Odd and Even Mathematical Functions

A mathematical function f is called even if

$$f(-x) = f(x) \tag{1.34}$$

and odd if

$$f(-x) = -f(x) \tag{1.35}$$

$f(x) = x^2$ is an even function, while $f(x) = x^3$ is an odd function.

Figure 1.9a represents the plot of $f(x) = f(-x)$; it can be clearly noticed from this that an even function is mirror symmetric about ordinate. $f(x) = -f(x)$ is plotted in Fig. 1.9b, where it can be marked that odd functions are antisymmetric about abscissa. The function $\cos x$ is an even function in the interval $[-\infty, \infty]$,

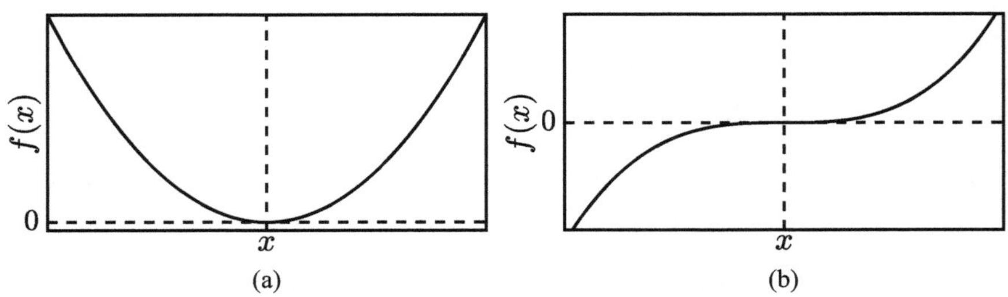

Fig. 1.9 (**a**) Even functions are symmetric about ordinate. (**b**) Odd functions are antisymmetric about abscissa

by definition $\cos(-x) = \cos x$. From Fig. 1.7b, it can be noticed that $\cos x$ is mirror symmetric around ordinate. $\sin x$ is an odd function, in $[-\infty, \infty]$ as $\sin(-x) = -\sin x$. Antisymmetric pattern of $\sin x$ can be noticed from Fig. 1.7a. There exists mathematical functions like e^x, which are neither even nor odd. Any mathematical functions in an interval $[-a, a]$ can be represented as a sum of odd and even function.

$$f(x) = \frac{1}{2}(f(x) + f(-x)) + \frac{1}{2}(f(x) - f(-x)) \tag{1.36}$$

with

$$g(x) = \frac{1}{2}(f(x) + f(-x)) \tag{1.37}$$

$$h(x) = \frac{1}{2}(f(x) - f(-x)) \tag{1.38}$$

The function $g(x)$ is even by definition $g(-x) = g(x)$, and $h(x)$ is odd as $h(-x) = -h(x)$. The product of two even or two odd functions is an even function, while the product of an even function and an odd function is odd. The sum of two even functions is even, while the sum of two odd functions is odd. If $f(x)$ is an even function, defined in an interval $[-a, a]$, then

$$\int_{-a}^{a} f(x)dx = 2\int_{0}^{a} f(x)dx \tag{1.39}$$

If $f(x)$ is an odd function, defined in an interval $[-a, a]$, then

$$\int_{-a}^{a} f(x)dx = 0 \tag{1.40}$$

It is important to notice from the odd and even properties of function that

$$\begin{aligned} \int_{-\pi}^{\pi} \cos x dx &= 2\int_{0}^{\pi} \cos x dx \\ \int_{-\pi}^{\pi} \sin x dx &= 0 \end{aligned} \tag{1.41}$$

1.5 Series Expansion of a Function

The idea of expanding a function in terms of series is a common practice in mathematics for a long time due to its advantages like term-by-term simple integration and differentiation of the function. The simplest series that represents the expansion of a function is power series expansion

$$f(x) = \sum_{n=0}^{\infty} a_n x^n \tag{1.42}$$

Here polynomial functions of x form the basis set $\{1, x, x^2, x^3, \cdot\}$. Exponential function can be represented using power series in the form

$$\begin{aligned} e^x &= 1 + x + \frac{x^2}{2!} + \frac{x^3}{3!} + \cdot + \frac{x^n}{n!} \\ &= \sum_{n=0}^{\infty} \frac{x^n}{n!} \end{aligned} \tag{1.43}$$

This power series expansion converges much faster with finite series (Fig. 1.10). But this power series expansion failed to represent any function having simple discontinuity. Other families of series expansion, analogous to power series like the Taylor series (named after English mathematician and barrister Brook Taylor (1685–1731)) and Maclaurin series (named after Scottish mathematician Colin Maclaurin (1698–1746)), also suffer from the same limitations. A complicated function often needs a relatively simple series representation that accommodates finite jump discontinuity in science and engineering problems.

The orthonormal basis of functions is very similar to the Cartesian basis vectors $\{\hat{i}, \hat{j}, \hat{k}\}$, which are used to define any vector $u_1\hat{i} + u_2\hat{j} + u_3\hat{k}$ in 3D Cartesian system. Analogous to vector space, any arbitrary function f, defined in

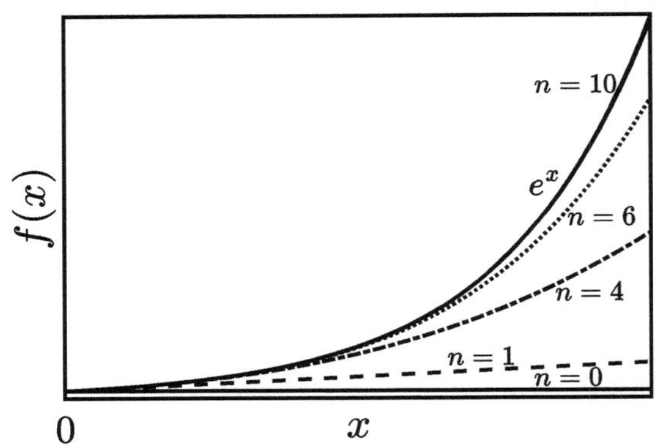

Fig. 1.10 Polynomial expansion of exponential function

an interval $[a, b]$, can be expanded using the infinite orthonormal functional basis $\left\{\tilde{f}_1, \tilde{f}_2, \tilde{f}_3, .., \tilde{f}_n, ..\right\}$ as follows:

$$\begin{aligned} f(x) &= c_0\tilde{f}_0(x) + c_1\tilde{f}_1(x) + c_2\tilde{f}_2(x) + \dots\dots\dots\dots + c_n\tilde{f}_n(x) + \dots\dots \\ &= \sum_{n=0}^{\infty} c_n\tilde{f}_n(x) \end{aligned} \tag{1.44}$$

The principle of orthogonality is used to determine the coefficients $\{c_0, c_1, c_2, .., c_n, ..\}$ in an interval $[a, b]$.

$$\begin{aligned} \int_a^b f(x)\,\tilde{f}_n(x)dx &= c_0\int_a^b \tilde{f}_0(x)\tilde{f}_n(x)dx + c_1\int_a^b \tilde{f}_1(x)\tilde{f}_n(x)dx + \dots. \\ &+ c_n\int_a^b \tilde{f}_n(x)\tilde{f}_n(x)dx + \dots\dots \\ &= c_n\int_a^b \tilde{f}_n^{\,2}(x)dx \end{aligned} \tag{1.45}$$

From Eq. (1.45)

$$c_n = \frac{\int_a^b f(x)\,\tilde{f}_n(x)dx}{\int_a^b \tilde{f}_n^{\,2}(x)dx} \tag{1.46}$$

Using Eq. (1.46), in Eq. (1.44) becomes

$$f(x) = \sum_{n=0}^{\infty} \frac{\int_a^b f(x)\,\tilde{f}_n(x)\,dx}{\int_a^b \tilde{f}_n^2(x)\,dx}\tilde{f}_n(x) \tag{1.47}$$

A set of real valued functions $\{f_1, f_2, f_3, \dots.\}$ in $\mathbb{R}^3$ is also said to be orthogonal in the interval $[a, b]$, with respect to a weight function $w(x)$.

$$\int_{a}^{b} w(x) f_m(x) f_n(x)\, dx = 0, m \neq n \tag{1.48}$$

Weight functions are usually positive $w(x) > 0$. In example 1.2, $w(x) = 1$. Equation (1.46) represents the orthogonal basis for expansion of function $f(x)$ in the interval $[a, b]$. Existence of integral in (1.46) is not yet discussed; hence, for the time being, the reader should assume that such orthonormal basis exists and function $f(x)$ uniformly converges in the interval $[a, b]$. The sufficient condition of convergence will be discussed in the next chapter.

1.6 Complex Variables

What exactly is a complex number, and when were they discovered and proven to exist? The history of complex numbers spans centuries. A square root of a negative number cannot be defined in the traditional sense, yet these numbers frequently arise in mathematical studies despite seeming nonsensical in everyday contexts. The earliest references to imaginary numbers are found in Italian cubic equations.

Niccolò Fontana Tartaglia (1500–1557), an Italian mathematician, was among the first to explore such numbers. Known as a mathematical prodigy, Tartaglia developed a secret technique to solve certain cubic equations using a Diophantine approach. His success in the 1535 mathematical contest at the University of Bologna, where he defeated senior Italian mathematician Scipione del Ferro (1465–1526), marked a significant achievement. Although Ferro had discovered a partial solution to the cubic equation earlier, Tartaglia provided the first general algebraic formula for solving these problems, including the square root of negative numbers. However, in 1539, Tartaglia shared his techniques with another Italian mathematician, Girolamo Cardano (1501–1576), who later published them without crediting Tartaglia in his book "Ars Magna". Cardano credited Ferro for a crude version of the solution, overshadowing Tartaglia's contribution, but Tartaglia remains recognized as the first to solve cubic equations involving negative square roots. In the competitive intellectual climate of sixteenth-century Italy, Tartaglia even encoded his solution in poetic form to prevent theft by rival mathematicians.

Another Italian mathematician, Rafael Bombelli (1526–1572), played a crucial role in the formal introduction of complex numbers. In his work on algebra, Bombelli embraced the concept of the square root of negative numbers and developed techniques for solving cubic equations involving complex numbers. He offered new insights into Cardano's formula, demonstrating how combinations of imaginary roots could yield real numbers, providing further evidence that any real number could be expressed as a complex number. Bombelli's research laid the groundwork for modern algebra and greatly influenced the study of cubic functions, although symbolic notation for imaginary numbers, as used today, had not yet been developed.

In the nineteenth century, French mathematician Augustin Louis Cauchy (1789–1857) and German mathematician Georg Friedrich Bernhard Riemann (1826–1866) advanced the analysis of complex numbers, shaping their role in modern mathematics. Bombelli's work, however, remains foundational, marking the birth of new algebra and expanding the mathematical understanding of cubic equations and complex numbers.

To understand the progression from natural numbers to complex numbers, let's break down the mathematical extensions introduced to solve specific equations and their implications.

Natural Numbers $\mathbb{N}$

From the principle of counting, the only known numbers are positive integers. The set of natural numbers, $\mathbb{N}$, includes positive integers:

$$\mathbb{N} = \{0, 1, 2, 3, \ldots\} \tag{1.49}$$

Integers $\mathbb{Z}$

To solve equations such as $x+1 = 0$, we need to introduce negative numbers, which extend the natural numbers to the set of integers:

$$\mathbb{Z} = \{\ldots, -3, -2, -1, 0, 1, 2, 3, \ldots\} \tag{1.50}$$

Rational Numbers $\mathbb{Q}$

For equations like $2x+3 = 0$, which simplify to $x = -\frac{3}{2}$, we need the set of rational numbers, which includes all fractions of the form $\frac{a}{b}$, where a and b are integers and $b \neq 0$:

$$\mathbb{Q} = \left\{\frac{a}{b} \mid a, b \in \mathbb{Z}, b \neq 0\right\} \tag{1.51}$$

Real Numbers $\mathbb{R}$

The set of real numbers $\mathbb{R}$ is required to solve equations like $x^2 - 2 = 0$, which simplifies to $x = \pm\sqrt{2}$. Real numbers include all the limits of convergent sequences of rational numbers and can be thought of as points on an infinitely long number line:

$$\mathbb{R} = \{x \mid x \text{ is a limit of a convergent sequence of rational numbers}\} \tag{1.52}$$

or

$$\mathbb{R} = \{x \mid x^2 \geq 0\} \tag{1.53}$$

Complex Numbers $\mathbb{C}$

To solve equations such as $x^2 + 1 = 0$, we need to extend real numbers to include solutions to polynomial equations that have no real solutions. The equation $x^2+1 = 0$ has solutions $x = \pm i$, where i is the imaginary unit defined by $i^2 = -1$. This leads to the set of complex numbers:

$$\mathbb{C} = \{z = x + iy \mid x, y \in \mathbb{R}, i^2 = -1\} \tag{1.54}$$

A complex number z consists of a real part x and an imaginary part y:

$$z = x + iy \tag{1.55}$$

French mathematician René Descartes named $i = \sqrt{-1}$ as an imaginary number. Realness is in the eyes of the beholder. Despite the historical nomenclature, imaginary and complex numbers have a mathematical existence as firm as that of real numbers. They are fundamental in the scientific description of the natural world, allowing solutions to all polynomial equations, even those without real solutions. Complex numbers are used extensively in engineering, physics, and applied mathematics, providing tools for solving a wide range of problems in these fields.

Often, a complex number is written as an ordered pair of real numbers (x, y), where the first number x is called the real part of the complex number and the second number y is called the imaginary part. Two complex numbers $z_1 = (x_1, y_1)$ and $z_2 = (x_2, y_2)$ are called equal $(z_1 = z_2)$ if and only if $x_1 = x_2$ and $y_1 = y_2$. The sum $(z_1 + z_2)$ and product $(z_1 z_2)$ are defined as

$$z_1 + z_2 = (x_1 + x_2, y_1 + y_2) \tag{1.56}$$

$$z_1 z_2 = (x_1 x_2 - y_1 y_2, x_1 y_2 + y_1 x_2) \tag{1.57}$$

Describing a complex number requires two real numbers, x and y, which are ordered, meaning (x, y) is not the same as (y, x). This ordering clarifies the definitions of equality, addition, and multiplication for complex numbers. Two complex numbers are equal if their respective real parts and imaginary parts are identical. For instance, $z_1 = (1, 2)$ and $z_2 = (2, 1)$ are not equal. To add two complex numbers, we add their real parts and their imaginary parts separately. For example, if $z_1 = (1, 2)$ and $z_2 = (6, 9)$, then

$$z_1 + z_2 = (7, 11)$$

Based on definitions of equality and addition, one might expect that multiplication of complex numbers involves multiplying the real parts and imaginary parts separately. However, this is not a practical definition for multiplication. Instead, we use a specific algorithm that combines the real and imaginary parts in a particular way. For example, if $z_1 = (1, 2)$ and $z_2 = (2, 1)$, then

$$z_1 z_2 = (0, 5)$$

Using the preceding definition of a complex number and the properties of real numbers, we can show that complex numbers obey the commutative, distributive, and associative laws. That is, for $z_1 = (x_1, y_1)$, $z_2 = (x_2, y_2)$, and $z_3 = (x_3, y_3)$, we have the following: Commutative law

$$\begin{aligned} z_1 + z_2 &= z_2 + z_1 \\ z_1 z_2 &= z_2 z_1 \end{aligned} \tag{1.58}$$

Associative law

$$\begin{aligned} z_1 + (z_2 + z_3) &= (z_1 + z_2) + z_3 \\ z_1 (z_2 z_3) &= (z_1 z_2) z_3 \end{aligned} \tag{1.59}$$

Distributive law

$$z_1 (z_2 + z_3) = z_1 z_2 + z_1 z_3 \tag{1.60}$$

The zero element (denoted as $\mathbf{0}$) is the complex number $(0, 0)$. This element has special properties. First, the sum of any complex number z and the zero element is z; that is,

$$z + \mathbf{0} = (x, y) + (0, 0) = (x + 0, y + 0) = (x, y) = z \tag{1.61}$$

Second, the product of any complex number and the zero element is the zero element; that is,

$$z\mathbf{0} = (x, y)(0, 0) = (x0 - y0, x0 + y0) = (0, 0) = 0 \tag{1.62}$$

The unit element (denoted as $\mathbf{1}$) is the complex number $(1, 0)$ and has the property that the product of any complex number z and the unit element is z; i.e.,

$$z1 = (x, y)(1, 0) = (x1 - y0, 1y + 0x) = (x, y) = z \tag{1.63}$$

Given any complex number $z = (x, y)$, the negative of z (denoted as $-z$) is defined as the complex number $(-x, -y)$ having the following property:

$$z + (-z) = (x, y) + (-x, -y) = (x - x, y - y) = (0, 0) = \mathbf{0} \tag{1.64}$$

Given any complex number $z = (x, y)$, not equal to $(0, 0)$, the inverse of z (denoted as z^{-1}) is defined to be the complex number

$$\left(\frac{x}{x^2+y^2}, \frac{-y}{x^2+y^2}\right) \tag{1.65}$$

This inverse element has the property

$$\begin{aligned} zz^{-1} &= (x, y)\left(\frac{x}{x^2+y^2}, \frac{-y}{x^2+y^2}\right) \\ &= \left(\frac{x^2}{x^2+y^2} + \frac{y^2}{x^2+y^2}, \frac{-xy}{x^2+y^2} + \frac{xy}{x^2+y^2}\right) \\ &= (1, 0) = 1 \end{aligned} \tag{1.66}$$

Using negative and inverse elements, we are able to define subtraction and division of complex numbers. Subtraction of the two complex numbers z_1 and z_2 (denoted as $z_1 - z_2$) is defined by

$$z_1 - z_2 = z_1 + (-z_2) \tag{1.67}$$

Similarly, division of the complex number z_1 by z_2, not equal to $(0, 0)$ (denoted as z_1/z_2), is defined by

$$\frac{z_1}{z_2} = z_1 z_2^{-1} \tag{1.68}$$

A real constant a can be written as the complex number $(a, 0)$. Thus, the multiplication of any complex number z by a real constant a results in

$$\mathrm{az} = (\mathrm{a}, 0)(x, y) = (\mathrm{a}x - 0y, 0x + \mathrm{a}y) = (\mathrm{a}x, \mathrm{a}y) \tag{1.69}$$

A term that frequently occurs in complex analysis is $1/i$. This can be rationalized as follows:

$$\frac{1}{i} = i^{-1} = \left(\frac{0}{0+1}, \frac{-1}{0+1}\right) = (0, -1) = -1(0, 1) = -i \tag{1.70}$$

The modulus, or absolute value, of a complex number $z = (x, y)$ is the non-negative real number given by

$$\|z\| = \left(x^2 + y^2\right)^{1/2} \tag{1.71}$$

The complex conjugate of a complex number $z = (x, y)$ is defined as $z^* = (x, -y)$. The complex number is geometrically represented in the Argand plane or Gauss plane. The real part of the complex number is projected in the horizontal axis (real axis), and the imaginary part is projected in the vertical axis (vertical axis). This rectangular form is a natural for a geometric, or vector, interpretation of a complex number. Under addition, they add like vectors.

A complex number $z = x + iy$ can be expressed as $z = r(\cos\theta + i\sin\theta)$

$$\begin{aligned} x &= r\cos\theta, \\ y &= r\sin\theta, \end{aligned} \tag{1.72}$$

where r is the absolute value or modulus of the complex number z

$$r^2 = x^2 + y^2 \tag{1.73}$$

and

$$\tan\theta = \frac{y}{x} \tag{1.74}$$

θ, the argument of z, is usually taken on the interval $0 \leq \theta \leq 2\pi$. This is known as the polar representation of a complex number. The multiplication of two complex numbers can be expressed more easily in polar coordinates—the magnitude or modulus of the product is the product of the two absolute values, or moduli, and the angle or argument of the product is the sum of the two angles, or arguments. In particular, multiplication by a complex number of modulus 1 acts as a rotation.

The complex conjugate z^* geometrically represents a mirror symmetric reflection of z with the real axis. z^* forms an angle $-\theta$ with the real axis that is mathematically expressed as

$$\begin{aligned} z &= x + iy = r(\cos\theta + i\sin\theta) \\ z^* &= x - iy = r[\cos(-\theta) + i\sin(-\theta)] \end{aligned} \tag{1.75}$$

Taylor series expansions of e^x, $\cos x$, and $\sin x$ are expressed as

$$e^x = 1 + x + \frac{x^2}{2!} + \frac{x^3}{3!} + \frac{x^4}{4!} + \frac{x^5}{5!} + \cdots \tag{1.76}$$

$$\cos x = 1 - \frac{x^2}{2!} + \frac{x^4}{4!} - \frac{x^6}{6!} + \cdots \tag{1.77}$$

$$\sin x = x - \frac{x^3}{3!} + \frac{x^5}{5!} - \frac{x^7}{7!} + \cdots \tag{1.78}$$

Substituting ix for x in Eq. (1.76)

$$\begin{aligned} e^{ix} &= 1 + ix - \frac{x^2}{2!} - \frac{ix^3}{3!} + \frac{x^4}{4!} + \frac{ix^5}{5!} + \cdots \\ &= \left(1 - \frac{x^2}{2!} + \frac{x^4}{4!} + \cdots\right) + i\left(x - \frac{x^3}{3!} + \frac{x^5}{5!} + \cdots\right) \end{aligned} \tag{1.79}$$

Using Eqs. (1.77) and (1.78) in Eq. (1.79), eighteenth-century Swiss mathematician Leonhard Euler (1707–1783) discovered one of the most profound relationship between the exponential function and trigonometric functions.

$$e^{ix} = \cos x + i \sin x \tag{1.80}$$

This elegant equation was first introduced by Euler in the 1740s. Euler's insight was to extend the exponential function, traditionally defined for real numbers, to complex numbers. By considering Taylor series expansions of e^x, $\cos x$, and $\sin x$, Euler showed that

$$e^z = \sum_{n=0}^{\infty} \frac{z^n}{n!} \tag{1.81}$$

where z is a complex number. Specifically, substituting $z = ix$ into the series and separating the real and imaginary parts yields the Euler's formula. Euler's formula not only unifies the exponential and trigonometric functions but also provides a deep connection between analysis and geometry. When $x = \pi$, Euler's formula leads to the celebrated Euler's identity:

$$e^{i\pi} + 1 = 0 \tag{1.82}$$

Euler's insight into the relationship between e, i, sin, cos, π, 1, and 0, in a single equation, remains a cornerstone of modern mathematical theory and a testament to the power of mathematical abstraction and generalization.

$$e^{-ix} = \cos x - i \sin x \tag{1.83}$$

In a polar coordinate notation, a complex number z and its conjugate z^* are written as

$$z = re^{i\theta} \tag{1.84}$$

$$z^* = re^{-i\theta} \tag{1.85}$$

The term $e^{i\theta}$ can be regarded as a unit vector rotated counterclockwise an angle θ from the real axis. Its real part is $\cos\theta$, and its imaginary part is $\sin\theta$.

$$\left\| e^{i\theta} \right\| = \left(\cos^2\theta + \sin^2\theta\right)^{1/2} = 1 \tag{1.86}$$

Euler's work on complex exponentials is often lauded for its beauty and has had far-reaching implications in many fields of mathematics and engineering, particularly in the study of differential equations, signal processing, and quantum mechanics. His discoveries have provided fundamental tools for understanding and describing oscillatory and wave phenomena. Euler's equations can be used to obtain a virtually unlimited number of useful results or identities. For example, addition and subtraction yield the following important equations:

$$\cos\theta = \frac{e^{i\theta} + e^{-i\theta}}{2}, \qquad \sin\theta = \frac{e^{i\theta} - e^{-i\theta}}{2i} \tag{1.87}$$

Euler's equations afford a simple technique for multiplication and division of two complex numbers. That is, if $z_1 = r_1 e^{(i\theta_1)}$ and $z_2 = r_2 e^{(i\theta_2)}$ ($r_2 \neq 0$), then

$$z_1 z_2 = r_1 r_2 e^{i(\theta_1+\theta_2)} \tag{1.88}$$

and

$$\frac{z_1}{z_2} = \frac{r_1}{r_2} e^{i(\theta_1-\theta_2)} \tag{1.89}$$

Thus, multiplying two complex numbers together in the exponential form is simply multiplying their magnitudes and adding their angles. Similarly, the division of two complex number is the division of their magnitudes and subtraction of their angles.

1.7 Differentiation and Integration of Complex Numbers

In complex analysis, differentiation and integration extend naturally to functions of a complex variable. The behavior of differentiation and integration in the complex plane is significantly different from the real case, particularly due to the properties of holomorphic (complex differentiable) functions. The derivative of a complex function $f(z)$, just like the derivative of a real function, represents the function's rate of change. However, complex differentiation introduces unique conditions and properties that differentiate it from real differentiation. Let's explore these differences and similarities in detail. A function $f(z)$ is said to be complex differentiable at a point z_0 if the following limit exists:

$$f'(z_0) = \lim_{\Delta z \to 0} \frac{f(z_0 + \Delta z) - f(z_0)}{\Delta z} \tag{1.90}$$

This resembles the real derivative definition

$$f'(x_0) = \lim_{\Delta x \to 0} \frac{f(x_0 + \Delta x) - f(x_0)}{\Delta x} \tag{1.91}$$

The derivative at a point measures the function's local behavior for both real and complex functions. The definition of the complex derivative is similar to the real derivative, but for a complex function to be differentiable, it must satisfy a stronger condition: the Cauchy-Riemann equation. The key difference arises like the complex plane, where Δz can approach zero from infinitely many directions (not just along the real axis). This makes complex differentiability more stringent than real differentiability. These are given by

$$\frac{\partial u}{\partial x} = \frac{\partial v}{\partial y}, \quad \frac{\partial u}{\partial y} = -\frac{\partial v}{\partial x} \tag{1.92}$$

where $f(z) = u(x, y) + iv(x, y)$ is expressed in terms of its real part $u(x, y)$ and imaginary part $v(x, y)$. A function satisfying these conditions is said to be holomorphic, and such functions possess smooth and well-behaved derivatives. Cauchy-Riemann equations are named after French mathematician Augustin Louis Cauchy (1789–1857) and German mathematician Georg Friedrich Bernhard Riemann (1826–1866). Cauchy-Riemann equations ensure that the limit defining the complex derivative is independent of the direction in which Δz approaches zero. If the Cauchy-Riemann equations are not satisfied, the complex derivative does not exist, even if partial derivatives of u and v exist. A holomorphic function, which is complex and differentiable, preserves angles and is locally conformal. This means that the function behaves in an angle-preserving manner in the complex plane. In contrast, real differentiability does not require this geometric property.

Contour integration evaluates integrals of complex functions over specific paths, called contours, in the complex plane. Unlike real integration, which is confined to the real number line, contour integration takes place over curves (or contours) in the complex plane, allowing for a broader analysis of functions that may exhibit singularities or more complicated behavior. A contour is a directed curve in the complex plane that evaluates an integral. If $f(z)$ is a complex-valued function and C is a contour (a smooth path or curve), the contour integral of $f(z)$ along C is defined as

$$\int_C f(z)\, dz \tag{1.93}$$

dz is the complex differential, representing movement along the contour. This is in contrast to real integrals, which only depend on limits of integration. For example, integrate the function $f(z) = 1/z$ around two different paths, Path 1, a line from $z = 1$ to $z = i$, then from $z = i$ to $z = -1$, and then from $z = -1$ back to $z = 1$; and Path 2, a circle around the origin. For path 1, because the contour does not enclose the origin, the integral is zero. However, for path 2, the contour does enclose the origin, and using the residue theorem (discussed later),

$$\int_{\Gamma_2} \frac{1}{z}\, dz = 2\pi i$$

Thus, the value of the integral depends on the path, showing the path dependence of complex integrals.

Cauchy's integral theorem is a critical theorem in complex analysis, and it states

$$\int_C f(z)\,dz = 0 \tag{1.94}$$

For any holomorphic function $f(z)$ and any closed curve C that does not enclose singularities of $f(z)$. This theorem is a consequence that holomorphic functions have no net change around a closed loop, provided they are analytic everywhere inside the loop. For example, applying Cauchy's integral theorem, if $f(z) = z^2 + 1$ is integrated around a closed contour C that doesn't enclose any singularities, Cauchy's theorem tells us that

$$\int_C (z^2 + 1)\,dz = 0$$

This is because $f(z)$ is holomorphic everywhere and the contour does not enclose any poles.

In practice, the contour is usually a closed path (such as a circle), and contour integration is often used in conjunction with the residue theorem and Cauchy's integral theorem to evaluate integrals of functions with singularities. Cauchy's residue theorem, often called the only residue theorem, generalizes Cauchy's integral theorem and is one of the most powerful tools in complex analysis. It allows the evaluation of contour integrals involving functions with singularities (poles). The theorem states that if a function $f(z)$ has isolated singularities inside a contour C, then

$$\int_C dz = 2\pi i \sum \text{Res}(f, z_k) \tag{1.95}$$

where the sum is over all singularities z_k inside Γ and $\text{Res}(f, z_k)$ is the residue of $f(z)$ at z_k, for example, the function $f(z) = \dfrac{1}{z(z-1)}$ and a contour Γ that encloses both $z = 0$ and $z = 1$. At $z = 0$, the residue is

$$\text{Res}(f, 0) = \frac{1}{1} = -1$$

and at $z = 1$,

$$\text{Res}(f, 1) = -\frac{1}{0} = 1.$$

Thus, by the residue theorem,

$$\int_\Gamma \frac{1}{z(z-1)}\,dz = 2\pi i(1-1) = 0$$

This shows how residues can be used to evaluate complex integrals, especially those involving singularities.

1.8 Useful Mathematical Formulae

1.8.1 Trigonometrical Relations

$$e^{\pm ix} = \cos x \pm i \sin x$$
$$\sin x = \frac{e^{ix} - e^{-iz}}{2i}$$
$$\cos x = \frac{e^{ix} + e^{-1z}}{2}$$
$$\sin^2 x + \cos^2 x = 1$$
$$\sec^2 x - \tan^2 x = 1$$
$$\mathrm{cosec}^2 x - \cot^2 x = 1$$
$$\sin(x \pm y) = \sin x \cos y \pm \cos x \sin y$$
$$\cos(x \pm y) = \cos x \cos y \mp \sin x \sin y$$
$$\sin x \cos y = \frac{1}{2}[\sin(x-y) + \sin(x+y)]$$
$$\sin x \sin y = \frac{1}{2}[\cos(x-y) - \cos(x+y)]$$
$$\cos x \cos y = \frac{1}{2}[(\cos(x-y) + \cos(x+y)]$$
$$\cos^2 x = \frac{1+\cos 2x}{2}$$
$$\sin^2 x = \frac{1-\cos 2x}{2}$$
$$\sin 2x = 2\sin x \cos x = \frac{2\tan x}{1+\tan^2 x}$$
$$\cos 2x = \cos^2 x - \sin^2 x = 2\cos^2 x - 1 = 1 - 2\sin^2 x$$
$$\tan(x \pm y) = \frac{\tan x \pm \tan y}{1 \mp \tan x \tan y}$$
$$1 - 2\sin^2 x = \frac{1-\tan^2 x}{1+\tan^2 x}$$
$$\tan 2x = \frac{2\tan x}{1-\tan^2 x}$$
$$\tan^2 x = \frac{1-\cos 2x}{1+\cos 2x}$$
$$\sin 3x = 3\sin x - 4\sin^3 x$$
$$\cos 3x = 4\cos^3 x - 3\cos x$$

$$\tan 3x = \frac{3\tan x - \tan^3 x}{1 - 3\tan^2 x}$$
$$\cot 3x = \frac{\cot^3 x - 3\cot x}{3\cot^2 x - 1}$$
$$\operatorname{cosec}^{-1} x = \sin^{-1}\left(\frac{1}{x}\right)$$
$$\cot^{-1} x = \tan^{-1}\left(\frac{1}{x}\right)$$
$$\sec^{-1} x = \cos^{-1}\left(\frac{1}{x}\right)$$
$$\sin^{-1} x + \cos^{-1} x = \frac{\pi}{2}$$
$$\tan^{-1} x + \cot^{-1} x = \frac{\pi}{2}$$
$$\operatorname{cosec}^{-1} x + \sec^{-1} x = \frac{\pi}{2}$$
$$\tan^{-1} x \pm \tan^{-1} y = \tan^{-1}\left(\frac{x \pm y}{1 \mp xy}\right)$$
$$\tan^{-1} x = \sin^{-1}\left(\frac{2x}{1 + x^2}\right)$$
$$3\tan^{-1} x = \tan^{-1}\left(\frac{3x - x^3}{1 - 3x^2} = \cos^{-1}\frac{1 - x^2}{1 + x^2}\right) = \tan^{-1}\left(\frac{2x}{1 - x^2}\right)$$
$$3\cos^{-1} x = \cos^{-1}\left(4x^3 - 3x\right)$$
$$3\sin^{-1} x = \sin^{-1}\left(3x - 4x^3\right)$$

1.8.2 Hyperbolic Relations

$$\sinh x = \frac{e^x - e^{-x}}{2}$$
$$\cosh x = \frac{e^x + e^{-x}}{2}$$
$$e^{\pm x} = \cosh x \pm \sinh x$$
$$\cosh^2 x - \sinh^2 x = 1$$
$$\operatorname{sech}^2 x + \tanh^2 x = 1$$
$$\coth^2 x - \operatorname{cosech}^2 x = 1$$
$$\sinh(x \pm y) = \sinh x \cosh y \pm \cosh x \sinh y$$
$$\cosh(x \pm y) = \cosh x \cosh y \pm \sinh x \sinh y$$
$$\sinh 2x = 2\sinh x \cosh x$$
$$\cosh 2x = \cosh^2 x + \sinh^2 x$$
$$\tanh 2x = \frac{2\tanh x}{1 + \tanh^2 x}$$
$$\cosh^2 x - 1 = 1 + 2\sinh^2 x$$
$$\sinh^{-1} x = \log\left|x + \sqrt{x^2 + 1}\right|$$

$$\cosh^{-1} x = \pm \log \left| x + \sqrt{x^2 - 1} \right|, \quad |x| \geq 1$$

$$\tanh^{-1} x = \frac{1}{2} \log \left| \frac{1+x}{1-x} \right|, \quad |x| < 1$$

$$\coth^{-1} x = \frac{1}{2} \log \left| \frac{x+1}{x-1} \right|, \quad |x| > 1$$

$$\operatorname{cosech}^{-1} x = \log \left| \frac{1 + \sqrt{1+x^2}}{x} \right|, \quad x \neq 0$$

$$\operatorname{sech}^{-1} x = \log \left| \frac{1 + \sqrt{1+x^2}}{x} \right|, \quad 0 < x \leq 1$$

1.8.3 Taylor Series Expansion

$$e^x = 1 + x + \frac{x^2}{2!} + \frac{x^3}{3!} + \cdots\cdots + \frac{x^n}{n!} + \cdots\cdots = \sum_{n=0}^{\infty} \frac{x^n}{n!}$$

$$\sin x = x - \frac{x^3}{3!} + \frac{x^5}{5!} - \frac{x^7}{7!} + \cdots + (-1)^{n-1} \frac{x^{2n-1}}{(2n-1)!} + \cdots = \sum_{n=0}^{\infty} (-1)^n \frac{x^{2n+1}}{(2n+1)!}$$

$$\cos x = 1 - \frac{x^2}{2!} + \frac{x^4}{4!} - \frac{x^6}{6!} + \cdots + (-1)^{n-1} \frac{x^{2n-2}}{(2n-2)!} + \cdots = \sum_{n=0}^{\infty} (-1)^n \frac{x^{2n}}{(2n)!}$$

$$\sinh x = x + \frac{x^3}{3!} + \frac{x^5}{5!} + \frac{x^7}{7!} + \cdots + \frac{x^{2n-1}}{(2n-1)!} + \cdots\cdots\cdots = \sum_{n=0}^{\infty} \frac{x^{2n+1}}{(2n+1)!}$$

$$\cosh x = 1 + \frac{x^2}{2!} + \frac{x^4}{4!} + \frac{x^6}{6!} + \cdots\cdots + \frac{x^{2n-2}}{(2n-2)!} + \cdots\cdots = \sum_{n=0}^{\infty} \frac{x^{2n}}{(2n)!}$$

$$\log(1+x) = x - \frac{x^2}{2} + \frac{x^3}{3} - \cdots + (-1)^{n-1} \frac{x^n}{n} + \cdots = \sum_{n=0}^{\infty} (-1)^{n+1} \frac{x^n}{n}, \quad -1 < x \leq 1$$

$$\log(1-x) = x - \frac{x^2}{2} - \frac{x^3}{3} - \cdots - \frac{x^n}{n} - \cdots = -\sum_{n=0}^{\infty} \frac{x^n}{n}, \quad -1 < x \leq 1$$

$$(1+x)^m = 1 + mx + \frac{m(m-1)}{2!} x^2 + \frac{m(m-1)(m-2)}{3!} x^3 + \cdots\cdots$$
$$+ \frac{m(m-1)\cdots(m-n+1)}{n!} x^n + \cdots\cdots = \sum_{n=0}^{\infty} \binom{m}{n} x^n, \quad -1 < x < 1$$

1.8.4 Standard Derivatives

$$\frac{d}{dx} x^n = n x^{n-1}, \quad n \text{ being a real}$$

$$\frac{d}{dx}\log x = \frac{1}{x}$$
$$\frac{d}{dx}e^{ax} = ae^{ax}$$
$$\frac{d}{dx}\log_a x = \frac{1}{x}\log_a e$$
$$\frac{d}{dx}a^{mx} = ma^{mx}\log_e a, \quad a > 0$$
$$\frac{d}{dx}a^{x} = a^{x}\log_c a, \quad a > 0$$
$$\frac{d}{dx}(cf(x)) = c\frac{d}{dx}f(x) = cf'(x)$$
$$\frac{d}{dx}f(ax+b) = af'(ax+b)$$
$$\frac{d}{dx}\sin x = \cos x$$
$$\frac{d}{dx}\cos x = -\sin x$$
$$\frac{d}{dx}\tan x = \sec^2 x$$
$$\frac{d}{dx}\cot x = -\operatorname{cosec}^2 x$$
$$\frac{d}{dx}\sec x = \sec x \tan x$$
$$\frac{d}{dx}\operatorname{cosec} x = -\operatorname{cosec} x \cot x$$
$$\frac{d}{dx}\sinh x = \cosh x$$
$$\frac{d}{dx}\cosh x = \sinh x$$
$$\frac{d}{dx}\tanh x = \operatorname{sech}^2 x$$
$$\frac{d}{dx}\coth x = -\operatorname{cosech}^2 x$$
$$\frac{d}{dx}\operatorname{sech} x = -\sec x \tan x$$
$$\frac{d}{dx}\operatorname{cosech} x = -\operatorname{cosec} x \cot x$$
$$\frac{d}{dx}\sin^{-1} x = \frac{1}{\sqrt{1-x^2}}, \quad |x| < 1$$
$$\frac{d}{dx}\cos^{-1} x = -\frac{1}{\sqrt{1-x^2}}, \quad |x| < 1$$
$$\frac{d}{dx}\tan^{-1} x = \frac{1}{1+x^2}$$
$$\frac{d}{dx}\cot^{-1} x = -\frac{1}{1+x^2}$$

$$\frac{d}{dx}\sec^{-1}x = \frac{1}{x\sqrt{x^2-1}}, \quad |x| > 1$$
$$\frac{d}{dx}\operatorname{cosec}^{-1}x = -\frac{1}{x\sqrt{x^2-1}}, |x| > 1$$
$$\frac{d}{dx}\sinh^{-1}x = \frac{1}{\sqrt{1+x^2}}$$
$$\frac{d}{dx}\cosh^{-1}x = \frac{1}{\sqrt{x^2-1}}, \quad |x| > 1$$
$$\frac{d}{dx}\tanh^{-1}x = \frac{1}{1-x^2}, \quad x < 1$$
$$\frac{d}{dx}\cot^{-1}x = -\frac{1}{x^2-1}, \quad x > 1$$
$$\frac{d}{dx}\operatorname{sech}^{-1}x = -\frac{1}{x\sqrt{1-x^2}}, \quad x < 1$$
$$\frac{d}{dx}\operatorname{cosech}^{-1}x = -\frac{1}{x\sqrt{x^2+1}}$$

1.8.5 Standard Integrals

$$\int x^n dx = \frac{x^{n+1}}{n+1}, \quad n \neq -1$$
$$\int \frac{1}{x}dx = \log|x|$$
$$\int f'(ax+b)dx = \frac{f(ax+b)}{a}, \quad a \neq 0$$
$$\int \frac{f'(x)}{f(x)}dx = \log|f(x)|$$
$$\int e^{ax+b}dx = \frac{e^{ax+b}}{a}, \quad a \neq 0$$
$$\int e^x dx = e^x$$
$$\int a^{mx}dx = \frac{a^{ma}}{m\log_e a}, \quad a > 0, a \neq 1$$
$$\int a^x dx = \frac{a^x}{\log_e a}, \quad a > 0, a \neq 1$$
$$\int \sin(ax+b)dx = -\frac{\cos(ax+b)}{a}, a \neq 0$$
$$\int \sin x dx = -\cos x$$
$$\int \cos(ax+b)dx = \frac{\sin(ax+b)}{a}, a \neq 0$$

$$\int \cos ax dx = \sin x$$
$$\int \sec^2 x dx = \tan x$$
$$\int \operatorname{cosec}^2 x dx = -\cot x$$
$$\int \sec x \tan x dx = \sec x$$
$$\int \operatorname{cosec} x \cot x dx = -\operatorname{cosec} x$$
$$\int \tan x dx = \log|\sec x|$$
$$\int \cot x dx = \log|\sin x|$$
$$\int \operatorname{cosec} x dx = \log\left|\tan\frac{x}{2}\right| = \log|\operatorname{cosec} x - \cot x|$$
$$\int \sec x dx = \log\left|\tan\left(\frac{\pi}{4}+\frac{x}{2}\right)\right| = \log|\sec x + \tan x|$$
$$\int \sinh x dx = \cosh x$$
$$\int \cosh x dx = \sinh x$$
$$\int \tanh x dx = \log|\cosh x|$$
$$\int \coth x dx = \log|\sinh x|$$
$$\int \operatorname{cosech} x dx = \log\left|\tanh\frac{x}{2}\right|$$
$$\int \operatorname{sech} x dx = 2\tan^{-1}(e^x)$$
$$\int \operatorname{sech}^2 x dx = \tanh x$$
$$\int \operatorname{cosech}^2 x dx = -\coth x$$
$$\int \operatorname{sech} x \tanh x dx = -\operatorname{sech} x$$
$$\int \operatorname{cosech} x \coth x dx = -\operatorname{cosech} x$$
$$\int \frac{dx}{x^2+a^2} = \frac{1}{a}\tan^{-1}\frac{x}{a}, \quad a \neq 0$$
$$\int \frac{dx}{1+x^2} = \tan^{-1} x$$
$$\int \frac{dx}{x^2-a^2} = \frac{1}{2a}\log\left|\frac{x-a}{x+a}\right|, \quad |x| > |a|$$

$$\int \frac{dx}{\sqrt{a^2 - x^2}} = \sin^{-1}\frac{x}{a} = -\cos^{-1}\frac{x}{a}, |x| < |a|$$

$$\int \frac{dx}{a^2 - x^2} = \frac{1}{2a}\log\left|\frac{a+x}{a-x}\right|, \quad |x| < |a|$$

$$\int \frac{dx}{\sqrt{x^2 + a^2}} = \sinh^{-1}\frac{x}{a} = \log\left|x + \sqrt{x^2 + a^2}\right|$$

$$\int \frac{dx}{\sqrt{x^2 - a^2}} = \cosh^{-1}\frac{x}{a} = \log\left|x + \sqrt{x^2 - a^2}\right|$$

$$\int \frac{dx}{x\sqrt{x^2 - a^2}} = \frac{1}{a}\sec^{-1}\frac{x}{a} = -\frac{1}{a}\operatorname{cosec}^{-1}\frac{x}{a}$$

$$\int \frac{dx}{\sqrt{1 - x^2}} = \sin^{-1} x$$

$$\int \frac{dx}{x\sqrt{x^2 - 1}} = \sec^{-1} x$$

$$\int \sqrt{x^2 + a^2}dx = \frac{x\sqrt{x^2 + a^2}}{2} + \frac{a^2}{2}\log\left|x + \sqrt{x^2 + a^2}\right| = \frac{x\sqrt{x^2 + a^2}}{2} + \frac{a^2}{2}\sinh^{-1}\frac{x}{a}$$

$$\int \sqrt{x^2 - a^2}dx = \frac{x\sqrt{x^2 - a^2}}{2} - \frac{a^2}{2}\log\left|x + \sqrt{x^2 - a^2}\right| = \frac{x\sqrt{x^2 - a^2}}{2} - \frac{a^2}{2}\cosh^{-1}\frac{x}{a}$$

$$\int \sqrt{a^2 - x^2}dx = \frac{a\sqrt{a^2 - x^2}}{2} + \frac{a^2}{2}\sin^{-1}\frac{x}{a}$$

$$\int e^{ax}\cos bx dx = \frac{e^{ax}(a\cos bx + b\sin bx)}{a^2 + b^2} = \frac{e^{ax}}{\sqrt{a^2 + b^2}}\cos\left(bx - \tan^{-1}\frac{b}{a}\right)$$

$$\int e^{ax}\sin bx dx = \frac{c^{ax}(a\sin bx - b\cos bx)}{a^2 + b^2} = \frac{e^{ax}}{\sqrt{a^2 + b^2}}\sin\left(bx - \tan^{-1}\frac{b}{a}\right)$$

Integral by parts

$$\int f_1(x)f_2(x)dx = f_1(x)\int f_2(x)dx - \int\left[\frac{d}{dx}f_1(x)\int f_2(x)dx\right]dx$$

1.9 Closure

In this chapter, several fundamental mathematical concepts that are useful for the study of Fourier analysis were presented. Initially, a basic definition of the trigonometric sine and cosine functions was provided, and several analytical properties of these functions were derived. These functions serve as fundamental building blocks of Fourier analysis. Since a significant portion of Fourier analysis is performed in the complex plane, several basic concepts and definitions from the field of complex variable theory were also presented.

Reference

Cadeddu, L., & Cauli, A. (2018). Wine and maths: Mathematical solutions to wine–inspired problems. *International Journal of Mathematical Education in Science and Technology, 49*(3), 459–469.

Chapter 2
Signals and Systems

The world is full of signals that we don't perceive.
—Stephen Jay Gould

Abstract This chapter covers the fundamental concepts of signals and systems, essential for applications across engineering and applied sciences. It begins by exploring various signal types and examples to build a solid understanding of signal behaviors and properties. Key signal types discussed include sinusoids, exponential and geometric signals, and standard functions like the box, Kronecker delta, Heaviside step, signum, sinc, Bessel, and Gaussian functions. The chapter then shifts focus to systems, examining critical properties such as linearity, time-invariance, causality, stability, and memory, which are essential for analyzing and predicting system behavior in signal processing. Finally, it addresses noise, an unavoidable element in real-world signals, and its impact on signal integrity and system performance.

Keywords Exponential signals · Geometric signals · Box and *rect* signals · Kronecker delta function · Heaviside step function · Signum function · *sinc* function · Gaussian function · Bessel function · Random signals · Linearity · Causality · Stability · Memory · Noise

2.1 Signals

A signal is a mathematical function $f(x)$ that imparts information regarding the behavior or attributes of a certain phenomenon. The set $\{f(x); x \in \mathbb{X}\}$, where $\mathbb{X}$ might be the entire real line or the positive real line $[0, \infty)$ or perhaps a more confined interval such as $[0, P)$ or $[-P/2, P/2]$. For instance, $\sin x$ or, more precisely, $\{\sin x; x \in (-\infty, \infty)\}$ constitutes a simple continuous and smooth signal. The signal is a function of an independent variable (or physical parameter), here termed x, which serves as the **space variable**; however, this independent variable could correspond to other physical quantities such as **time**. $\mathbb{X}$ delineates

S. K. Jena, *Fourier, Laplace, and the Tangled Love Affair with Transforms*,
https://doi.org/10.1007/978-3-031-80165-5_2

the permissible values of the parameter and is referred to as the domain of the signal. The minimum and maximum values of $g(x)$ are known as the range of the signal. A finite duration signal is defined within a finite extent domain of definition, and the signal assumes a zero value outside this domain. For example, the signal $\{\sin x; x \in [0, 2\pi)\}$ is of finite duration and is defined as follows:

$$f(x) = \begin{cases} \sin x & x \in [0, 2\pi) \\ 0 & x \in \mathbb{R}, x \notin [0, 2\pi) \end{cases} \tag{2.1}$$

Such finite-duration signals are also called space-/time-limited signals. Signals can be classified into two main categories. When the set $\mathbb{X}$ is continuous, the signal is continuous, a continuous parameter signal, or simply a waveform. Examples include analog audio signals and ECG signals. The functional notation $f(x)$ is often used for the signal $\{f(x); x \in \mathbb{X}\}$, but this can cause confusion as $f(x)$ could either represent the value of the signal at a specific x or the entire waveform $f(x)$ for all $x \in \mathbb{X}$. The whole signal is denoted by simply the name of the function f to avoid this ambiguity.

$$f = \{f(x); x \in \mathbb{X}\} \tag{2.2}$$

It is also fairly common practice to use boldface to denote the entire signal; that is, $\mathbf{f} = \{f(x); x \in \mathbb{X}\}$.

Temperature information recorded at a point for a month over 1-hour intervals is represented through some finite data points, and such information is called discrete signals. Discrete signals are defined only at finite discrete points in the domain and are usually represented as a sequence of numbers. The sampled sinusoid $\{\sin(nT); n \in N\}$, where N is the set of all integers $\{\ldots, -2, -1, 0, 1, 2, \ldots\}$, a geometric progression $\{r^n; n = 0, 1, 2, \ldots\}$, is an example of discrete representation of a signal. Analogous to the continuous case $\{f(x); x \in \mathbb{X}\}$, a discrete signal is denoted with subscripts $f_n = f(x_n)$ rather than functional notation. Thus, a sequence is denoted by $\{f_n; n \in \mathbb{X}\}$. $\mathbb{X}$ may have infinite duration (e.g., all integers) or finite duration (e.g., the integers from 0 to $N - 1$).

In the previous examples, $\mathbb{X}$ is one-dimensional, consisting of some real or complex input x. Often, signals are multidimensional, as exemplified by a photograph. A two-dimensional square sampled image intensity raster could be represented as $\left\{f_{n,m}; n = 1, \ldots, N; m = 1, \ldots, M\right\}$, where each $f_{n,k}$ denotes the intensity (a non-negative number) of a single picture element or pixel within the image, specifically located in the nth column and kth row of the square image. In the RGB (red-green-blue) color model, any hue can be achieved by amalgamating red, green, and blue light, each encoded as a value from 0 (off) to 255 (brightest). A megapixel (MP) image contains a million pixels, and for each combination of n and k, $f_{n,k}$ stores a value ranging from 0 to 255. For instance, a camera producing a 2048×1536 pixel image (3,145,728 final image pixels) typically employs a few additional rows and columns of sensor elements and is commonly referred to as having 3.2 MP. A series of two-dimensional images captured over time and concatenated form a three-dimensional movie clip. Thus, a movie clip constitutes a three-dimensional discrete signal encompassing both space and time. In summary, a signal is merely a function whose domain of definition is $\mathbb{X}$ and whose range lies within the space of real or complex numbers. The nature of $\mathbb{X}$ dictates whether the signal is continuous or discrete and is finite or infinite. The signal is deemed real-valued or complex-valued contingent upon the possible values of $f(x)$.

The energy content of a signal is defined as

$$E = \int_{-\infty}^{\infty} |f(x)|^2 \tag{2.3}$$

If E is finite, then the signal is an energy signal. For practical signals, the energy of the signal can be approximated using a finite number of terms of the summation.

$$E = \sum_{n=-\infty}^{\infty} |f[n]|^2 \tag{2.4}$$

For signals not having finite energy, average power may be finite. The average power is defined as

$$P = \lim_{N \to \infty} \frac{1}{2N+1} \sum_{n=-N}^{N} |f[n]|^2 \tag{2.5}$$

Such signals are known as the power signal.

2.2 Basic Signal Examples

Signals convey information in various domains, such as communications, control systems, and data processing. Signals can be broadly categorized into two types: analog and digital signals. Analog signals are continuous and vary smoothly over time, representing information through amplitude, frequency, or phase variations. In contrast, digital signals are discrete, consisting of binary values (0s and 1s), and are used in modern computing and digital communication systems. They offer advantages in terms of noise resistance and ease of processing. Both types of signals play crucial roles in different applications, with analog signals essential for representing real-world phenomena and digital signals providing robust data storage, transmission, and manipulation solutions.

2.2.1 Sinusoids

Sinusoids are continuous, smooth, periodic signals representing oscillating phenomena, such as sound waves, alternating current (AC) in electrical circuits, radio waves, etc. A sinusoidal signal is represented as

$$f(x) = A\sin(\omega x + \phi) \tag{2.6}$$

A is the amplitude, ω is the frequency, and ϕ is the phase. A sinusoid can also be considered as a discrete signal by sampling the continuous version.

$$f[n] = A\sin[\omega n + \phi];\ n \in \mathbb{Z} \tag{2.7}$$

Figure 2.1 represents the continuous and discrete representation of sinusoids.

2.2.2 Exponential and Geometric Signals

Exponential signal is fundamental in signal processing and systems theory and is characterized by its distinctive mathematical form.

$$f(x) = Ae^{\lambda x} \tag{2.8}$$

A is the amplitude, and e is the base of the natural logarithm. λ is a constant that determines the growth or decay rate of the signal. For $\lambda > 0$, the signal represents exponential growth. For $\lambda < 0$, the signal represents exponential decay, and for $\lambda = 0$, the signal reduces to a constant value A. A decaying coefficient $\lambda < 0$ is used for a finite energy-containing signal. An exponential signal in signal processing is often expressed as

$$f(x) = \begin{cases} Ae^{-\lambda x} & x \geq 0 \\ 0 & \text{otherwise} \end{cases}. \tag{2.9}$$

Exponential and geometric signals are intrinsically linked, with the former representing continuous signals and the latter representing discrete counterparts. For example, consider the signal given by the finite sequence.

$$f[n] = \{r^n; n = 0, 1, \ldots, N-1\} = \left\{1, r, \ldots, r^{N-1}\right\} \tag{2.10}$$

This is the discrete and finite duration geometric signal because it has a finite-length piece of a geometric progression. Figure 2.2 represents the exponential and geometric signals.

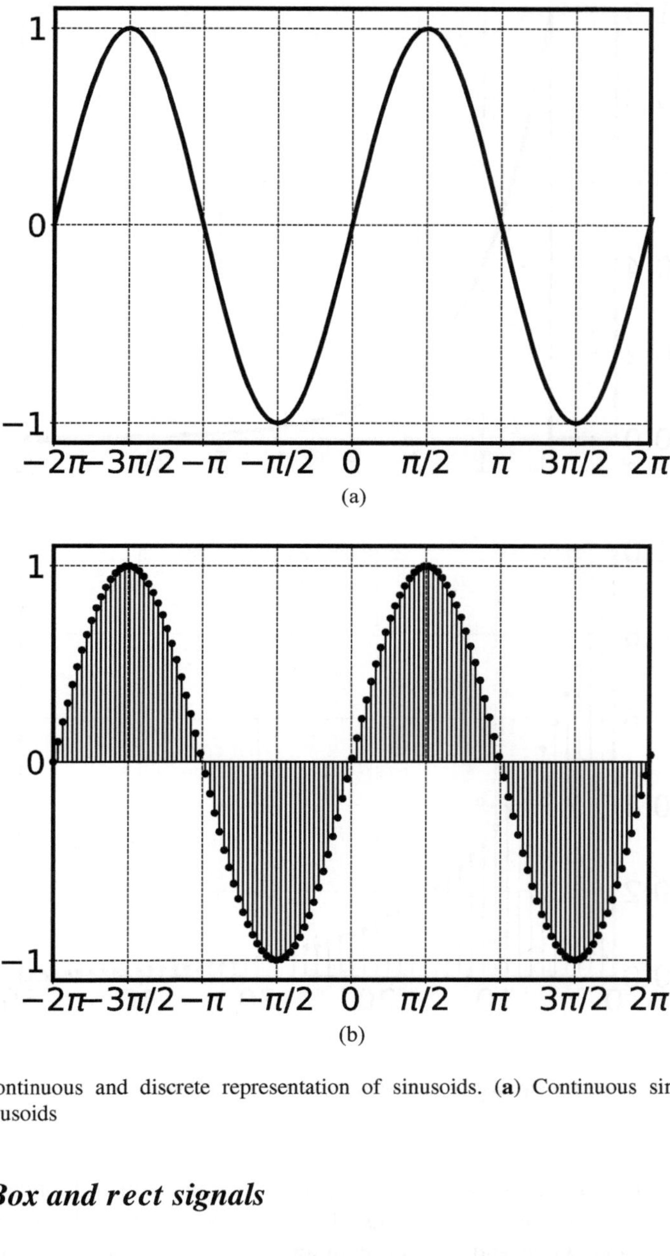

Fig. 2.1 Continuous and discrete representation of sinusoids. (**a**) Continuous sinusoids. (**b**) Discrete sinusoids

2.2.3 *Box and rect signals*

The box signal's history is intertwined with the evolution of signal processing, from early mathematical foundations to modern digital applications. The box signal, often called a rectangular pulse or a rectangular window function, is a fundamental concept in signal processing and analysis. Its simple shape and constant amplitude

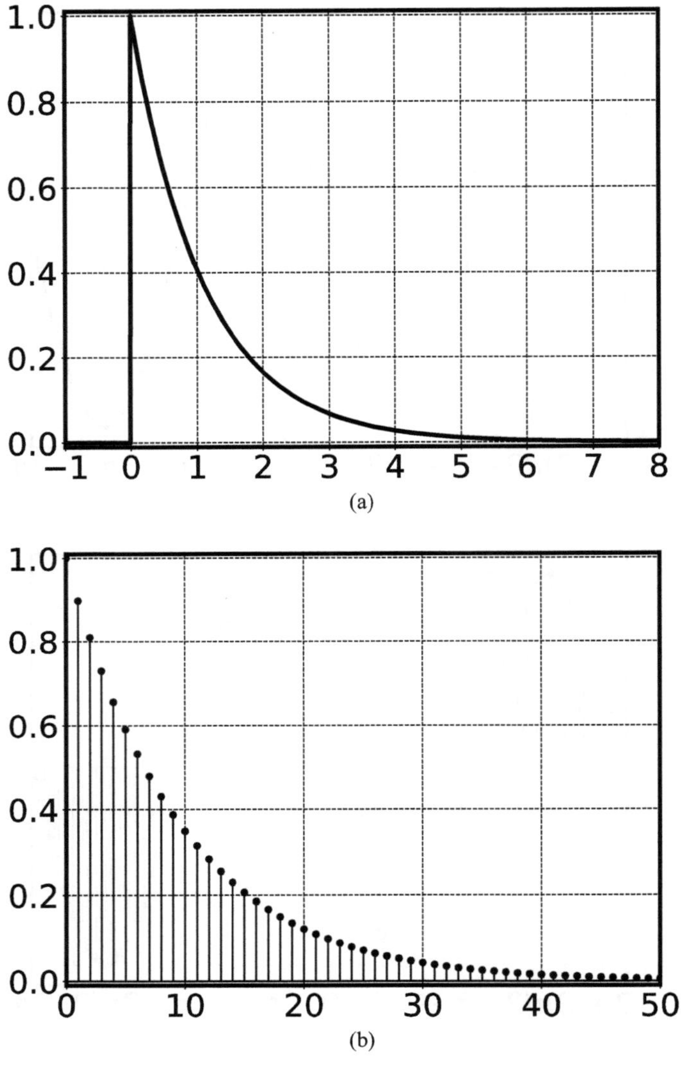

Fig. 2.2 (**a**) Exponential signals for $\lambda = -0.9$ (**b**) Geometric signals for $r = 0.9$

over a specified duration make it an essential tool in various applications. Box signals became a standard tool in digital signal processing (DSP) for tasks such as filtering, windowing, and time-domain analysis. For any real $T > 0$, the box function $\square_T(x)$ is defined for any real x by

$$\square_T(x) = \begin{cases} 1 & |x| \leq T \\ 0 & \text{otherwise} \end{cases}. \tag{2.11}$$

This $\square_T(x)$ is non-smooth and requires special care for Fourier inversion. The rectangle function considered by Bracewell (1965) is a variant of the box function and is reproduced exactly after a sequence of a Fourier transform and an inverse Fourier transform.

$$\Pi(x) = \begin{cases} 1 & |x| < \frac{1}{2} \\ \frac{1}{2} & |x| = \frac{1}{2} \\ 0 & \text{otherwise} \end{cases}. \tag{2.12}$$

The notation $rect(x)$ and top hat function is also used for $\Pi(x)$. Figure 2.3 represents the continuous and discrete form of Bracewell (1965)'s rectangle signal.

2.2.4 Kronecker Delta Function

The Kronecker delta function is named after the German mathematician Leopold Kronecker (1823–1891), who significantly contributed to number theory, algebra, and analysis in the nineteenth century. In discrete signal processing, the Kronecker delta function represents the discrete unit impulse signal δ_t and is defined for any real x by

$$\delta_t(x) = \begin{cases} 1 & x = 0 \\ 0 & x \neq 0 \end{cases}. \tag{2.13}$$

This function is used to analyze and design discrete systems and is a basis for building and decomposing signals. Kronecker delta function is an ordinary signal and should not be confused with the Dirac delta function, which is a generalized function and will be introduced later. Figure 2.4 represents the Kronecker delta function.

2.2.5 Heaviside Step Function

The Heaviside step function, named after the British engineer and mathematician Oliver Heaviside (1850–1925), is a piecewise function widely used in mathematics, engineering, physics, and various scientific disciplines. It is denoted as $\mathcal{H}(x)$ and is defined as

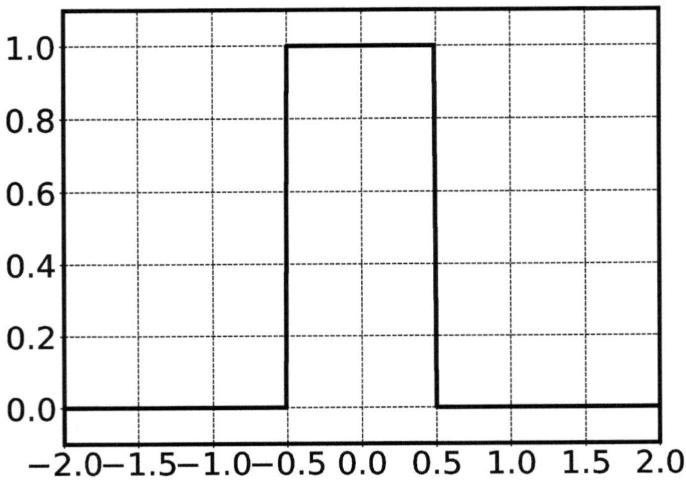

(a)

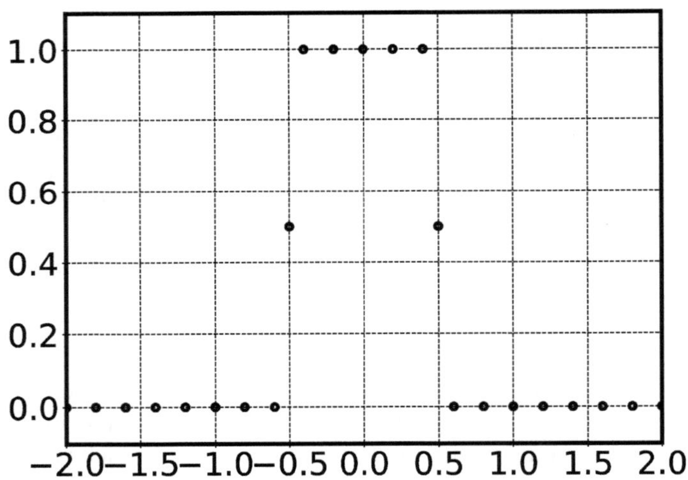

(b)

Fig. 2.3 Continuous and discrete form of Bracewell (1965)'s rectangle signal. (**a**) Continuous rectangle signal. (**b**) Discrete rectangle signal

$$\mathcal{H}(x) = \begin{cases} 1 & x > 0 \\ \dfrac{1}{2} & x = 0 \\ 0 & x < 0 \end{cases}. \tag{2.14}$$

This simple definition of the Heaviside function allows it to represent the switch-like behavior where a system transitions instantaneously from one state to another.

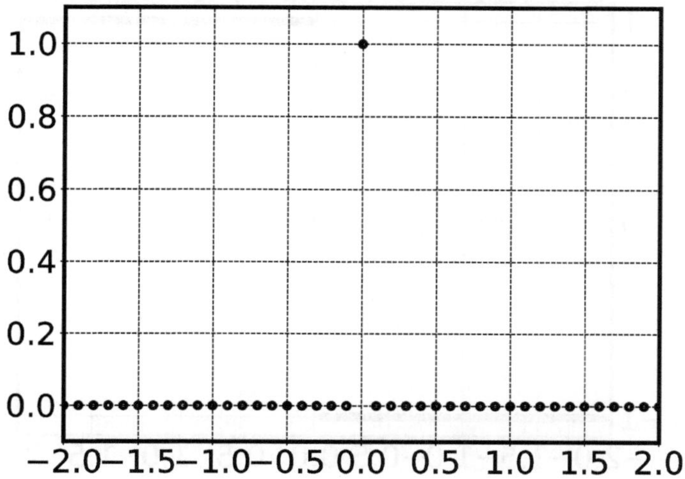

Fig. 2.4 Kronecker delta function

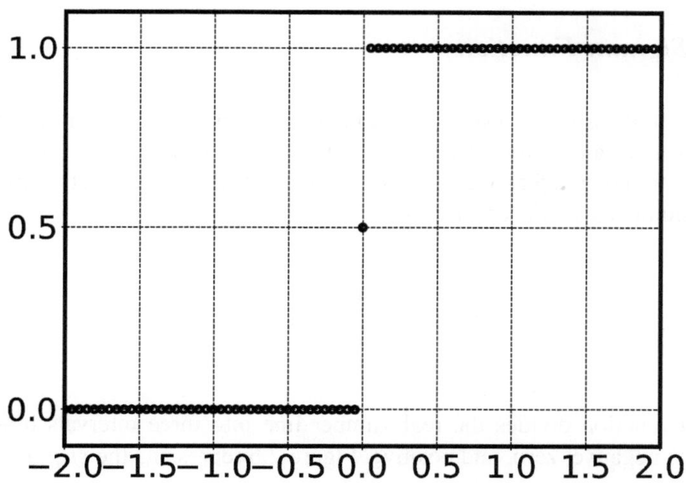

Fig. 2.5 Heaviside step function

The Heaviside step function is used for the same purpose as the rectangle function; defining its value at discontinuities as the midpoint between the values above and below the discontinuity is helpful in forming Fourier transform pairs. Both step functions share a common Fourier transform, but the inverse continuous Fourier transform yields the Heaviside step function. Figure 2.5 represents the Heaviside step function.

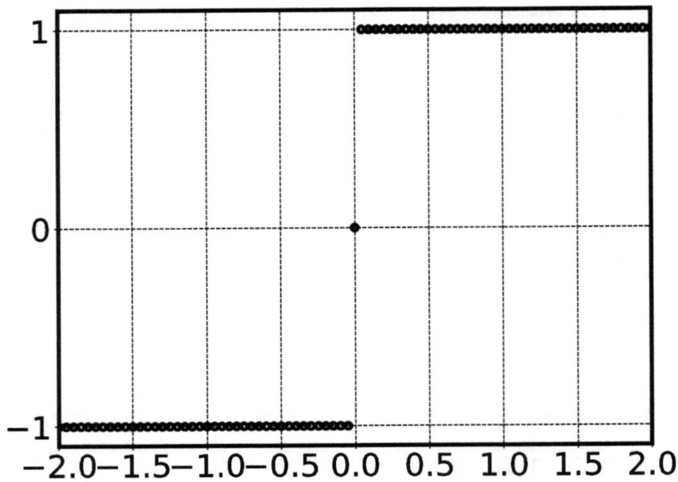

Fig. 2.6 Signum function

2.2.6 *Signum Function*

The concept of assigning a sign to a number based on its positive or negative nature is ancient and can be traced back to early mathematical civilizations. The $sign$ function, often denoted as $sign(x)$, is a mathematical function that returns the sign of a real number x and is defined as

$$sgn(x) = \begin{cases} +1 & \text{if } x > 0 \\ 0 & \text{if } x = 0 \\ -1 & \text{if } x < 0 \end{cases} \tag{2.15}$$

The $sign$ function divides the real number line into three intervals based on the value of x: negative, zero, and positive. Figure 2.6 represents the $sign$ function.

2.2.7 *sinc Function*

The $sinc$ function is the short term for "sine cardinal", reflecting its connection to the sine function and its cardinality in Fourier analysis. The $sinc$ function is defined as

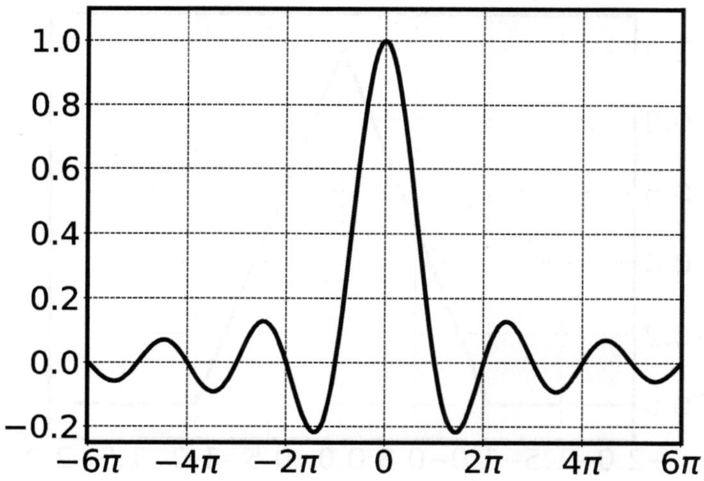

Fig. 2.7 *sinc* function

$$sinc(\omega x) = \begin{cases} \dfrac{\sin \omega x}{\omega x} & x \neq 0 \\ 1 & x = 0 \end{cases}. \tag{2.16}$$

The *sinc* function stands as a testament to the profound impact of Fourier analysis on modern mathematics and engineering. Its ubiquity in signal processing and its theoretical underpinnings in Fourier transforms underscore its importance in analyzing and manipulating signals in analog and digital domains. Figure 2.7 represents the *sinc* function.

2.2.8 $\wedge(x)$ Function

The triangle or wedge $\wedge(x)$ function is another common signal used in engineering. $\wedge(x)$ is defined for all x

$$\wedge(x) = \begin{cases} 1 - |x| & \text{if } |x| < 1 \\ 0 & \text{otherwise.} \end{cases} \tag{2.17}$$

Figure 2.8 represents the $\wedge$ function.

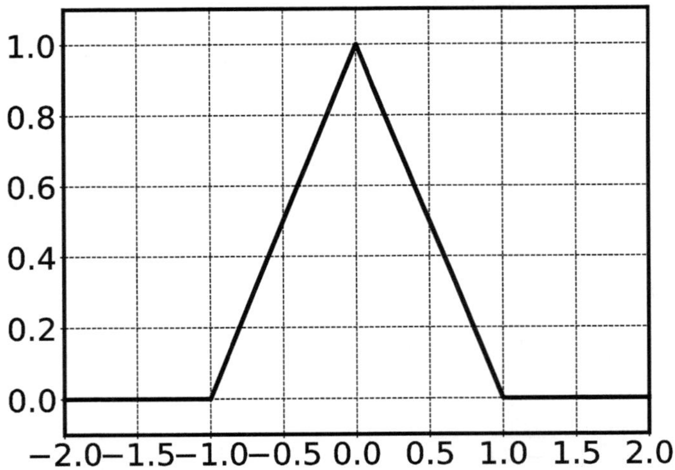

Fig. 2.8 $\wedge(x)$ function

2.2.9 Gaussian Function

The Gaussian function, often referred to as the normal distribution or bell curve in the context of probability and statistics, is a fundamental function in mathematics and science. Its historical roots trace back to the work of prominent German mathematician and physicist Carl Friedrich Gauss (1777–1855). In 1809, while studying the method of least squares for astronomical data, Gauss published his influential work "Theoria Motus Corporum Coelestium in Sectionibus Conicis Solem Ambientium", which laid the groundwork for the Gaussian distribution.

$$f(x) = \frac{1}{\sqrt{2\pi}\sigma} e^{-\frac{(x-\mu)^2}{2\sigma^2}} \tag{2.18}$$

μ is the mean value and σ is the standard deviation. Figure 2.9 represents the Gaussian function.

2.2.10 Bessel Function

The Bessel function of the first kind, denoted as $J_n(x)$, originates in the work of German mathematician and astronomer Friedrich Wilhelm Bessel (1784–1846). Bessel first introduced these functions in the early nineteenth century while studying solutions to Kepler's equations of planetary motion. His work was pivotal

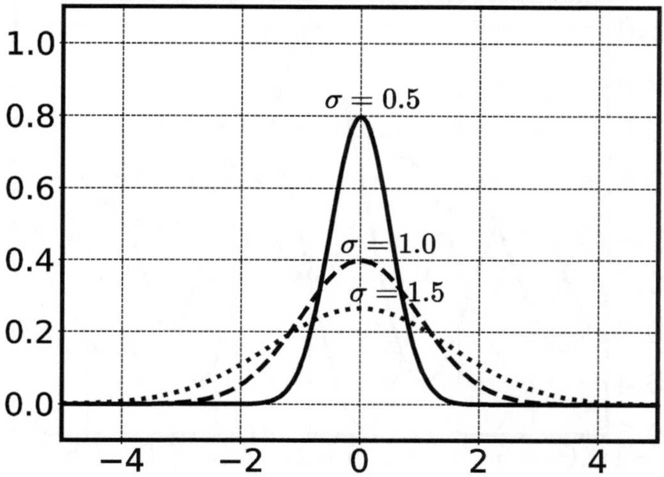

Fig. 2.9 Gaussian function for $\mu = 0$

in addressing problems in mathematical physics involving cylindrical symmetry, such as heat conduction, wave propagation, and electromagnetic fields. Bessel functions, which are solutions to Bessel's differential equation, have since become fundamental in various branches of applied mathematics, physics, and engineering, particularly in problems involving radial symmetry, frequency modulation (FM) and quantization. Bessel's meticulous analysis and extensive application of these functions have cemented his legacy in the mathematical sciences, where Bessel functions remain indispensable tools in theoretical and applied contexts. The nth order ordinary Bessel function of the first kind is defined for real x, and any integer n is denoted as

$$J_n(x) = \frac{1}{2\pi} \int_{-\pi}^{\pi} e^{ix\sin\phi - in\phi} d\phi. \tag{2.19}$$

Bessel function is discrete series form expressed as

$$J_n(x) = \sum_{k=0}^{\infty} \frac{(-1)^k}{k!(n+k)!} \left(\frac{x}{2}\right)^{2k+n} \tag{2.20}$$

For integer order n, the following relationship is valid:

$$J_{-n}(x) = (-1)^n J_n(x) \tag{2.21}$$

Figure 2.10 represents the Bessel function for different n.

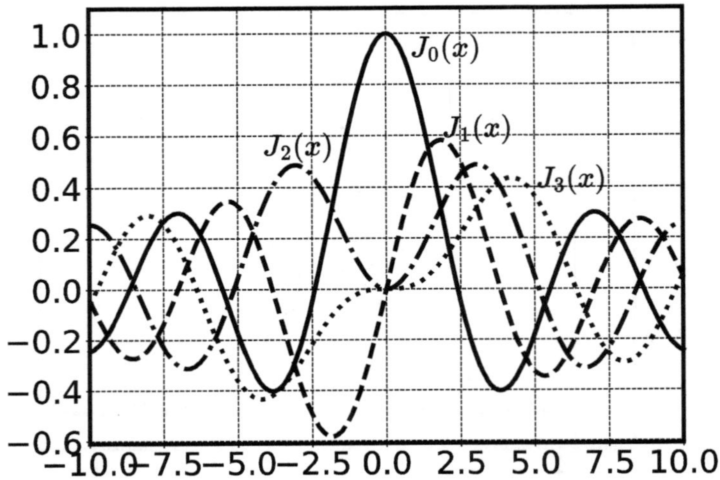

Fig. 2.10 Bessel function

2.2.11 Random Signals

All of the signals discussed above have specific mathematical forms and structures. Often, real-life signals are not well understood and are not constructible in any useful and simple mathematical form. One way of modelling unknown signals is to consider them as having been produced by a random process. Random signals are a fundamental concept in signal processing and communications, characterized by their unpredictable and stochastic nature. Unlike deterministic signals, random signals are best understood through statistical properties and probabilistic descriptions. Figure 2.11 represents a discrete random signal.

2.3 Systems

A system, in the context of signal processing, is an entity that transforms input signals into output signals. A system acts as a mapping function $\mathcal{H}$, receiving an input signal $v = \{v(x); x \in \mathbb{X}_i\}$ and generating a corresponding output signal $w = \{w(x); x \in \mathbb{X}_0\} = \mathcal{H}(v)$.

$$w(x) = \mathcal{H}_x(v) \tag{2.22}$$

Systems can be classified based on the type of signals they handle into continuous systems and discrete systems. The behavior of a system is defined by the relationship between its input and output signals. For instance, the simplest form of a system

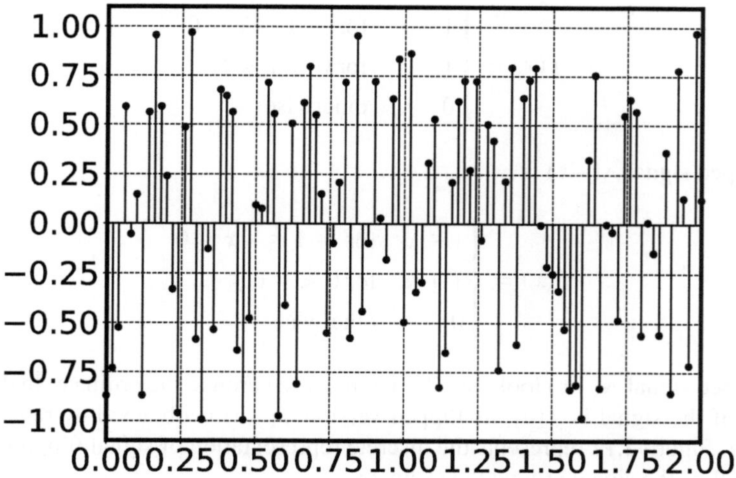

Fig. 2.11 Random signal

is the identity system, which directly transmits the input signal to the output without any modification $\mathcal{H}(v) = v$, akin to an ideal "wire". Another elementary system is the null system, which outputs a zero signal regardless of the input $\mathcal{H}(v) = 0$, resembling an ideal "ground". More complex systems perform a range of linear or nonlinear operations on the input signal to produce the desired output. These operations can involve amplification, filtering, modulation, or other signal processing techniques, enabling systems to execute sophisticated tasks across various applications.

In many applications, input and output signals are of the same type, meaning that $\mathbb{X}_\mathrm{i}$ and $\mathbb{X}_\mathrm{o}$ are identical. However, this is not always the case. The output of a system can theoretically depend on the entire history and future of the input signal, particularly if x corresponds to time. Although this might seem unphysical, it serves as a valuable abstraction for understanding properties of systems in their most general form. This generalization allows for more comprehensive system analysis, including causality, memory, and predictability considerations. It encompasses many potential system behaviors, from simple instantaneous mappings to complex processes incorporating extensive temporal information. This abstraction is instrumental in developing robust theoretical frameworks and practical applications in signal processing, control systems, and other fields where understanding the intricate relationship between input and output signals is crucial.

Flipping a signal, also known as time reversal or reflection, involves reversing the order of its samples. This operation is common in digital signal processing (DSP) for various analyses and transformations. For a given signal $f(x)$, its flipped version $f(-x)$ is obtained by replacing x with $-x$. For example, consider the $\wedge(x)$ function.

$$f(x) = \begin{cases} 1 + x & \text{for } -1 \leq x \leq 0 \\ 1 - x & \text{for } 0 < x \leq 1 \\ 0 & \text{otherwise} \end{cases} \tag{2.23}$$

The flipped signal $\wedge(-x)$ is

$$f(x) = \begin{cases} 1 - x & \text{for } -1 \leq -x \leq 0 \\ 1 + x & \text{for } 0 < -x \leq 1 \\ 0 & \text{otherwise} \end{cases} \tag{2.24}$$

The flipped signal $\wedge(-x)$ looks similar but mirrored around the ordinate. In discrete signals, if the signal is $f[n]$, its flipped version $f[-n]$ is obtained by replacing n with $-n$. Figure 2.12 represents the discrete representation of signal flipping. Time reversal flips the signal around the ordinate.

Scaling a signal by a constant is one of the most straightforward operations in signal processing. Consider a signal $f = \{f(x); x \in \mathbb{X}\}$ and a constant a. A new signal $\{af(x); x \in \mathbb{X}\}$ can be defined by multiplying every value of the original signal by a. This process is called amplitude scaling and exemplifies a simple system. In this context, the system $\mathcal{H}$ can be described by the operation $\mathcal{H}_x(f) = af(x); x \in \mathbb{X}$. Such a system is fundamental in signal processing, illustrating the concept of linear transformation. By scaling the amplitude of the signal uniformly across its domain, this operation preserves the shape and characteristics of the original signal while adjusting its magnitude. The second type of scaling $g(x) = \{f(ax); x \in \mathbb{X}\}$ is called the width or time scaling that compresses or dilates a signal by multiplying the time variable by some nonzero constant a. For $a > 1.0$, the signal becomes narrower, and the operation is called compression, and for $a < 0$, the signal becomes wider and is called dilation. Figure 2.13a represents the magnitude scaling of the $\wedge(x)$ signal, while Fig. 2.13b represents the time scaling of the $\wedge(x)$ signal. This basic yet essential example underlines how signals can be manipulated and transformed, forming the foundation for more complex signal processing techniques.

In signal processing, shifting a signal is a fundamental operation that involves displacing the entire signal in the time domain (or another relevant domain, such as space). Shifting a signal changes the position of the signal's values without altering their magnitude or shape. Alternatively, shifting a signal means redefining the time origin. If the independent variable corresponds to space instead of time, shifting the signal corresponds to moving the signal in space without changing its shape. This operation is crucial in various applications, including communication, control, and data analysis. There are two primary types of signal shifting: time shifting and space shifting.

Right Shift (Delay $f(t - t_0)$) When a signal is delayed, each value of the signal is shifted to a later point in time. Mathematically, if $f(t)$ represents the original

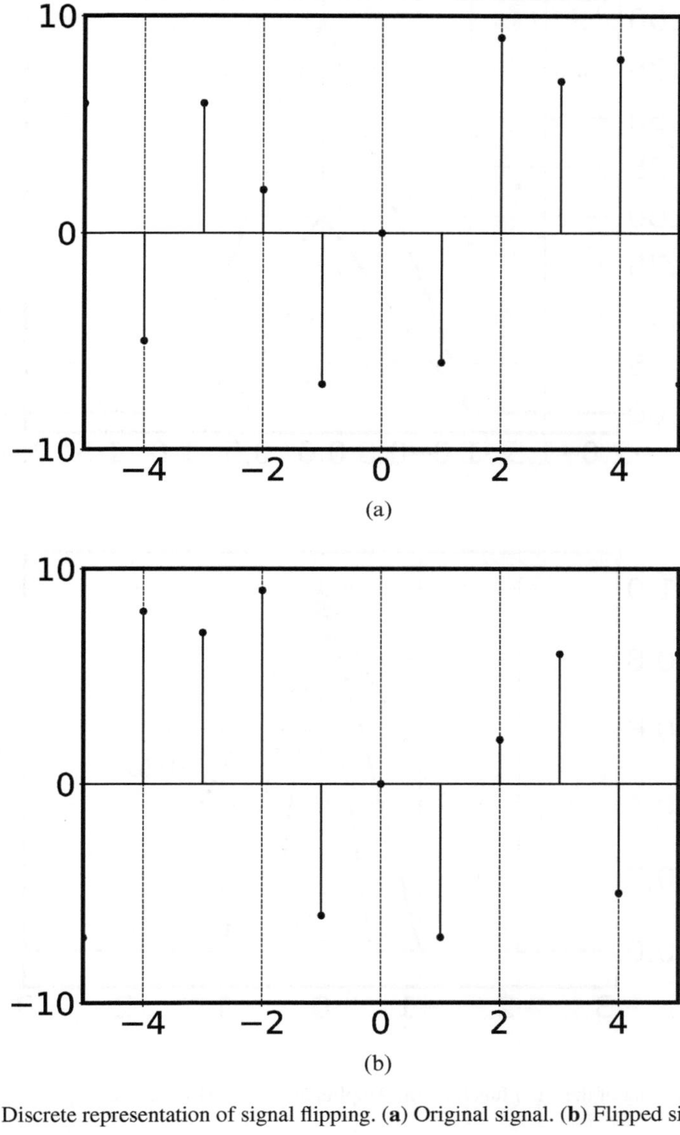

Fig. 2.12 Discrete representation of signal flipping. (**a**) Original signal. (**b**) Flipped signal

signal, a right shift by t_0 units results in the shifted signal $f(t - t_0)$. The shifted signal can be thought of as a signal that starts t_0 seconds after $f(t)$ does and then mimics it.

Left Shift (Advance $f(t + t_0)$) When a signal is advanced, each value of the signal is shifted to an earlier point in time. This is represented as $f(t + t_0)$.

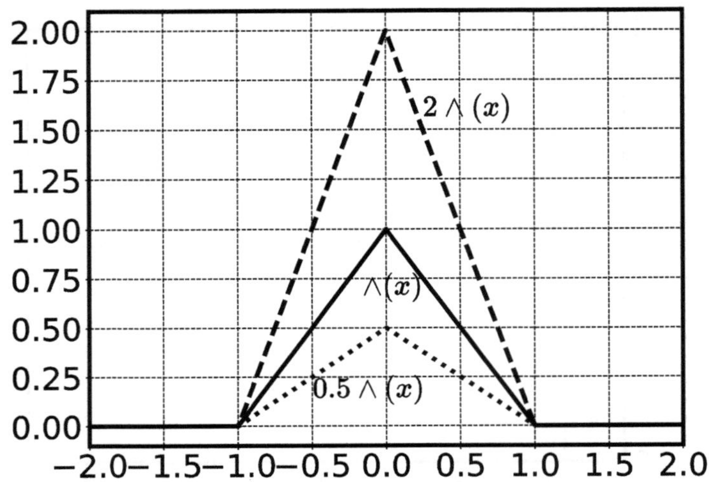

(a)

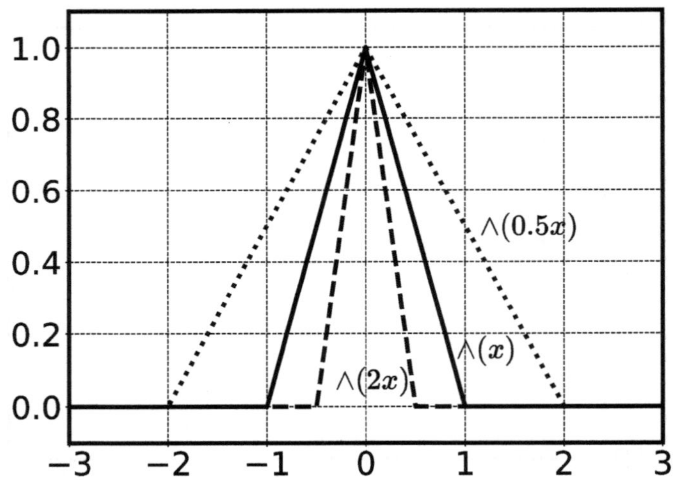

(b)

Fig. 2.13 Scaling of the $\wedge(x)$ function. (**a**) Amplitude scaling. (**b**) Time scaling

Space Shifting Similar to time shifting, signals can also be shifted in space for spatial signals (like images). For example, if $f(x)$ is a spatial signal, a shift by x_0 units results in $f(x - x_0)$ for a right shift and $f(x + x_0)$ for a left shift.

Figure 2.14 represents shifting of the $\wedge(x)$ signal. Cyclic shifting, also known as circular or periodic shifting, is an operation where signal elements are shifted and values that fall off one end of the signal are wrapped around the other. Cyclic shifting involves rotating the elements of a signal in a circular manner. This operation

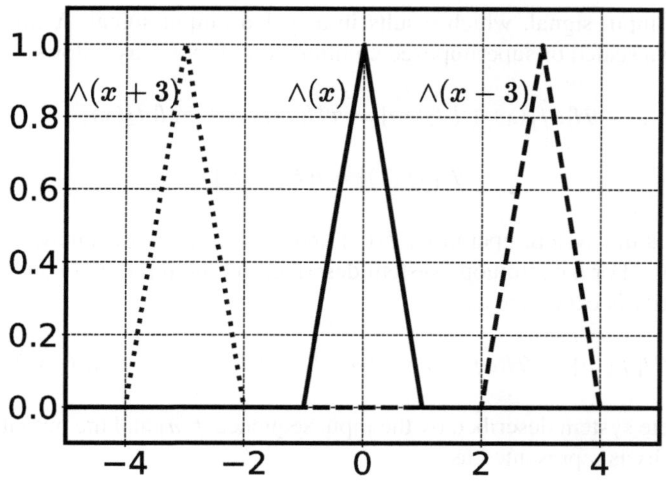

Fig. 2.14 Shifting of the $\wedge(x)$ function

ensures that no information is lost; instead, values are simply repositioned. Cyclic shifting can be performed both to the left and to the right. This type of shifting is particularly useful in digital signal processing and applications involving periodic signals. A right cyclic shift by k positions for a discrete signal $f[n]$ with a finite length N results in a signal $g[n] = f[(n-k) \mod N]$. This means each element of f is moved k positions to the right, and elements that exceed the length N wrap around to the beginning. Similarly, a left cyclic shift by k positions for a discrete signal $f[n]$ results in the signal $h[n] = f[(n+k) \mod N$. Here, each element of f is moved k positions to the left, and elements that fall below the starting index wrap around to the end.

In communication systems, signals often need to be synchronized. Shifting allows aligning signals in time, ensuring that transmitted data matches the receiver's timing. In audio processing, creating echoes or delay effects involves shifting the audio signal in time and combining it with the original signal. In time series analysis, shifting can help compare data points from different time periods, such as computing differences or correlations between past and current values. In control theory, shifting is used to predict the future states of a system based on past behavior, aiding in system stabilization and optimization.

2.3.1 Linearity

A system is linear if it satisfies principles of superposition and homogeneity. Superposition means that if an input signal is a sum of multiple signals, the output is the sum of the system's responses to each individual signal. Homogeneity means scaling the input signal, which results in a scaled output signal. A linear system responds to a scaled or superimposed combination of inputs as follows:

$$\mathcal{H}[af_1(x) + bf_2(x)] = a\mathcal{H}[f_1(x)] + b\mathcal{H}[f_2(x)] \tag{2.25}$$

$$\mathcal{H}[af(x)] = a\mathcal{H}[f(x)] \tag{2.26}$$

$\mathcal{H}[\cdot]$ denotes the system operation, $f_1(x)$ and $f_2(x)$ are input signals, and a and b are constants. For a continuous system described by the input $f(x)$ and the output $g(x)$, linearity is expressed as

$$g(x) = \mathcal{H}[f(x)] \Rightarrow \mathcal{H}[af_1(x) + bf_2(x)] = a\mathcal{H}[f_1(x)] + b\mathcal{H}[f_2(x)] \tag{2.27}$$

For a discrete system described by the input sequence $f[n]$ and the output sequence $g[n]$, linearity is represented as

$$g[n] = \mathcal{H}[f[n]] \Rightarrow \mathcal{H}[af_1[n] + bf_2[n]] = a\mathcal{H}[f_1[n]] + b\mathcal{H}[f_2[n]] \tag{2.28}$$

Linearity allows us to analyze a system's response to complex inputs by considering simpler components separately and combining their responses. Linearity simplifies the analysis of systems using tools such as Fourier transforms and Laplace transforms.

In contrast to linear systems, nonlinear systems do not satisfy the superposition and scaling properties. They exhibit responses that are not proportional to the inputs or do not add linearly. The following systems are easily seen to be nonlinear:

$$\begin{aligned} \mathcal{H}(f) &= f^2(x) \\ \mathcal{H}(f) &= a + bf(x) \\ \mathcal{H}(f) &= sgn(f(x)) \\ \mathcal{H}(f) &= \sin(f(x)) \\ \mathcal{H}(f) &= e^{-i2\pi f(x)} \end{aligned} \tag{2.29}$$

2.3.2 *Time-Invariant Systems*

A system is time invariant if its response to an input signal shifted in time is the same as the response to the original signal, shifted in time. Mathematically, a continuous time invariant-system with input $f(t)$ and the output $g(t)$ is expressed as

$$g(t) = \mathcal{H}[f(t)] \quad \Rightarrow \quad g(t - t_0) = \mathcal{H}[f(t - t_0)] \tag{2.30}$$

t_0 is the time shifting. Delaying the input $f(t)$ by t_0 delays the output t_0 as well. A discrete time-invariant system described by the input sequence $f[n]$ and the output sequence $g[n]$ is expressed as

$$g[n] = \mathcal{H}[f[n]] \quad \Rightarrow \quad g[n - n_0] = \mathcal{H}[f[n - n_0]] \tag{2.31}$$

n_0 is any integer shift. This implies that shifting the input sequence $f[n]$ by n_0 samples results in the output sequence $g[n]$ being shifted by n_0 samples as well.

Time-invariant systems are often known as shift invariant or stationary systems and are crucial in signal processing and control theory due to their predictable and consistent behavior over time shifts. Time-invariant systems are used extensively in digital signal processing for filtering, modulation, and other operations where consistent and predictable responses are required.

2.3.3 *Causality*

Causality is a fundamental concept in system theory and signal processing. A causal (physical or non-anticipative system) system depends on the current and past values of the input signal to determine the current output. This means the system does not anticipate future inputs; it responds only to present and past inputs. Causality is essential in real-world systems because it aligns with the physical principle that causes precede effects.

Mathematically, a continuous system with input signal $f(t)$ and the output signal $g(t) = \mathcal{H}[f(t)]$ is causal if for any two input signals $f_1(t)$ and $f_2(t)$ and for any time t_0, the following condition holds:

$$f_1(t) = f_2(t) \text{ for all } t \leq t_0 \Rightarrow g_1(t) = g_2(t) \text{ for all } t \leq t_0 \tag{2.32}$$

where $g_1(t) = \mathcal{H}[f_1(t)]$ and $g_2(t) = \mathcal{H}[f_2(t)]$. Two input signals are identical up to a certain time t_0; their corresponding outputs must also be identical up to that same time. In discrete form, for $g[n] = \mathcal{H}[f[n]]$, the causal system is mathematically expressed as

$$f_1[n] = f_2[n] \text{for all} n \leq n_0 \Rightarrow g_1[n] = g_2[n] \text{ for all } n \leq n_0 \tag{2.33}$$

$g(x) = f(x - x_0)$ for $x_0 \geq 0$ and $g(x) = \int_{-\infty}^{x} e^{-(x-\tau)} f(\tau)\, d\tau$ are examples of causal systems. On the contrary, $g(x) = f(x + x_0)$ for $x_0 \geq 0$ is the anticipatory or non-causal system. Understanding and identifying causality in systems is crucial for a foundational framework for comprehending the temporal relationships within systems.

2.3.4 Stability

Signal stability refers to a system's property whereby bounded input signals result in bounded output signals. This property is essential for ensuring that a system's response does not grow uncontrollably, which is crucial for practical applications like communications, control systems, and electronics. Mathematically, for a bounded input signal $f(x)$ for all x and some finite constant M_x, the output signal $g(x) = \mathcal{H}[f(x)]$ is stable for the following conditions:

$$|f(x)| \leq M_x \Rightarrow |g(x)| \leq M_y \tag{2.34}$$

M_y is a finite constant. For a discrete system, the same is expressed as

$$|f[n]| \leq M_x \Rightarrow |g[n]| \leq M_y \tag{2.35}$$

In signal processing, this is called bounded-input, bounded-output (BIBO) stability. For linear time-invariant (LTI) systems, stability can be analyzed through the system's impulse response. The output $g(t)$ of a continuous time LTI system can be expressed as the convolution of the input $f(t)$ with the impulse response $h(t)$.

$$g(t) = \int_{-\infty}^{\infty} f(\tau)h(t-\tau)\,d\tau \tag{2.36}$$

The system is BIBO stable if and only if the impulse response $h(t)$ is absolutely integrable.

$$\int_{-\infty}^{\infty} |h(t)|\,dt < \infty \tag{2.37}$$

For a discrete LTI system, the output $g[n]$ of a discrete-time LTI system can be expressed as the convolution of the input $f[n]$ with the impulse response $h[n]$.

$$g[n] = \sum_{k=-\infty}^{\infty} f[k]h[n-k] \tag{2.38}$$

The discrete system is BIBO stable if and only if the impulse response $h[n]$ is absolutely summable.

$$\sum_{k=-\infty}^{\infty} |h[k]| < \infty \tag{2.39}$$

2.3.5 Memory

Systems can be classified based on their memory into memoryless, finite, and infinite memory systems. A memoryless system is a type of system where the output at any given time depends solely on the input at that same time. In other words, the system has no memory of past inputs, and future inputs do not influence the current output. For a continuous system,

$$g(x) = \mathcal{H}[f(x)] \tag{2.40}$$

For a discrete system,

$$g[n] = \mathcal{H}[f[n]] \tag{2.41}$$

A finite memory system is a system where the output at any time depends only on a finite number of past input values. The system remembers a fixed window of input values, not the entire past history. For a continuous system,

$$g(x) = \mathcal{H}[f(x), f(x - \Delta x), f(x - 2\Delta x), \ldots, f(x - N\Delta x)] \tag{2.42}$$

For a discrete system,

$$g[n] = \mathcal{H}[f[n], f[n-1], f[n-2], \ldots, f[n-N]] \tag{2.43}$$

A moving average filter of length N computes the output as the average of the last N input samples; $g\,[n] = \frac{1}{N}\sum_{k=0}^{N-1} f\,[n-k]$ is a finite memory system.

An infinite memory system is a system where the output at any time depends on all past input values. These systems have an infinite memory length, meaning the entire history of the input influences the current output.

$$g(x) = \int_{-\infty}^{x} h(x-\tau) f(\tau)\, d\tau \tag{2.44}$$

$g(x)$ depends on the entire past input $f(\tau)$ for $-\infty < \tau \leq x$. $h(x)$ is the impulse response of the system. For a discrete infinite memory system,

$$g[n] = \sum_{k=-\infty}^{n} h[n-k] f[k] \tag{2.45}$$

Infinite impulse response (IIR) filter $g[n] = \sum_{k=0}^{N} b_k f_1[n-k] + \sum_{j=1}^{M} a_j f_2[n-j]$ is an infinite memory system. b_k and a_j are filter coefficients.

2.4 Noise

In signal processing, "noise" refers to any unwanted, unpredictable, or random disturbance that interferes with the original signal, degrading its quality and making it more difficult to interpret or transmit the information it carries accurately. Noise is an omnipresent challenge in signal processing, affecting the accuracy and reliability of data transmission, storage, and interpretation.

Noise can originate from various sources, including environmental factors, electronic components, and even inherent thermal fluctuations in the system. Thermal noise, also known as Johnson-Nyquist noise, arises from the random motion of electrons in conductive materials due to thermal energy. Thermal noise is inherent in all electronic devices and becomes more pronounced as temperature increases. Shot noise is another form generated by the discrete nature of electric charge in devices like transistors and diodes, where the flow of electrons or holes occurs in small, quantized packets. Environmental noise encompasses various disturbances, from electromagnetic interference (EMI) caused by nearby electronic devices to acoustic noise from mechanical vibrations or airflow. For instance, in wireless communication systems, signals can be corrupted by interference from other wireless networks, leading to data loss or errors. In digital signal processing, noise can also result from quantization, the process of mapping a large set of input values to a smaller set, such as when converting an analog signal to a digital one. This introduces quantization noise, which is a by-product of rounding or truncating errors during the conversion process.

Dealing with noise is crucial for maintaining signal integrity. Various techniques are employed to mitigate its effects, such as filtering, which removes or reduces unwanted components from a signal. Noise reduction algorithms, such as averaging or adaptive filtering, are also commonly used to enhance the signal-to-noise ratio (SNR), thereby improving the clarity and usability of the signal.

2.5 Closure

This chapter embarked on an in-depth exploration of fundamental concepts of signals and systems. We began by defining various types of signals, including continuous-time and discrete-time signals, and examined their different representations. We covered basic properties and operations of signals, such as scaling, shifting, and flipping, which are essential for effective signal manipulation and analysis. Following this, we investigated different types of systems, including linear and non-linear, time-invariant and time-variant, causal and non-causal, and stable and unstable systems. Understanding these properties is crucial for predicting system behavior and ensuring the reliability and efficiency of signal-processing applications. Foundational operations and properties of signals and systems discussed in this chapter will be instrumental in subsequent chapters of this book.

Reference

Bracewell, R. (1965). *The fourier transform and its applications* (1st ed.). McGraw Hill.

Chapter 3
The Fourier Series

Primary causes are unknown to us; but are subject to simple and constant laws, which may be discovered by observation, the study of them being the object of natural philosophy. Heat, like gravity, penetrates every substance of the universe, its rays occupy all parts of space. The object of our work is to set forth the mathematical laws which this element obeys. The theory of heat will hereafter form one of the most important branches of general physics

—Joseph Fourier

Abstract This chapter offers both theoretical and practical perspectives on the Fourier series. It begins by drawing inspiration from periodic behaviors observed in nature, illustrating the motivation for using the Fourier series. The chapter then explores the Fourier series expansion of odd and even functions, which enables efficient decomposition of signals with particular symmetries and simplifies complex signal representations. Moving further, it introduces the complex number representation of the Fourier series, a powerful method for compactly expressing periodic functions. This approach is extended with the double Fourier series, facilitating the representation of functions that depend on two variables and are useful in multidimensional signal processing. Practical examples of Fourier series expansions for commonly encountered signals are provided, demonstrating the versatility and wide applications of Fourier analysis.

Keywords Fourier series · Superposition · Euler formula · Gibbs phenomenon · Parseval identity · Half-range expansions · Double Fourier series · Sawtooth wave · Triangular wave · Forward pulse

3.1 Observation from Nature

The evolution of science and mathematics is a testament to humanity's enduring fascination with the natural world and the quest to decipher its mysteries. From the

S. K. Jena, *Fourier, Laplace, and the Tangled Love Affair with Transforms*,
https://doi.org/10.1007/978-3-031-80165-5_3

dawn of civilization, keen observers of the heavens noted the regularity of celestial movements, which fostered the nascent fields of astronomy and timekeeping. Ancient Greeks, inspired by the geometrical elegance observed in nature, such as the symmetry of leaves and the harmony of proportions in the human body, laid the foundations of geometry and number theory. Through rigorous contemplation and empirical scrutiny, these early scholars formulated principles that sought to explain the natural order. With its spirit of inquiry and rediscovery, the Renaissance era witnessed a profound scientific and mathematical exploration resurgence. Visionaries like Italian physicist and astronomer Galileo Galilei (1564–1642) and German astronomer Johannes Kepler (1571–1630), driven by meticulous observations of planetary motions, revolutionized our understanding of the cosmos and cemented the heliocentric model. Simultaneously, the advent of calculus by Issac Newton and German mathematician Gottfried Wilhelm Leibniz (1646–1716) provided the mathematical tools necessary to describe motion and change, further unifying the realms of physics and mathematics. Over time, the curiosity to understand the underlying principles of these patterns drove the formulation of hypotheses and experimental methods, giving birth to the scientific method. With its boundless complexity and inherent order, nature inspires contemporary research. This iterative process of observation, theory, and experimentation allowed for the refinement of knowledge, leading to the discovery of the sophisticated science and mathematics we have today. Thus, the relentless pursuit of understanding the natural world through meticulous observation has been the cornerstone of the evolution of these disciplines.

Before delving into the rigorous mathematical description of the Fourier series, let us first appreciate the serene image of mother swans swimming with their cygnets in a pond (Fig. 3.1). Each time a swan flaps its wings, a simple water wave is created. The amplitude and frequency of these waves differ significantly between the mother

Fig. 3.1 Superimposing water waves formed behind swans during swimming

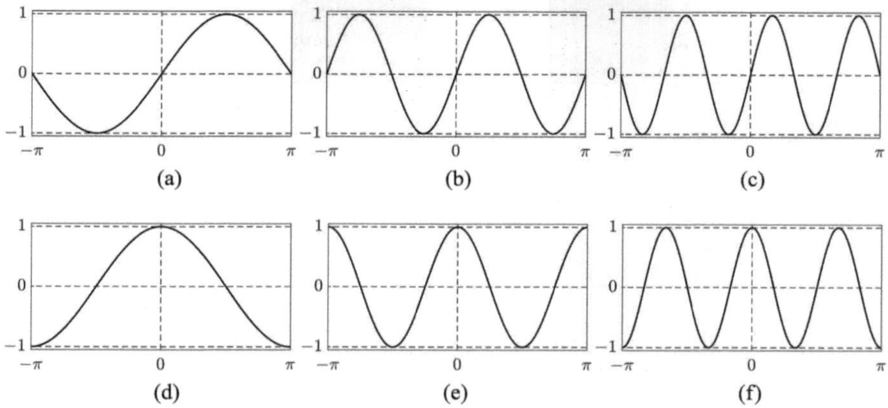

Fig. 3.2 Sine and cosine and functions over a period 2π. (**a**) $\sin x$. (**b**) $\sin 2x$. (**c**) $\sin 3x$. (**d**) $\cos x$. (**e**) $\cos 2x$. (**f**) $\cos 3x$

swan and her babies. Eventually, all these individual wave patterns combine to form complex wave patterns seen in the pond. Conversely, any intricate pattern in nature results from the superimposition of several simple phenomena. Mathematics plays a crucial role in unravelling the mysteries of these beautiful natural occurrences.

Figure 3.2 presents simple sinusoids over a period of 2π. Figure 3.3 represents the structure forms from the superimposition of these simple sinusoids. It is quite flabbergasting and interesting to notice how these simple sinusoids superimpose together in some mysterious conspiracy to express any functions in the interval $[-\pi, \pi]$, even with jump discontinuity. In Chap. 1, it has been discussed how a function $f(x)$ can be expanded using an orthonormal basis in an interval $[-L, L]$. $\{1, \cos x_i, \cos 2x_i, \cdots \cos nx_i, \sin x_i, \sin 2x_i, \cdots \sin nx_i\}$ forms an orthogonal basis in the interval $[-\pi, \pi]$ having a period of 2π (fundamental period). The first term of the basis 1 is periodic with any period.

Any periodic function $f(x)$ with period 2π in the interval $x \in [-\pi, \pi]$ formed due to superimposition of an infinite number of harmonics is mathematically expressed as

$$f(x) = \sum_{n=-\infty}^{\infty} (A_n \cos nx + B_n \sin nx) \tag{3.1}$$

The sum is taken from $-\infty$ to ∞ because of mathematical symmetry. This process of constructing a function by adding harmonics of various amplitudes is called Fourier synthesis. Using the odd-even properties of harmonics $\cos(-x) = \cos(x)$ and $\sin(-x) = -\sin(x)$, the series above series can be rewritten in the form

$$f(x) = \frac{a_0}{2} + \sum_{n=1}^{\infty} (a_n \cos nx + b_n \sin nx) \tag{3.2}$$

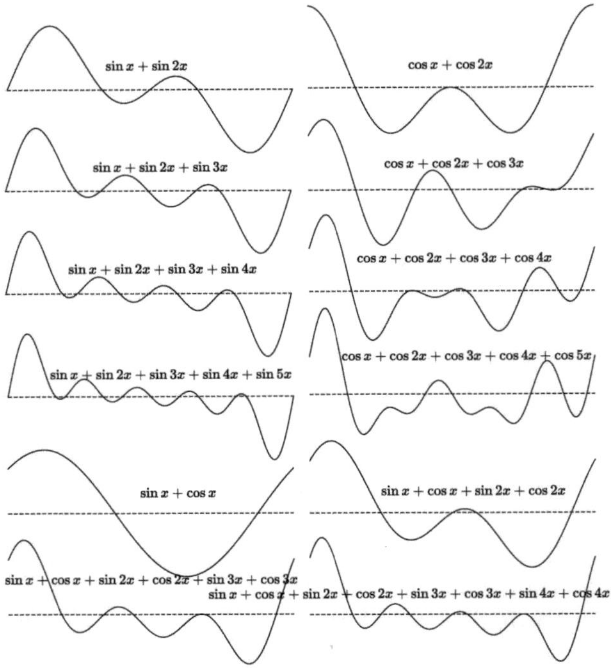

Fig. 3.3 Structures from the superimposition of sinusoids in 2π

where

$$
\begin{aligned}
a_n &= \frac{1}{2}\left(A_{-n} + A_n\right) \\
b_n &= \frac{1}{2}\left(B_{-n} + B_n\right)
\end{aligned}
\tag{3.3}
$$

Equation (3.2) is known as the Fourier series. This series is the crowning glories of eighteenth-century mathematics. The terms a_n and b_n represent the amplitude of the harmonics, and a_0 is called the DC or RMS value of the function $f(x)$ (named from its use in signal analysis). The term a_0 is divided by two to avoid counting it twice as a_0 is found from the same formula used to calculate a_n. Some books like Kreyszig et al. (2011) and Jeffrey (2002) used the notation a_0 instead of $\frac{a_0}{2}$ to represent the Fourier series. Despite the difference in notations, the series remains the same and should not be confused. Here we adhere to the classical definitions as Simmons (2017) and Zill et al. (2013) and use the notation $\frac{a_0}{2}$ instead of a_0.

Joseph Fourier is not the original inventor of the series in Eq. (3.2). Neugebauer (1952) has reported that centuries before the birth of Joseph Fourier, a similar primitive series was used in Babylonian civilization to calculate the astronomical movement of celestial bodies. In modern history, the use of this series arose when solving the classical wave equation generated by oscillating violin strings. D'Alembert (1747a,b) derived the deflection of violin string as a function of time and space using the superposition of two travelling waves in two opposite directions. Euler (1748) proposed the solution of the vibrating string problem as the summation of a series of sine waves. Bernoulli (1755) derived the solution of a vibrating string through an infinite trigonometric series form. Controversy arises regarding the convergence of the series proposed by Bernoulli. Euler, D'Alembert, and Laplace were not convinced about the existence of Bernoulli's series solution. In 1759, young mathematical scientist Lagrange joined in this controversial debate and gave an entirely new treatment for the solution of vibrating string. Both Euler and D'Alembert criticized Lagrange for his solution. In 1768, D'Alembert wrote a series of notes in which he attacked Euler's views on the problem's solution. In 1779, Laplace entered into this debate and supported D'Alembert's views. Different solutions arose from different scientists out of the controversy. For a long time, the prime issue regarding the convergence of such series for any arbitrary function was not answered. Joseph Fourier was engaged in the task of cellar design to preserve wine for Napoleon Bonaparte. During the cellar design process, Fourier discovered the one-dimensional heat equation, and in 1807, he used the controversial series expansion to solve the steady-state heat equation. The idea was that the temperature at a particular place on Earth is periodic, with a time period of 1 year. He found heat conduction in an object like a circular disc is periodic due to rotational symmetry. To solve this kind of periodic phenomenon that appeared in his design problem, Fourier postulated that any arbitrary periodic function $f(x)$ could be represented using the trigonometric series. He communicated a paper to the Academy of Sciences in Paris, mentioning the applicability of such a series in solving heat transfer. The idea was initially questioned and opposed by contemporary mathematicians Laplace, Poisson, and Lagrange due to a lack of rigor. Fourier (1822) was finally published with the mathematical treatment of problems involving heat conduction, and eventually, the series became famous as the "Fourier Series". The argument of Fourier regarding the convergence of series expansion was not mathematically rigorous; subsequently, Dirichlet (1829) and Riemann (1867) did rigorous mathematical underpinning to represent a convergent Fourier series. Development of the Fourier series opened the gate to many other branches of mathematics and is used in many applied fields like mathematical physics, probability, mathematical statistics, mathematical economics, harmonic analysis, signal analysis, wave propagation, wavelet analysis, and solution of boundary value partial differential equation. The Fourier series is more universal than the Taylor or Maclaurin series as it accommodates the jump discontinuities, where the other series fails.

The amplitude of the harmonics in Eq. 3.3 is also known as Fourier coefficients and is determined using the principle of orthogonality. To obtain the value of $\frac{a_0}{2}$, let us integrate the series in Eq. (3.2), in the interval $[-\pi, \pi]$. Term-by-term integration results

$$\int_{-\pi}^{\pi} f(x)\,dx = a_0\pi$$
$$\frac{a_0}{2} = \frac{1}{2\pi}\int_{-\pi}^{\pi} f(x)\,dx \tag{3.4}$$

To obtain the amplitude of the cosines a_n, let us multiply $\cos nx$ with both sides of Eq. (3.2) and integrate term by term in the interval $[-\pi, \pi]$. Using orthogonality principle

$$\int_{-\pi}^{\pi} f(x)\cos nx dx = \int_{-\pi}^{\pi} a_n \cos^2 nx dx = a_n\pi$$
$$a_n = \frac{1}{\pi}\int_{-\pi}^{\pi} f(x)\cos nx dx \tag{3.5}$$

Similarly to evaluate the amplitude of the sine components, $\sin nx$ is multiplied in both sides of Eq. (3.2) and integrated in the interval $[-\pi, \pi]$

$$\int_{-\pi}^{\pi} f(x)\sin nx dx = \int_{-\pi}^{\pi} b_n \sin^2 nx dx = b_n\pi$$
$$b_n = \frac{1}{\pi}\int_{-\pi}^{\pi} f(x)\sin nx dx \tag{3.6}$$

Derivations in Eqs. (3.4)–(3.6) first appeared in the paper by Euler (1777) and are hence well known as the Euler formula.

Example 3.1 (Periodic Rectangular Wave) Find the Fourier series expansion of a pulsating square wave expressed as

$$f(x) = \begin{cases} -k & \text{if} \quad -\pi < x < 0 \\ k & \text{if} \quad 0 < x < \pi \end{cases}$$

and

$$f(x \pm 2\pi) = f(x)$$

(continued)

Example 3.1 (continued)
Solution: The shape of the square wave function is given in Fig. 3.4a. The first Fourier coefficient or mean value of the square wave is zero.

$$\frac{a_0}{2} = \frac{1}{2\pi}\left[\int_{-\pi}^{0} -k dx + \int_{0}^{\pi} k dx\right] = 0$$

Fourier cosine coefficient is too zero (even function, symmetric about ordinate)

$$\begin{aligned} a_n &= \frac{1}{\pi}\left[\int_{-\pi}^{0} -k \cos nx dx + \int_{0}^{\pi} k \cos nx dx\right] \\ &= \frac{1}{\pi}\left[-k \left.\frac{\sin nx}{n}\right|_{-\pi}^{0} + k \left.\frac{\sin nx}{n}\right|_{0}^{\pi}\right] \\ &= 0 \end{aligned}$$

The Fourier sine coefficient is

$$\begin{aligned} b_n &= \frac{1}{\pi}\left[\int_{-\pi}^{0} -k \sin nx dx + \int_{0}^{\pi} k \sin nx dx\right] \\ &= \frac{1}{\pi}\left[k \left.\frac{\cos nx}{n}\right|_{-\pi}^{0} - k \left.\frac{\cos nx}{n}\right|_{0}^{\pi}\right] \\ &= \frac{2k}{n\pi}(1 - \cos n\pi) \end{aligned}$$

For $n = 1, 2, 3, \cdots$, hence

$$b_1 = \frac{4k}{\pi}, b_2 = 0, b_3 = \frac{4k}{3\pi}, b_4 = 0, b_5 = \frac{4k}{5\pi}, \cdots$$

The Fourier series expansion of the square wave is expressed as

$$f(x) = \frac{4k}{\pi}\left(\sin x + \frac{1}{3}\sin 3x + \frac{1}{5}\sin 5x + \frac{1}{7}\sin 7x + \cdots\right)$$

(continued)

Example 3.1 (continued)
Figure 3.4b presents the $f(x)$ for $n = 1$ and Fig. 3.4c for $n = 2$. From Fig. 3.4e and f, it's clear that as $n \to \infty$, the series takes the shape of the square wave, except at the point of discontinuity $\{-\pi, 0, \pi\}$, where a ripple (pronounced spikes or wiggles) is forming. This overshooting by the partial sum from the expanded Fourier series near-jump-discontinuity location does not smooth out but remains fairly constant, even when the value n is taken to be large (Fig. 3.4f). When the Fourier series approximates a function with a jump discontinuity, the partial sums of the series exhibit overshoot and oscillations near the discontinuity. This behavior is known as the Gibbs phenomenon. This phenomenon is named after American physicist Josiah Willard Gibbs (1839–1903). American physicists Albert Abraham Michelson (1852–1931) and Samuel Wesley Stratton (1861–1931) experimentally observed the Gibbs phenomenon while designing a mechanical device to draw finite Fourier series. Michelson and Stratton assumed these extra wiggles at jump locations were due to mechanical problems with the machine. However, for the first time, Gibbs stated that these phenomena are real and do not go away. The oscillation is gradually compressed with an increasing value of n and approaches a finite value typically around 9% of the jump's magnitude. However, they contribute zero in the limit of the L^2 norm of the difference between the function and its Fourier series.

Value of the expanded function at $x = \dfrac{\pi}{2}$ becomes

$$f\left(\frac{\pi}{2}\right) = \frac{4k}{\pi}\left(1 - \frac{1}{3} + \frac{1}{5} - \frac{1}{7} + \cdots\right) = \frac{4k}{\pi}\frac{\pi}{4} = k$$

This exactly equals its analytical description. Leibniz got this famous result in 1673 from geometric consideration. Figure 3.5 represents the amplitude of each harmonic in the expanded Fourier series. For high-frequency oscillations, the amplitude gradually diminishes to zero; hence, with a finite value of n, the Fourier series of square wave converges to its functional value.

Equation (3.2) represents a function having a period 2π. However, all real-valued periodic functions in $\mathbb{R}^3$ may not have a period of 2π; there might be some periodic function $f(x)$ defined in the interval $[-L, L]$ having the period of $2L$. In order to expand such function using sinusoids basis, a scaled-up method is used. Let y be any new variable such that the same function $(f(y))$ has a period of 2π with respect to the new variable y. The function f remains unaltered by changing the variable x to y; hence,

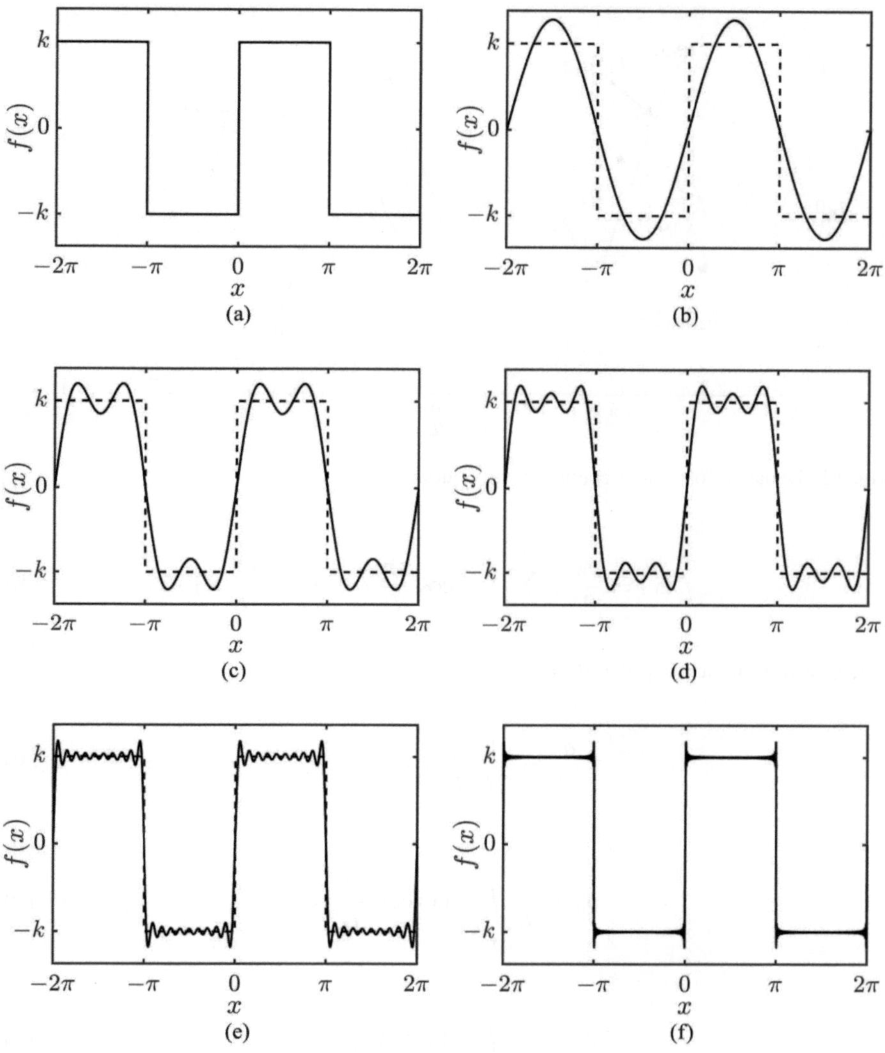

Fig. 3.4 Fourier series expansion of periodic square wave. (**a**) $f(x)$. (**b**) $n = 1$. (**c**) $n = 2$. (**d**) $n = 3$. (**e**) $n = 10$. (**f**) $n = 100$

$$
\begin{aligned}
2Ly &= 2\pi x \\
y &= \frac{2\pi}{x}
\end{aligned}
\tag{3.7}
$$

$f(y)$ at $y = \pm\pi$ corresponds to $x = \pm L$. Symbolically replacing y with x, Fourier series for any arbitrary function $f(x)$ having a period of $2L$ is represented by

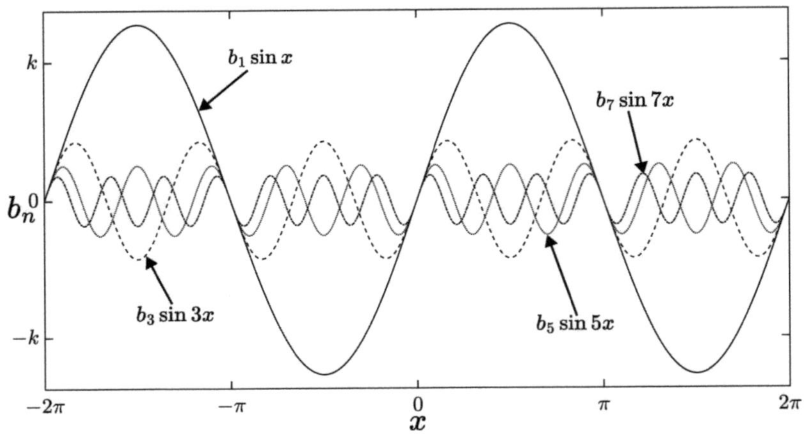

Fig. 3.5 Dynamics of various coefficients of Fourier series

$$f(x) = \frac{a_0}{2} + \sum_{n=1}^{\infty} \left(a_n \cos \frac{n\pi}{L} x + b_n \sin \frac{n\pi}{L} x \right) \tag{3.8}$$

The coefficients are represented as

$$\frac{a_0}{2} = \frac{1}{2L} \int_{-L}^{L} f(x_i)\, dx \tag{3.9}$$

$$a_n = \frac{1}{L} \int_{-L}^{L} f(x) \cos \frac{n\pi}{L} x dx \tag{3.10}$$

$$b_n = \frac{1}{L} \int_{-L}^{L} f(x) \sin \frac{n\pi}{L} x dx \tag{3.11}$$

$$n = 1, 2, 3, \cdots$$

Viability of the Fourier series expansion of a function in Eq. (3.8) depends on the existence and convergence of the integrals in Eqs. (3.9)–(3.11). Dirichlet (1829) showed that the function f needs to be bounded for the existence and convergence of the above integrals in Eqs. (3.9)–(3.11). Mathematically, the boundedness of a function in $[-L, L]$ is expressed as

$$|f(x)| \leq M \tag{3.12}$$

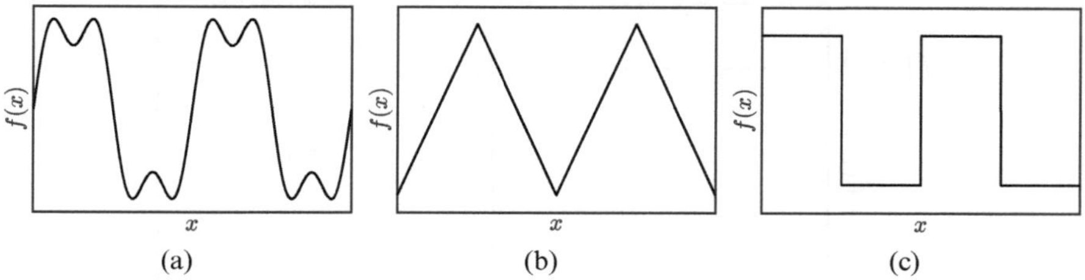

Fig. 3.6 Types of functions

M is some finite constant for all x in $[-L, L]$. The total energy content of the function (system/signal) in $[-L, L]$ is expressed as

$$E = \frac{1}{2L} \int_{-L}^{L} |f(x)|^2 dx \tag{3.13}$$

An important point to notice here is $|f(x)|$ and $|f(x)|^2$ are used instead of $f(x)$ and $f^2(x)$ to represent a complex valued function. For real valued function, $|f(x)|^2 = f^2(x)$. A finite energy-containing function $E < \infty$ in $[-L, L]$ is also called a square-integrable function in L^2 space. A function f needs to be square integrable in $[-L, L]$ for the existence of its Fourier series. For a square-integrable function, the Fourier series is said to be converged if the amplitude of higher-frequency terms in the Fourier series gradually approaches zero (shown in Fig. 3.5). Physically, the same signifies that the Fourier series ideally replicates the original function. To clarify the concept of convergence, let us classify the functions into three categories. Figure 3.6a represents a smooth function, i.e., everywhere continuous and differentiable. The Fourier series converges at every point in x for smooth functions. The finite sum of the Fourier series replicates the original function, and the convergence rate depends upon the smoothness. Figure 3.6b represents the category of continuous functions everywhere in x but are non-smooth at some discrete points. For these categories of functions, relatively high-frequency terms are required for convergence at non-smooth points. Figure 3.6c represents the function having finite jump discontinuity. From Example 3.1, it has already been noticed, irrespective of any high-frequency term consideration, that the series does not converge at jump discontinuity location (Gibbs phenomenon). It is convenient to replace $=$ with $\approx$ to accommodate the Gibbs phenomenon. If $f(x)$ and $f'(x)$ are piecewise continuous functions in $[-L, L]$, then for all x_i in the interval $[-L, L]$,

$$f(x) \approx \frac{a_0}{2} + \sum_{n=1}^{\infty} \left(a_n \cos \frac{n\pi}{L} x + b_n \sin \frac{n\pi}{L} x \right) \tag{3.14}$$

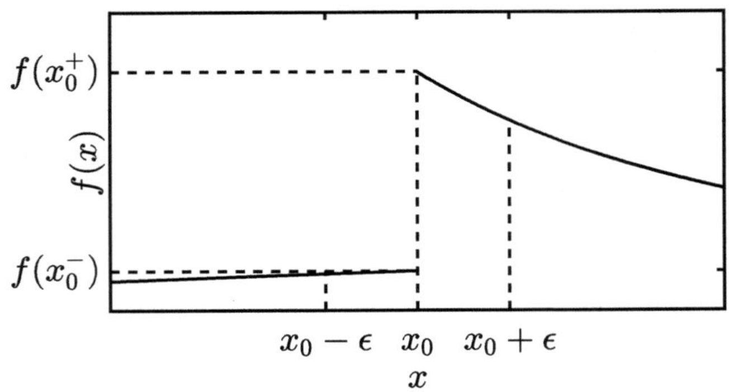

Fig. 3.7 Function with simple discontinuity

The Fourier series of the function converges to its functional value $f(x)$ at the continuity points. At finite jump locations, the Fourier series converges to the average of the limiting functional value. This is known as the Dirichlet theorem for the Fourier series. For a physical understanding of the Dirichlet theorem, consider Fig. 3.7. At a point x_0, the function has jump discontinuity. That means the finite left-hand limit and right-hand limit are unequal. At this jump discontinuity location, Fourier series expansion $S_N(x)$ (S_N is finite term summation of the Fourier series) converges to a value having an average of left and right hand limiting value of original function $f(x)$. For any arbitrary small positive value ε, mathematically, the same is expressed as

$$S_N(x_0) = \frac{f(x_0 - \varepsilon) + f(x_0 + \varepsilon)}{2} = \frac{f\left(x_0^-\right) + f\left(x_0^+\right)}{2} \tag{3.15}$$

Equation (3.15) is known as the Dirichlet condition for convergence of Fourier series at finite jump location. The Dirichlet condition is sufficient but not necessary for the convergence. If the condition is satisfied, then the convergence is guaranteed; however, the series may or may not converge if it is not satisfied.

Mean square convergences of the Fourier series often make more sense than point-wise convergence. The mean square convergence is expressed as

$$\lim_{N\to\infty} \int_{-L}^{L} [f(x) - S_N(x)]^2 dx = 0 \tag{3.16}$$

Simplifying Eq. (3.16) using the principle of orthogonality leads to an expression.

$$\frac{1}{2L} \int_{-L}^{L} |f(x)|^2 dx = \frac{1}{2} a_0^2 + \sum_{n=1}^{\infty} \left(a_n^2 + b_n^2\right) \tag{3.17}$$

Equation (3.17) is known as the Parseval relation (named after French mathematician Marc-Antoine Parseval des Chênes (1755–1836)) for the Fourier series or Rayleigh Identity. Rayleigh's identity is an essential aspect of the Fourier series. Physically, it signifies the conservation of energy. The right side of Eq. (3.17) is analogous to obtaining the magnitude of a vector in Cartesian space from the summation of the square of its components. In infinite-dimensional Fourier space, the sum of the square of individual components gives the function's total energy content. The function's total energy remains the same in its functional and Fourier series forms. In the subsequent section, we see that Rayleigh's identity will appear as Parseval identity for the Fourier integral transform. Parseval identity permits variables to transfer from the spatial domain to the wavenumber domain and time domain to the frequency domain without violating the fundamental principle of energy conservation. The same holds for electrical signals and optics.

Fourier series can be written in form of a single sinusoid by considering $a_n = R_n \cos\phi_n$ and $b_n = R_n \sin\phi_n$. Then the Fourier series in Eq. (3.8) takes the form

$$f(x_i) = \frac{a_0}{2} + \sum_{n=1}^{\infty} R_n \cos\left(\frac{n\pi}{L}x_i + \phi_n\right) \tag{3.18}$$

where $R_n = \sqrt{\left(a_n^2 + b_n^2\right)}$ and $\phi_n = \tan^{-1}\left(\frac{b_n}{a_n}\right)$ are the amplitude and phase of nth harmonic. $|R_n|^2$ gives the energy of the nth harmonic mode. Two harmonics are said to be in phase, if both their crest or trough are reaching at a single point together (zero phase difference) and out of phase when trough of one harmonic is arriving with crest of other harmonic at a point together (π phase difference). Any phase angle in between 0 and π is the phase lag of one harmonic from other.

Fourier coefficients of the sum of f_1 and f_2 is equal to sum of the corresponding Fourier coefficients of f_1 and f_2. For any constant c, the Fourier coefficient of cf equals to c times the corresponding Fourier coefficients of f.

3.2 Fourier Series Expansion of Odd and Even Functions

In the previous section, we have noticed that the product of two even or two odd functions is even, while the product of one odd and one even function is odd. So if $f(x)$ is even, then $f(x)\sin nx$ is odd and its definite integral in the period $[-L, L]$ is zero and the Fourier series reduced to the form

$$f(x) = \frac{a_0}{2} + \sum_{n=1}^{\infty} a_n \cos\frac{n\pi}{L}x \tag{3.19}$$

This is known as the Fourier cosine series, and the coefficients are expressed as

$$\frac{a_0}{2} = \frac{1}{L}\int_0^L f(x)\,dx \tag{3.20}$$

$$a_n = \frac{2}{L}\int_0^L f(x)\cos\frac{n\pi}{L}x_i dx \tag{3.21}$$

If $f(x)$ is an odd function, then $f(x)\cos nx$ is too an odd function, and its definite integral in the limit $[-L, L]$ is zero. The Fourier series reduced to the form.

$$f(x) = \sum_{n=1}^{\infty} b_n \sin\frac{n\pi}{L}x \tag{3.22}$$

This is known as the Fourier sine series, having the coefficient

$$b_n = \frac{2}{L}\int_0^L f(x)\sin\frac{n\pi}{L}x dx \tag{3.23}$$

Half-range expansions are another simple and useful application Fourier series. Figure 3.8 represents a function $f(x)$ in the interval L. Fourier series expansion of the above function in the interval $[-L/2, L/2]$ can be done in terms of either Fourier sine or cosine series. But in order to simplify the problem, $f(x)$ can artificially extend to the interval $f[-L, 0]$, such that it can be represented either by an even function (Fig. 3.9a) or an odd function (Fig. 3.9b). Once the function is replicated by an even function in $[-L, L]$, then it can be expanded through a relatively simple Fourier cosine series. If the function is replicated by an odd function in $[-L, L]$, then it can be expanded through the Fourier sine series. Half-range expansion reduces the computational effort.

Fig. 3.8 The original function of period L

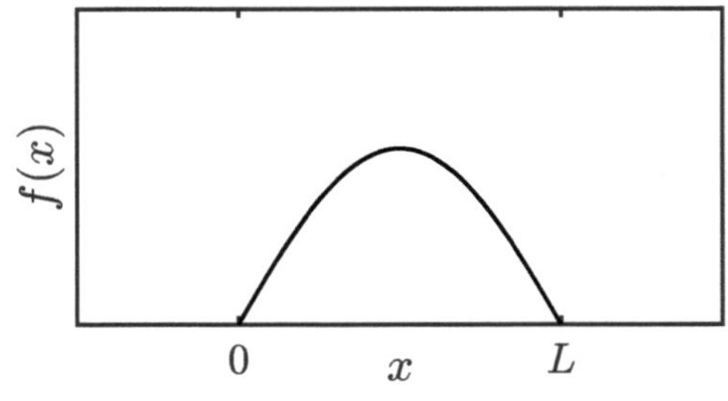

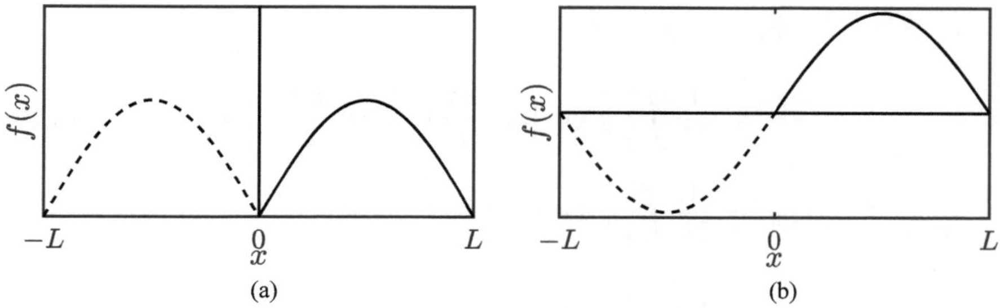

Fig. 3.9 Half-range expansion. (**a**) Even transformation. (**b**) Odd transformation

3.3 Complex Number Representation of Fourier Series

Using Euler formula, exponential functions $e^{i\frac{n\pi}{L}x}$ and $e^{-i\frac{n\pi}{L}x}$ are represented as

$$e^{i\frac{n\pi}{L}x} = \cos\frac{n\pi}{L}x + i\sin\frac{n\pi}{L}x \tag{3.24}$$

$$e^{-i\frac{n\pi}{L}x} = \cos\frac{n\pi}{L}x - i\sin\frac{n\pi}{L}x \tag{3.25}$$

where i is called iota and defined as $i = \sqrt{-1}$ and $\frac{1}{i} = -i$ (i should not be confused with subscript i, used for index notation). Conversely,

$$\cos\frac{n\pi}{L}x = \frac{1}{2}\left(e^{i\frac{n\pi}{L}x} + e^{-i\frac{n\pi}{L}x}\right) \tag{3.26}$$

$$\begin{aligned}\sin\frac{n\pi}{L}x &= \frac{1}{2i}\left(e^{i\frac{n\pi}{L}x} - e^{-i\frac{n\pi}{L}x}\right)\\ &= -\frac{i}{2}\left(e^{i\frac{n\pi}{L}x} - e^{-i\frac{n\pi}{L}x}\right)\end{aligned} \tag{3.27}$$

Using Eqs. (3.26) and (3.27),

$$\begin{aligned}a_n\cos\frac{n\pi}{L}x + b_n\sin\frac{n\pi}{L}x &= \frac{1}{2}a_n\left(e^{i\frac{n\pi}{L}x} + e^{-i\frac{n\pi}{L}x}\right) - \frac{i}{2}b_n\left(e^{i\frac{n\pi}{L}x} - e^{-i\frac{n\pi}{L}x}\right)\\ &= \frac{1}{2}(a_n - ib_n)\,e^{i\frac{n\pi}{L}x} + \frac{1}{2}(a_n + ib_n)\,e^{-i\frac{n\pi}{L}x}\\ &= \bar{c}_n e^{i\frac{n\pi}{L}x} + c_n e^{-i\frac{n\pi}{L}x}\end{aligned} \tag{3.28}$$

where

$$\begin{aligned}\bar{c}_n &= \frac{1}{2}(a_n - ib_n)\\ &= \frac{1}{2L}\int_{-L}^{L} f(x)\left(\cos\frac{n\pi}{L}x - i\sin\frac{n\pi}{L}x\right)dx \\ &= \frac{1}{2L}\int_{-L}^{L} f(x)\,e^{-i\frac{n\pi}{L}x}dx\end{aligned} \tag{3.29}$$

$$\begin{aligned}c_n &= \frac{1}{2}(a_n + ib_n)\\ &= \frac{1}{2L}\int_{-L}^{L} f(x)\left(\cos\frac{n\pi}{L}x + i\sin\frac{n\pi}{L}x\right)dx \\ &= \frac{1}{2L}\int_{-L}^{L} f(x)\,e^{i\frac{n\pi}{L}x}dx\end{aligned} \tag{3.30}$$

From Eqs. (3.29) and (3.30), $\bar{c}_{-n} = c_n$ (complex conjugates). For $n = 0$, Eq. (3.2) becomes

$$\frac{a_0}{2} = \frac{1}{2L}\int_{-L}^{L} f(x)\,e^{i\frac{n\pi}{L}x}dx \tag{3.31}$$

Combining Eqs. (3.29)–(3.31), the complex notation for Fourier series is expressed as

$$f(x) = \sum_{-\infty}^{\infty} c_n e^{i\frac{n\pi}{L}x} \tag{3.32}$$

where harmonic amplitude or Fourier coefficient in complex form c_n is expressed as

$$c_n = \frac{1}{2L}\int_{-L}^{L} f(x)\,e^{-i\frac{n\pi}{L}x}dx \tag{3.33}$$

$$n = 0, \pm 1, \pm 2, \cdot$$

The summation of the complex Fourier series is taken from $-\infty$ to ∞, as per the original Fourier series definition in Eq. (3.2). The real Fourier series considers summation from 1to ∞ due to symmetric properties.

3.4 The Double Fourier Series

As discussed earlier, some phenomena like waves are periodic in both time and space and represented as $f(x,t)$. Fourier series expansion of such bi-variate function is called double Fourier series. The principle of expansion remains the same as of single variable expansion. The Fourier series for the first variable is taken, keeping the second variable constant, and then the Fourier series of the second variable is taken, keeping the first variable constant. Let $f(x,t)$ be a function having spatial periodicity of $2L$ with $-L \leq x \leq L$ and temporal periodicity of $2T$ with $0 \leq t \leq 2T$. Let us consider Fourier series of $f(x,t)$ with respect to x, keeping t constant

$$f(x,t) = \sum_{m=0}^{\infty} \left(A_m(t) \cos \frac{m\pi x}{L} + B_m(t) \sin \frac{m\pi x}{L} \right) \tag{3.34}$$

Then allow t to vary by replacing Fourier coefficients $A_m(t)$ and $B_m(t)$ by Fourier series representation

$$A_m(t) = \sum_{n=0}^{\infty} \left(a_{mn} \cos \frac{n\pi t}{T} + b_{mn} \sin \frac{n\pi t}{T} \right) \tag{3.35}$$

$$B_m(t) = \sum_{n=0}^{\infty} \left(c_{mn} \cos \frac{n\pi t}{T} + d_{mn} \sin \frac{n\pi t}{T} \right) \tag{3.36}$$

Using Eqs. (3.35) and (3.36) in Eq. (3.34), the double Fourier series expansion of $f(x,t)$ in the interval $-L \leq x \leq L$ and $0 \leq t \leq 2T$ is expressed as

$$\begin{aligned} f(x,t) = {} & \sum_{m=0}^{\infty} \sum_{n=0}^{\infty} \left(a_{mn} \cos \frac{m\pi x}{L} \cos \frac{n\pi t}{T} + b_{mn} \cos \frac{m\pi x}{L} \sin \frac{n\pi t}{T} \right) \\ & + \sum_{m=0}^{\infty} \sum_{n=0}^{\infty} \left(c_{mn} \sin \frac{m\pi x}{L} \cos \frac{n\pi t}{T} + d_{mn} \sin \frac{m\pi x}{L} \sin \frac{n\pi t}{T} \right) \end{aligned} \tag{3.37}$$

Fourier coefficients in Eq. (3.37) are evaluated using the principle of orthogonality and expressed as

$$a_{mn} = \frac{1}{LT} \int_{L}^{L} \int_{0}^{2T} f(x,t) \cos \frac{m\pi x}{L} \cos \frac{n\pi t}{T} dt dx \tag{3.38}$$

$$b_{mn} = \frac{1}{LT}\int\limits_{L}^{L}\int\limits_{0}^{2T} f(x,t)\cos\frac{m\pi}{L}\sin\frac{n\pi t}{T}dtdx \tag{3.39}$$

$$c_{mn} = \frac{1}{LT}\int\limits_{L}^{L}\int\limits_{0}^{2T} f(x,t)\sin\frac{m\pi x}{L}\cos\frac{n\pi t}{T}dtdx \tag{3.40}$$

$$d_{mn} = \frac{1}{LT}\int\limits_{L}^{L}\int\limits_{0}^{2T} f(x,t)\sin\frac{m\pi x}{L}\sin\frac{n\pi t}{T}dtdx \tag{3.41}$$

for $m, n = 1, 2, 3, \ldots$. The constant coefficient is expressed as

$$a_{00} = \frac{1}{4LT}\int\limits_{L}^{L}\int\limits_{0}^{2T} f(x,t)\,dtdx \tag{3.42}$$

With other first cross coefficients

$$a_{m0} = \frac{1}{2LT}\int\limits_{L}^{L}\int\limits_{0}^{2T} f(x,t)\cos\frac{m\pi x}{L}dtdx \tag{3.43}$$

$$a_{0n} = \frac{1}{2LT}\int\limits_{L}^{L}\int\limits_{0}^{2T} f(x,t)\cos\frac{n\pi t}{T}dtdx \tag{3.44}$$

$$b_{m0} = 0 \tag{3.45}$$

$$c_{0n} = 0 \tag{3.46}$$

$$d_{0n} = 0 \tag{3.47}$$

$$d_{m0} = 0 \tag{3.48}$$

Because the index zero causes the sine function to vanish in the integrals' integrands defining these constants. For details derivation of double integral expansion, interested readers are encouraged to follow Jeffrey (2002).

3.5 Fourier Series Expansion of Some Common Signals

Example 3.2 (Full Sawtooth Wave)

$$f(x) = x \quad -\pi < x < \pi$$

$$\frac{a_0}{2} = \frac{1}{2\pi}\left[\int_{-\pi}^{\pi} x dx\right] = \frac{1}{2\pi}\frac{x^2}{2}\bigg|_{-\pi}^{\pi} = 0$$

$$a_n = \frac{1}{\pi}\left[\int_{-\pi}^{\pi} x \cos nx dx\right] = \frac{1}{\pi n^2}(nx \sin nx + \cos nx)|_{-\pi}^{\pi} = 0$$

$$\begin{aligned} b_n &= \frac{1}{\pi}\left[\int_{-\pi}^{\pi} x \sin nx dx\right] \\ &= \frac{1}{\pi n^2}(-nx \cos nx + \sin nx)|_{-\pi}^{\pi} \\ &= -\frac{2}{n}\cos n\pi = -\frac{2}{n}(-1)^n \end{aligned}$$

$$f(x) = \sum_{n=1}^{\infty} \frac{-2}{n}(-1)^n \sin nx$$

Figure 3.10 graphically represents the Fourier series expansion of a full sawtooth wave.

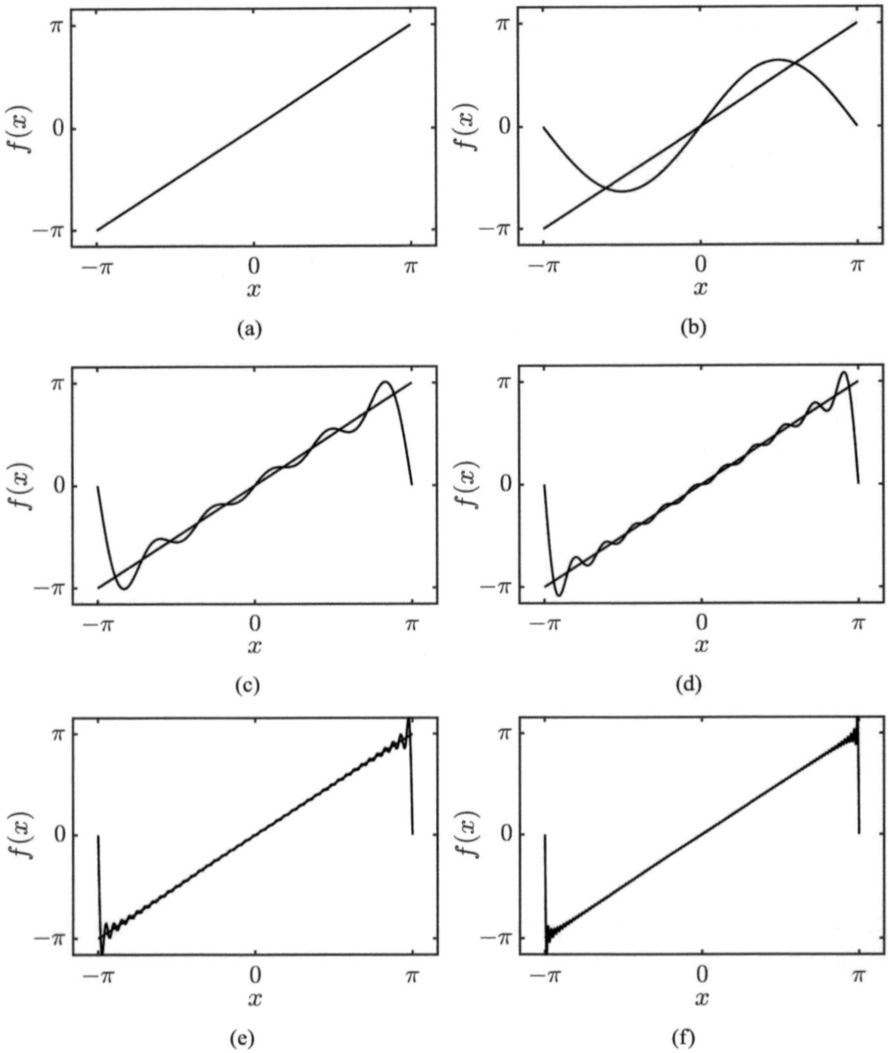

Fig. 3.10 Fourier series expansion of a full sawtooth wave. (**a**) $f(x)$. (**b**) $n = 1$. (**c**) $n = 5$. (**d**) $n = 10$. (**e**) $n = 40$. (**f**) $n = 100$

Example 3.3 (Half Sawtooth Wave)

$$f(x) = \begin{cases} x & 0 < x < \pi \\ 0 & -\pi < x < 0 \end{cases}$$

$$\frac{a_0}{2} = \frac{1}{2\pi}\left[\int_0^{\pi} x dx\right]$$

$$= \frac{1}{2\pi}\left.\frac{x^2}{2}\right|_0^{\pi} = \frac{\pi}{4}$$

$$a_n = \frac{1}{\pi}\left[\int_0^{\pi} x \cos nx dx\right]$$

$$= \frac{1}{\pi n^2}\left(nx \sin nx + \cos nx\right)|_0^{\pi}$$

$$= \frac{1}{\pi n^2}\left(\cos n\pi - 1\right)$$

$$= \frac{1}{\pi n^2}\left((-1)^n - 1\right)$$

$$b_n = \frac{1}{\pi}\left[\int_0^{\pi} x \sin nx dx\right]$$

$$= \frac{1}{\pi n^2}\left(-nx \cos nx + \sin nx\right)|_0^{\pi}$$

$$= \frac{1}{\pi n^2}\left(-nx \cos nx\right)|_0^{\pi}$$

$$= -\frac{1}{n}\cos n\pi = -\frac{(-1)^n}{n}$$

$$f(x) = \frac{\pi}{4} + \sum_{n=1}^{\infty} \frac{1}{\pi n^2}\left((-1)^n - 1\right)\cos nx - \frac{(-1)^n}{n}\sin nx$$

Figure 3.11 graphically represents the Fourier series expansion of a half sawtooth wave.

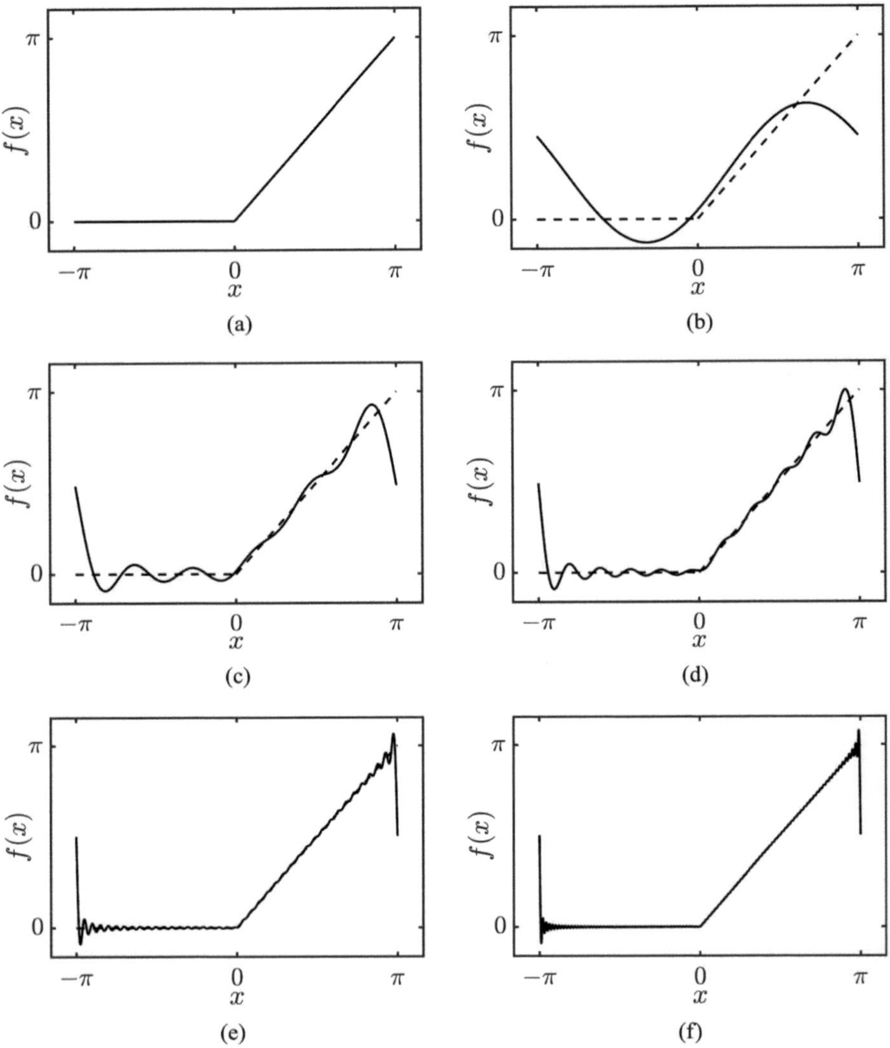

Fig. 3.11 Fourier series expansion of a half sawtooth wave. (**a**) $f(x)$. (**b**) $n = 1$. (**c**) $n = 5$. (**d**) $n = 10$. (**e**) $n = 40$. (**f**) $n = 100$

Example 3.4 (Lower Triangular Wave)

$$f(x) = \begin{cases} x & 0 < x < \pi \\ -x & -\pi < x < 0 \end{cases}$$

$$\frac{a_0}{2} = \frac{1}{2\pi}\left[\int_{-\pi}^{0} -xdx + \int_{0}^{\pi} xdx\right] = \frac{1}{2\pi}\left[-\left.\frac{x^2}{2}\right|_{-\pi}^{0} + \left.\frac{x^2}{2}\right|_{0}^{\pi}\right] = \frac{\pi}{2}$$

$$\begin{aligned} a_n &= \frac{1}{\pi}\left[\int_{-\pi}^{0} -x\cos nxdx + \int_{0}^{\pi} x\cos nxdx\right] \\ &= \frac{1}{\pi n^2}\left[(nx\sin nx + \cos nx)|_{0}^{-\pi} + (nx\sin nx + \cos nx)|_{0}^{\pi}\right] \\ &= \frac{1}{\pi n^2}\left[2(\cos n\pi - 1)\right] = \frac{2}{n^2\pi}\left[(-1)^n - 1\right] \end{aligned}$$

$$\begin{aligned} b_n &= \frac{1}{\pi}\left[\int_{-\pi}^{0} -x\sin nxdx + \int_{0}^{\pi} x\sin nxdx\right] \\ &= \frac{1}{\pi n^2}\left[(nx\cos nx - \sin nx)|_{-\pi}^{0} + (-nx\cos nx + \sin nx)|_{0}^{\pi}\right] \\ &= \frac{1}{\pi n^2}\left[n\pi\cos n\pi - n\pi\cos n\pi\right] = 0 \end{aligned}$$

$$f(x) = \frac{\pi}{2} + \sum_{n=1}^{\infty} \frac{2}{n^2\pi}\left[(-1)^n - 1\right]\cos nx$$

Figure 3.12 graphically represents the Fourier series expansion of a lower triangular wave.

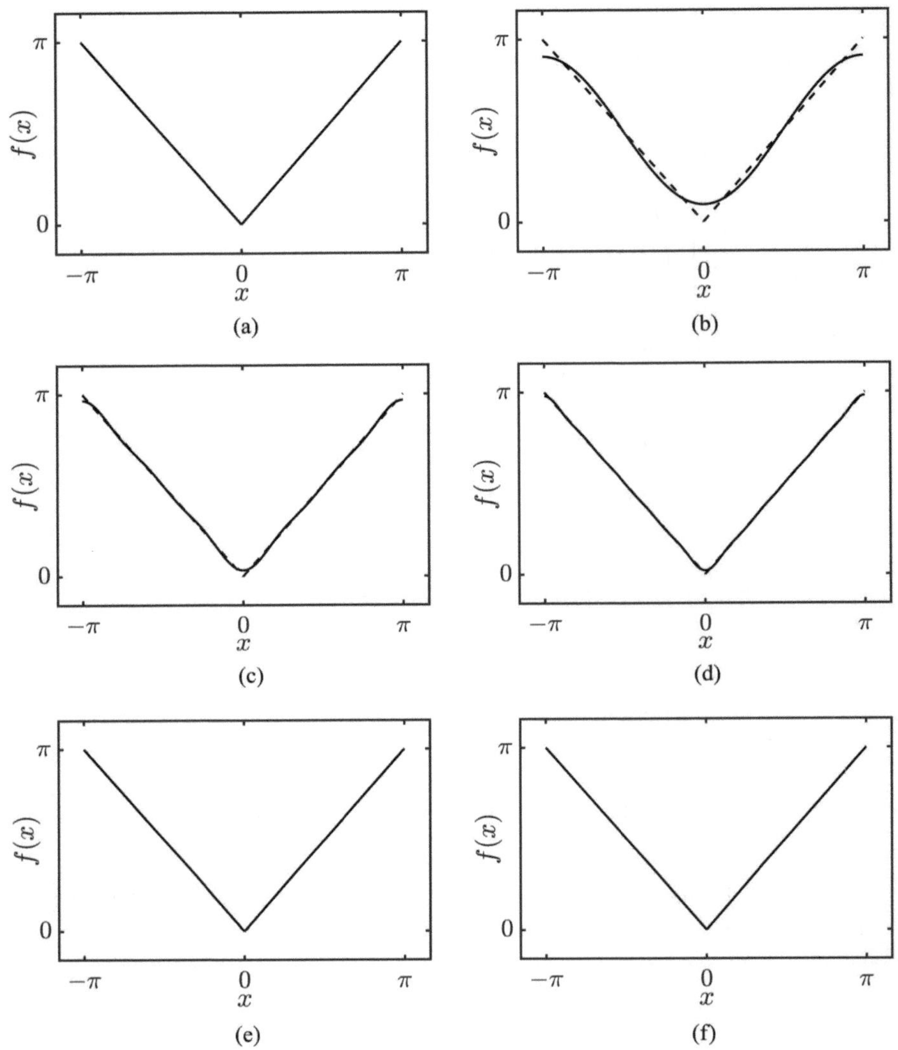

Fig. 3.12 Fourier series expansion of a lower triangular wave. (**a**) $f(x)$. (**b**) $n = 1$. (**c**) $n = 5$. (**d**) $n = 10$. (**e**) $n = 40$. (**f**) $n = 100$

Example 3.5 (Upper Triangular Wave)

$$f(x) = \begin{cases} \pi + x, & -\pi < x < 0 \\ \pi - x, & 0 \le x < \pi \end{cases}$$

(continued)

Example 3.5 (continued)

$$\frac{a_0}{2} = \frac{1}{2\pi}\left[\int_{-\pi}^{0}(\pi + x)\,dx + \int_{0}^{\pi}(\pi - x)\,dx\right]$$

$$= \frac{1}{2\pi}\left[\int_{-\pi}^{\pi}\pi dx + \int_{-\pi}^{0}xdx - \int_{0}^{\pi}xdx\right]$$

$$= \frac{1}{2\pi}\left[\pi x|_{-\pi}^{\pi} + \left.\frac{x^2}{2}\right|_{-\pi}^{0} - \left.\frac{x^2}{2}\right|_{0}^{\pi}\right] = \frac{\pi}{2}$$

$$a_n = \frac{1}{\pi}\left[\int_{-\pi}^{0}(\pi + x)\cos nxdx + \int_{0}^{\pi}(\pi - x)\cos nxdx\right]$$

$$= \frac{1}{\pi}\left[\pi\int_{-\pi}^{\pi}\cos nxdx - \int_{0}^{-\pi}x\cos nxdx - \int_{0}^{\pi}x\cos nxdx\right]$$

$$= \frac{1}{\pi}\left[\pi\left.\frac{\sin nx}{n}\right|_{-\pi}^{\pi} - \frac{1}{n^2}\left(nx\sin nx + \cos nx\right)\Big|_{0}^{-\pi}\right.$$

$$\left. - \frac{1}{n^2}\left(nx\sin nx + \cos nx\right)\Big|_{0}^{\pi}\right]$$

$$= \frac{2}{n^2\pi}(1 - \cos n\pi) = \frac{2}{n^2\pi}\left(1 - (-1)^n\right)$$

$$b_n = \frac{1}{\pi}\left[\int_{-\pi}^{0}(\pi + x)\sin nxdx + \int_{0}^{\pi}(\pi - x)\sin nxdx\right]$$

$$= \frac{1}{\pi}\left[\pi\int_{-\pi}^{\pi}\sin nxdx - \int_{0}^{-\pi}x\sin nxdx - \int_{0}^{\pi}x\sin nxdx\right]$$

$$= \frac{1}{\pi}\left[-\pi\left.\frac{\cos nx}{n}\right|_{-\pi}^{\pi} - \frac{1}{n^2}\left(nx\cos nx + \sin nx\right)\Big|_{0}^{-\pi} - \frac{1}{n^2}\left(nx\cos nx + \sin nx\right)\Big|_{0}^{\pi}\right]$$

$$= 0$$

$$f(x) = \frac{\pi}{2} + \sum_{n=1}^{\infty}\left\{\frac{2\left(1 - (-1)^n\right)}{n^2\pi}\cos nx\right\}$$

(continued)

Example 3.5 (continued)
Figure 3.13 graphically represents the Fourier series expansion of an upper triangular wave.

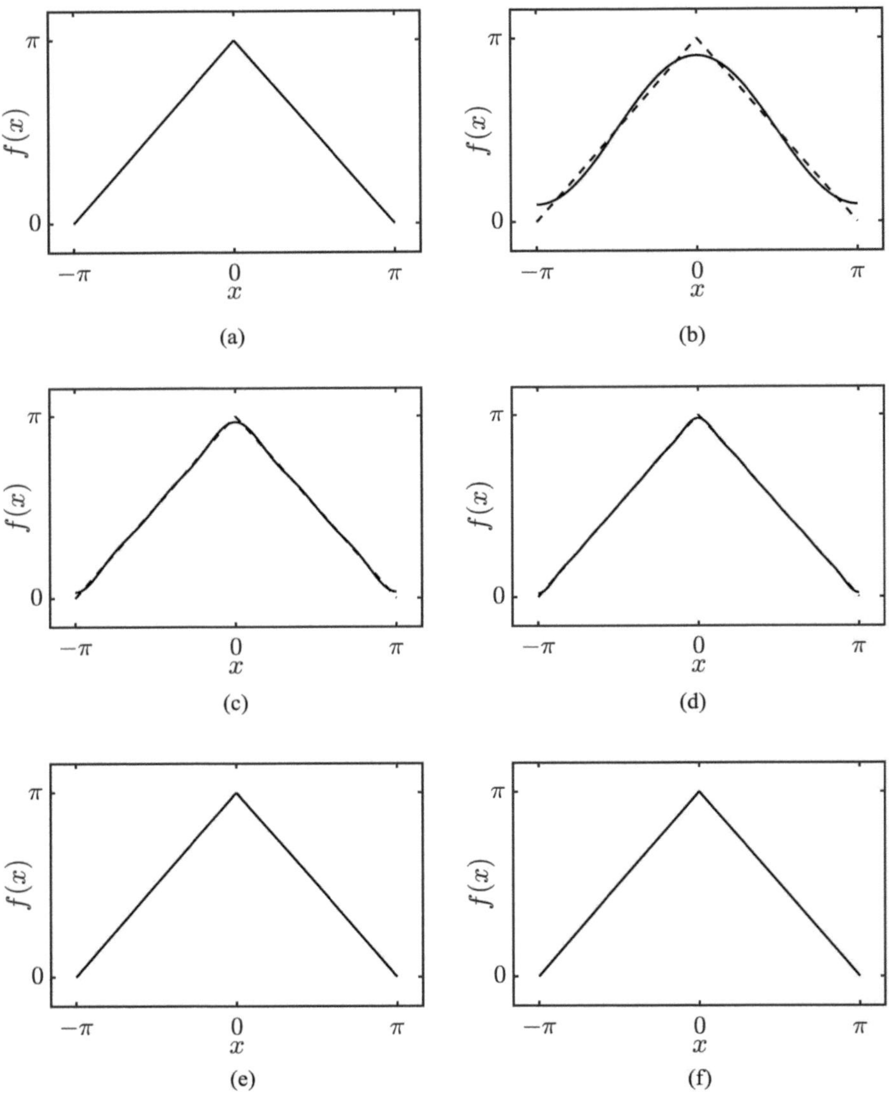

Fig. 3.13 Fourier series expansion of an upper triangular wave. **(a)** $f(x)$. **(b)** $n = 1$. **(c)** $n = 5$. **(d)** $n = 10$. **(e)** $n = 40$. **(f)** $n = 100$

Example 3.6 (Inverted Sawtooth Wave)

$$f(x) = \begin{cases} \pi - x & 0 < x < \pi \\ -(\pi + x) & -\pi < x < 0 \end{cases}$$

$$\frac{a_0}{2} = \frac{1}{2\pi}\left[-\int_{-\pi}^{0}(\pi + x)\,dx + \int_{0}^{\pi}(\pi - x)\,dx\right]$$

$$= \frac{1}{2\pi}\left[-\int_{-\pi}^{0}\pi dx - \int_{-\pi}^{0}x dx + \int_{0}^{\pi}\pi dx - \int_{0}^{\pi}x dx\right]$$

$$= 0$$

$$a_n = \frac{1}{\pi}\left[-\int_{-\pi}^{0}(\pi + x)\cos nx dx + \int_{0}^{\pi}(\pi - x)\cos nx dx\right]$$

$$= \frac{1}{\pi}\left[-\pi\int_{-\pi}^{0}\cos nx dx + \pi\int_{0}^{\pi}\cos nx dx - \int_{-\pi}^{\pi}x\cos nx dx\right]$$

$$= \frac{1}{\pi}\left[-\pi\left.\frac{\sin nx}{n}\right|_{-\pi}^{0} + \pi\left.\frac{\sin nx}{n}\right|_{0}^{\pi} - \left.\frac{1}{n^2}(nx\sin nx + \cos nx)\right|_{-\pi}^{\pi}\right]$$

$$= 0$$

$$b_n = \frac{1}{\pi}\left[-\int_{-\pi}^{0}(\pi + x)\sin nx dx + \int_{0}^{\pi}(\pi - x)\sin nx dx\right]$$

$$= \frac{1}{\pi}\left[-\pi\int_{-\pi}^{0}\sin nx dx + \pi\int_{0}^{\pi}\sin nx dx - \int_{-\pi}^{\pi}x\sin nx dx\right]$$

$$= \frac{1}{\pi}\left[\pi\left.\frac{\cos nx}{n}\right|_{-\pi}^{0} - \pi\left.\frac{\cos nx}{n}\right|_{0}^{\pi} - \left.\frac{1}{n^2}(-nx\cos nx + \sin nx)\right|_{-\pi}^{\pi}\right]$$

$$= \frac{1}{\pi}\left[\frac{\pi}{n}(1 - \cos n\pi) - \frac{\pi}{n}(\cos n\pi - 1) + \frac{2\pi}{n}\cos n\pi\right] = \frac{2}{n}$$

$$f(x) = \sum_{n=1}^{\infty}\frac{2}{n}\sin nx$$

(continued)

Example 3.6 (continued)
Figure 3.14 graphically represents the Fourier series expansion of an inverted sawtooth wave.

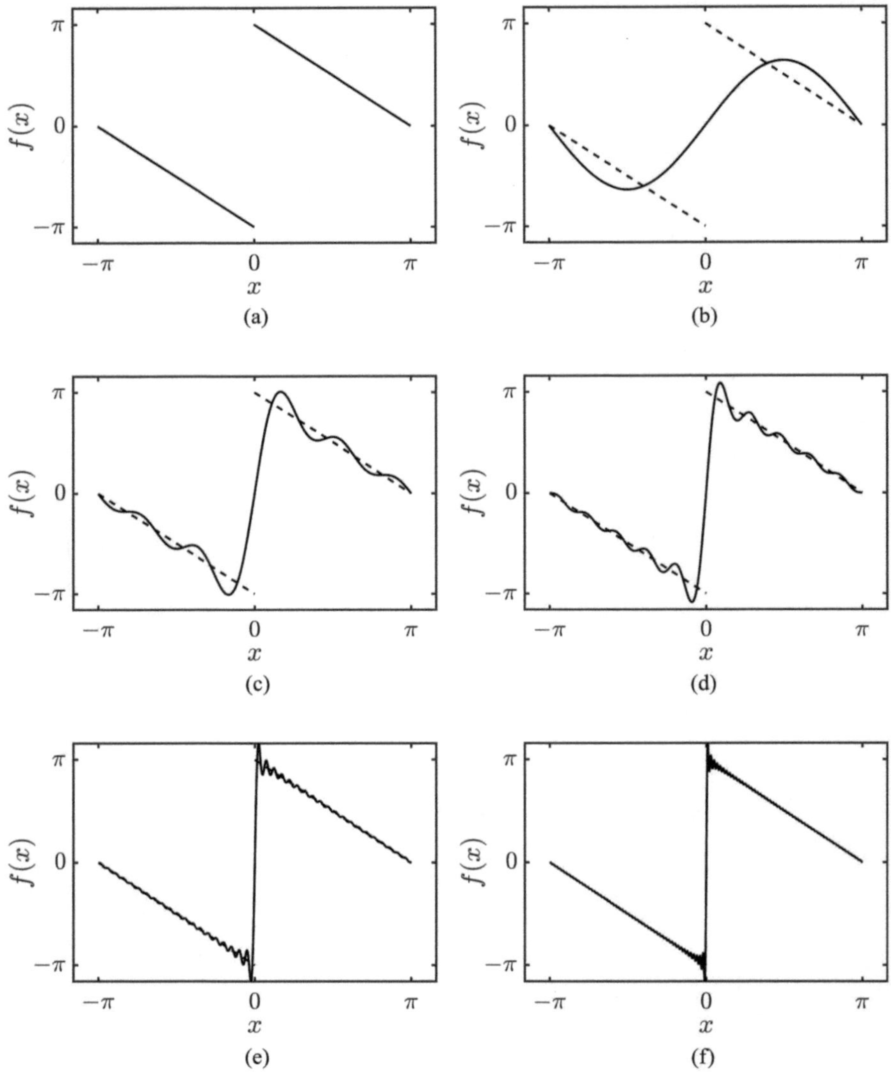

Fig. 3.14 Fourier series expansion of an inverted sawtooth wave. (**a**) $f(x)$. (**b**) $n = 1$. (**c**) $n = 5$. (**d**) $n = 10$. (**e**) $n = 40$. (**f**) $n = 100$

Example 3.7 (Trapezoidal Wave)

$$f(x) = \begin{cases} \frac{2}{\pi}(x+\pi), & -\pi \le x \le -\frac{\pi}{2} \\ 1, & -\frac{\pi}{2} \le x \le \frac{\pi}{2} \\ \frac{2}{\pi}(\pi - x), & \frac{\pi}{2} \le x \le \pi \end{cases}$$

$$\frac{a_0}{2} = \frac{1}{2\pi}\left(\int_{-\pi}^{-\frac{\pi}{2}} \frac{2}{\pi}(x+\pi)\,dx + \int_{-\frac{\pi}{2}}^{\frac{\pi}{2}} 1\,dx + \int_{\frac{\pi}{2}}^{\pi} \frac{2}{\pi}(\pi - x)\,dx\right)$$

$$= \frac{1}{2\pi}\left(\left(\frac{x^2}{\pi} + 2x\right)\Big|_{-\pi}^{-\pi/2} + (x)|_{-\pi/2}^{\pi/2} + \left(2x - \frac{x^2}{\pi}\right)\Big|_{\pi/2}^{\pi}\right)$$

$$= \frac{1}{2\pi}\left(\frac{\pi}{4} + \pi + \frac{\pi}{4}\right) = \frac{3}{4}$$

$$a_n = \frac{1}{\pi}\left(\int_{-\pi}^{-\frac{\pi}{2}} \frac{2}{\pi}(x+\pi)\cos(nx)\,dx + \int_{-\frac{\pi}{2}}^{\frac{\pi}{2}} \cos(nx)\,dx\right.$$

$$\left. + \int_{\frac{\pi}{2}}^{\pi} \frac{2}{\pi}(\pi - x)\cos(nx)\,dx\right)$$

$$= \frac{1}{\pi}\left[\left(\frac{2\left(n\left(x+\pi\right)\sin\left(nx\right)+\cos\left(nx\right)\right)}{\pi n^2}\right)\Big|_{-\pi}^{-\pi/2} + \left(\frac{\sin\left(nx\right)}{n}\right)\Big|_{-\pi/2}^{\pi/2}\right.$$

$$\left. + \left(-\frac{2\left(n\left(x-\pi\right)\sin\left(nx\right)+\cos\left(nx\right)\right)}{\pi n^2}\right)\Big|_{\pi/2}^{\pi}\right]$$

$$= \frac{1}{\pi}\left[-\frac{2\cos\left(\pi n\right) + \pi n \sin\left(\frac{\pi n}{2}\right) - 2\cos\left(\frac{\pi n}{2}\right)}{\pi n^2}\right.$$

$$\left. + \frac{2\sin\left(\frac{\pi n}{2}\right)}{n} - \frac{2\cos\left(\pi n\right) + \pi n \sin\left(\frac{\pi n}{2}\right) - 2\cos\left(\frac{\pi n}{2}\right)}{\pi n^2}\right]$$

$$= \frac{4\cos\left(\frac{\pi n}{2}\right) - 4\left(-1\right)^n}{\pi^2 n^2}$$

$$b_n = 0$$

$$f\left(x\right) = \frac{3}{4} + \sum_{n=1}^{\infty} \frac{4\cos\left(\frac{\pi n}{2}\right) - 4\left(-1\right)^n}{\pi^2 n^2}\cos(nx)$$

(continued)

Example 3.7 (continued)

Figure 3.15 graphically represents the Fourier series expansion of a trapezoidal wave.

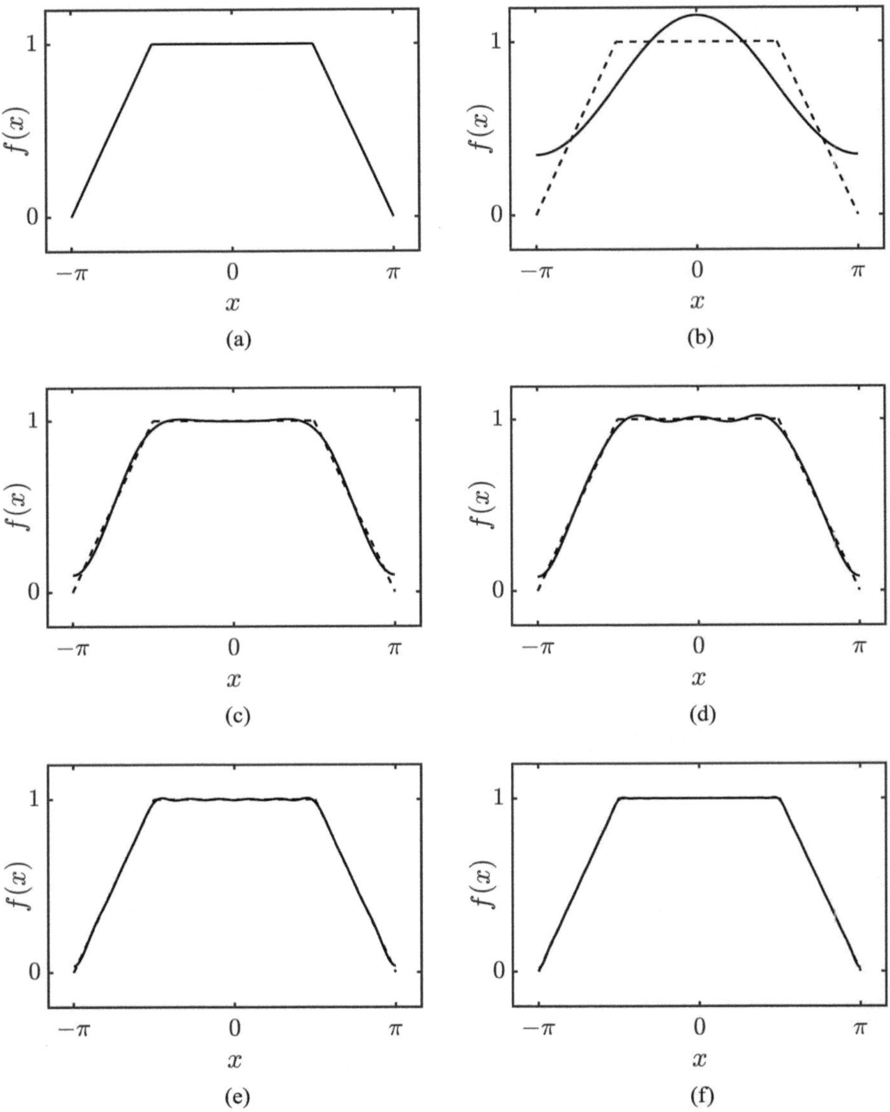

Fig. 3.15 Fourier series expansion of a trapezoidal wave. (**a**) $f(x)$. (**b**) $n = 1$. (**c**) $n = 3$. (**d**) $n = 5$. (**e**) $n = 10$. (**f**) $n = 20$

Example 3.8 (Triangular Wave)

$$f(x) = \begin{cases} -(\pi + x), & -\pi \leq x < -\frac{\pi}{2}, \\ x, & -\frac{\pi}{2} \leq x < \frac{\pi}{2}, \\ (\pi - x), & \frac{\pi}{2} \leq x \leq \pi \end{cases}$$

$$\frac{a_0}{2} = \frac{1}{2\pi}\left[-\int_{-\pi}^{-\frac{\pi}{2}} (\pi + x)\, dx + \int_{-\frac{\pi}{2}}^{\frac{\pi}{2}} x\, dx + \int_{\frac{\pi}{2}}^{\pi} (\pi - x)\, dx\right]$$
$$= 0$$

$$a_n = 0$$

$$b_n = \frac{1}{\pi}\left[-\int_{-\pi}^{-\frac{\pi}{2}} -(\pi + x)\sin nx\, dx + \int_{-\frac{\pi}{2}}^{\frac{\pi}{2}} x \sin nx\, dx + \int_{\frac{\pi}{2}}^{\pi} (\pi - x)\sin nx\, dx\right]$$
$$= \frac{1}{\pi}\left[\left(\frac{(x+\pi)\cos(nx)}{n} - \frac{\sin(nx)}{n^2}\right)\Bigg|_{-\pi}^{-\pi/2} + \left(\frac{\sin(nx) - nx\cos(nx)}{n^2}\right)\Bigg|_{-\pi/2}^{\pi/2}\right.$$
$$\left. + \left(\frac{(x-\pi)\cos(nx)}{n} - \frac{\sin(nx)}{n^2}\right)\Bigg|_{\pi/2}^{\pi}\right]$$
$$= \frac{1}{\pi}\left[\left(\frac{2\sin\left(\frac{\pi n}{2}\right) + \pi n\cos\left(\frac{\pi n}{2}\right)}{2n^2} - \frac{\sin(\pi n)}{n^2}\right) + \left(\frac{2\sin\left(\frac{\pi n}{2}\right) - \pi n\cos\left(\frac{\pi n}{2}\right)}{n^2}\right)\right.$$
$$\left. + \left(\frac{2\sin\left(\frac{\pi n}{2}\right) + \pi n\cos\left(\frac{\pi n}{2}\right)}{2n^2} - \frac{\sin(\pi n)}{n^2}\right)\right]$$
$$= \frac{4\sin\left(\frac{\pi n}{2}\right)}{\pi n^2}$$

$$f(x) = -\sum_{n=1}^{\infty} \frac{4\sin\left(\frac{\pi n}{2}\right)}{\pi n^2} \sin nx$$

Figure 3.16 graphically represents the Fourier series expansion of a full triangular wave.

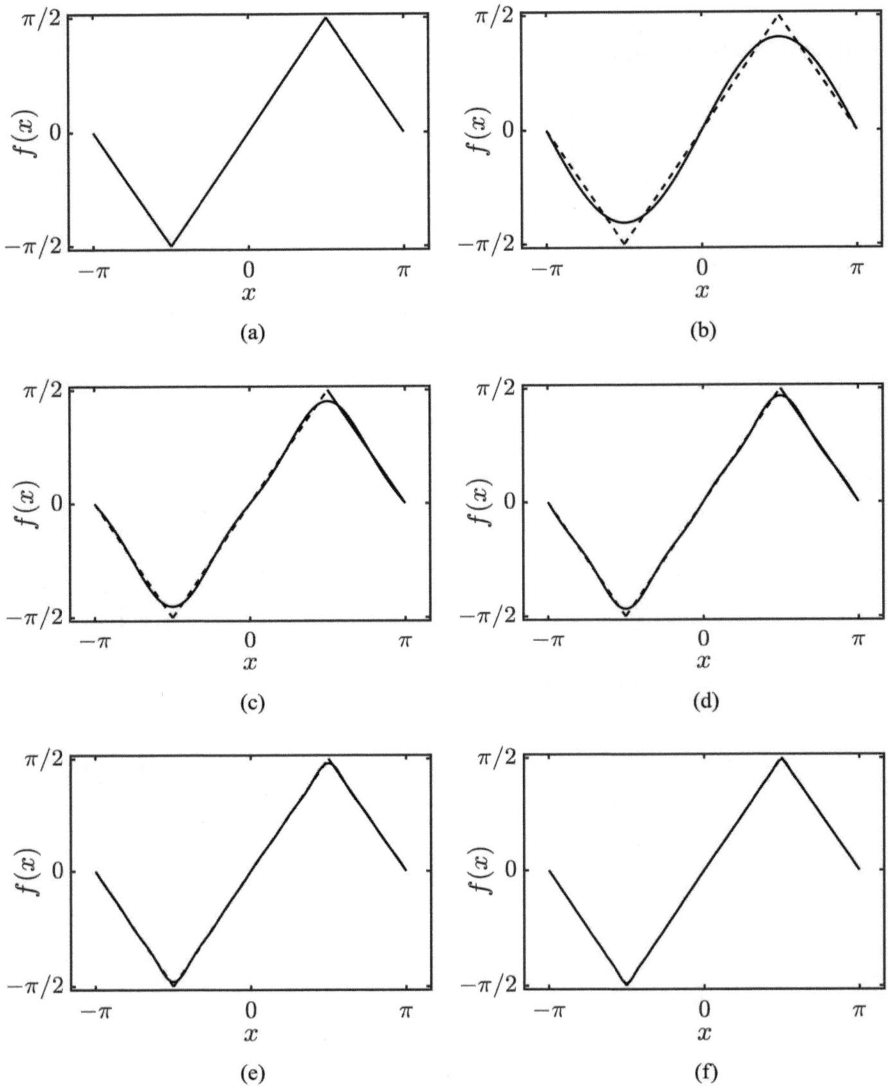

Fig. 3.16 Fourier series expansion of a full triangular wave. (**a**) $f(x)$. (**b**) $n = 1$. (**c**) $n = 3$. (**d**) $n = 5$. (**e**) $n = 10$. (**f**) $n = 20$

Example 3.9 (Half Square Wave)

$$f(x) = \begin{cases} 0, & -\pi \leq x < 0, \\ 1, & 0 \leq x \leq \pi, \end{cases}$$

$$\begin{aligned} \frac{a_0}{2} &= \frac{1}{2\pi} \left[\int_0^{\pi} 1\, dx \right] \\ &= \frac{1}{2\pi} (x)|_0^{\pi} = \frac{1}{2} \end{aligned}$$

$$\begin{aligned} a_n &= \frac{1}{\pi} \left[\int_0^{\pi} \cos nx\, dx \right] \\ &= \frac{1}{\pi} \left(\frac{\sin(nx)}{n} \right) \Big|_0^{\pi} \\ &= \frac{\sin(\pi n)}{\pi n} \end{aligned}$$

$$\begin{aligned} b_n &= \frac{1}{\pi} \left[\int_0^{\pi} \sin nx\, dx \right] \\ &= \frac{1}{\pi} \left(-\frac{\cos(nx)}{n} \right) \Big|_0^{\pi} \\ &= -\frac{\cos(\pi n) - 1}{\pi n} \end{aligned}$$

$$f(x) = \frac{1}{2} + \sum_{n=1}^{\infty} \frac{1 - \cos(\pi n)}{\pi n} \sin nx$$

Figure 3.17 graphically represents the Fourier series expansion of a half square wave.

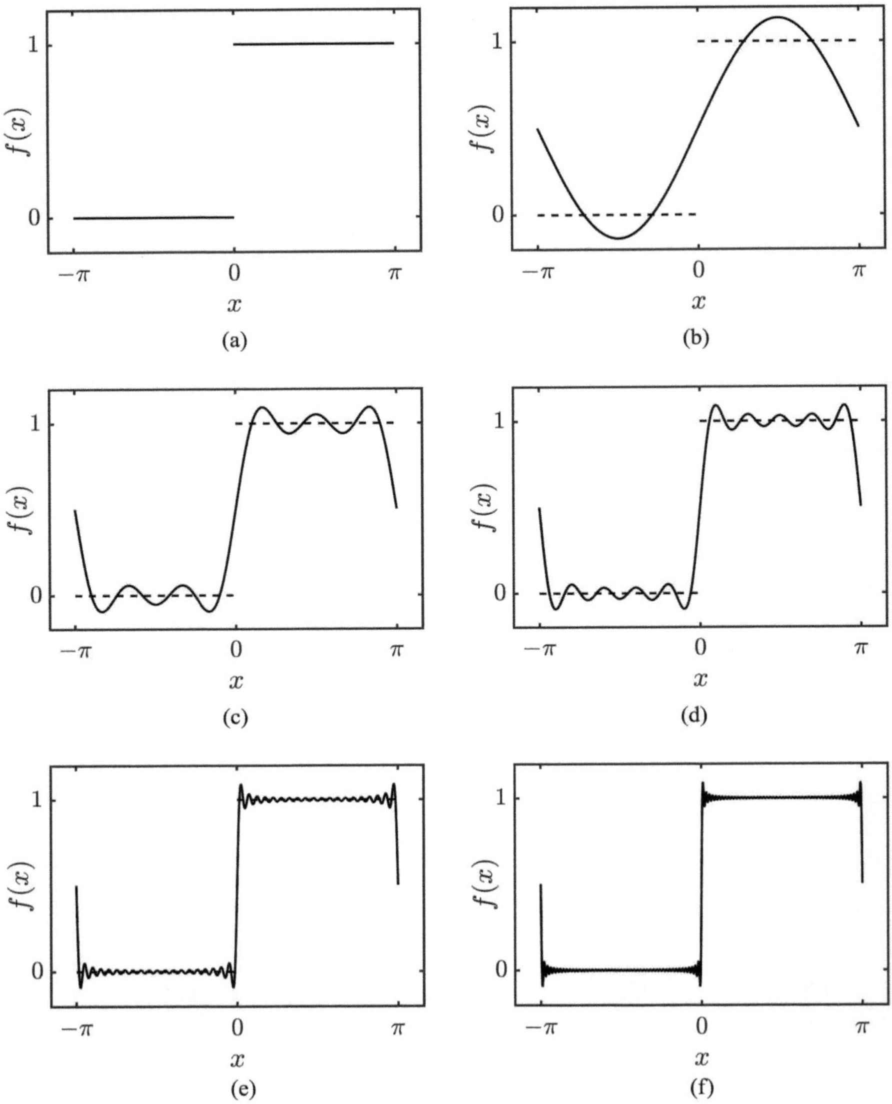

Fig. 3.17 Fourier series expansion of a half square wave. (**a**) $f(x)$. (**b**) $n = 1$. (**c**) $n = 5$. (**d**) $n = 10$. (**e**) $n = 40$. (**f**) $n = 100$

Example 3.10 (Full Square Wave)

$$f(x) = \begin{cases} -1, & -\pi \le x < 0, \\ 1, & 0 \le x \le pi, \end{cases}$$

$$\frac{a_0}{2} = \frac{1}{2\pi}\left[\int_{-\pi}^{0} -1\,dx + \int_{0}^{\pi} 1\,dx\right]$$
$$= \frac{1}{2\pi}\left[(-x)|_{-\pi}^{0} + (x)|_{0}^{\pi}\right] = 0$$

$$a_n = \frac{1}{\pi}\left[\int_{-\pi}^{0} -\cos nx\,dx + \int_{0}^{\pi} \cos nx\,dx\right] = 0$$

$$b_n = \frac{1}{\pi}\left[\int_{-\pi}^{0} -\sin nx\,dx + \int_{0}^{\pi} \sin nx\,dx\right]$$
$$= \frac{1}{\pi}\left[\left(\frac{\cos(nx)}{n}\right)\Big|_{-\pi}^{0} + \left(-\frac{\cos(nx)}{n}\right)\Big|_{0}^{\pi}\right]$$
$$= -\frac{\cos(\pi n) - 1}{\pi n} - \frac{\cos(\pi n) - 1}{\pi n} = 2\left(\frac{1 - \cos(\pi n)}{\pi n}\right)$$

$$f(x) = 2\sum_{n=1}^{\infty} \frac{1 - \cos(\pi n)}{\pi n} \sin nx$$

Figure 3.18 graphically represents the Fourier series expansion of a full square wave.

Example 3.11 (Forward Pulse)

$$f(x) = \begin{cases} -\frac{\pi}{2}, & -\pi \le x < -\frac{\pi}{2}, \\ x, & -\frac{\pi}{2} \le x < \frac{\pi}{2}, \\ \frac{\pi}{2}, & \frac{\pi}{2} \le x \le \pi \end{cases}$$

(continued)

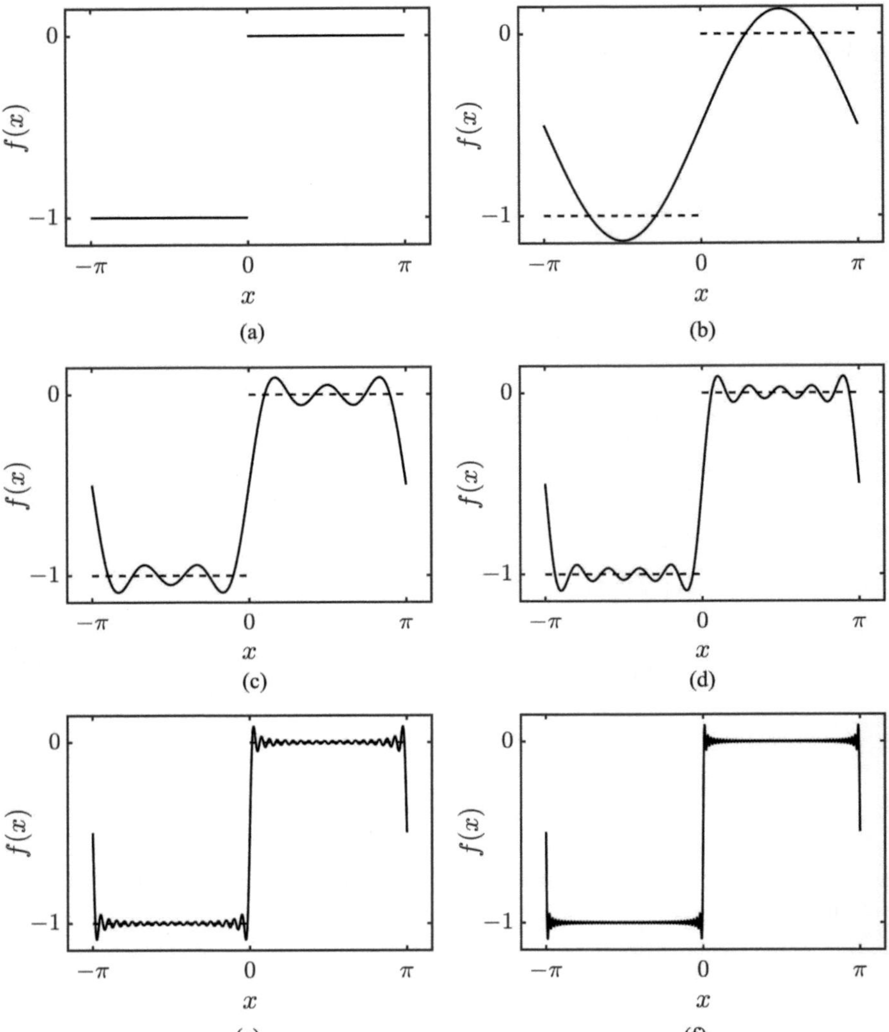

Fig. 3.18 Fourier series expansion of a full square wave. (**a**) $f(x)$. (**b**) $n = 1$. (**c**) $n = 5$. (**d**) $n = 10$. (**e**) $n = 40$. (**f**) $n = 100$

Example 3.11 (continued)

$$\frac{a_0}{2} = \frac{1}{2\pi}\left[-\int_{-\pi}^{-\frac{\pi}{2}} -\frac{\pi}{2}\,dx + \int_{-\frac{\pi}{2}}^{\frac{\pi}{2}} x\,dx + \int_{\frac{\pi}{2}}^{\pi} \frac{\pi}{2}\,dx\right]$$

$$= 0$$

$$a_n = \frac{1}{\pi}\left[\int_{-\pi}^{-\frac{\pi}{2}} -\frac{\pi}{2}\cos nx\,dx + \int_{-\frac{\pi}{2}}^{\frac{\pi}{2}} x\cos nx\,dx + \int_{\frac{\pi}{2}}^{\pi} \frac{\pi}{2}\cos nx\,dx\right]$$

$$= \frac{1}{\pi}\left[\left(-\frac{\pi\sin(nx)}{2n}\right)\Bigg|_{-\pi}^{-\pi/2} + \left(\frac{nx\sin(nx)+\cos(nx)}{n^2}\right)\Bigg|_{-\pi/2}^{\pi/2}\right.$$

$$\left. + \left(\frac{\pi\sin(nx)}{2n}\right)\Bigg|_{-\pi/2}^{\pi}\right]$$

$$= \frac{1}{\pi}\left[\left(-\frac{\pi\left(\sin(\pi n)-\sin\left(\frac{\pi n}{2}\right)\right)}{2n}\right) + (0) + \left(\frac{\pi\left(\sin(\pi n)-\sin\left(\frac{\pi n}{2}\right)\right)}{2n}\right)\right]$$

$$= 0$$

$$b_n = \frac{1}{\pi}\left[-\int_{-\pi}^{-\frac{\pi}{2}} -\frac{\pi}{2}\sin nx\,dx + \int_{-\frac{\pi}{2}}^{\frac{\pi}{2}} x\sin nx\,dx + \int_{\frac{\pi}{2}}^{\pi} \frac{\pi}{2}\sin nx\,dx\right]$$

$$= \frac{1}{\pi}\left[\left(\frac{\pi\cos(nx)}{2n}\right)\Bigg|_{-\pi}^{-\pi/2} + \left(\frac{\sin(nx)-nx\cos(nx)}{n^2}\right)\Bigg|_{-\pi/2}^{\pi/2}\right.$$

$$\left. + \left(-\frac{\pi\cos(nx)}{2n}\right)\Bigg|_{-\pi/2}^{\pi}\right]$$

$$= \frac{1}{\pi}\left[\left(-\frac{\pi\left(\cos(\pi n)-\cos\left(\frac{\pi n}{2}\right)\right)}{2n}\right) + \left(\frac{2\sin\left(\frac{\pi n}{2}\right)-\pi n\cos\left(\frac{\pi n}{2}\right)}{n^2}\right)\right.$$

$$\left. + \left(-\frac{\pi\left(\cos(\pi n)-\cos\left(\frac{\pi n}{2}\right)\right)}{2n}\right)\right]$$

$$= \frac{2}{\pi n^2}\sin\left(\frac{\pi n}{2}\right) - \frac{(-1)^n}{n}$$

$$f(x) = \sum_{n=1}^{\infty}\left(\frac{2}{\pi n^2}\sin\left(\frac{\pi n}{2}\right) - \frac{(-1)^n}{n}\right)\sin nx$$

(continued)

Example 3.11 (continued)
Figure 3.19 graphically represents the Fourier series expansion of a forward pulse.

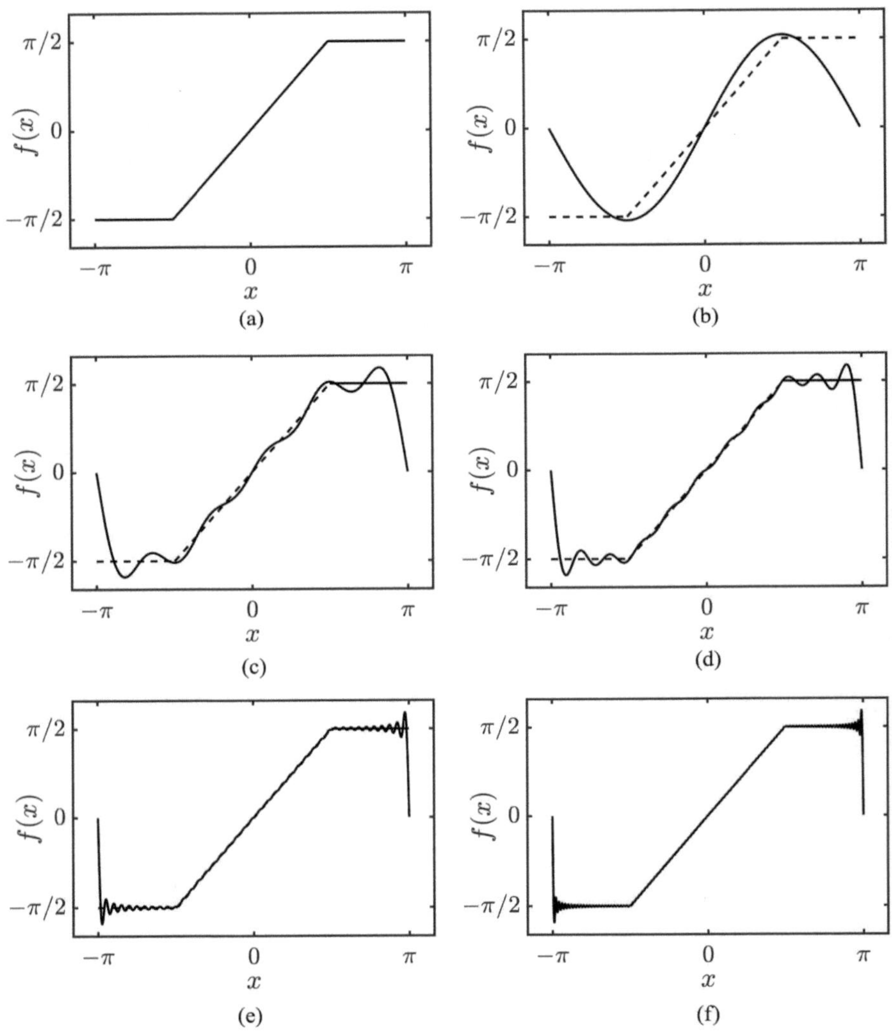

Fig. 3.19 Fourier series expansion of a forward pulse. **(a)** $f(x)$. **(b)** $n = 1$. **(c)** $n = 5$. **(d)** $n = 10$. **(e)** $n = 40$. **(f)** $n = 100$

Example 3.12 (Half Sine-Wave Rectifier)

$$f(x) = \begin{cases} 0, & -\pi \le x < 0, \\ \sin x, & 0 \le x \le pi, \end{cases}$$

$$\begin{aligned} \frac{a_0}{2} &= \frac{1}{2\pi}\left[\int_0^{\pi} \sin x \; dx\right] \\ &= \frac{1}{2\pi}\left[(-\cos x)|_0^{\pi}\right] = \frac{1}{\pi} \end{aligned}$$

$$\begin{aligned} a_n &= \frac{1}{\pi}\left[\int_0^{\pi} \sin x \cos nx \; dx\right] \\ &= \frac{1}{\pi}\left[\left(\frac{n\sin(x)\sin(nx) + \cos(x)\cos(nx)}{n^2-1}\right)\Big|_0^{\pi}\right] \\ &= \frac{\cos(\pi n) + 1}{\pi(1-n^2)} \\ &= \frac{2}{\pi(1-n^2)} \quad \text{for } n = 2, 4, 6, \ldots, \end{aligned}$$

$$\begin{aligned} b_n &= \frac{1}{\pi}\left[\int_0^{\pi} \sin x \sin nx \; dx\right] \\ &= \frac{1}{\pi}\left[\left(\frac{\cos(x)\sin(nx) - n\sin(x)\cos(nx)}{n^2-1}\right)\Big|_0^{\pi}\right] \\ &= \frac{\sin(\pi n)}{\pi(1-n^2)} \\ &= 0 \quad \text{for } n \ne 1 \end{aligned}$$

for $n = 1$

$$\begin{aligned} b_1 &= \frac{1}{\pi}\left[\int_0^{\pi} \sin x \sin x \; dx\right] \\ &= \frac{1}{\pi}\left[\left(-\frac{\sin(2x) - 2x}{4}\right)\Big|_0^{\pi}\right] = \frac{1}{2} \end{aligned}$$

$$f(x) = \frac{1}{\pi} + \frac{1}{2}\sin x - \sum_{n=2,4,6,\ldots}^{\infty} \frac{2}{\pi(1-n^2)}\cos nx$$

(continued)

Example 3.12 (continued)

Figure 3.20 graphically represents the Fourier series expansion of a half-sine wave rectifier.

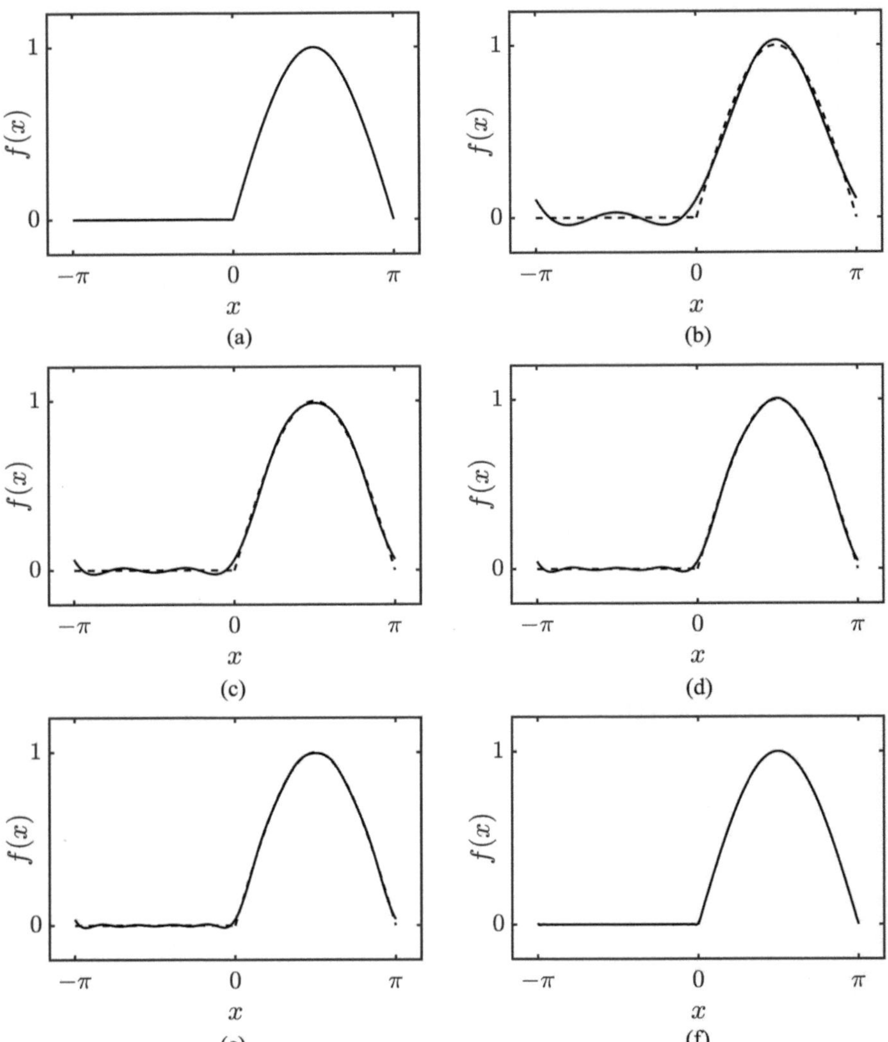

Fig. 3.20 Fourier series expansion of a half-sine wave rectifier. **(a)** $f(x)$. **(b)** $n = 2$. **(c)** $n = 4$. **(d)** $n = 6$. **(e)** $n = 8$. **(f)** $n = 40$

Example 3.13 (Inverted Parabolic Wave Rectifier)

$$f(x) = (\pi^2 - x^2) \quad -\pi \le x \le pi$$

$$\frac{a_0}{2} = \frac{1}{2\pi}\left[\int_{-\pi}^{\pi} (\pi^2 - x^2)\, dx\right]$$

$$= \frac{1}{2\pi}\left[\left(\pi^2 x - \frac{x^3}{3}\right)\Big|_{-\pi}^{\pi}\right] = \frac{2\pi^2}{3}$$

$$a_n = \frac{1}{\pi}\left[\int_{-\pi}^{\pi} (\pi^2 - x^2)\cos nx\, dx\right]$$

$$= \frac{1}{\pi}\left[\left(-\frac{\left(n^2 (x-\pi)(x+\pi) - 2\right)\sin(nx) + 2nx\cos(nx)}{n^3}\right)\Big|_{-\pi}^{\pi}\right]$$

$$= \frac{4\left(\sin(\pi n) - \pi n\cos(\pi n)\right)}{\pi n^3}$$

$$b_n = \frac{1}{\pi}\left[\int_{-\pi}^{\pi} (\pi^2 - x^2)\sin nx\, dx\right]$$

$$= \frac{1}{\pi}\left[\left(\frac{\left(n^2 (x-\pi)(x+\pi) - 2\right)\cos(nx)}{n^3} - \frac{2x\sin(nx)}{n^2}\right)\Big|_{-\pi}^{\pi}\right]$$

$$= 0$$

$$f(x) = \frac{2\pi^2}{3} - \sum_{n=1}^{\infty} \frac{4}{n^2}\cos(\pi n)\cos nx$$

Figure 3.21 graphically represents the Fourier series expansion of an inverted parabolic wave rectifier.

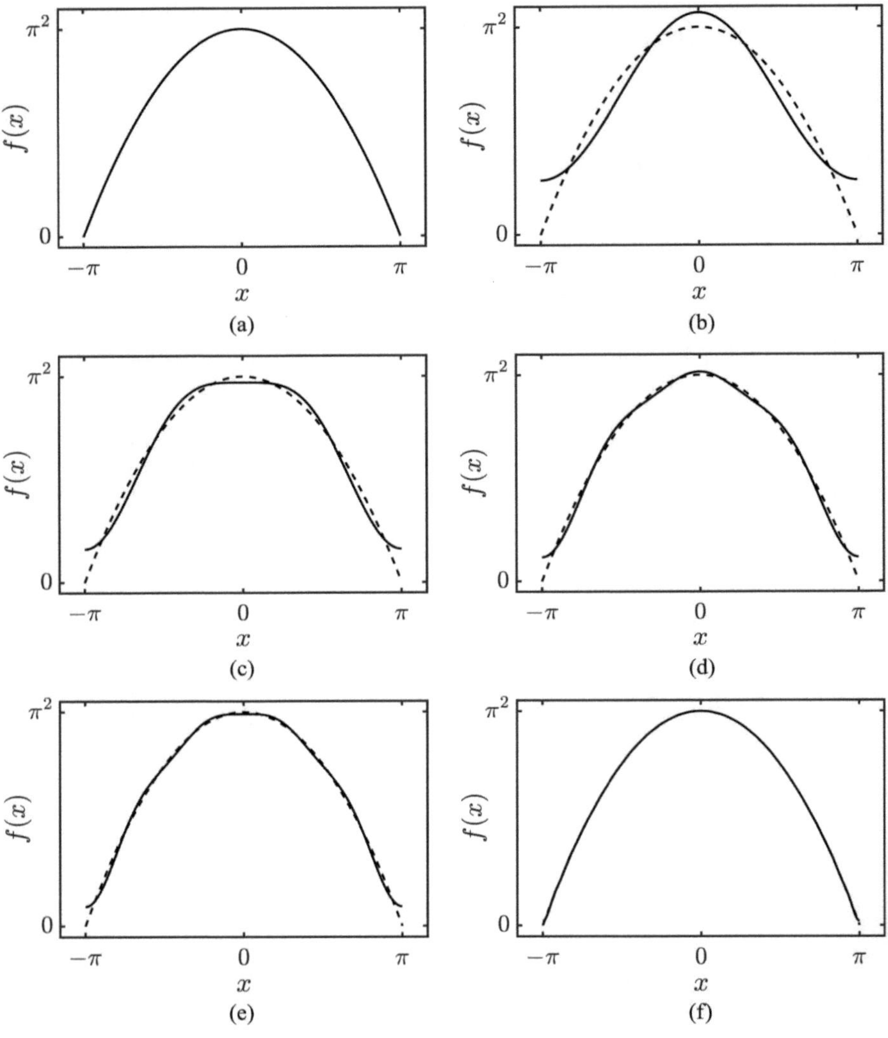

Fig. 3.21 Fourier series expansion of an inverted parabolic wave rectifier. **(a)** $f(x)$. **(b)** $n = 1$. **(c)** $n = 2$. **(d)** $n = 3$. **(e)** $n = 4$. **(f)** $n = 20$

Example 3.14 (Cubic Wave)

$$f(x) = x^3 \quad -\pi \leq x \leq pi$$

$$\frac{a_0}{2} = \frac{1}{2\pi}\left[\int_{-\pi}^{\pi} x^3\,dx\right]$$

$$= \frac{1}{2\pi}\left[\left(\frac{x^4}{4}\right)\Big|_{-\pi}^{\pi}\right] = 0$$

$$a_n = \frac{1}{\pi}\left[\int_{-\pi}^{\pi} x^3 \cos nx\,dx\right]$$

$$= \frac{1}{\pi}\left[\frac{\left(n^3x^3 - 6nx\right)\sin(nx) + \left(3n^2x^2 - 6\right)\cos(nx)}{n^4}\Bigg|_{-\pi}^{\pi}\right]$$

$$= 0$$

$$b_n = \frac{1}{\pi}\left[\int_{-\pi}^{\pi} x^3 \sin nx\,dx\right]$$

$$= \frac{1}{\pi}\left[\left(\frac{3\left(n^2x^2 - 2\right)\sin(nx) - nx\left(n^2x^2 - 6\right)\cos(nx)}{n^4}\right)\Bigg|_{-\pi}^{\pi}\right]$$

$$= \frac{2\left(\left(3\pi^2n^2 - 6\right)\sin(\pi n) + \left(6\pi n - \pi^3n^3\right)\cos(\pi n)\right)}{\pi n^4}$$

$$f(x) = \sum_{n=1}^{\infty} \frac{2\left(6 - \pi^2n^2\right)}{n^3}\cos(\pi n)\sin nx$$

Figure 3.22 graphically represents the Fourier series expansion of a cubic wave.

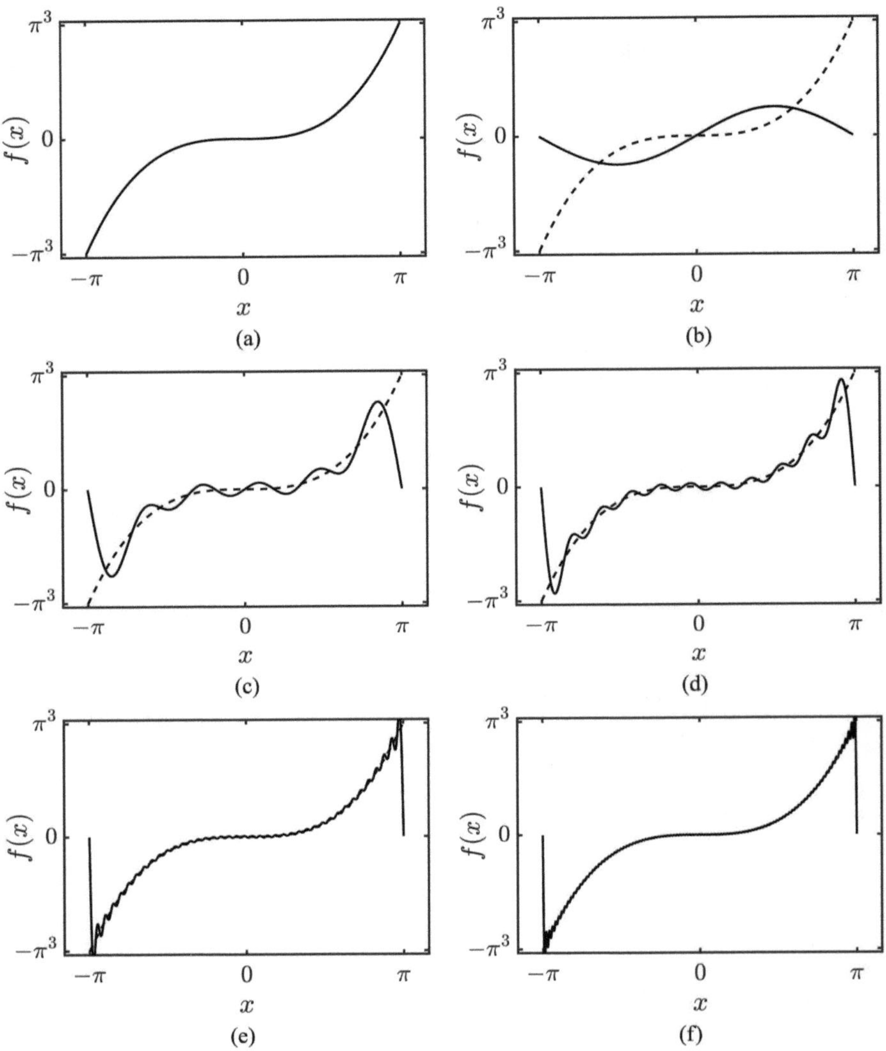

Fig. 3.22 Fourier series expansion of a cubic wave. **(a)** $f(x)$. **(b)** $n = 1$. **(c)** $n = 5$. **(d)** $n = 10$. **(e)** $n = 40$. **(f)** $n = 100$

This section expands many continuous and non-smooth functions as smooth and continuous functions of simple sinusoids using the Fourier series. Expressing these complex functions in terms of sinusoids achieves several benefits. First, such expansions allow for easier analysis and manipulation of the functions, as sinusoids are well understood and mathematically tractable. Complex behaviors like sharp transitions or discontinuities, which may be difficult to study directly, are broken down into smoother components, simplifying calculations. Another key advantage lies in applications to signal processing, where even non-smooth signals can be efficiently represented and reconstructed. The Fourier series expansion enables filtering and noise reduction by isolating specific frequency components. Moreover, this approach provides a way to approximate functions with a finite number of terms, striking a balance between accuracy and computational efficiency.

3.6 Closure

This chapter explores the rich and far-reaching concept of the Fourier series, beginning with its foundational philosophy and advancing through various applications and extensions. The Fourier series, rooted in the idea that any periodic function can be expressed as a sum of sines and cosines, is a profound mathematical tool that bridges continuous and discrete domains. This principle allows even highly complex or irregular functions to be broken down into simpler harmonic components, making the Fourier series indispensable in physics, engineering, and beyond.

The study then extended to special cases of odd and even functions, where the symmetry of a function determines which terms in its Fourier series survive. Due to their symmetry about the origin, odd functions have Fourier series composed solely of sine terms, while even functions are represented only by cosine terms. These properties provide powerful simplifications in many practical applications, from wave analysis to signal processing.

Next, the complex Fourier series was introduced, where traditional real sine and cosine terms were replaced by exponential functions involving complex numbers. This reformulation not only simplifies mathematics but also deepens the understanding of the Fourier series, connecting it to complex analysis and broader areas of mathematics.

The chapter also ventured into the double Fourier series, which extends the concept to functions of two variables, such as in analyzing surfaces or 2D waveforms. This powerful extension allows the decomposition of periodic functions across two dimensions, revealing applications in fields like image processing, fluid dynamics, and heat distribution.

Concrete examples of Fourier series expansions for common functions, such as square waves, sawtooth waves, and piecewise functions, were provided to demonstrate the practical utility of the theory. These examples highlighted how the Fourier series approximates and can exactly represent functions under certain conditions, emphasizing their robustness in applications across continuous and discrete domains.

This chapter's comprehensive exploration of the Fourier series has laid a solid foundation. The historical context also reinforces Fourier's profound impact on the mathematical and scientific worlds. As the journey continues into more advanced areas, powerful concepts of the Fourier series will continue to unfold in various mathematical and physical applications.

References

Bernoulli, D. (1755). Histoire de l'académie r. In *Royale des sciences et belles lettres de Berlin*, p. 175.

D'Alembert, J. L. R. (1747a). Recherches sur la courbe que forme une corde tenduë mise en vibration. *Histoire de l'académie Royale des Sciences et Belles Lettres de Berlin, 3*, 214–219.

D'Alembert, J. L. R. (1747b). Suite des recherches sur la courbe que forme une corde tenduë mise en vibration. *Histoire de l'académie Royale des Sciences et Belles Lettres de Berlin, 3*, 220–249.

Dirichlet, L. G. (1829). Sur la convergence des séries trigonométriques qui servent à représenter une fonction arbitraire entre des limites données. *Journal für die reine und angewandte Mathematik, 4*, 157–169.

Euler, L. (1748). *Introductio in Analysin Infinitorum* (1st ed.). Marcum-Michaelem Bosquet & Socios.

Fourier, J. (1822). *Théorie analytique de la chaleur* (1st ed.). Firmin Dido.

Jeffrey, A. (2002). *Advanced engineering mathematics* (1st ed.). Harcourt /Academic Press.

Kreyszig, E., Kreyszig, H., & Norminton, E. (2011). *Advanced engineering mathematics* (10th ed.). Wiley.

Neugebauer, O. (1952). *The exact sciences in antiquity* (2nd ed.). Dover Publications.

Riemann, B. (1867). *Über die Darstellbarkeit einer Function durch eine trigonometrische Reihe*, vol. 13. Dieterichsche Buchhandlung.

Simmons, G. F. (2017). *Differential equations with applications and historical notes* (3rd ed.). CRC Press.

Zill, D. G., Wright, W. S., & Cullen, M. R. (2013). *Differential equations with boundary-value problems* (8th ed.). Brooks/Cole Cengage Learning.

Chapter 4
The Fourier Analysis

After the Fourier series, other series have entered the domain of analysis; they entered by the same door; they have been imagined in view of applications.

—Henri Poincaré

Abstract This chapter extends from the Fourier series to a broader framework in Fourier analysis. It begins by introducing the Fourier integral, which generalizes the Fourier series to accommodate non-periodic functions, broadening the applicability of Fourier methods to a wider range of signals. The chapter then delves into the properties of the Fourier transform, the core element of Fourier analysis, and examines essential features such as transform pair duality, linearity, constant multiplication, and symmetry. These properties enable critical manipulations for signal processing. Additionally, topics such as frequency shifts, scaling, and the differentiation of Fourier pairs provide greater flexibility for analyzing complex signals. These properties make the Fourier transform an invaluable tool across physics, engineering, and applied sciences. This chapter establishes a solid foundation in Fourier analysis, preparing readers for more advanced techniques in signal transformation and analysis.

Keywords Fourier analysis · Non-periodic function · Duality · Linearity · Time delay · Space shift · Wavenumber shift · Stretching and shrinking · Heisenberg uncertainty

4.1 Fourier Series to Fourier Analysis

In Chap. 3, meticulous detailing is provided regarding the expansion of intricate, non-smooth periodic functions via the Fourier series. It is well established that any non-smooth periodic function can be represented as a sum of simple sine and cosine functions. However, it is recognized that not all functions are inherently periodic. A reasonable and natural question may arise in the inquisitive reader's mind: whether

S. K. Jena, *Fourier, Laplace, and the Tangled Love Affair with Transforms*,
https://doi.org/10.1007/978-3-031-80165-5_4

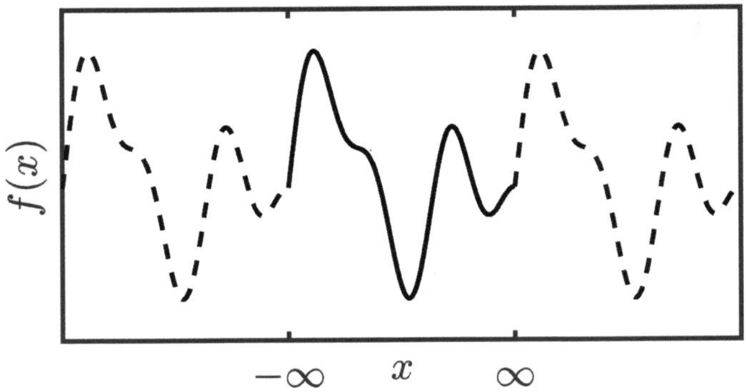

Fig. 4.1 Replicating a non-periodic function by virtual periodic function

a similar method exists for expanding complex non-periodic functions using simple harmonic components. This intriguing question is answered affirmatively. The extended form of the Fourier series, applicable to non-periodic functions, is referred to as the Fourier integral. Through the Fourier integral, any non-periodic function with finite energy can be expanded in terms of simple sinusoids.

To grasp the fundamental philosophy behind the Fourier integral, the notion that "every mathematical function is periodic" must momentarily be entertained. Although this statement may initially seem surprising, it should not be dismissed as a whimsical conjecture akin to the flat earth theory but rather regarded as a concept grounded in a rational and compelling analogy. This idea is illustrated through Fig. 4.1, where an arbitrary non-periodic function (depicted by solid lines) is considered. By virtually placing identical functions on either side (as indicated by the dotted lines in Fig. 4.1), the resulting representation closely approximates a periodic function, with its periodicity extended hypothetically to infinity.

Let us now contemplate the Fourier series expansion of such a virtually periodic function with a period of $2L$ as $L \to \infty$. This approach reveals the profound insight that any non-periodic function can be analyzed and expressed through the elegant framework of Fourier integrals, extending the versatility and applicability of Fourier analysis beyond periodic functions.

$$f_\infty(x) = \frac{a_0}{2} + \sum_{n=1}^{\infty} (a_n \cos \omega x + b_n \sin \omega x) \tag{4.1}$$

where $\omega = \dfrac{n\pi}{L}$. As L approaches infinity, ω is no longer confined to discrete integral multiples of $\dfrac{n\pi}{L}$; instead, it extends continuously across all real values. Any infinitesimal variation in ω can be expressed as

$$\Delta\omega = \frac{(n+1)\pi}{L} - \frac{n\pi}{L} = \frac{\pi}{L} \tag{4.2}$$

$$\frac{1}{L} = \frac{\Delta\omega}{\pi} \tag{4.3}$$

Putting the values of the Fourier coefficients in Eq. (4.1)

$$f_\infty(x) = \frac{1}{2L}\int_{-L}^{L} f_\infty(v)\,dv + \frac{1}{L}\sum_{n=1}^{\infty}\left[\cos\omega x\int_{-L}^{L} f_\infty(v)\cos\omega v dv + \sin\omega x\int_{-L}^{L} f_\infty(v)\sin\omega v dv\right] \tag{4.4}$$

An absolutely integrable function in L^2 space has a finite value of the integral of its absolute value. Mathematically, this means that a function $f(x)$ is absolutely integrable on a given interval $[-L, L]$ if

$$\int_{-L}^{L} |f(x)|\,dx < \infty \tag{4.5}$$

Absolute integrability is a stronger condition than simple integrability because it ensures that the total magnitude of the function's values (without regard to sign) is finite. In practical terms, absolute integrability ensures that a function behaves well enough over its domain, neither growing too rapidly nor oscillating too wildly, allowing for useful applications in signal processing, physics, and other fields. An absolute integrating function physically represents the finite energy content of the function signal. As $L \to \infty$, $\frac{1}{2L} \to 0$, hence the product of $\frac{1}{2L}$ times the definite integral of an infinite periodic absolute integrable function is zero. The first term of Eq. (4.4) equals to zero. Fourier series expansion of the infinite periodic function becomes

$$\begin{aligned} f_\infty(x) &= \frac{1}{L}\sum_{n=1}^{\infty}\left[\cos\omega x\int_{-L}^{L} f_\infty(v)\cos\omega v dv + \sin\omega x\int_{-L}^{L} f_\infty(v)\sin\omega v dv\right] \\ &= \frac{\Delta\omega}{\pi}\sum_{n=1}^{\infty}\left[\cos\omega x\int_{-L}^{L} f_\infty(v)\cos\omega v dv + \sin\omega x\int_{-L}^{L} f_\infty(v)\sin\omega v dv\right] \end{aligned} \tag{4.6}$$

As $L \to \infty$, it is plausible to replace the infinite series $\sum_{n=1}^{\infty}$ by $\int_{0}^{\infty}$ and $\Delta\omega$ becomes $d\omega$. Hence, Eq. (4.6) becomes

$$f_{\infty}(x) = \frac{1}{\pi}\int_{0}^{\infty}\left[\cos\omega x\int_{-\infty}^{\infty} f_{\infty}(v)\cos\omega v dv + \sin\omega x\int_{-\infty}^{\infty} f_{\infty}(v)\sin\omega v dv\right]d\omega \tag{4.7}$$

As a standard notation moving forward, the subscript from the non-periodic function $f_{\infty}(x)$ is omitted, and the infinite periodic function will be denoted by $f(x)$. The simplified version of Eq. (4.7) is expressed as

$$f(x) = \int_{0}^{\infty}[A(\omega)\cos\omega x + B(\omega)\sin\omega x]\,d\omega \tag{4.8}$$

$$A(\omega) = \frac{1}{\pi}\int_{-\infty}^{\infty} f(v)\cos\omega v dv \tag{4.9}$$

$$B(\omega) = \frac{1}{\pi}\int_{-\infty}^{\infty} f(v)\sin\omega v dv \tag{4.10}$$

Equation (4.8) is known as the Fourier integral of the function $f(x)$. Equations (4.9) and (4.10) are Fourier cosine and sine integral coefficients of $f(x)$, respectively. Substituting $A(\omega)$ and $B(\omega)$ into Eq. (4.8)

$$f(x) = \frac{1}{\pi}\int_{0}^{\infty}\int_{-\infty}^{\infty} f(v)[\cos\omega v\cos\omega x + \sin\omega v\sin\omega x]\,dv d\omega \tag{4.11}$$

Using the trigonometric identities for cosine addition, Eq. (4.11) becomes

$$f(x) = \frac{1}{\pi}\int_{0}^{\infty}\left[\int_{-\infty}^{\infty} f(v)\cos\omega(x - v)dv\right]d\omega \tag{4.12}$$

$\cos\omega(x - v)$ is an even function; the integral inside the bracket in Eq. (4.12) is even with respect to ω. Using the symmetric property of definite integrals for even function, Eq. (4.12) becomes

$$f(x) = \frac{1}{2\pi}\int_{-\infty}^{\infty}\left[\int_{-\infty}^{\infty} f(v)\cos\omega(x - v)dv\right]d\omega \tag{4.13}$$

$$0 = \frac{1}{2\pi}\int_{-\infty}^{\infty}\left[\int_{-\infty}^{\infty} f(v)\sin\omega(x - v)dv\right]d\omega \tag{4.14}$$

The symmetric integral in Eq. (4.14) is zero because the integrand $\sin\omega\,(x - v)$ is an odd function of $d\omega$. Multiplying i with Eq. (4.14) and adding it with Eq. (4.13) results in

$$f\,(x) = \frac{1}{2\pi}\int_{-\infty}^{\infty}\left[\int_{-\infty}^{\infty} f\,(v)\cos\omega\,(x - v) + i\,\sin\omega\,(x - v)dv\right]d\omega \tag{4.15}$$

Using Euler's formula,

$$e^{i\omega(x-v)} = \cos\omega\,(x - v) + i\omega\sin\,(x - v) \tag{4.16}$$

Equation (4.15) becomes

$$\begin{aligned} f\,(x) &= \frac{1}{2\pi}\int_{-\infty}^{\infty}\int_{-\infty}^{\infty} f\,(v)\,e^{i\omega(x-v)}dvd\omega \\ &= \int_{-\infty}^{\infty}\left[\frac{1}{2\pi}\int_{-\infty}^{\infty} f\,(v)\,e^{-i\omega v}dv\right]e^{i\omega x}d\omega \end{aligned} \tag{4.17}$$

The bracketed term in Eq. (4.17) is a function of ω and is known as the Fourier transform of the function $f\,(x)$, replacing the dummy integrand variable v with x

$$\mathscr{F}\,\{f\,(x)\} = F\,(\omega) = \frac{1}{2\pi}\int_{-\infty}^{\infty} f\,(x)\,e^{-i\omega x}dx \tag{4.18}$$

Using Eq. (4.18) and (4.17) becomes

$$f\,(x) = \int_{-\infty}^{\infty} F\,(\omega)\,e^{i\omega x}d\omega = \mathscr{F}^{-1}\,\{F\,(\omega)\} \tag{4.19}$$

$\mathscr{F}^{-1}\,\{F\,(\omega)\}$ is known as the inverse Fourier transform. Equation (4.18) uses intricate function/signal and breaks it into simpler constituent parts, known as analysis. In contrast, Eq. (4.19) reconstructs an intricate function/signal by reassembling its simple constituent components, known as synthesis. A non-periodic has periodicity infinite, and fundamental frequency tends to be zero. In the Fourier transform, the harmonics are more and more closely spaced; hence, the summation sign of the Fourier series is replaced by an integral sign.

Remarkably, the inverse Fourier transform functions in precisely the same manner as the synthesis process. If $f(x)$ is a real-valued function, then $F(\omega)$ exhibits symmetry, and conversely, if $f(x)$ is symmetric, $F(\omega)$ will be real. The forward Fourier transform (Eq. (4.18)) and the inverse Fourier transform (Eq. (4.19)) are collectively referred to as a "Fourier pair".

$$\begin{aligned} &\mathscr{F}\{f(x)\} \rightleftharpoons \mathscr{F}^{-1}\{F(\omega)\} \\ &f(x) \rightleftharpoons F(\omega) \end{aligned} \tag{4.20}$$

The universe's greatest secret is unveiled through this pair of equations. Both mathematicians and physicists acknowledged the profound importance of the Fourier integral in the nineteenth and twentieth centuries. This transform is now recognized as one of the most fundamental achievements of modern mathematical analysis, with extensive applications in both physical sciences and engineering. So impressed were British physicist William Thomson (1824–1907, also known as Lord Kelvin) and Scottish mathematical physicist Peter Guthrie Tait (1831–1901) by the generality and significance of the Fourier transform that they declared, "Fourier's theorem is not only one of the most beautiful results of modern analysis but may be said to furnish an indispensable instrument in the treatment of nearly every recondite question in modern physics. To mention only sonorous vibrations, the propagation of electric signals along a telegraph wire, and the conduction of heat by the earth's crust, as subjects in their generality intractable without it, is to give but a feeble idea of its importance". Fourier analysis is regarded as central and crucial in mathematical physics, probability and statistics, mathematical economics, geometry, harmonic analysis, signal processing, wave propagation, wavelet analysis, Brownian motion, stochastic processes, and distribution functions.

Given the wide range of applications for Fourier analysis, a single set of notations is not universally available. Representations in the frequency and time domains, as well as in the wavenumber and spatial domains, are interrelated reciprocally. In certain physics and engineering contexts, measurements are expressed in cycles per second or repetitions per meter instead of using angular frequency and wavenumber. The variable ω is equivalent to the wavenumber k for spatial periodicity. The spectroscopic wavenumber $\tilde{\nu}$ is converted to angular wavenumber k by multiplying it by 2π, so $k = 2\pi\tilde{\nu}$, and therefore, $dk = 2\pi d\tilde{\nu}$. Substituting the spectroscopic wavenumber into Eqs. (4.18) and (4.19) modifies the Fourier forward and inverse transforms to the following forms:

$$\mathscr{F}\{f(x)\} = F(\tilde{\nu}) = \int_{-\infty}^{\infty} f(x)\, e^{-2\pi i \tilde{\nu} x} dx \tag{4.21}$$

$$\mathscr{F}^{-1}\{F(\tilde{\nu})\} = f(x) = \int_{-\infty}^{\infty} F(\tilde{\nu})\, e^{2\pi i \tilde{\nu} x} d\tilde{\nu} \tag{4.22}$$

In this form, the Fourier forward transform pair and inverse Fourier transform pair form a symmetric twin without multiplication factor of $\frac{1}{2\pi}$ as in Eq. (4.18). Few textbooks like Lighthill (1970); Dym and McKean (1972); Osgood (2007); James (2011b) used Eqs. (4.21) and (4.22) as Fourier forward and inverse transform pair. However, many mechanical vibrations, turbomachinery and turbulence problems are measured in angular frequency and wavenumber. For these applications, $\frac{1}{2\pi}$ multiplier is required in either of the integral transformation processes as in Eqs. (4.18) and (4.19), and the symmetry between the pairs breaks. To preserve symmetry for those applications, few books like Jeffrey (2002); Kreyszig et al. (2011); Lokenath and Dambaru (2015) puts the $\frac{1}{\sqrt{2\pi}}$ multiplication in front of both forward and inverse transform. Books like Spiegel (1974); O'Neil (2007); James (2011a); Zill et al. (2013) puts the multiplication factor one in front of forward transform and a factor $\frac{1}{2\pi}$ in front of inverse transform. A few books like Duffy (2017) put a multiplication factor $\frac{1}{2\pi}$ for forward transform and one for inverse transform. Books like Spiegel (1974); Jeffrey (2002); O'Neil (2007); James (2011a); Kreyszig et al. (2011) and Duffy (2017) used $e^{-i\omega x}$ kernel for forward Fourier transform and $e^{i\omega x}$ kernel for inverse Fourier transform, while Zill et al. (2013) used $e^{i\omega x}$ kernel for forward transform and $e^{-i\omega x}$ kernel for inverse Fourier Transform. Changing the sign of the kernel in the Fourier pair calculation of a given function results in the complex conjugate. Different notations and expressions are prime sources of confusion among readers. Despite diversified definitions, the central theme of the Fourier and inverse Fourier transforms remains unique.

Depending on the application, users have the freedom to select different definitions for the Fourier transform. Any chosen notation functions correctly as long as the established rules are consistently adhered to. However, deviations from these rules during a lengthy calculation can lead to confusion. In this text, the forward transform employs a multiplicative factor of $\frac{1}{2\pi}$, while the inverse transform uses no such factor. The forward transform uses the kernel $e^{-i\omega x}$, whereas $e^{i\omega x}$ is used for the inverse transform. The notation $\mathscr{F}$ denotes the forward transform, and $\mathscr{F}^{-1}$ represents the inverse transform. Additionally, in the time or spatial domain, the function is denoted by $f(x)$; in the spectral domain, it is represented by $F(\omega)$. Both notational forms are applicable and will be discussed in detail in a subsequent section.

Example 4.1 (Fourier Transform of Rectangular Pulse) Without procrastinating further, let us find out the Fourier transfer of a rectangular pulse function $\Pi(x)$ (Fig. 4.2a), defined as follows:

$$\Pi(x) = \begin{cases} 1 & |x| < b \\ 0 & |x| \geq b \end{cases} \tag{4.23}$$

(continued)

Example 4.1 (continued)
$\Pi(x)$ function in Eq. (4.23) is also known as the top hat function or indicator function. Using Eq. (4.18), Fourier transform of Eq. (4.23) becomes

$$\begin{aligned}\mathscr{F}\{\Pi(x)\} = \hat{\Pi}(\omega) &= \frac{1}{2\pi}\int_{-\infty}^{\infty}\Pi(x)\,e^{-i\omega x}dx = \frac{1}{2\pi}\int_{-b}^{b}e^{-i\omega x}dx \\ &= -\frac{1}{2\pi i\omega}\left[e^{-i\omega x}\right]_{-b}^{b} = \frac{1}{\pi\omega}\frac{\left(e^{i\omega b}-e^{-i\omega b}\right)}{2i} \\ &= \frac{b}{\pi}\left(\frac{\sin\omega b}{\omega b}\right) = \frac{b}{\pi}sinc\,(\omega b)\end{aligned} \tag{4.24}$$

where $sinc\,(\omega b) = \dfrac{\sin\omega b}{\omega b}$. $sinc$ function has a value equal to 1 when $\omega \to 0$ (shown in Fig. 4.2b). In the limiting case of never repeating rectangular pulse function (infinite period and zero frequency),

$$\lim_{\omega\to 0}\hat{\Pi}(\omega) = \frac{b}{\pi} \tag{4.25}$$

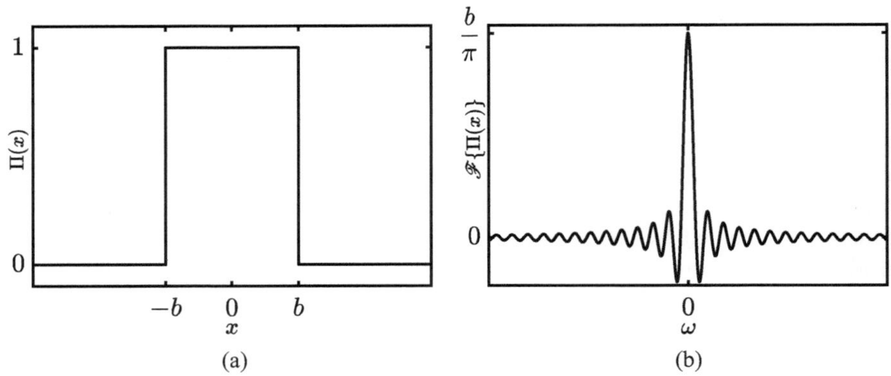

Fig. 4.2 (**a**) Rectangular pulse function. (**b**) Fourier transform of rectangular pulse function

Example 4.2 (Fourier Transform of Triangle Function) A non-periodic triangle function ($\Lambda(x)$) is defined as follows:

$$\Lambda(x) = \begin{cases} 1 - \dfrac{|x|}{b} & |x| \le b \\ 0 & |x| > b \end{cases} \tag{4.26}$$

Figure 4.3a represents the triangle function. Fourier transform of Λ function becomes

$$\begin{aligned}
\mathscr{F}\{\Lambda(x)\} = \hat{\Lambda}(\omega) &= \frac{1}{2\pi}\int_{-\infty}^{\infty} \Lambda(x)\, e^{-i\omega x} dx \\
&= \frac{1}{2\pi}\left[\int_{-b}^{0}\left(1+\frac{x}{b}\right)e^{-i\omega x}dx + \int_{0}^{b}\left(1-\frac{x}{b}\right)e^{-i\omega x}dx\right] \\
&= \frac{1}{2\pi}\left[-\frac{1}{i\omega}\left(1+\frac{x}{b}\right)e^{-i\omega x} + \frac{1}{b\omega^2}e^{-i\omega x}\right]_{-b}^{0} \\
&+ \frac{1}{2\pi}\left[-\frac{1}{i\omega}\left(1-\frac{x}{b}\right)e^{-i\omega x} - \frac{1}{b\omega_i^2}e^{-i\omega x}\right]_{0}^{b} \\
&= \frac{1}{2\pi}\frac{2}{b\omega_i^2}\left[1 - \frac{\left(e^{i\omega_i x_i} + e^{-i\omega x}\right)}{2}\right] \\
&= \frac{1}{2\pi}\frac{2}{b\omega^2}\left[1 - \cos\omega b\right] \\
&= \frac{b}{2\pi}\frac{\sin^2\left(\dfrac{\omega b}{2}\right)}{\left(\dfrac{\omega b}{2}\right)^2} = \frac{b}{2\pi} sinc^2\left(\frac{\omega b}{2}\right)
\end{aligned} \tag{4.27}$$

Fourier transfer of triangular function appears in the form of $\frac{b}{2\pi} sinc^2\left(\frac{\omega b}{2}\right)$ (Fig. 4.3b). In the limiting case of $\omega = 0$, the Fourier transfer of the triangle function becomes

$$\lim_{\omega \to 0} \hat{\Lambda}(\omega) = \frac{b}{2\pi} \tag{4.28}$$

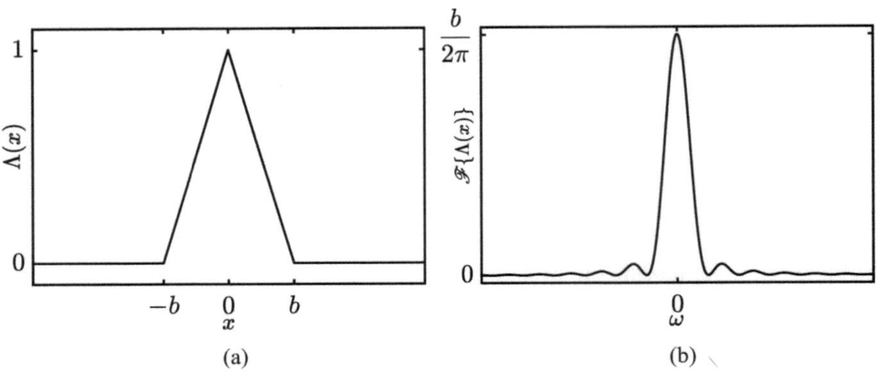

Fig. 4.3 (**a**) Triangle function. (**b**) Fourier transform of triangle function

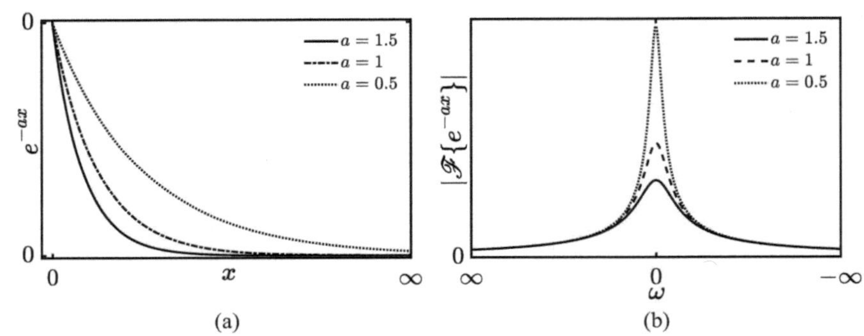

Fig. 4.4 (**a**) Exponential decaying function. (**b**) Fourier transform of exponential decaying function

Example 4.3 (Fourier Transform of Exponential Decaying Function) Consider the Fourier transform of the most commonly used decaying exponential function (Fig. 4.4a). This function is used to model certain types of chemical reactions (those depend on the concentration of one or another reactant), electrostatics, luminescence, radioactivity, thermo-electricity, and finance. The exponentially decaying function is defined as

$$f(x) = \begin{cases} 0 & x < 0 \\ e^{-ax} & x \geq 0 \end{cases} \tag{4.29}$$

(continued)

Example 4.3 (continued)
where a is any positive constant. This function models the signal that starts at the origin and decays gradually with time. A graphical representation of the function for the various constant parameters is presented in Fig. 4.4a. The Fourier transform of the function is expressed as

$$\begin{aligned}\mathcal{F}\{f(x)\} = F(\omega) &= \frac{1}{2\pi}\int_{-\infty}^{\infty} f(x)\, e^{-i\omega x} dx \\ &= \frac{1}{2\pi}\int_{0}^{\infty} e^{-(a+i\omega)x} dx \\ &= -\frac{1}{2\pi(a+i\omega)}\left[e^{-(a+i\omega)x}\right]_0^{\infty} \\ &= \frac{1}{2\pi(a+i\omega)}\end{aligned} \tag{4.30}$$

Unlike the Π or Λ function, the Fourier transform of the exponential function is not real but rather a complex-valued function. The power spectrum of the exponential decay function is expressed as

$$|F(\omega)|^2 = \frac{1}{|2\pi(a+i\omega)|^2} = \frac{1}{4\pi^2\left(a^2+\omega^2\right)} \tag{4.31}$$

The Fourier transform of an exponentially decaying function results in a complex-valued function of ω. The power spectrum shown in Fig. 4.4b represents the magnitude of this complex function, with its square corresponding to the total energy content in the spectral domain. The characteristic bell-shaped curve visible in Fig. 4.4b is identified as the Lorenz profile, which frequently arises in the analysis of transition probabilities and the lifetimes of excited atomic states. It is important to note that the Lorenz profile, though similar in appearance, is distinct from a Gaussian probability distribution bell curve. The Lorenz profile has heavier tails and describes resonance phenomena in various fields, making it essential in spectroscopy and the study of atomic line shapes, where it reflects the broadening of spectral lines due to natural and collisional effects.

4.2 Existence of Fourier Integral

In the preceding section, we examined how Fourier analysis facilitates the decomposition of any arbitrary non-periodic function. However, an essential question still lingers: does the Fourier transform ($\mathscr{F}$) of a given function $f(x)$ truly exist? To explore this, let us substitute $\omega = 0$ into Eq. (4.18).

$$F(\omega = 0) = \frac{1}{2\pi} \int_{-\infty}^{\infty} f(x)\, dx \tag{4.32}$$

This is the first Fourier coefficient. The Fourier transform's plausibility requires the integral's finite value in Eq. (4.32). If the function $f(x)$ is an absolute integrable (defined in Eq. (4.5)) and piecewise continuous on every finite interval, then both $\mathscr{F}\{f(x)\}$ and $\mathscr{F}^{-1}\{g(\omega)\}$ do exist. An absolute integrable function $f(x)$ indicates the square integrability in L^2 space (finite energy content system). The function must intercept the x axis as $x \to \infty$ in a thumb rule. The same is expressed in mathematical form as

$$\lim_{|x| \to \infty} f(x) = 0 \tag{4.33}$$

Both sine and cosine functions are infinitely periodic and open in the x direction without finite energy content. The Fourier transform of sine and cosine functions do not exist in the classical definition of Fourier analysis.

Analogous to the Fourier series, the Fourier integral in Eq. (4.18) converges to functional values $f(x)$ everywhere in the continuous domain, except at the finite jump location, where Fourier integral gives the arithmetic average of left- and right-hand limiting value of the function (Dirichlet condition for Fourier analysis).

$$\frac{1}{2}\left[f\left(x^{+}\right) + f\left(x^{-}\right)\right] = \frac{1}{2\pi} \int_{-\infty}^{\infty} \left[\int_{-\infty}^{\infty} f(v)\, e^{-i\omega v} dv\right] e^{i\omega x} d\omega \tag{4.34}$$

Equation (4.34) is widely recognized as the Fourier integral theorem. For smooth and continuous segments of x, the function satisfies $f\left(x^{+}\right) = f\left(x^{-}\right) = f(x)$, ensuring the function's values from either direction match at x. The Fourier integral theorem was first articulated in Fourier's seminal work, "La Théorie Analytique de la Chaleur" (Fourier, 1822). Mathematicians and physicists quickly acknowledged its profound importance throughout the nineteenth century. This theorem remains one of the most significant and groundbreaking achievements in modern mathematical analysis, providing the foundation for numerous applications across science and engineering.

The exponential factor $e^{[i\omega(x-v)]}$ in Eq. (4.34) can be expanded in terms of trigonometric functions. Using the even and odd nature of cosine and sine functions of ω, Eq. (4.34) is written as

$$f(x) = \frac{1}{\pi}\int_0^{\infty} d\omega \int_{-\infty}^{\infty} f(v)\cos\omega(x-v)dv \tag{4.35}$$

Π and Λ are even functions of x; hence, the Fourier sine integral vanishes during the Fourier transform, and the transform function becomes a real-valued function. The Fourier transform of even functions is known as the Fourier cosine transform.

$$f(x) = f(-x) = \frac{2}{\pi}\int_0^{\infty} \cos(\omega x)d\omega \int_0^{\infty} f(v)\cos(\omega v)dv \tag{4.36}$$

But when $f(x)$ is odd, the Fourier cosine integral vanishes, and the Fourier transform function has a purely imaginary component. Such transforms are known as Fourier sine transform.

$$f(x) = -f(-x) = \frac{2}{\pi}\int_0^{\infty} (\omega x)d\omega \int_0^{\infty} f(v)\sin(\omega v)dv \tag{4.37}$$

These integral formulas were discovered independently by Cauchy in his work on the propagation of waves on the surface of water (Cauchy, 1825). But the function which is neither even nor odd like decaying exponential function e^{-ax}, both sine and cosine component persists during the Fourier transforms, and the transform function is a complex-valued function having both non-zero real and imaginary part.

The square of a transform function $|F(\omega)|^2$ represents its magnitude and is called the energy spectrum. In fluid turbulence, it represents spectral turbulence kinetic energy; in communication engineering, it represents the power spectrum; and in optics, it represents spectral power density. The energy spectrum establishes an important relation between the function in both spatial/time and spectral domains.

$$\int_{-\infty}^{\infty} |F(\omega)|^2 d\omega = \frac{1}{2\pi}\int_{-\infty}^{\infty} |f(x)|^2 dx \tag{4.38}$$

Equation (4.38) is known as Parseval's identity for Fourier transforms (Plancherel formula), which facilitates transform of function/signal from time/spatial domain to spectral domain without violating the conservation of energy principle. Swiss mathematician Michel Plancherel (1885–1967) proved Eq. (4.38) (Plancherel, 1910). By intuition, it is obvious that $\mathscr{F}\{f(x)\} = F(\omega) \to 0$ as $\omega \to \infty$, which is called the Riemann-Lebesgue lemma. $\mathscr{F}\{f(x)\}$ is usually a contentious function of ω.

4.3 Properties of Fourier Transform

Properties of the Fourier transform must be comprehended thoroughly and precisely from all perspectives to ensure a fair and accurate understanding and practical application.

4.3.1 *Physical Realization of Fourier Transform*

Hamming (1977) proposed an analogy between the Fourier transform and a glass prism, illustrated in Fig. 4.5. In this comparison, the function $f(x)$ is likened to white light, composed of a mixture of waves at different wavelengths. White light disperses into its constituent frequencies when it passes through a prism. Similarly, the Fourier transform decomposes the function $f(x)$ into a spectrum of frequencies ω, with each frequency having an associated intensity $F(\omega)$. The complete range of these frequencies, for which both the Fourier transform and its inverse exist, is referred to as the spectrum. The procedure of converting variables from the spatial or temporal domain to the frequency or wavenumber domain is known as spectral analysis. The term "spectrum" originates from optics, describing the array of colors produced when a prism disperses light. In this context, the spectral density $F(\omega)$ quantifies light intensity within a specific frequency range $\omega + d\omega$. The total energy contribution within a frequency band $a \leq \omega \leq b$ is expressed as

$$E(\omega) = \int_a^b |F(\omega)|^2 d\omega \tag{4.39}$$

Fourier analysis is indispensable in quantum mechanics, particularly for understanding wave functions and their characteristics. The position and momentum of particles are linked via Fourier transforms, where the wave function in position space, $\psi(x)$, and momentum space, $\phi(p)$ forms a Fourier pair. This connection sheds light on the Heisenberg uncertainty principle, illustrating the inherent limitations in precisely determining a particle's position and momentum simultaneously.

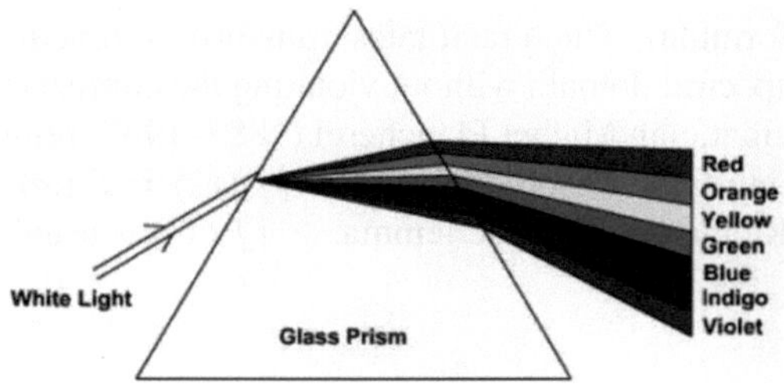

Fig. 4.5 Dispersion in a prism

In the realm of optics, Fourier analysis plays a crucial role in describing the behavior of light. It is employed to examine diffraction patterns and light propagation through different media. The Fourier pair relationship in this context exists between the aperture angle x and the sine of the diffraction angle divided by the wavelength, expressed as $p = \sin\theta/\lambda$. By transforming the aperture function of an optical system into its frequency components, physicists can predict and interpret the resulting diffraction patterns. This is vital for designing optical instruments such as microscopes and telescopes, where accurate control of light behavior is essential. Fourier analysis is equally valuable in condensed matter physics, where it is applied to examine the periodic structures of crystals through X-ray diffraction patterns. By analyzing these patterns, scientists can determine the atomic arrangement within a crystal, providing key insights into its physical properties and potential technological applications.

In the realm of fluid dynamics, Fourier analysis provides a framework for understanding turbulence. Turbulent flows consist of eddies of various sizes and time scales, making direct analysis challenging. By transforming the turbulent flow data into the frequency and wavenumber domain, physicists can study the energy distribution across different scales, known as the turbulent spectrum. This approach is essential for developing models and simulations that predict fluid behavior in engineering and natural systems. Fourier analysis is also crucial in studying wave phenomena, including sound, electromagnetic, and water waves. By decomposing these waves into their frequency components, physicists can analyze their behavior and interactions. For instance, in acoustics, Fourier transforms help in analyzing sound signals, leading to advancements in sound engineering, noise reduction, and audio compression technologies. Fourier transforms are employed in signal processing, a fundamental aspect of experimental physics. They enable the extraction of meaningful information from noisy data, facilitating the analysis of signals from various detectors and sensors. This capability is crucial in particle physics, where detecting rare events amidst background noise is a significant challenge.

In summary, Fourier analysis is a versatile and indispensable tool in physics, offering a means to transform complex, time-dependent, or spatially varying phenomena into a more manageable frequency domain. Its applications span numerous fields, from quantum mechanics and optics to fluid dynamics and condensed matter physics, underscoring its fundamental role in advancing our understanding of the physical world. Furthermore, the inverse Fourier transform is crucial for recombining these different spectral components, effectively reconstructing the original function or signal from its transformed state. This blending process allows for a comprehensive analysis and understanding of complex systems, such as fluid turbulence, by providing a clearer picture of the underlying energy distributions.

4.3.2 Transform Pair and Duality in Fourier Transform

Duality in Fourier analysis is a significant property that dramatically simplifies the process of performing Fourier transforms. Duality refers to the symmetry between the forward Fourier transform and its inverse, often leading to misunderstandings when not appropriately applied or interpreted. The forward and inverse Fourier transforms are mathematically expressed as follows:

$$\mathscr{F}\{f(x)\} = F(\omega) = \frac{1}{2\pi}\int_{-\infty}^{\infty} f(x)\, e^{-i\omega x} dx \tag{4.40}$$

$$\mathscr{F}^{-1}\{F(\omega)\} = f(x) = \int_{-\infty}^{\infty} F(\omega)\, e^{i\omega x} d\omega \tag{4.41}$$

where $f(x) \rightleftharpoons F(\omega)$ are the Fourier pair. $f(-x)$ is the signal reversal for a signal $f(x)$. For convenience, let us write $f(-x) = f^{-}(x)$. $F(-\omega)$ is the reverse transfer signal for the original Fourier transfer signal $F(\omega)$. For convenience, let us use the notation $F(-\omega) = F^{-}(\omega)$. Fourier integral of reversal signal $f(-x)$ is expressed as

$$\mathscr{F}\{f(-x)\} = \mathscr{F}\left\{f^{-}(x)\right\} = \frac{1}{2\pi}\int_{-\infty}^{\infty} f(-x)\, e^{-i\omega x} dx \tag{4.42}$$

Taking $y = -x$, Eq. (4.42) becomes

$$\mathscr{F}\{f(-x)\} = -\frac{1}{2\pi}\int_{\infty}^{-\infty} f(y)\, e^{i\omega y} dy = \frac{1}{2\pi}\int_{-\infty}^{\infty} f(y)\, e^{-i(-\omega)y} dy = F(-\omega) \tag{4.43}$$

Equation (4.43) is known as the reversal property of the Fourier transform. Fourier transform of the reverse signal equals the reverse of the transformed signal.

$$\mathscr{F}\{f(-x)\} = \mathscr{F}\left\{f^{-}(x)\right\} = F^{-}(\omega) = F(-\omega) \tag{4.44}$$

The symmetry between the Fourier and inverse Fourier transform is the striking feature. Replacing ω with $-\omega$ in Eq. (4.40) leads to

$$F(-\omega) = \frac{1}{2\pi}\int_{-\infty}^{\infty} f(x)\, e^{i\omega x} dx = \frac{1}{2\pi}\mathscr{F}^{-1}\{f(x)\} \tag{4.45}$$

Diddling with symbols is inevitable in this subject. From Eqs. (4.44) and (4.45),

$$\mathscr{F}\{f(-x)\} = F(-\omega) = \frac{1}{2\pi}\mathscr{F}^{-1}\{f(x)\} \tag{4.46}$$

The Fourier transform of a reversed signal is $\frac{1}{2\pi}$ times its inverse Fourier transform. In this formulation, both functions exist within the same domain, making the previous discussions on transforming signals between the time/spatial and frequency/wavenumber domains somewhat less intuitive from a physical perspective. It is crucial to recognize that the same mathematical expression can have multiple interpretations and should not be viewed in a singular, rigid way. The Fourier transform acts as a mathematical kernel, providing two distinct representations, $f(x)$ and $F(\omega)$, for the same function. Equation (4.46) is referred to as the "first duality principle of Fourier transforms". This principle asserts that the Fourier transform of a reversed signal is equivalent to the reversed Fourier transform, which is $\frac{1}{2\pi}$ times the inverse Fourier transform of the original signal. For this reason, the inverse Fourier transform is often called the reverse or back transform. The equation can also take on the following form:

$$\mathscr{F}\{f(x)\} = \frac{1}{2\pi}\mathscr{F}^{-1}\{f(-x)\} \tag{4.47}$$

Second duality in Fourier transform $f(x) \rightleftharpoons F(\omega)$ pair is expressed as follows: Let the forward transform be expressed as

$$\mathscr{F}\{f(x)\} = F(\omega) \tag{4.48}$$

From inverse transform formula (Eq. (4.41)),

$$f(x) = \int_{-\infty}^{\infty} F(\omega)\, e^{i\omega x} d\omega \tag{4.49}$$

Replacing x with $-y$, in Eq. (4.49), and multiplying both sides with $\frac{1}{2\pi}$ results in

$$\frac{1}{2\pi} f(-y) = \frac{1}{2\pi}\int_{-\infty}^{\infty} F(\omega)\, e^{-i\omega y} d\omega = \mathscr{F}\{F(\omega)\} = \mathscr{F}\{\mathscr{F}\{f(y)\}\} \tag{4.50}$$

Equation (4.50) represents the second duality statement of Fourier transform, which states Fourier transform of the Fourier transform of a function equals $\frac{1}{2\pi}$ times reverse of the function.

$$\mathscr{F}\{\mathscr{F}\{f(x)\}\} = \frac{1}{2\pi} f(-x) \tag{4.51}$$

Any effort to ascribe a physical interpretation to duality inevitably results in ambiguity and misunderstanding. By regarding the Fourier transform as an abstract mathematical construct, the duality principles become far clearer, dispelling any risk of misinterpretation. These inherent dualities within Fourier transforms serve to greatly simplify the application of Fourier analysis, rendering the process more intuitive and less prone to confusion.

Example 4.4 (Fourier Transfer of $sinc(x)$) The Fourier transfer of $sinc = \frac{\sin x}{x}$ function from the definition takes the form

$$\mathscr{F}\{sinc(x)\} = \frac{1}{2\pi} \int_{-\infty}^{\infty} \frac{\sin x}{x} e^{-i\omega x} dx \tag{4.52}$$

Earlier, we discussed that $\sin x$ and $\cos x$ are non-bounding functions in the x axis and do not possess any finite amount of energy. Finding any analytical solution of the integral in Eq. (4.52) is challenging. However, the principle of duality makes the job much simpler.

Fourier transfer of a rectangular pulse function $\mathscr{F}\{\Pi(x)\}$ equals to $\frac{b}{\pi} sinc(\omega b)$; hence, $\mathscr{F}\{\Pi(bx)\} = \frac{1}{\pi} sinc(\omega)$. The Fourier transform of the $sinc(\omega)$ becomes

$$\mathscr{F}\{sinc(x)\} = \mathscr{F}\{\pi\mathscr{F}\{\Pi(b\omega)\}\} = \frac{1}{2}\Pi(-b\omega) \tag{4.53}$$

In Eq. (4.53), the second principle of duality (Eq. (4.51)) is used. Π is an even function, $\Pi(-b\omega) = \Pi(b\omega)$; hence,

$$\mathscr{F}\{sinc(x)\} = \frac{1}{2}\Pi(b\omega) \tag{4.54}$$

Example 4.5 (Fourier Transfer of $sinc^2(x)$) The Fourier transform of $sinc^2(x)$ does not exist from the ab initio definition, as the analytic integral does not converge. From Example 4.2, $\mathscr{F}\{\Lambda(x)\} = \frac{b}{2\pi} sinc^2\left(\frac{\omega b}{2}\right)$. Replacing the variable x with $\frac{b}{2}x$ results in $\mathscr{F}\left\{\Lambda\left(\frac{b}{2}x\right)\right\} = \frac{1}{\pi} sinc^2(\omega)$.

(continued)

Example 4.5 (continued)

Applying the second duality principle,

$$\mathscr{F}\left\{sinc^2(x)\right\} = \mathscr{F}\left\{\pi\mathscr{F}\left\{\Lambda\left(\frac{b}{2}x\right)\right\}\right\} = \frac{1}{2}\Lambda\left(-\frac{b}{2}x\right) \tag{4.55}$$

Λ is an even function; hence,

$$\mathscr{F}\left\{sinc^2(x)\right\} = \frac{1}{2}\Lambda\left(\frac{b}{2}x\right) \tag{4.56}$$

4.3.3 Linearity in Fourier Transform

Fourier transform is a linear operation. For any two Fourier pair $f_1(x) \rightleftharpoons F_1(\omega)$ and $f_2(x) \rightleftharpoons F_2(\omega)$, Fourier transformation of linear combination of $f_1(x)$ and $f_2(x)$ equals to linear combination of their individual Fourier transform.

$$\begin{aligned}\mathscr{F}\{\alpha f_1(x) + \beta f_2(x)\} &= \frac{1}{2\pi}\int_{-\infty}^{\infty}\{\alpha f_1(x) + \beta f_2(x)\}e^{-i\omega x}dx \\ &= \alpha\frac{1}{2\pi}\int_{-\infty}^{\infty} f_1(x)e^{-i\omega x}dx + \beta\frac{1}{2\pi}\int_{-\infty}^{\infty} f_2(x)e^{-i\omega x}dx \\ &= \alpha\mathscr{F}f_1(x) + \beta\mathscr{F}f_2(x) = \alpha F_1(\omega) + \beta F_2(\omega)\end{aligned} \tag{4.57}$$

α and β are constants.

Example 4.6 (Fourier Transform of Two-Sided Decaying Exponential Function) A two-sided decaying exponential function is expressed as

$$f(x) = e^{-a|x|} \tag{4.58}$$

With any $a > 0$. For $x \geq 0$, the Fourier transform for the exponential function is already derived in Eq. (4.30). Using the reversal property of the transform $\mathscr{F}(f(-x)) = F(-\omega)$, the solution of Eq. (4.58) for $x < 0$ becomes

(continued)

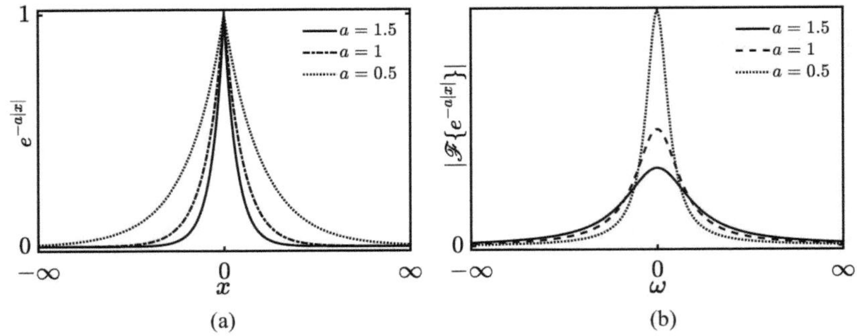

Fig. 4.6 (**a**) The two-sided decaying exponential function. (**b**) Fourier transform of the two-sided decaying exponential function

Example 4.6 (continued)

$$\mathscr{F}\{f(x)\} = \frac{1}{2\pi(a - i\omega)} \tag{4.59}$$

Using the principle of linearity, the solution for $-\infty \leq x \leq \infty$ is the linear combination for $x \geq 0$ and $x < 0$.

$$\mathscr{F}\{f(x)\} = \frac{1}{2\pi(a + i\omega)} + \frac{1}{2\pi(a - i\omega)} = \frac{a}{\pi\left(a^2 + \omega^2\right)} \tag{4.60}$$

The Fourier transform of the two-sided decaying exponential function is real in ω due to its symmetry about ordinate. Fourier transform is complex for a one-sided decaying exponential function. Figure 4.6 represents a two-sided exponential function's time and spectral domain graph.

4.3.4 Constant Multiplication with Fourier Transform

If $f(x) \rightleftharpoons F(\omega)$ is the Fourier pair, then the Fourier transform of a constant multiple a times $f(x)$ equals to a times Fourier transform of its functional value. Mathematically, the same is expressed as

$$\mathscr{F}\{af(x)\} = \frac{1}{2\pi}\int_{-\infty}^{\infty} af(x)e^{-i\omega x}dx = a\mathscr{F}\{f(x)\} = aF(\omega) \tag{4.61}$$

Physically, a acts as a multiplier of the intensity of the function/signal $|F(\omega)|$. Beethoven's symphony or Kishore Kumar songs are played at slow volume in a radio room, but when in a concert, the same is played with a loudspeaker or microphone. Microphone and loudspeaker act as a, which amplify the intensity of $|F(\omega)|$ to $a\,|F(\omega)|$, without any alteration in its frequency ω.

4.3.5 Symmetric Properties of Fourier Pair

Any function $f(x)$ can be written as a sum of even function $g(x)$ and odd function $h(x)$. So the Fourier transform of $f(x)$ takes the form

$$\begin{aligned}
\mathscr{F}\{f(x)\} &= \frac{1}{2\pi}\int_{-\infty}^{\infty} g(x)\, e^{-i\omega x} dx + \frac{1}{2\pi}\int_{-\infty}^{\infty} h(x)\, e^{-i\omega x} dx \\
&= \frac{1}{2\pi}\int_{-\infty}^{\infty} g(x)\,(\cos(\omega x) - i\sin(\omega x))\, dx \\
&\quad + \frac{1}{2\pi}\int_{-\infty}^{\infty} h(x)\,(\cos(\omega x) - i\sin(\omega x))\, dx \\
&= \frac{1}{2\pi}\int_{-\infty}^{\infty} g(x)\cos(\omega x)\, dx - i\frac{1}{2\pi}\int_{-\infty}^{\infty} h(x)\sin(\omega x)\, dx
\end{aligned} \tag{4.62}$$

The Fourier transform of any arbitrary functions $f(x)$ takes the following forms because of the above symmetric property. Fourier transform properties of a function $f(x)$ is provided in Table 4.1.

4.3.6 Time Delay or Space Shift in Fourier Transform

The property of a Fourier pair $f(x) \rightleftharpoons F(\omega)$, during the delay of time or shift in space, is interesting. Any delay or shifting for function $f(x)$ an amount b is expressed as $f(x \pm b)$. The Fourier transform for $f(x \pm b)$ from its definition is

$$\mathscr{F}\{f(x \pm b)\} = \frac{1}{2\pi}\int_{-\infty}^{\infty} f(x \pm b)\, e^{-i\omega x} dx \tag{4.63}$$

Table 4.1 Fourier transform properties of a function $f(x)$

Original function $f(x)$	Transfer function $F(\omega)$	Comments
Real	$F(-\omega) = F^*(\omega)$	Conjugate symmetry
Imaginary	$F(-\omega) = -F^*(\omega)$	Conjugate antisymmetric
Even	$F(-\omega) = F(\omega)$	Even
Odd	$F(-\omega) = -F(\omega)$	Odd
Real and even	$\Re\{F(-\omega)\} = \Re\{F(\omega)\}$, $\Im\{F(\omega)\} = 0$	Real and even
Real and odd	$\Re\{F(\omega)\} = 0$, $\Im\{F(-\omega)\} = \Im\{-F(\omega)\}$	Imaginary and odd
Imaginary and even	$\Re\{F(\omega)\} = 0$, $\Im\{F(-\omega)\} = \Im\{F(\omega)\}$	Imaginary and even
Imaginary and odd	$\Re\{F(-\omega)\} = \Re\{-F(\omega)\}$, $\Im\{F(\omega)\} = 0$	Real and odd

By setting $y = x \pm b$, the Fourier transform becomes (limits of the integration remains unchanged)

$$\begin{aligned}
\mathscr{F}\{f(x \pm b)\} &= \frac{1}{2\pi}\int_{-\infty}^{\infty} f(y)\, e^{-i\omega(y \mp b)} dy \\
&= \frac{1}{2\pi}\int_{-\infty}^{\infty} f(y)\, e^{-i\omega \mp b} e^{-i\omega y} dy \\
&= e^{\pm i\omega b} \frac{1}{2\pi}\int_{-\infty}^{\infty} f(y)\, e^{-i\omega y} dy \\
&= e^{\pm i\omega b} \mathscr{F}\{f(x)\} = e^{\pm i\omega b} F(\omega)
\end{aligned} \tag{4.64}$$

Any delay in time or shift of space $\pm b$ results in shifting the phase $e^{\pm i\omega b}$ in its transform pair $F(\omega)$.

Figure 4.7a represents b amount of space shift or time delay in rectangular pulse function. This results in a $e^{i\omega b}$ phase shift in the spectral domain. Due to the even property, the spectral transformation of $\Pi(x)$ is real and symmetric. Any amount of shift b breaks that symmetric property of Π function. This results in a complex-valued transfer function in ω, with a phase lag of $e^{\pm i\omega b}$. Figure 4.7b represents both real and imaginary parts of the time delayed for unit pulse signal. Despite phase shift, total energy content of $\Pi(x)$ and $\Pi(x+b)$ remains the same.

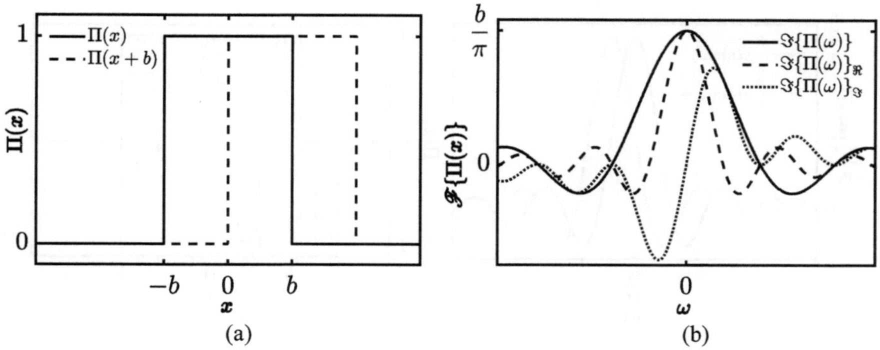

Fig. 4.7 Fourier transform for a delay or shift in function. (**a**) delay/shift in time/space. (**b**) phase change in spectrum

4.3.7 Frequency or Wavenumber Shift in Fourier Transform

For a Fourier pair $f(x) \rightleftharpoons F(\omega)$, any shifting of frequency ω_0 in $F(\omega)$ is represented as $F(\omega + \omega_0)$. Taking inverse Fourier transform of $F(\omega + \omega_0)$ from its definition,

$$\mathscr{F}^{-1}\{F(\omega \pm \omega_0)\} = \int_{-\infty}^{\infty} F(\omega \pm \omega_0)\, e^{i\omega x} d\omega \tag{4.65}$$

Taking $\omega_d = \omega \pm \omega_0$, Eq. (4.65) becomes

$$\begin{aligned}\mathscr{F}^{-1}\{F(\omega \pm \omega_0)\} &= \int_{-\infty}^{\infty} F(\omega_d)\, e^{i(\omega_d \mp \omega_0)x} d\omega_d \\ &= e^{\mp i\omega_0 x} \int_{-\infty}^{\infty} F(\omega_d)\, e^{i\omega_d x} d\omega_d \\ &= e^{\mp i\omega_0 x} f(x)\end{aligned} \tag{4.66}$$

Any amount of frequency or wavenumber shifts ω_0 results in $e^{\mp i\omega_0 x}$ phase shift of function/signal in time or spatial domain. Figure 4.8a represents ω_0 shift in $\frac{b}{\pi} sinc(\omega)$ function. Inverse Fourier transfer of $\frac{b}{\pi} sinc(\omega)$ equals to $\Pi(x)$ function with width $2b$. Inverse Fourier transfer of $\frac{b}{\pi} sinc(\omega \pm \omega_0)$ becomes $\Pi(x)\, e^{\mp i\omega_0 x}$. For simplicity, let's call it $\tilde{\Pi}(x)$. Till now, the discussion was about real functions in

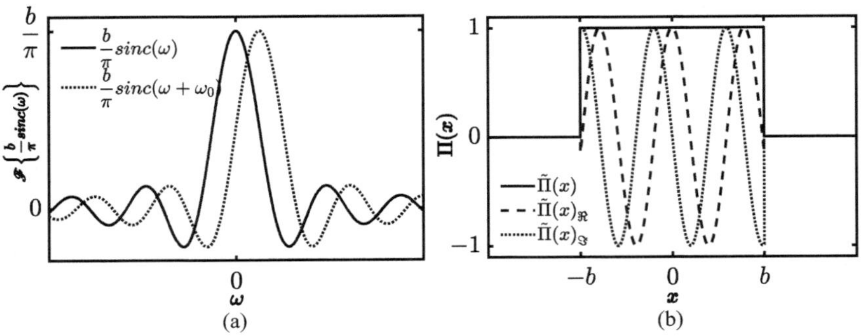

Fig. 4.8 (**a**) Frequency shift in *sinc* function. (**b**) Corresponding Fourier transform due to frequency shift

the time/spatial domain and complex functions in the spectral domain. But $\tilde{\Pi}(x)$ is a complex function in time/spatial domain, having a magnitude of $|\Pi(x)|$ and a phase shift of $e^{\mp i\omega_0 x}$. Figure 4.8b represents the real and imaginary part of the function in the time/spatial domain. Point to notice that the energy of the function/signal is conserved during frequency shift.

Example 4.7 (Fourier Transform of Gaussian Distribution) The probability density function for normal distribution with mean μ and standard deviation σ is

$$f(y) = \frac{1}{\sqrt{2\pi}\sigma} e^{-\frac{(y-\mu)^2}{2\sigma^2}} \tag{4.67}$$

Substituting $x = y - \mu$, $b = \frac{1}{2\sigma^2}$, and $a = \frac{1}{\sqrt{2\pi}\sigma}$ in Eq. (4.67) for simplification results,

$$f(x+\mu) = g(x) = ae^{-bx^2} \tag{4.68}$$

$g(x)$ is the general Gaussian function of x, with $a, b > 0$. Fourier transform for a Gaussian distribution exists as it contains finite energy ($|x| \to \infty$, $g(x) \to 0$). However, in single-variable calculus, an analytic expression for the Fourier transform of the Gaussian function does not exist from the ab initio principle.

(continued)

Example 4.7 (continued)

Gaussian integral is defined as

$$I = \int_{-\infty}^{\infty} g(x)\, dx = \int_{-\infty}^{\infty} ae^{-bx^2} dx \tag{4.69}$$

Squaring both sides of Eq. (4.69),

$$I^2 = \int_{-\infty}^{\infty} ae^{-bx^2} dx \int_{-\infty}^{\infty} ae^{-bx^2} dx = \int_{-\infty}^{\infty} ae^{-bx^2} dx \int_{-\infty}^{\infty} ae^{-by^2} dy \tag{4.70}$$

In a definite integral, variables inside the integrand are dummy; hence, variable y appears for x in Eq. (4.70). On simplification of Eq. (4.70),

$$\begin{aligned} I^2 &= a^2 \int_{-\infty}^{\infty} \int_{-\infty}^{\infty} e^{-b(x^2+y^2)} dxdy = a^2 \int_{0}^{\infty} \int_{0}^{2\pi} e^{-br^2} rdrd\theta \\ &= a^2 \int_{0}^{\infty} \left[e^{-br^2} r \int_{0}^{2\pi} d\theta dr \right] = 2\pi a^2 \int_{0}^{\infty} e^{-br^2} rdr \\ &= -\frac{\pi a^2}{b} \left[e^{-br^2} \right]_0^{\infty} = \frac{\pi a^2}{b} \end{aligned} \tag{4.71}$$

Variables in $x - y$ coordinate are transferred into $r - \theta$ coordinate. $dxdy$ takes the form $rdrd\theta$ (due to Jacobi transformation, named after German mathematician Carl Gustav Jacob Jacobi (1804–1851)). The solution of the Gaussian integral becomes

$$I = a\sqrt{\frac{\pi}{b}} \tag{4.72}$$

In Eq. (4.72), substitution of values for a and b results in $I = 1$. This is the requirement for any probability density function. Fourier transfer of the Gaussian function becomes

(continued)

Example 4.7 (continued)

$$\mathscr{F}\{g(x)\} = G(\omega) = \frac{1}{2\pi}\int_{-\infty}^{\infty} g(x)\, e^{-i\omega x} dx = \frac{1}{2\pi}\int_{-\infty}^{\infty} ae^{-bx^2} e^{-i\omega x} dx$$

$$= \frac{1}{2\pi}\int_{-\infty}^{\infty} ae^{-(bx^2+i\omega x)} dx = \frac{1}{2\pi}\int_{-\infty}^{\infty} ae^{-b\left(x+i\frac{\omega}{2b}\right)^2 - \frac{\omega^2}{4b}} dx$$

$$= \frac{1}{2\pi}e^{-\frac{\omega^2}{4b}}\int_{-\infty}^{\infty} ae^{-b\left(x+i\frac{\omega}{2b}\right)^2} dx \tag{4.73}$$

Substitution of $z = \left(x + \mathrm{i}\frac{\omega}{2b}\right)$ with $dz = dx$ in Eq. (4.73) results in

$$\mathscr{F}\{g(x)\} = G(\omega) = \frac{1}{2\pi}e^{-\frac{\omega^2}{4b}}\int_{-\infty}^{\infty} ae^{-bz_i^2} dz_i = \frac{1}{2\pi}e^{-\frac{\omega^2}{4b}} a\sqrt{\frac{\pi}{b}}$$

$$= \frac{a}{2\sqrt{\pi b}}e^{-\frac{\omega^2}{4b}} = \frac{1}{2\pi}e^{-\frac{\sigma^2}{2}\omega^2} \tag{4.74}$$

Fourier transfer of a Gaussian function is another Gaussian function. The integrating limit remains unchanged due to the nature of the substitution in Eq. (4.74). Fourier transfer of original Gaussian PDF $f(x + \mu) = g(x)$ can be determined using the shifting principle.

$$F(\omega) = \frac{1}{2\pi}e^{-\frac{\sigma^2}{2}\omega^2 + i\omega\mu} \tag{4.75}$$

Figure 4.9 represents both the forward and backward transform of the Gaussian function for different values of standard deviation σ. Increasing the standard deviation (σ) in the original Gaussian function squeezes the standard deviation in the transferred Gaussian pair. A decrease in standard deviation in the original Gaussian function spreads the standard deviation in the transferred Gaussian pair. The subsequent section will cover the physical interpretation of the same. The Fourier transfer is real and even for a standard normal distribution ($\mu = 0$ and $\sigma = 1$).

$$F(\omega) = \frac{1}{2\pi}e^{-\frac{\omega^2}{2}} \tag{4.76}$$

(continued)

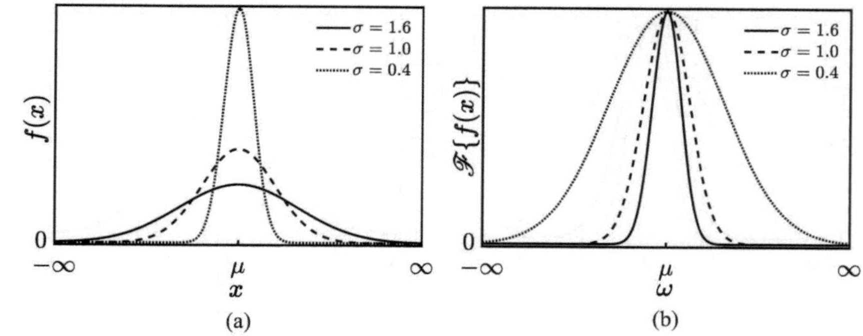

Fig. 4.9 (**a**) Gaussian distribution function. (**b**) Fourier transform of Gaussian distribution function

Example 4.7 (continued)
For any non-zero mean $\mu \neq 0$, the symmetry around the ordinate is lost, and a phase component $e^{i\omega\mu}$ is multiplied with the standard normal distribution.

4.3.8 Stretching and Shrinking Phenomenon in Fourier Transform

When a variable x is stretched or shrunk by a factor ax (stretched for $a > 1$, shrink for $a < 1$, and unchanged for $a = 1$), its effect is impassioned on the Fourier transfer of the function. From the ab initio principle, the Fourier transform of the function with stretched variable ax is

$$\mathscr{F}\{f(ax)\} = \frac{1}{2\pi}\int_{-\infty}^{\infty} f(ax)\, e^{-i\omega x} dx \tag{4.77}$$

Substitution of $y = ax$ results in

$$\mathscr{F}\{f(ax)\} = \frac{1}{a}\frac{1}{2\pi}\int_{-\infty}^{\infty} f(y)\, e^{-i\omega \frac{y}{a}} dy = \frac{1}{a}F\left(\frac{y}{a}\right) \tag{4.78}$$

A function/signal cannot be localized both in the time and frequency domain. Figure 4.10 represents the stretching/shrinking effect on the Fourier transform of a Gaussian pair. For $a < 1$, the variable ax shrinks, and the Gaussian function

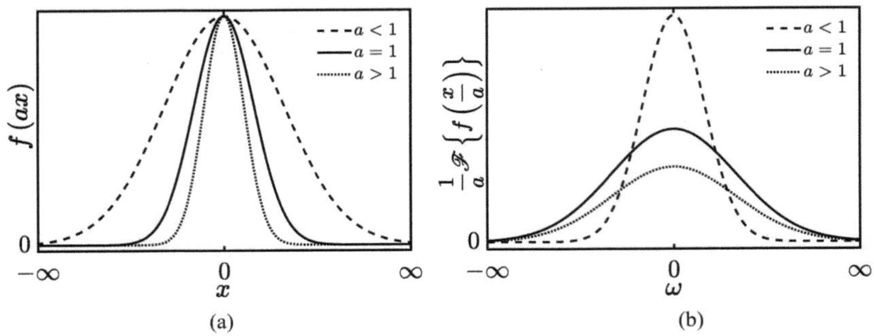

Fig. 4.10 (**a**) Stretching/Shrinking of the original function. (**b**) Corresponding response in Fourier transform

is spread over x (being the negative square of exponential) (Fig. 4.10a). At the same time, $\frac{\omega}{a}$ spreads for $a < 1$, and the transformed Gaussian pair shrinks in spectral space (Fig. 4.10b). For $a > 1$, the phenomenon is reversed. Simultaneous localization of variables in both the time and frequency domain is impossible. From Fig. 4.10, it is clear that the standard deviation in the Gaussian function works as a stretching parameter. Increasing the standard deviation in one space reduces the standard deviation in the transform space. The product of standard deviation in f and $\mathscr{F}\{f\}$ becomes

$$\sigma(f)\,\sigma(\mathscr{F}\{f\}) \geq \frac{1}{4\pi} \tag{4.79}$$

Equation (4.79) is the famous Heisenberg uncertainty principle in quantum mechanics. Fourier transform in quantum mechanics works as a transform pair between position space x and momentum space $\frac{p}{\hbar}$ ($\hbar$ Planck's constant). In 1927, German physicist Werner Heisenberg (1901–1976) found the above inequality and stated that "The more precisely the position of some particle is determined, the less precisely its momentum can be known, and vice versa" (Heisenberg, 1927). Mathematically, the same is expressed as

$$\Delta x \cdot \Delta p \geq \frac{\hbar}{2} \tag{4.80}$$

Δx is the uncertainty in position, and Δp is the uncertainty in momentum. The Heisenberg uncertainty principle has profound implications for understanding the quantum world. It challenges the classical notion that particles always have well-defined positions and momenta. Instead, it suggests that at a quantum level, nature is inherently probabilistic, and one can only predict the likelihood of finding a particle in a particular state. A detailed description of this brilliant finding is beyond the purview of the present book. In fluid turbulence, Tennekes (1976) stated, "Since eddies in turbulent flows are a dense collection of very short wave packets, their position, and wavenumber(or lifetime and frequency) cannot be determined without ambiguity."

4.3.9 Differentiation of Fourier Transform Pair

$f'(x)$ is the differentiation of a continuous function $f(x)$ with respect to x. The Fourier transform of $f'(x)$ becomes

$$\mathscr{F}\left\{f'(x)\right\} = \frac{1}{2\pi}\int_{-\infty}^{\infty} f'(x)\, e^{-i\omega x} dx \tag{4.81}$$

Integrating the right-hand side of Eq. (4.81) by parts results in

$$\mathscr{F}\left\{f'(x)\right\} = \frac{1}{2\pi}\left[e^{-i\omega x} f(x)\Big|_{-\infty}^{\infty} - (-i\omega)\int_{-\infty}^{\infty} f(x)\, e^{-i\omega x} dx \right] \tag{4.82}$$

The first term of the right side of Eq. (4.82) vanishes, and the energy content of the system is finite ($|x| \to \infty$, $f(x) \to 0$).

$$\mathscr{F}\left\{f'(x)\right\} = (i\omega)\frac{1}{2\pi}\int_{-\infty}^{\infty} f(x)\, e^{-i\omega x} dx = (i\omega)\,\mathscr{F}\{f(x)\} \tag{4.83}$$

Differentiating Eq. (4.83), k times results in

$$\mathscr{F}\left\{f^{k}(x)\right\} = (i\omega)^{k}\,\mathscr{F}\{f(x)\} \tag{4.84}$$

Equation (4.83) shows that differentiation in one domain corresponds to $i\omega$ multiplication in the transform domain. This property simplifies the solution of differential equations, analyzing systems, and processing signals by converting complex differential operations into more manageable algebraic ones. Details about the same are described in Chap. 12.

4.4 Closure

This chapter delved into the profound and versatile field of Fourier analysis, exploring its foundational concepts and key properties. Fourier analysis provides a powerful mathematical kernel for transforming functions between two domains, enabling a deeper understanding and manipulation of signals and systems. Key properties of the Fourier transform are thoroughly examined, including linearity, time and frequency shifting, scaling, and the differentiation property. Each of these properties is illustrated with mathematical expressions and graphical illustrations. The famous Heisenberg uncertainty principle, a cornerstone of quantum mechanics, is also explored in the context of Fourier analysis. Fourier analysis serves as an indispensable tool in both theoretical and applied disciplines. Its ability to transform and analyze functions across domains opens up a multitude of possibilities for innovation and discovery. By mastering the concepts and properties of Fourier analysis, one gains a deeper insight into the nature of signals and systems, paving the way for advancements in science, engineering, and technology. The advanced concepts and applications of Fourier analysis are provided in the subsequent chapters as we move forward.

References

Cauchy, A.-L. (1825). *Mémoire sur les intégrales définies, prises entre des limites imaginaires.* chez De Bure fréres, rue Serpente.

Duffy, D. G. (2017). *Advanced engineering mathematics with MATLAB* (4th ed.). CRC Press.

Dym, H., & McKean, H. (1972). *Fourier series and integrals* (1st ed.). Academic Press.

Fourier, J. (1822). *Théorie analytique de la chaleur* (1st ed.). Firmin Dido.

Hamming, R. (1977). *Digital filters* (1st ed.) Prentice-Hall.

Heisenberg, W. (1927). Über den anschaulichen Inhalt der quantentheoretischen Kinematik und Mechanik. *Zeitschrift für Physik, 43*(3), 172–198.

James, G. (2011a). *Advanced modern engineering mathematics* (4th ed.). Pearson Education Limited.

James, J. (2011b). *A student's guide to fourier transforms with applications in physics and engineering* (3rd ed.). Cambridge University Press.

Jeffrey, A. (2002). *Advanced engineering mathematics* (1st ed.). Harcourt /Academic Press.

Kreyszig, E., Kreyszig, H., & Norminton, E. (2011). *Advanced engineering mathematics* (10th ed.). Wiley.

Lighthill, M. (1970). *Introduction to fourier analysis and generalized functions* (1st ed.). Cambridge University Press.

Lokenath, D., & Dambaru, B. (2015). *Integral transforms and their applications* (3rd ed.). CRC Press.

O'Neil, P. V. (2007). *Advanced engineering mathematics* (1st ed.). Thomson Canada Limited.

Osgood, B. (2007). *Lecture notes for the fourier transform and its applications* (1st ed.). Stanford University.

Plancherel, M. (1910). Contribution à l'étude de la représentation d'une fonction arbitraire par des intégrales définies. *Rendiconti del Circolo Matematico di Palermo (1884-1940), 30*(1), 289–335.

Spiegel, M. (1974). *Theory and problems of fourier analysis with applications to boundary value problems* (1st ed.). Schaum'S Outline Series.

Tennekes, H. (1976). Fourier-transform ambiguity in turbulence dynamics. *Journal of the Atmospheric Sciences, 33*, 1660–1663.

Zill, D. G., Wright, W. S., & Cullen, M. R. (2013). *Differential equations with boundary-value problems* (8th ed.). Brooks/Cole Cengage Learning.

Chapter 5
Fourier Transform of Generalized Functions

Mathematics is the tool specially suited for dealing with abstract concepts of any kind and there is no limit to its power in this field..

—Paul Adrien Maurice Dirac

Abstract This chapter extends the Fourier transform to the realm of generalized functions or distributions, enhancing its ability to handle a wider range of mathematical constructs. It begins by highlighting the Fourier transform's universality and application to functions that decrease rapidly, paving the way for a deeper understanding of how generalized functions behave under transformation. The chapter introduces Dirac's delta function, examining its essential properties and use in modelling impulse responses. It further explores how functions can induce distributions and cover the derivatives of distributions, allowing the Fourier transform to capture complex, non-standard behaviors often encountered in both theory and practical applications. Key distribution properties are discussed, including the Dirac comb (or Sha function), which models periodic sampling and is crucial for understanding the discrete aspects of sampled signals.

Keywords Rapidly decreasing function · Generalized functions · Lebesgue's integral · Schwartz class of function · Dirac's delta function · Sha function · Poisson summation · Distributions

5.1 The Universality of the Fourier Transform

Fourier analysis is a rich and rigorous subject that provides detailed insight into various physical phenomena. The most significant disadvantage of Fourier analysis in its classical form is that the solution of the integral does not exist for many classes of simple functions like sine, cosine, and constant functions. The second limitation of Fourier transfer is its silent nature toward the inverse Fourier transform.

S. K. Jena, *Fourier, Laplace, and the Tangled Love Affair with Transforms*,
https://doi.org/10.1007/978-3-031-80165-5_5

$$\mathscr{F}^{-1}\left(\mathscr{F}\left\{f\left(x\right)\right\}\right) = f\left(x\right) \tag{5.1}$$

The forward Fourier transform of $\Pi(x)$ is $sinc(x)$. By principle, the inverse Fourier transform of $sinc(x)$ should be $\Pi(x)$.

$$\mathscr{F}^{-1}\left\{sinc\left(x\right)\right\} = \int_{-\infty}^{\infty} \frac{\sin x}{x} e^{i\omega x} dx \tag{5.2}$$

Unfortunately, the integral equation (5.2) does not satisfy the integrability criterion (Eq. 4.5). Using the principle of duality, we have established the inverse Fourier transform of $sinc(x)$ as $\Pi(x)$. The method is quite restrictive and not useful for all classes of functions we encounter in physics or engineering. Despite being a well-established branch of mathematics, this limitation existed with inverse Fourier analysis until the fourth decade of the twentieth century, until the development of generalized functions. Generalized functions are especially useful in making discontinuous functions more like smooth functions and describing discrete physical phenomena.

The informal techniques of summing infinitesimally small quantities, which had been employed for centuries, necessitated a more robust mathematical framework. This need prompted significant advancements in the eighteenth and nineteenth centuries. A pivotal contribution came from the German mathematician Georg Friedrich Bernhard Riemann (1826–1866), who, in his 1854 habilitation dissertation, "Über die Darstellbarkeit einer Function durch eine trigonometrische Reihe" (On the Representability of a Function by a Trigonometric Series), introduced what is now known as the Riemann integral (Riemann, 1867). Riemann's method involved partitioning a function's domain into smaller subintervals and then summing the areas of rectangles formed by evaluating the function at specific points within these subintervals. Riemann provided a rigorous definition of the integral by taking the infinitely small limit for these subintervals.

Although not immediately recognized for its significance, the Riemann integral gradually gained widespread acceptance, profoundly influencing the development of modern mathematical analysis. This approach offered a precise way to handle functions that were not necessarily continuous but still integrable. However, Riemann's method also revealed its own limitations, particularly when dealing with functions exhibiting numerous discontinuities or complex behaviors over small intervals. These shortcomings led to the search for more general integration techniques.

In the late nineteenth and early twentieth centuries, French mathematicians such as Marie Ennemond Camille Jordan (1838–1922) introduced the concept of the Jordan measure, while Félix Édouard Justin Émile Borel (1871–1956) developed measure theory to address these challenges. Building upon Borel's foundational work, another French mathematician, Henri Léon Lebesgue (1875–1941), introduced the Lebesgue integral in 1902. This method offered a more versatile and general approach to integration, accommodating a broader range of functions, including those with dense sets of discontinuities. In his doctoral thesis, Lebesgue redefined the integration process using measure theory, allowing for precisely defining the "size" or "measure" of sets of points where functions assume particular values. This new approach transformed the understanding of integration, providing powerful convergence theorems and expanding the class of integrable functions. Mathematicians solved many problems using Lebesgue's integral that were previously unsolved using the Riemann integral. While complementary to Riemann's work, Lebesgue's integral addressed many of its limitations without discarding its core principles.

Around 1930, British physicist Paul Adrien Maurice Dirac (1902–1984) introduced a special class of functions, termed "improper functions", to solve the Schrödinger wave equation in quantum mechanics (Dirac, 1927, 1930). These functions, later known as Dirac's delta function, initially faced skepticism as they did not conform to the classical understanding of functions. In truth, Dirac's delta function was not a traditional function but a distribution, a conceptual extension of functions that captures the idea of an infinitely concentrated point mass. Despite initial reservations, the delta function was eventually embraced by mathematicians and physicists alike, becoming formalized within the broader framework of distribution theory.

Before Dirac, an English self-taught electrical engineer, Oliver Heaviside (1850–1925) laid the groundwork for generalized functions in his influential 1899 treatise on operational calculus for electromagnetic theory (Heaviside, 1893b,a). Later, Russian mathematician Sergei Lvovich Sobolev (1908–1989) developed a comprehensive theory of generalized functions for weak solutions of partial differential equations (Sobolev, 1992). Finally, French mathematician Laurent-Moïse Schwartz (1915–2002) provided a rigorous mathematical definition of Dirac's delta function and expanded the theory to include various classes of generalized functions (Schwartz, 1947).

While an in-depth discussion of Lebesgue's integral theory of distributions is beyond the scope of this book, these concepts play a crucial role in applications involving Fourier analysis. For further exploration, readers are encouraged to consult works by Lighthill (1958); Strichartz (1994); Hoskins (2009).

5.1.1 Rapidly Decreasing Function

Existence of $\mathscr{F}\{f(x)\}$ requires absolute integrability of $f(x)$. From derivative formula in Eq. (4.83),

$$\mathscr{F}\{f(x)\} = \frac{1}{(i\omega)^k}\mathscr{F}\left\{\frac{d^k f(x)}{dx^k}\right\} \tag{5.3}$$

As per Riemann-Lebesgue lemma, $\mathscr{F}\{f(x)\} \to 0$ when $\omega \to \pm\infty$. The above is possible if the function is decaying fast. But how fast $\mathscr{F}\{f(x)\}$ approaches zero, that needs additional assumption apart from absolute integrability of $f(x)$. $\mathscr{F}\{f(x)\}$ is smooth (order of smoothness equals to the number of times $\mathscr{F}\{f(x)\}$ is differentiable), despite discontinuity in $f(x)$. If $f(x)$ is absolutely integrable, then $xf(x)$ is too absolutely integrable.

$$\int_{-\infty}^{\infty} |xf(x)|\,dx < \infty \tag{5.4}$$

Taking the Fourier transform of $-ixf(x)$,

$$\begin{aligned}\mathscr{F}\{-ixf(x)\} &= \frac{1}{2\pi}\int_{-\infty}^{\infty} -ixf(x)\,e^{-i\omega x}dx = \frac{1}{2\pi}\int_{-\infty}^{\infty}\left(\frac{de^{-i\omega x}}{d\omega}\right)f(x)\,dx \\ &= \frac{d}{d\omega}\left(\frac{1}{2\pi}\int_{-\infty}^{\infty} f(x)\,e^{-i\omega x}dx\right) = \frac{d\mathscr{F}\{f(x)\}}{d\omega} = \frac{dF(\omega)}{d\omega}\end{aligned} \tag{5.5}$$

Multiplying with $(-ix)^2$ and taking the Fourier transform result in

$$\mathscr{F}\left\{(-ix)^2 f(x)\right\} = \frac{d^2\mathscr{F}\{f(x)\}}{d\omega^2} = \frac{d^2F(\omega)}{d^2\omega} \tag{5.6}$$

Multiplying with $(-ix)^m$ and taking the Fourier transform result in

$$\mathscr{F}\left\{(-ix)^m f(x)\right\} = \frac{d^m\mathscr{F}\{f(x)\}}{d\omega^m} = \frac{d^mF(\omega)}{d^m\omega} \tag{5.7}$$

Equation (5.7) gives the condition for m order smoothness of $\mathscr{F}\{f(x)\}$. The faster the decay of $f(x)$, the more smooth is the $\mathscr{F}\{f(x)\}$ (because x^m will always increase). A function is said to be rapidly decreasing if it is infinitely differentiable, and for any positive power m of x, times and order of derivative tend to zero as $x \to \pm\infty$.

$$\left|x^m \frac{d^k f(x)}{dx^k}\right| \to 0 \qquad x \to \pm\infty \tag{5.8}$$

French mathematician Laurent-Moïse Schwartz (1915–2002) discovered these rapidly decreasing functions for application in the Fourier series. Functions satisfying the above criteria are called Schwartz class of function ($\mathcal{S}$). They are too known as the best class of functions. $\Pi \notin \mathcal{S}$, because of discontinuity; $\Lambda \notin \mathcal{S}$, because it is not smooth. sin, cos $\notin \mathcal{S}$, because they are periodic functions (not rapidly decaying). It is quite surprising to find a function that meets these stringent regulations. Gaussian function $e^{-x^2} \in \mathcal{S}$. Many other rapidly decreasing functions do exist. From Eq. (5.7), if a function rapidly decreases, then its Fourier transform too rapidly decreases. Inverse Fourier transform of $\mathscr{F}\{f(x)\}$ $(f(x))$ exists for Schwartz class of functions $\mathcal{S}$ as $\mathscr{F}\{f(x)\}$ decays fast.

Generalized functions or distribution $\mathscr{D}(x)$ are not classical mathematical functions; they operate or are paired with the classical legitimate function $\varphi(x)$ (often called as test function) and extracts value of the function at a point. The pairing is a linear operation of integral from $-\infty$ to ∞.

$$\langle \mathscr{D}(x), \varphi(x) \rangle = \int_{-\infty}^{\infty} \mathscr{D}(x)\,\varphi(x)\,dx \tag{5.9}$$

The physical interpretation of this pairing is quite remarkable. Consider using a thermometer to measure the temperature inside an oven with a wide range of temperature distribution. The mercury column in the thermometer displays the temperature as an average over the bulb's surface. To achieve a precise temperature reading at a specific point, one would need to shrink the thermometer's bulb to a single point (or zero area), which is impossible in practice. Hence, the thermometer reflects the average temperature over the area of the bulb. If the temperature is measured over an extended period, the thermometer reading corresponds to the time-averaged temperature at that location.

Similarly, the function $\mathscr{D}(x)$ plays a role akin to temperature, while $\varphi(x)$ functions like a thermometer, providing a precise reading of $\mathscr{D}(x)$ at a specific point. A generalized function $\mathscr{D}(x)$ is one that, when applied to a well-defined function $\varphi(x)$, extracts the value of that function at a particular point, a. Therefore, the definite integral in Eq. (5.9) represents a single value at a specific point rather than describing a continuous function over an interval.

5.1.2 Dirac's Delta Function

A unit step function $\mathcal{H}$ (often called Heaviside function or impulse response) is defined as

$$\mathcal{H}(x) = \begin{cases} 0 & x < 0 \\ 1 & x \geq 0 \end{cases} \tag{5.10}$$

Figure 5.1a shows the Heaviside step function. The slope to this unit step function is defined as

$$\delta_0(x) = \frac{d\mathcal{H}(x)}{dx} = \begin{cases} 0 & x \neq 0 \\ \infty & x = 0 \end{cases} \tag{5.11}$$

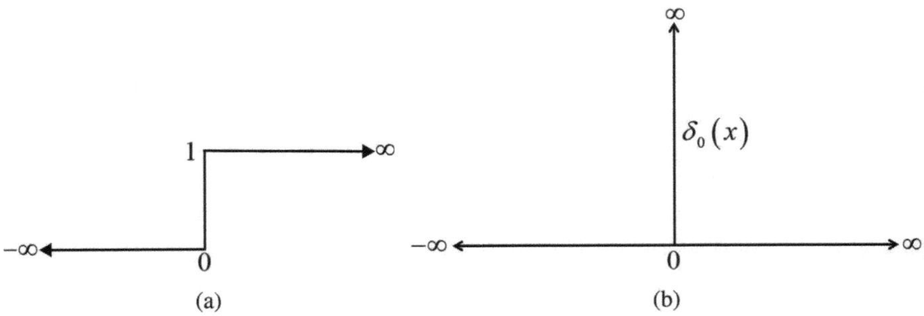

Fig. 5.1 A step function and its slope. (**a**) Unit step function. (**b**) Dirac delta function

Figure 5.1b shows the slope to this unit step function, which represents Dirac's delta function. Integration of Eq. (5.11) results in

$$\int_{-\infty}^{\infty} \delta_0(x)\, dx = \mathcal{H}(x)|_{-\infty}^{\infty} = 1 \tag{5.12}$$

Dirac's delta function is one of the most well-known and pioneering examples of a generalized function, first introduced by the physicist Paul Dirac (1902–1984) during his research on canonical transformations in quantum mechanics. His work involved key components such as the Schrödinger wave equation (named after Austrian physicist Erwin Rudolf Josef Alexander Schrödinger (1887–1961)) and Heisenberg's matrix formulations. Abstract definitions of the delta function are provided by Eqs. (5.11) and (5.12), but they offer little practical insight into the function itself. Here, the symbol δ_0 is used instead of the classical δ, as it is focused at the point 0. Similarly, δ_a denotes concentration at a point a.

Notably, no ordinary function satisfies Eqs. (5.11) and (5.12) in the conventional sense. To visualize the concept, Fig. 5.2a shows how a sequence of functions $\Pi_\varepsilon(x)$, with Fig. 5.2b a sequence of Gaussian functions $g(x)$, are progressively compressed around the origin $x = 0$, in an attempt to meet the conditions of these equations. This squeezing effect demonstrates how these sequences approach the behavior described by Dirac's delta function, though no ordinary function truly satisfies the abstract definitions.

For the physical interpretation of Dirac's delta function, let us consider pairing the rectangular pulse function $\Pi_\varepsilon(x)$ with a legitimate function $\varphi(x)$ under the limiting condition $\varepsilon \to 0$.

$$\lim_{\varepsilon\to 0} \langle \Pi_\varepsilon(x), \varphi(x)\rangle = \lim_{\varepsilon\to 0} \int_{-\infty}^{\infty} \Pi_\varepsilon(x)\,\varphi(x)\, dx = \lim_{\varepsilon\to 0} \frac{1}{\varepsilon} \int_{-\frac{\varepsilon}{2}}^{\frac{\varepsilon}{2}} \varphi(x)\, dx \tag{5.13}$$

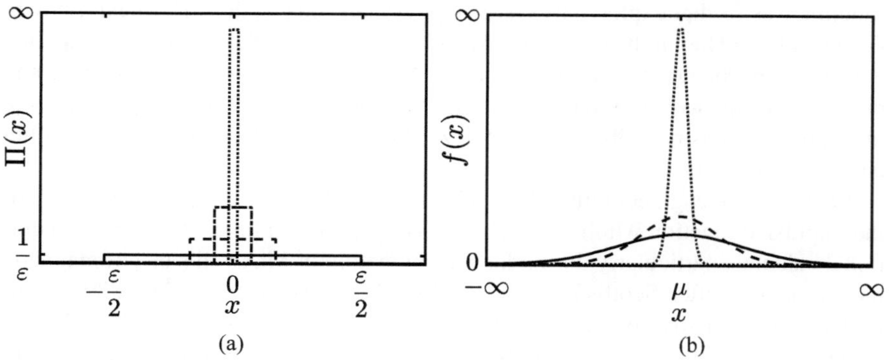

Fig. 5.2 Dirac of delta function in operation. (**a**) Rectangular pulse. (**b**) Gaussian pulse

Applying Taylor series expansion for $\delta_0(x)$ at 0

$$\lim_{\varepsilon\to 0}\langle \Pi_\varepsilon(x), \varphi(x)\rangle = \lim_{\varepsilon\to 0}\frac{1}{\varepsilon}\int_{-\frac{\varepsilon}{2}}^{\frac{\varepsilon}{2}}\left(\varphi(0)+\varphi'(0)\,x+\frac{1}{2}\varphi''(0)\,x^2+\cdots\right)dx \tag{5.14}$$

Term-by-term integration results in

$$\lim_{\varepsilon\to 0}\langle \Pi_\varepsilon(x), \varphi(x)\rangle = \lim_{\varepsilon\to 0}\frac{1}{\varepsilon}\left[\varphi(0)\,x+\frac{1}{2}\varphi'(0)\,x^2+\frac{1}{6}\varphi''(0)\,x^3+\cdots\right]_{-\frac{\varepsilon}{2}}^{\frac{\varepsilon}{2}} = \varphi(0) \tag{5.15}$$

Higher-order terms of ε are neglected for being extremely small. Under the limiting condition $\varepsilon\to 0$.

$$\lim_{\varepsilon\to 0}\Pi_\varepsilon(x) = \delta_0(x) \tag{5.16}$$

Dirac's delta is an impulsive function. When it pairs with any legitimate test function $\varphi(x)$, it pulls out the concentrated value of the test function at a point $\varphi(0)$.

$$\langle \delta_0(x), \varphi(x)\rangle = \int_{-\infty}^{\infty}\delta_0(x)\,\varphi(x)\,dx = \varphi(0) \tag{5.17}$$

The concept of an impulsive function existed long before Dirac's formal application. As early as 1827, the mathematician Augustin-Louis Cauchy (1789–1857) introduced the use of an impulsive function, treating it as an infinitesimal version of the Cauchy distribution. Around the same period, Siméon Denis Poisson (1781–1840) utilized a similar concept in his studies on wave propagation. The significance of this function was further explored by German physician and physicist Hermann Ludwig Ferdinand von Helmholtz (1821–1894) and Lord Kelvin (1824–1907), who applied it in fluid mechanics to represent point sources and sinks. Additionally, German physicist Gustav Robert Kirchhoff (1824–1887) used the idea of an impulsive force in electromagnetism, applying it to analyze instantaneous changes in current within an electrical circuit.

Later, British engineer Oliver Heaviside (1850–1925) advanced the formalism of the impulsive function within the realms of both space and time. His work in operational calculus, particularly during his reexamination of Maxwell's electromagnetic theory (named after Scottish physicist James Clerk Maxwell (1831–1879)), contributed significantly to the development of this concept. Heaviside's influence on Dirac was profound, as Dirac himself remarked, "All electrical engineers are familiar with the idea of a pulse, and the δ_0 function is just a way of expressing a pulse mathematically". This impulsive function, in turn, was a generalization of Kronecker's delta function, δ_{ij}, and it was this connection that inspired Dirac to adopt the term "delta function" for the mathematical impulse. Kronecker's delta function is named after German mathematician Leopold Kronecker (1823–1891).

Dirac made the systematic application of this impulsive function in 1926 during his groundbreaking work in quantum mechanics, which later became known as Dirac's delta function. The historical evolution of this concept is beautifully summarized by Debnath (2013), who provides an insightful account of its discovery and development over time.

Often shifted Dirac's delta function ($\delta_{\pm a}$) is used in engineering and physics, which is defined at some point $\pm a$ away from the origin and denoted as (a is positive).

$$\delta_{\pm a}(x) = \begin{cases} 0 & x \neq \pm a \\ \infty & x = \pm a \end{cases} \tag{5.18}$$

Figure 5.3 represents a shifted delta function $\delta_{\pm a}$. When a shifted Dirac's delta function ($\delta_{\pm a}$) paired with a test function $\varphi(x)$, it pulls out the concentrated value of $\varphi(x)$ at $\pm a$.

$$\langle \delta_{\pm a}(x), \varphi(x) \rangle = \int_{-\infty}^{\infty} \delta_0(x \pm a)\, \varphi(x)\, dx = \varphi(\mp a) \tag{5.19}$$

Equation (5.19) is also known as the shifting property of Dirac's delta function, which shifts the value of $\varphi(\pm x)$ to $\pm a$. The shifting property is precisely expressed as

$$\langle \delta_{\pm a}(x), \varphi(x) \rangle = \int_{\pm a - \varepsilon_1}^{\pm a + \varepsilon_2} \delta_0(x \pm a)\, \varphi(x)\, dx = \varphi(\mp a) \tag{5.20}$$

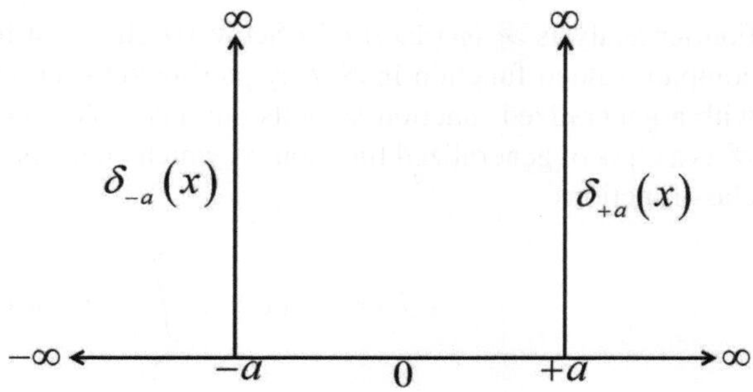

Fig. 5.3 Dirac's shifted delta function

Integral in Eq. (5.20) does not exist for $\varepsilon_1 = \varepsilon_2$ or for any discontinuity in $\varphi(x)$ at $\pm a$.

5.1.3 Distribution Induced by Functions

Integral pairing of any function $f(x)$ with $\varphi(x)$ induces a distribution.

$$\langle f(x), \varphi(x)\rangle = \int_{-\infty}^{\infty} f(x)\,\varphi(x)\,dx \tag{5.21}$$

For example,

$$\langle 1, \varphi(x)\rangle = \int_{-\infty}^{\infty} \varphi(x)\,dx$$

Distribution induced by constant function.

$$\left\langle e^{iax}, \varphi(x)\right\rangle = \int_{-\infty}^{\infty} e^{iax}\varphi(x)\,dx$$

Distribution induced by an exponential function.

For a test function $\varphi(x) \in \mathcal{S}$, its Fourier transform $\mathscr{F}\{\varphi(x)\} \in \mathcal{S}$ and inverse Fourier transform $\mathscr{F}^{-1}\{\varphi(x)\} \in \mathcal{S}$.

$$\begin{aligned} \mathscr{F}^{-1}\{\mathscr{F}\{\varphi(x)\}\} &= \varphi(x) \\ \mathscr{F}\left\{\mathscr{F}^{-1}\{\varphi(x)\}\right\} &= \varphi(x) \end{aligned} \tag{5.22}$$

Fourier analysis $\mathscr{F}\{\varphi(\pm x)\}$ of a Schwartz class test function $\varphi(x)$ often results a complex valued function in $\mathcal{S}$. Any pairing between this complex-valued function with a generalized function $\mathscr{D}$ pulls out a complex number. Tempered distribution $\mathscr{T}$ is a class of generalized function $\mathscr{D}$, which operates on complex-valued Schwartz class functions.

$$\langle \mathscr{T}(x), \varphi(x)\rangle = \int_{-\infty}^{\infty} \mathscr{T}(x)\,\varphi(x)\,dx \tag{5.23}$$

The above integral pairing is a linear operation. For two different test functions $\varphi_1(x)$ and $\varphi_2(x_2)$, the pairing with $\mathscr{T}$ becomes

$$\langle \mathscr{T}(x), \alpha_1\varphi_1(x) + \alpha_2\varphi_2(x)\rangle = \alpha_1\langle \mathscr{T}(x), \varphi_1(x)\rangle + \alpha_2\langle \mathscr{T}(x), \varphi_2(x)\rangle \tag{5.24}$$

α_1 and α_2 are constants. Fourier transform of a tempered distribution $\mathscr{F}\{\mathscr{T}(x)\}$ is also a tempered distribution and operates on a test function $\varphi(\omega)$ as follows:

$$\begin{aligned}
\langle \mathscr{F}\{\mathscr{T}(x)\}, \varphi(\omega)\rangle &= \int_{-\infty}^{\infty} \mathscr{F}\{\mathscr{T}(x)\}\,\varphi(\omega)\,d\omega \\
&= \int_{-\infty}^{\infty}\left[\frac{1}{2\pi}\int_{-\infty}^{\infty} \mathscr{T}(x)e^{-i\omega x}dx\right]\varphi(\omega)\,d\omega \\
&= \int_{-\infty}^{\infty}\left[\frac{1}{2\pi}\int_{-\infty}^{\infty} \varphi(\omega)e^{-i\omega x}d\omega\right]\mathscr{T}(x)\,dx \\
&= \int_{-\infty}^{\infty} \mathscr{F}\{\varphi(\omega)\}\,\mathscr{T}(x)\,dx \\
&= \langle \mathscr{T}(x), \mathscr{F}\{\varphi(\omega)\}\rangle
\end{aligned} \tag{5.25}$$

The right side of Eq. (5.25) represents the pairing of a tempered distribution with the Fourier transform of a test function (which is classically defined). Equation (5.25) establishes an important relationship. From the symmetric nature of the Fourier transform, a similar relationship for the inverse Fourier transform holds true.

$$\left\langle \mathscr{F}^{-1}\{\mathscr{T}(\omega)\}, \varphi(x)\right\rangle = \left\langle \mathscr{T}(\omega), \mathscr{F}^{-1}\{\varphi(x)\}\right\rangle \tag{5.26}$$

Example 5.1 (Fourier Transfer of Dirac's Delta Function) Using the relationship in Eq. (5.25), Fourier transfer of the Dirac's delta function becomes $\mathscr{F}\{\delta_0(x)\}$

$$\begin{aligned}\langle \mathscr{F}\{\delta_0(x)\}, \varphi(\omega)\rangle &= \langle \delta_0(x), \mathscr{F}\{\varphi(\omega)\}\rangle \\ &= \frac{1}{2\pi}\int_{-\infty}^{\infty} \varphi(\omega)\, e^{-i0\omega} d\omega \\ &= \frac{1}{2\pi}\int_{-\infty}^{\infty} \varphi(\omega)\, d\omega\end{aligned} \tag{5.27}$$

The pairing between $\frac{1}{2\pi}$ with the test function $\varphi(\omega)$ becomes

$$\left\langle \frac{1}{2\pi}, \varphi(\omega)\right\rangle = \frac{1}{2\pi}\int_{-\infty}^{\infty} \varphi(\omega)\, d\omega \tag{5.28}$$

Equating Eqs. (5.27) and (5.28), the Fourier transfer of Dirac's delta function becomes

$$\mathscr{F}\{\delta_0(x)\} = \frac{1}{2\pi}(\omega) \tag{5.29}$$

where $\frac{1}{2\pi}(\omega)$. Notation represents the constant value $\frac{1}{2\pi}$ in ω space. $\delta_0(x)$ function is concentrated at x domain, and $\mathscr{F}\{\delta_0(x)\}$ uniformly spreads in ω domain.

Fourier transform of a shifted Dirac's delta function $\mathscr{F}\{\delta_{\pm a}(x)\}$ is

$$\begin{aligned}\langle \mathscr{F}\{\delta_{\pm a}(x)\}, \varphi(\omega)\rangle &= \langle \delta_{\pm a}(x), \mathscr{F}\{\varphi(\omega)\}\rangle \\ &= \frac{1}{2\pi}\int_{-\infty}^{\infty} \varphi(\omega)\, e^{\mp ia\omega} d\omega \\ &= \left\langle \frac{1}{2\pi} e^{\mp ia\omega}, \varphi(\omega)\right\rangle\end{aligned} \tag{5.30}$$

From Eq. (5.30), the Fourier transform of a shifted delta function is expressed as

(continued)

Example 5.1 (continued)

$$\mathscr{F}\left\{\delta_{\pm a}(x)\right\} = \frac{1}{2\pi} e^{\mp i a \omega} \tag{5.31}$$

Example 5.2 (Fourier Transfer of Few Classical Functions) Pairing of Fourier transfer of a complex exponential function $\mathscr{F}\left\{e^{iax}\right\}$ with a test function $\phi(x)$ is expressed as

$$\begin{aligned} \left\langle \mathscr{F}\left\{e^{iax}\right\}, \varphi(x) \right\rangle &= \left\langle \mathscr{F}\left\{\varphi(x)\right\}, e^{iax} \right\rangle \\ &= \int_{-\infty}^{\infty} e^{iax} \mathscr{F}\left\{\varphi(x)\right\} dx \end{aligned} \tag{5.32}$$

The integral in the right-hand side of Eq. (5.32) represents the inverse Fourier transform definition. Using the property in Eq. (5.32), the above integral becomes

$$\begin{aligned} \left\langle \mathscr{F}\left\{e^{iax}\right\}, \varphi(x) \right\rangle &= \varphi(a) \\ &= \left\langle \delta_a(x), \varphi(x) \right\rangle \end{aligned} \tag{5.33}$$

From Eq. (5.33), the Fourier transform of the complex exponential becomes

$$\mathscr{F}\left\{e^{iax}\right\} = \delta_a \tag{5.34}$$

Putting $a = 0$, the Fourier transfer of a constant function becomes

$$\mathscr{F}\{1\} = \delta_0(\omega) \tag{5.35}$$

(ω) in right side of Eq. (5.35) represents δ_0 operation in ω space. A constant function is uniformly distributed in the x domain, hence concentrated at the ω domain.

The Fourier transform of the trigonometric functions can be derived as

$$\mathscr{F}\{\cos(ax)\} = \frac{\mathscr{F}\left\{e^{iax}\right\} + \mathscr{F}\left\{e^{-iax}\right\}}{2} = \frac{\delta_a + \delta_{-a}}{2}(\omega) \tag{5.36}$$

$$\mathscr{F}\{\sin(ax)\} = \frac{\mathscr{F}\left\{e^{iax}\right\} - \mathscr{F}\left\{e^{-iax}\right\}}{2i} = \frac{\delta_a - \delta_{-a}}{2i}(\omega) \tag{5.37}$$

Equations (5.35)–(5.37) represent generalized Fourier transform of some commonly used functions, for which Fourier transfer does not exist with its classical definition.

5.1.4 Derivative of Distributions

Distributions are concentrated at specific points and may not appear differentiable in the conventional sense. Nevertheless, derivatives of distributions are defined through their action on a test function, allowing them to possess well-defined derivatives despite their lack of traditional differentiability. The derivative $\mathscr{T}'(x)$ of a distribution $\mathscr{T}(x)$ operates on a test function $\varphi(x)$ as follows:

$$\begin{aligned}\langle \mathscr{T}'(x), \varphi(x)\rangle &= \int_{-\infty}^{\infty} \mathscr{T}'(x)\,\varphi(x)\,dx \\ &= [\mathscr{T}(x)\varphi(x)]_{-\infty}^{\infty} - \int_{-\infty}^{\infty} \mathscr{T}(x)\,\varphi'(x)\,dx\end{aligned} \tag{5.38}$$

Being a rapidly decreasing function, the product $\mathscr{T}(x)\,\varphi(x) \to 0$ as $x \to \infty$. Equation (5.38) becomes

$$\langle \mathscr{T}'(x), \varphi(x)\rangle = -\int_{-\infty}^{\infty} \mathscr{T}(x)\,\varphi'(x)\,dx = \langle \mathscr{T}(x), \varphi'(x)\rangle \tag{5.39}$$

Being a rapidly decreasing function, the derivative $\varphi'(x)$ exists. The pairing of $\mathscr{T}'(x)$ with $\varphi(x)$ produces the same effect as negative of the pairing of $\mathscr{T}(x)$ with $\varphi'(x)$. Using Eq. (4.83), Fourier transfer of $\mathscr{T}'(x)$ is expressed as

$$\mathscr{F}\{\mathscr{T}'(x)\} = i\omega\mathscr{F}\{\mathscr{T}(x)\} \tag{5.40}$$

Example 5.3 (Derivative of Heaviside Function) Derivative of the Heaviside function of unit step function defined in Eq. (5.10) can be calculated using Eq. (5.39) as follows:

(continued)

Example 5.3 (continued)

$$\begin{aligned}\left\langle \mathcal{H}'(x), \varphi(x)\right\rangle &= -\left\langle \mathcal{H}(x), \varphi'(x)\right\rangle = \int_{-\infty}^{\infty} \mathcal{H}(x)\varphi'(x)\,dx \\ &= -\int_{0}^{\infty} \varphi'(x)\,dx = -\left[\varphi(x)\right]_0^{\infty} = \varphi(0)\end{aligned} \tag{5.41}$$

From Eq. (5.17), $\langle \delta_0(x), \varphi(x)\rangle = \varphi(0)$. Hence,

$$\mathcal{H}'(x) = \delta_0(x) \tag{5.42}$$

Example 5.4 (Signum Function and its Fourier Transform) The signum function $sgn\,(x)$ is defined as

$$sgn\,(x) = \begin{cases} 1 & x > 0 \\ 0 & x = 0 \\ -1 & x < 0 \end{cases} \tag{5.43}$$

From Eq. (5.39),

$$\begin{aligned}\left\langle sgn'(x), \varphi(x)\right\rangle &= -\left\langle sgn\,(x), \varphi'(x)\right\rangle \\ &= -\int_{-\infty}^{\infty} sgn\,(x)\varphi'(x)\,dx \\ &= \int_{-\infty}^{0} \varphi'(x)\,dx - \int_{0}^{\infty} \varphi'(x)\,dx \\ &= 2\int_{0}^{\infty} \varphi'(x)\,dx = 2\varphi(0) = 2\,\langle 2\delta_0(x), \varphi(x)\rangle\end{aligned} \tag{5.44}$$

So the derivative of the signum function is

$$sgn'(x) = 2\delta_0(x) \tag{5.45}$$

(continued)

Example 5.4 (continued)
Fourier transform of $sgn'(x)$ is expressed as

$$\mathscr{F}\left\{sgn'(x)\right\} = i\omega\mathscr{F}\left\{sgn(x)\right\} \tag{5.46}$$

Fourier transform of $sgn(x)$ becomes

$$\begin{aligned} \mathscr{F}\left\{sgn(x)\right\} &= \frac{1}{i\omega}\mathscr{F}\left\{sgn'(x)\right\} \\ &= \frac{2}{i\omega}\mathscr{F}\left\{\delta_0(x)\right\} \\ &= \frac{1}{\pi i\omega} \end{aligned} \tag{5.47}$$

$\mathscr{F}\left\{sgn(x)\right\}$ as a singularity at $\omega = 0$

Example 5.5 (Fourier Transfer of the Heaviside Function) The Heaviside function can be expressed in terms of signum function as

$$\mathcal{H}(x) = \frac{1}{2}\left(1 + sgn'(x)\right) \tag{5.48}$$

Taking Fourier transform of both sides of Eq. (5.48),

$$\begin{aligned} \mathscr{F}\left\{\mathcal{H}(x)\right\} &= \frac{1}{2}\mathscr{F}\left\{1\right\} + \frac{1}{2}\mathscr{F}\left\{sgn(x)\right\} \\ &= \frac{1}{2}\delta_0(\omega) + \frac{1}{2\pi i\omega} \end{aligned} \tag{5.49}$$

5.1.5 Few More Properties of Distributions

Most of the algebraic operations between two functions are applicable for distributions, except multiplication. Multiplication between two distributions is not defined. But product of a classical function $f(x)$ with a distribution $\mathscr{T}(x)$ is defined and expressed as follows:

$$\langle f(x)\mathscr{T}(x), \varphi(x)\rangle = \int_{-\infty}^{\infty} f(x)\mathscr{T}(x)\varphi(x)\,dx = \langle \mathscr{T}(x), f(x)\varphi(x)\rangle \tag{5.50}$$

Equation (5.50) makes sense only if the product $f(x)\varphi(x)$ is another test function, which scaled up by $f(x)$. For the special case when distribution is Dirac's delta function, the product is defined as

$$\begin{aligned}\langle f(x)\delta_0(x), \varphi(x)\rangle &== \langle \delta_0(x), f(x)\varphi(x)\rangle = f(0)\varphi(0)\\ &= \langle f(0)\delta_0(x), \varphi(x)\rangle\end{aligned} \tag{5.51}$$

So the product of a classical function $f(x)$ with Dirac's delta function is defined as

$$f(x)\delta_0(x) = f(0)\delta_0(x) \tag{5.52}$$

Equation (5.52) is known as the sampling property of Dirac's delta function. For a shifted delta function, the sampling property is expressed as

$$f(x)\delta_{\pm a}(x) = f(\pm a)\delta_{\pm a}(x) \tag{5.53}$$

Similar to multiplication, convolution between two distributions is not defined in general. For a special case when the distribution is Dirac's delta function, its convolution with a classical function $f(x)$ is expressed as

$$\begin{aligned}f(x) * \delta_0(x) &= \int_{-\infty}^{\infty} f(y_i)\delta_0(x - y_i)\,dy_i\\ &= \int_{-\infty}^{\infty} f(x - y_i)\delta_0(y_i)\,dy_i = f(x)\end{aligned} \tag{5.54}$$

Convolution of a classical function $f(x)$ with a shifted delta function $\delta_{\pm a}(x)$ is expressed as

$$\begin{aligned}f(x) * \delta_{\pm a}(x) &= \int_{-\infty}^{\infty} f(y_i)\delta_0(x - y_i \pm a)\,dy_i\\ &= \int_{-\infty}^{\infty} f(x - y_i \pm a)\delta_0(y_i)\,dy_i = f(x \pm a)\end{aligned} \tag{5.55}$$

Convolution between two shifted delta functions δ_a and δ_b is defined as

$$\begin{aligned}\delta_a(x) * \delta_b(x) &= \int_{-\infty}^{\infty} \delta_0(x-a)\,\delta_0(x-b-y_i)\,dy_i \\ &= \delta_0(x-a-b) = \delta_{a+b}\end{aligned} \tag{5.56}$$

Convolution between two Dirac's delta functions is a Dirac's delta function.

$$\delta_0(x) * \delta_0(x) = \delta_0(x) \tag{5.57}$$

Dirac's delta function is even and symmetric.

$$\delta_0(-x) = \delta_0(x) \tag{5.58}$$

Variable scaling ax is another important property of Dirac's delta function. $\delta_0(ax)$ works on a test function as

$$\langle \delta_0(ax), \phi(x) \rangle = \int_{-\infty}^{\infty} \delta_0(ax)\,\phi(x)\,dx \tag{5.59}$$

Taking $y = ax$, Eq. (5.59) becomes

$$\begin{aligned}\langle \delta_0(ax), \phi(x) \rangle &= \frac{1}{a}\int_{-\infty}^{\infty} \delta_0(y)\,\phi\left(\frac{y}{a}\right)dy = \frac{1}{a}\phi(0) \\ &= \frac{1}{a}\langle \delta_0(x), \phi(x) \rangle\end{aligned} \tag{5.60}$$

Equation (5.60) is valid for any $a > 0$ and $a < 0$. Variable scaled Dirac's delta function is expressed as

$$\delta_0(ax) = \frac{1}{|a|}\delta_0(x) \tag{5.61}$$

5.1.6 Dirac's Comb or Sha Function

Any function $f(x)$ can be periodized with a period p by summation.

$$f_p(x) = \sum_{k=-\infty}^{\infty} f(x-kp) \tag{5.62}$$

Using the sampling property of the delta function in Eq. (5.55), the periodic form is expressed as

$$\begin{aligned} f_p(x) &= \sum_{k=-\infty}^{\infty} f(x) * \delta(x - kp) \\ &= f(x) * \sum_{k=-\infty}^{\infty} \delta(x - kp) \\ &= f(x) * Ш_p(x) \end{aligned} \tag{5.63}$$

Dirac's comb or Sha function is expressed as

$$Ш_p(x) = \sum_{k=-\infty}^{\infty} \delta(x - kp) \tag{5.64}$$

This periodic tempered distribution is also called an impulse train or sampling function. It is constructed by repeating a series of Dirac's delta functions, as shown in Fig. 5.4.

Multiplication of Sha function $Ш_p(x)$ with any classical function $f(x)$, sampled the value of the classical function at discrete points, with periodicity p.

$$f(x)\, Ш_p(x) = \sum_{k=-\infty}^{\infty} f(kp)\, \delta(x - kp) \tag{5.65}$$

Equation (5.65) is known as the sampling property of the Sha function. Fourier transfer of the periodic function $f_p(x)$ results in

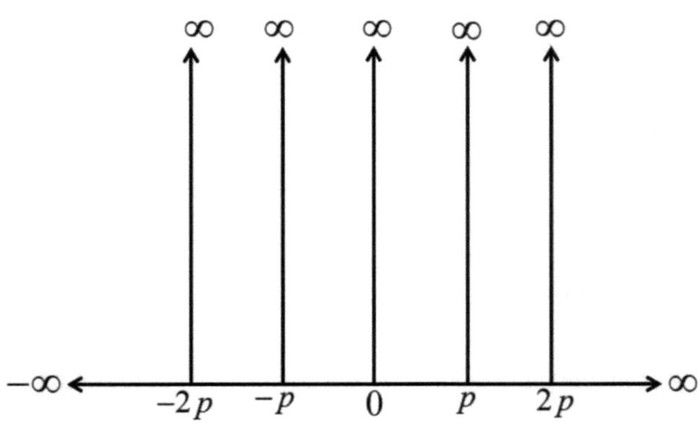

Fig. 5.4 Sha function of period p

$$\mathscr{F}\left\{f_p(x)\right\} = 2\pi \mathscr{F}\{f(x)\} \mathscr{F}\left\{\text{Ш}_p(x)\right\} \tag{5.66}$$

$\text{Ш}_p(x)$ make sense as a periodic distribution, and hence its Fourier transform $\mathscr{F}\left\{\text{Ш}_p(x)\right\}$ too. For simplicity, let us consider a Sha function that has a period of $p = 1$.

$$\text{Ш}(x) = \sum_{k=-\infty}^{\infty} \delta(x-k) \tag{5.67}$$

Fourier transfer of Eq. (5.67) results in

$$\begin{aligned} \mathscr{F}\{\text{Ш}(x)\} &= \sum_{k=-\infty}^{\infty} \mathscr{F}\{\delta(x-k)\} \\ &= \frac{1}{2\pi} \sum_{k=-\infty}^{\infty} e^{-ik\omega} \end{aligned} \tag{5.68}$$

The sum of complex exponential in the right side of Eq. (5.68) does not converge; hence, a circular route is required to compute the Fourier transfer of Dirac's comb. Sha functions and its Fourier transfer are a legitimate tempered distribution. Pairing of $\text{Ш}(x)$ with any test function $\varphi(x)$ is expressed as

$$\langle \text{Ш}(x), \varphi(x) \rangle = \sum_{k=-\infty}^{\infty} \varphi(k) \tag{5.69}$$

The sum is the right-hand side of Eq. (5.68), which converges as $|\varphi(k)| \to 0$ as $k \to \infty$. $\mathscr{F}\{\text{Ш}(x)\}$ operates on $\varphi(x)$ as

$$\begin{aligned} \langle \mathscr{F}\{\text{Ш}(x)\}, \varphi(x) \rangle &= \langle \text{Ш}(x), \mathscr{F}\{\varphi(x)\} \rangle \\ &= \sum_{k=-\infty}^{\infty} \mathscr{F}\{\varphi(k)\} \end{aligned} \tag{5.70}$$

From Poisson summation formula (John J. Benedetto, 1997; Strichartz, 1994),

$$\sum_{k=-\infty}^{\infty} \mathscr{F}\{\varphi(k)\} = \frac{1}{2\pi} \sum_{k=-\infty}^{\infty} \varphi(k) \tag{5.71}$$

Equating Eqs. (5.69) and (5.70) using Eq. (5.71),

$$\mathscr{F}\{\text{Ш}(x)\} = \frac{1}{2\pi} \text{Ш}(x) \tag{5.72}$$

Taking the inverse Fourier transfer of Eq. (5.72),

$$\mathscr{F}^{-1}\{\text{Ш}(x)\} = 2\pi\,\text{Ш}(x) \tag{5.73}$$

$\text{Ш}_p(x)$ is expressed using the variable scaling property of the delta function as

$$\begin{aligned}\text{Ш}_p(x) &= \sum_{k=-\infty}^{\infty} \delta(x-kp) \\ &= \sum_{k=-\infty}^{\infty} \delta\left(p\left(\frac{x}{p}-k\right)\right) = \frac{1}{p}\sum_{k=-\infty}^{\infty} \delta\left(\frac{x}{p}-k\right) \\ &= \frac{1}{p}\text{Ш}\left(\frac{x}{p}\right)\end{aligned} \tag{5.74}$$

Taking the Fourier transfer of Eq. (5.74),

$$\begin{aligned}\mathscr{F}\left\{\text{Ш}_p(x)\right\} &= \frac{1}{p}\mathscr{F}\left\{\text{Ш}\left(\frac{x}{p}\right)\right\} = \mathscr{F}\{\text{Ш}(px)\} = \frac{1}{2\pi}\text{Ш}(px) \\ &= \frac{1}{2\pi}\sum_{k=-\infty}^{\infty} \delta(px-k) = \frac{1}{2\pi}\sum_{k=-\infty}^{\infty} \delta\left(p\left(x-\frac{k}{p}\right)\right) \\ &= \frac{1}{2\pi p}\sum_{k=-\infty}^{\infty} \delta\left(x-\frac{k}{p}\right) = \frac{1}{2\pi p}\text{Ш}_{\frac{1}{p}}(x)\end{aligned} \tag{5.75}$$

Spacing between the impulse in $\text{Ш}_p(x)$ is p, while the spacing between the impulse in transferred domain is $\frac{1}{p}$ (shown in Fig. 5.5). This is obvious from the reciprocal

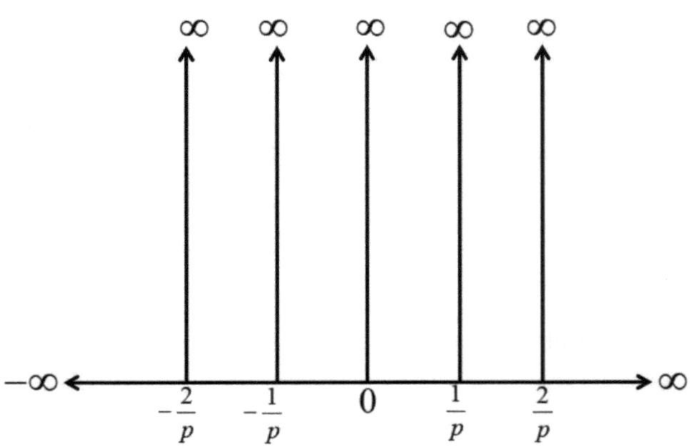

Fig. 5.5 Fourier transfer of Sha function of period p

relationship between the two domains. The inverse Fourier transform of the Sha function is expressed as

$$\mathscr{F}^{-1}\left\{\text{Ш}_p(x)\right\} = \frac{2\pi}{p}\text{Ш}_{\frac{1}{p}}(x) \tag{5.76}$$

Example 5.6 (Poisson Summation Formula) Periodic train $\Phi(x)$ of a test function $\varphi(x)$ with periodicity one is expressed as

$$\Phi(x) = \varphi(x) * \text{Ш}(x) = \sum_{k=-\infty}^{\infty} \varphi(x-k) \tag{5.77}$$

The periodic train $\Phi(x)$ can also be expressed in terms of Fourier series as

$$\Phi(x) = \sum_{n=-\infty}^{\infty} \widehat{\Phi}_n e^{-2\pi i n x} = \sum_{\omega_n=-\infty}^{\infty} \widehat{\Phi}_n e^{-i\omega_n x} \tag{5.78}$$

where $\omega_n = 2\pi n$ and $\widehat{\Phi}_n$ is the Fourier coefficient and is expressed as

$$\widehat{\Phi}_n = \int_{-\frac{1}{2}}^{\frac{1}{2}} e^{-2\pi i n x} \sum_{k=-\infty}^{\infty} \varphi(x-k)dx = \sum_{k=-\infty}^{\infty} \int_{-\frac{1}{2}}^{\frac{1}{2}} e^{-2\pi i n x} \varphi(x-k)\, dx \tag{5.79}$$

Substituting $y_i = x - k$ and $dy_i = dx$,

$$\begin{aligned}
\widehat{\Phi}_n &= \sum_{k=-\infty}^{\infty} \int_{-k-\frac{1}{2}}^{-k+\frac{1}{2}} e^{-2\pi i n (y_i+1)} \varphi(y_i)\, dy_i = \sum_{k=-\infty}^{\infty} \int_{-k-\frac{1}{2}}^{-k+\frac{1}{2}} e^{-2\pi i n y_i} e^{-2\pi i n k} \varphi(y_i)\, dy_i \\
&= \int_{-\infty}^{\infty} e^{-2\pi i n y_i} \varphi(y_i)\, dy_i = 2\pi \frac{1}{2\pi} \int_{-\infty}^{\infty} e^{-i\omega_n y_i} \varphi(y_i)\, dy_i \\
&= 2\pi \mathscr{F}\{\varphi(y_i)\} = 2\pi \widehat{\varphi}(\omega_n)
\end{aligned} \tag{5.80}$$

$e^{-2\pi i n k} = 1$ and $\widehat{\varphi}(\omega_n) = \mathscr{F}\{\varphi(x)\}$. Equation (5.78) becomes

$$\Phi(x) = 2\pi \sum_{\omega_n=-\infty}^{\infty} \widehat{\varphi}(\omega_n)\, e^{-i\omega_n x} \tag{5.81}$$

(continued)

Example 5.6 (continued)
Now, two different definitions of periodic train $\Phi(x)$ are available. Plucking the value $x = 0$ in Eqs. (5.77) and (5.81), and equating results in

$$\begin{aligned}\Phi(0) &= \sum_{k=-\infty}^{\infty} \varphi(-k) = \sum_{k=-\infty}^{\infty} \varphi(k) \\ &= 2\pi \sum_{\omega_n=-\infty}^{\infty} \widehat{\varphi}(\omega_n) = 2\pi \sum_{k=-\infty}^{\infty} \mathscr{F}\{\varphi(k)\}\end{aligned} \tag{5.82}$$

This impulse train is used for continuous function sampling and is covered in detail in Chap. 7.

5.2 Fourier Transform of Some Standard Functions

Few standard functions and their Fourier transform are provided in Table 5.1

Table 5.1 Fourier transform of some standard functions

	$f(x)$	$F(\omega) = \frac{1}{2\pi} \int_{-\infty}^{\infty} f(x) e^{-i\omega x}\, dx$
1.	$e^{-a\lvert x\rvert}, \quad a > 0$	$\frac{a}{\pi\left(a^2+\omega^2\right)}$
2.	$xe^{-a\lvert x\rvert}, \quad a > 0$	$\frac{-2ai\omega}{\pi\left(a^2+\omega^2\right)}$
3.	$e^{-ax^2}, \quad a > 0$	$\frac{1}{2\sqrt{\pi a}} e^{\left(-\frac{\omega^2}{4a}\right)}$
4.	$\frac{1}{\left(x^2+a^2\right)}, \quad a > 0$	$\frac{e^{-a\lvert\omega\rvert}}{2a}$
5.	$\frac{x}{\left(x^2+a^2\right)}, \quad a > 0$	$\left(\frac{i\omega}{4a}\right) e^{-a\lvert\omega\rvert}$
6.	$\begin{Bmatrix} c, & a \le x \le b \\ 0, & \text{outside} \end{Bmatrix}$	$\frac{ic}{2\pi}\frac{1}{\omega}\left(e^{-ib\omega} - e^{-ia\omega}\right)$
7.	$\lvert x\rvert e^{-a\lvert x\rvert} \quad a > 0$	$\frac{\left(a^2-\omega^2\right)}{\pi\left(a^2+\omega^2\right)^2}$
8.	$\frac{\sin ax}{x}$	$\frac{1}{2}\mathcal{H}(a - \lvert\omega\rvert)$
9.	$e^{-x(a-ik)}\mathcal{H}(x)$	$\frac{i}{2\pi(k-\omega+ia)}$

(continued)

Table 5.1 (continued)

10.	$\frac{1}{\left(a^2-x^2\right)^{1/2}}\mathcal{H}(a-\vert x\vert)$	$\frac{J_0(a\omega)}{2}$
11.	$\frac{\sin\left[b\left(x^2+a^2\right)^{\frac{1}{2}}\right]}{\left(x^2+a^2\right)^{\frac{1}{2}}}$	$\frac{1}{2}J_0\left(a\sqrt{b^2-\omega^2}\right)\mathcal{H}(b-\vert\omega\vert)$
12.	$\frac{\cos\left(b\sqrt{a^2-x^2}\right)}{\left(a^2-x^2\right)^{\frac{1}{2}}}\mathcal{H}(a-\vert x\vert)$	$\frac{1}{2}J_0\left(a\sqrt{b^2+\omega^2}\right)$
13.	$e^{-ax}\mathcal{H}(x),\quad a>0$	$\frac{(a-i\omega)}{2\pi\left(a^2+\omega^2\right)}$
14.	$\frac{1}{\sqrt{\vert x\vert}}e^{-a\vert x\vert}$	$\sqrt{\frac{\left[a+\left(a^2+\omega^2\right)^{\frac{1}{2}}\right]}{2\pi\left(a^2+\omega^2\right)}}$
15.	$\delta(x)$	$\frac{1}{2\pi}$
16.	$\delta^{(n)}(x)$	$\frac{1}{2\pi}(i\omega)^n$
17.	$\delta(x-a)$	$\frac{1}{2\pi}e^{-ia\omega}$
18.	$\delta^{(n)}(x-a)$	$\frac{1}{2\pi}(i\omega)^n e^{-ia\omega}$
19.	e^{iax}	$\delta(\omega-a)$
20.	1	$\delta(\omega)$
21.	x	$i\delta'(\omega)$
22.	x^n	$i^n\delta^{(n)}(\omega)$
23.	$\mathcal{H}(x)$	$\frac{1}{2}\left[\frac{1}{i\pi\omega}+\delta(\omega)\right]$
24.	$\mathcal{H}(x-a)$	$\frac{1}{2}\left[\frac{e^{-i\omega a}}{\pi i\omega}+\delta(\omega)\right]$
25.	$\mathcal{H}(x)-\mathcal{H}(-x)$	$-\frac{i}{\pi\omega}$
26.	$x^n e^{iax}$	$i^n\delta^{(n)}(\omega-a)$
27.	$\frac{1}{\vert x\vert}$	$\frac{1}{2\pi}(A-2\log\vert\omega\vert)\quad A$ is a constant
28.	$\log(\vert x\vert)$	$-\frac{1}{2\vert\omega\vert}$
29.	$\mathcal{H}(a-\vert x\vert)$	$\left(\frac{\sin a\omega}{\pi\omega}\right)$

(continued)

Table 5.1 (continued)

30.	$\lvert x\rvert^{\alpha}$ $\alpha < 1$, not a negative integer	$\frac{1}{\pi}\Gamma(\alpha+1)\lvert\omega\rvert^{-(1+\alpha)}\cos\left[\frac{\pi}{2}(\alpha+1)\right]$
31.	$\operatorname{sgn} x$	$\frac{1}{(i\pi\omega)}$
32.	$x^{-(n+1)}\operatorname{sgn} x$	$\frac{(-i\omega)^n}{2\pi n!}(A-2\log\lvert\omega\rvert)$
33.	$\frac{1}{x}$	$-\frac{i}{2}\operatorname{sgn}\omega$
34.	$\frac{1}{x^n}$	$-\frac{i}{2}\left[\frac{(-i\omega)^{n-1}}{(n-1)!}\operatorname{sgn}\omega\right]$
35.	$x^n e^{iax}$	$i^n\delta^{(n)}(\omega-a)$
36.	$x^{\alpha}\mathcal{H}(x)$ (α not an integer)	$\frac{\Gamma(\alpha+1)}{2\pi}\lvert\omega\rvert^{-(\alpha+1)}e^{\left[-\left(\frac{\pi i}{2}\right)(\alpha+1)\operatorname{sgn}\omega\right]}$
37.	$x^n e^{iax}\mathcal{H}(x)$	$\frac{1}{2}\left[\frac{n!}{i\pi(\omega-a)^{n+1}}+i^n\delta^{(n)}(\omega-a)\right]$
38.	$e^{iax}\mathcal{H}(x-b)$	$\frac{1}{2}\left[\frac{e^{[-ib(\omega-a)]}}{i\pi(\omega-a)}+\delta(\omega-a)\right]$
39.	$\frac{1}{x-a}$	$-\frac{i}{2}e^{-ia\omega}\operatorname{sgn}\omega$
40.	$\frac{1}{(x-a)^n}$	$-\frac{i}{2}e^{-ia\omega}\frac{(-i\omega)^{n-1}}{(n-1)!}\operatorname{sgn}\omega$
41.	$\frac{e^{iax}}{(x-b)}$	$\frac{i}{2}e^{[ib(a-\omega)]}\left[1-2\mathcal{H}(\omega-a)\right]$
42.	$\frac{e^{iax}}{(x-b)^n}$	$\frac{i}{2}\left[1-2\mathcal{H}(\omega-a)\right]$ $\times\frac{e^{ib(a-\omega)}}{(n-1)!}[-i(\omega-a)]^{n-1}$
43.	$\lvert x\rvert^{\alpha}\operatorname{sgn} x$ (α not integer)	$\frac{-i}{\pi}\frac{\Gamma(\alpha+1)}{\lvert\omega\rvert^{\alpha+1}}\cos\left(\frac{\pi\alpha}{2}\right)\operatorname{sgn}\omega$
44.	$x^n f(x)$	$\frac{(-i)^n}{\sqrt{2\pi}}\frac{d^n}{d\omega^n}F(\omega)$
45.	$\frac{d^n}{dx^n}f(x)$	$\frac{(i\omega)^n}{\sqrt{2\pi}}F(\omega)$
46.	$e^{iax}f(bx)$	$\frac{1}{\sqrt{2\pi}b}F\left(\frac{\omega-a}{b}\right)$
47.	$\sin\left(ax^2\right)$	$\frac{1}{2\sqrt{a\pi}}\sin\left(\frac{\omega^2}{4a}-\frac{\pi}{4}\right)$
48.	$\cos\left(ax^2\right)$	$\frac{1}{2\sqrt{a\pi}}\cos\left(\frac{\omega^2}{4a}-\frac{\pi}{4}\right)$

Two Bessel and gamma functions are used in the table. Bessel's functions are

$$J_0(x) = \sum_{r=0}^{\infty} \frac{(-1)^r}{(r!)^2} \left(\frac{x}{2}\right)^{2r} \tag{5.83}$$

$$J_1(x) = \sum_{r=0}^{\infty} \frac{(-1)^r}{r!(r+1)!} \left(\frac{x}{2}\right)^{2r+1} \tag{5.84}$$

The gamma function is

$$\Gamma(x) = \int_0^{\infty} e^{-t} t^{x-1} dt, \quad x > 0 \tag{5.85}$$

With

$$\Gamma(x+1) = x\Gamma(x) \tag{5.86}$$

5.3 Closure

This chapter introduced the fundamental concepts of rapidly decreasing functions, generalized functions, and their relationship through Fourier transforms. Rapidly decreasing functions were characterized as smooth and well behaved, decaying faster than any polynomial as their argument approaches infinity. The importance of these functions in mathematical physics and signal processing was highlighted due to their advantageous properties under Fourier transformation, resulting in similarly well-behaved transforms.

Next, the discussion was extended to generalized functions or distributions, allowing objects like the Dirac delta function and impulse trains to be rigorously handled. These concepts are essential for modelling point sources, singularities, and periodic signals. The Fourier transform of such distributions is shown to broaden the classical framework of Fourier analysis, providing a method to analyze functions that do not fit within the standard categories of integrable functions.

The impulse train, introduced as a periodic extension of the Dirac delta function, is an example of the practical utility of generalized functions in Fourier analysis. It was demonstrated that its Fourier transform results in another impulse train in the frequency domain, illustrating the duality between time and frequency domains and emphasizing the significance of periodicity.

By examining rapidly decreasing functions, generalized functions, and their Fourier transforms, a cohesive framework for addressing problems in practical applications ranging from signal processing to quantum mechanics was established. With this foundational understanding set, more advanced topics can now be approached, and the applicability of these mathematical structures will continue to be explored.

References

Debnath, L. (2013). A short biography of Paul A. M. Dirac and historical development of Dirac delta function. *International Journal of Mathematical Education in Science and Technology, 44*, 1201–1223.

Dirac, P. A. M. (1927). The physical interpretation of the quantum dynamics. *Proceedings of the Royal Society of London. Series A, Containing Papers of a Mathematical and Physical Character, 113*(765), 621–641.

Dirac, P. A. M. (1930). *The principles of quantum mechanics*. Oxford University Press.

Heaviside, O. (1893a). *Electromagnetic theory*, vol. i London. The Electrician Publishing.

Heaviside, O. (1893b). On operators in physical mathematics: Part I. *Proceedings of the Royal Society of London, 52*(315-320), 504–529.

Hoskins, R. F. (2009). *Delta functions: introduction to generalised functions* (2nd ed.). Woodhead Publishing.

John J. Benedetto, G. Z. (1997). Sampling multipliers and the Poisson Summation Formula. *Journal of Fourier Analysis and Applications, 3*, 505–523.

Lighthill, M. J. (1958). *Introduction to fourier analysis and generalized functions* (1st ed.). Cambridge University.

Riemann, B. (1867). *Über die Darstellbarkeit einer function durch eine trigonometrische Reihe*, vol. 13. Dieterichsche Buchhandlung.

Schwartz, L. (1947). Théorie générale des fonctions moyenne-périodiques. *Annals of Mathematics, 48*(4), 857–929.

Sobolev, S. L. (1992). *Differential equations and function spaces: collection of papers dedicated to the memory of academician sergei lvovich sobolev*. American Mathematical Society, p. 3.

Strichartz, R. S. (1994). *A guide to distribution theory and fourier transform* (1st ed.). CRC Press.

Chapter 6
The Laplace Transform

What we know is not much. What we do not know is immense.
— Pierre-Simon Laplace

Abstract This chapter thoroughly explores the Laplace transform, a critical tool for analyzing complex systems in engineering and applied sciences. It opens with a historical background, establishing the transform's role in simplifying differential equations and dynamic system analysis. The chapter defines the Laplace transform and explains the conditions for its existence and uniqueness, ensuring its applicability across diverse functions. Key properties like linearity, shifting, and scaling are detailed, offering efficient methods for manipulating transformed functions. The coverage of the inverse Laplace transform follows, with techniques like partial fraction decomposition and contour integration for reconstructing original functions. Heaviside's expansion theorem enhances inversion methods, while differentiation and integration in the Laplace domain allow for refined modeling. Initial and final value theorems provide insights into system behavior at endpoints, with applications in system modeling underscoring the transform's engineering utility. The chapter concludes with the bilateral Laplace transform and standard function examples, equipping readers with a solid grounding in both theory and practical applications of the Laplace transform.

Keywords Laplace transform · Heaviside's first shifting theorem · Second shifting theorem · Impulse response · Partial fraction decomposition · Bromwich contour integration · Heaviside's expansion theorem · Initial value theorem · Final value theorem · Nyquist plot · Bode plot · Nichols plot · Bilateral Laplace transform

6.1 History of Laplace Transform

German mathematician Gottfried Wilhelm Leibniz (1646–1716) was the pioneer in introducing the concept of symbolic methods in calculus. Following his

S. K. Jena, *Fourier, Laplace, and the Tangled Love Affair with Transforms*,
https://doi.org/10.1007/978-3-031-80165-5_6

groundbreaking work, Italian mathematician Joseph-Louis Lagrange (1736–1813) and French mathematician Pierre-Simon, Marquis de Laplace (1749–1827) made substantial contributions that evolved these symbolic methods into what is known today as operational calculus. In his seminal work on probability theory in the 1780s, Laplace introduced the integral transform now bearing his name. His classic book "*La Théorie Analytique des Probabilités*" contains foundational results of the Laplace transform, one of the earliest and most widely utilized integral transforms in the mathematical literature. This transform has proven to be a powerful tool in solving linear differential and integral equations. The formalism and utility of Laplace integral transforms were not yet fully appreciated among their contemporaries. French general Napoléon Bonaparte was the student of Laplace. Laplace was Napoléon's examiner when Napoléon graduated from the École Militaire in Paris in 1785. After the Bourbon Restoration, Laplace became a count of the Empire in 1806 and was named a marquis in 1817.

Interest in the Laplace transform saw a revival in the early twentieth century, driven by engineers and scientists who recognized its potential in electrical engineering and control theory. Although the Laplace transform was initially discovered earlier, it was British electrical engineer Oliver Heaviside (1850–1925) who popularized it by applying it to solve ordinary differential equations related to electrical circuits and systems, thereby advancing modern operational calculus. In his celebrated papers, "On Operational Methods in Physical Mathematics", Parts I and II, published in The Proceedings of the Royal Society, London, in 1892 and 1893, Heaviside developed operational methods that leveraged this transform. He used operational calculus to solve the telegraph equation and second-order hyperbolic partial differential equations with constant coefficients. Heaviside's development of operational methods paid scant attention to mathematical rigor, which led to controversy. The widespread use of his methods, before being formally justified by the Fourier or Laplace transform theory, drew considerable skepticism. The classical works of Laplace, Fourier, and Cauchy inspired these operational methods. Despite Heaviside's calculus being one of the most useful mathematical methods, his contemporary mathematicians largely overlooked his contributions due to their lack of rigor.

In the 1920s and 1930s, the Laplace transform gained further prominence with the formalization of control theory and signal processing. American electrical engineer Harold Stephen Black (1898–1983) and Swedish-American electronic engineer Harry Nyquist (1889–1976) utilized the transform to analyze feedback control systems, thus laying the foundation for modern control engineering. The transform's capability to handle initial conditions and its compatibility with the convolution operation made it an indispensable tool for studying system behavior and stability. By the mid-twentieth century, the Laplace transform had become a standard technique in the analysis of linear systems across various disciplines.

For nearly two centuries, Laplace transformations have been instrumental in addressing numerous challenges in applied mathematics, mathematical physics, and engineering. The Laplace transform has proven invaluable for solving partial differential equations and boundary value problems in applied mathematics. Its ability to simplify complex boundary conditions into manageable algebraic forms has enabled mathematicians and engineers to tackle a broad range of practical problems, from heat conduction to wave propagation. Today, the Laplace transform remains a fundamental technique in analyzing and designing linear systems. Its enduring relevance is a testament to the foresight of Pierre-Simon Laplace and the successive generations of mathematicians and engineers who recognized and expanded upon its potential. The Laplace transform continues to be a cornerstone of modern engineering, seamlessly bridging the gap between mathematical theory and practical application.

6.2 Definition of the Laplace Transform

It is worth noting that the Laplace transform is a special case of the Fourier transform for causal functions, i.e., the functions defined on the positive real axis. However, the Laplace transform is simpler than the Fourier transform for several reasons. The Fourier transform is primarily used to analyze energy signals, so it does not converge if the function is not rapidly decreasing. Unlike the Fourier transform, the Laplace transform is a more generalized mathematical tool and can handle both energy signals (finite energy) and power signals (finite power but possibly infinite energy). The convergence issue is less problematic with the Laplace transform due to its exponentially decaying kernel, e^{-sx}. Here, s is a complex variable with a real part greater than zero ($\Re(s) > 0$), and $x > 0$. This exponential decay ensures that the transform converges more readily compared to the Fourier transform. Second, the Laplace transform is an analytic function of the complex variable s. This analytic nature allows its properties to be studied using the well-established theory of complex variables, making it more accessible and straightforward to analyze. Third, the definitions of the Laplace transform and its inverse can be derived from the Fourier integral formula. This connection involves expressing these transforms as complex contour integrals, which can be evaluated using the Cauchy residue theorem and the deformation of contours in the complex plane. These techniques from complex analysis facilitate a deeper understanding and easier computation of the Laplace transform.

Fourier integral of a function $f(x)$ is defined as

$$f(x) = \frac{1}{2\pi} \int_{-\infty}^{\infty} \left[\int_{-\infty}^{\infty} f(v)\, e^{-i\omega v} dv \right] e^{i\omega x} d\omega \tag{6.1}$$

where the domain of function $f(x)$ is $-\infty < x < \infty$. When this function $f(x)$ is multiplied with the Heaviside step function $H(x)$, then its domain is limited within $0 \leq x < \infty$. e^{-cx}, for any positive value of $c > 0$ asymptotically approaches zero as $x \to \infty$. A function $f(x)$ defined over the set of positive real numbers is said to be exponential order for large x if there exist real constants $k, M > 0$ such that

$$e^{-kx}|f(x)| \leq M \tag{6.2}$$

$f(x)$ is of exponential order if there exists a real constant k such that $e^{-kx}|f(x)|$ is bounded for sufficiently large values of x. $f(x)$ is of the order e^{kx} and can be expressed as $|f(x)| \sim O\left(e^{kx}\right)$. This means that as x increases, the modulus of $f(x)$ increases not faster than an exponential function e^{kx}. Multiplying the function with e^{-cx} for $c > 0$ ensures the function's finite energy content and guarantees the Fourier integral's convergence. So, the function is redefined as

$$g(x) = f(x)\mathcal{H}(x)e^{-cx} = f(x)e^{-cx} \quad x \geq 0 \tag{6.3}$$

The Fourier integral of this modified function $g(x)$ becomes

$$f(x) = \frac{e^{-cx}}{2\pi}\int_{-\infty}^{\infty}\left[\int_{0}^{\infty} f(v)\,e^{-(c+i\omega)v}dv\right]e^{i\omega x}d\omega \tag{6.4}$$

Substituting the variable $c+i\omega = s$ results in $id\omega = ds$. Equation (6.4) is expressed as

$$f(x) = \frac{e^{-cx}}{2\pi i}\int_{c-i\infty}^{c+i\infty}\left[\int_{0}^{\infty} f(v)\,e^{-(s)v}dv\right]e^{(s-c)x}ds \tag{6.5}$$

The Laplace transform is formally defined as

$$\mathscr{L}\{f(x)\} = F(s) = \int_{0}^{\infty} f(x)\,e^{-sx}dx \quad \Re(s) > 0 \tag{6.6}$$

e^{-sx} is the kernel of the transform and s is the complex transform variable with $\Re(s) > 0$. Under broad conditions on $f(x)$, its transform $F(s)$ is analytic in s in the half-plane, where $\Re(s) > 0$. From Eq. (6.5), the inverse Laplace transform is defined as

$$\mathscr{L}^{-1}\{F(s)\} = f(x) = \frac{1}{2\pi i}\int_{c-i\infty}^{c+i\infty} F(s)\,e^{sx}ds \quad \Re(s) > 0 \tag{6.7}$$

$\mathscr{L}$ is the Laplace operator. Equations (6.6) and (6.7) together imply

$$\begin{aligned} \mathscr{L}^{-1}\{\mathscr{L}\{f(x)\}\} &= f(x) \\ \mathscr{L}\{\mathscr{L}^{-1}\{F(s)\}\} &= F(s) \end{aligned} \tag{6.8}$$

The Laplace transform is called an integral transform because it transforms (changes) a function in one space to a function in another space by a process of integration that involves the kernel e^{-sx}. The kernel is a function of the variables in the two spaces and defines the integral transform. In probability theory, the Laplace transform may be thought of as an expectation value of a random variable. If x is a random variable with probability density function f defined over the set of positive real numbers (i.e., $f(x) = 0$ for $x < 0$), then the Laplace transform of $f(x)$ is given by the expectation

$$\mathscr{L}\{f(x)\}(s) = E\left[e^{-sx}\right] \tag{6.9}$$

To avoid further confusion, now onward $\mathscr{L}\{f(x)\}$ will be written instead of $\mathscr{L}\{f(x)\}(s)$. The same is true for the inverse Laplace transform. The original functions are denoted by lowercase letters f, and their transformations are made using the same letters in capital F.

6.3 Existence and Uniqueness of Laplace Transforms

A function $f(x)$ possesses a Laplace transform if its growth is sufficiently controlled. Specifically, for all $x \geq 0$ and given constants M and k, the function must satisfy the "growth restriction":

$$|f(x)| \leq Me^{kx} \tag{6.10}$$

The growth restriction in Eq. (6.10) is often referred to as "growth of exponential order". However, this term can be misleading as it implies that the exponent must be kx, not kx^2 or a similar term. The function $f(x)$ does not need to be continuous but should not exhibit overly erratic behavior. The technical term commonly used in mathematics is piecewise continuity. A function $f(x)$ is considered piecewise continuous on a finite interval $a \leq x \leq b$ if this interval can be divided into a finite number of subintervals, within each of which f is continuous and has finite limits as x approaches either endpoint of each subinterval from within the interval (Fig. 6.1).

If $f(x)$ is defined and piecewise continuous on every finite interval along the semiaxis $x \geq 0$ and satisfies equation (6.10) for all $x \geq X$ with certain constants M and k, then the Laplace transform $\mathscr{L}(f)$ exists for all $\Re(s) > k$. This states that the graph of $f(x)$ on the interval $[X, \infty)$ does not grow faster than the graph of the exponential function Me^{kx}, where k is a positive constant (Fig. 6.2).

Because $f(x)$ is piecewise continuous, $e^{-sx} f(x)$ is integrable over any finite interval on the x axis. Using Eq. (6.10) and assuming $\Re(s) > k$ (which is necessary for the existence of the following integrals), we can demonstrate the existence of $\mathscr{L}(f)$ from this:

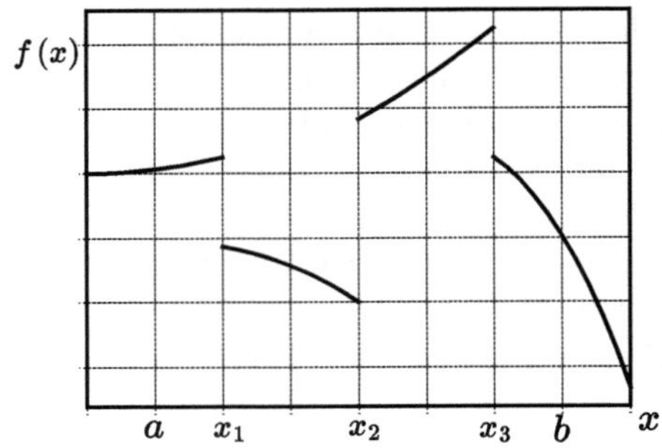

Fig. 6.1 Piecewise continuous function

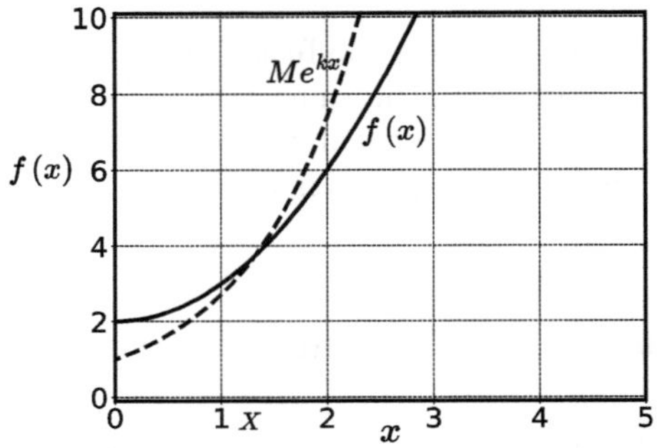

Fig. 6.2 $f(x)$ is of exponential order

$$
\begin{aligned}
|\mathscr{L}\{f\}| = \left|\int_0^{\infty} e^{-sx} f(x)dx\right| &\leqq \int_0^{\infty} |f(x)|e^{-sx}dx \\
&\leqq \int_0^{\infty} Me^{kx}e^{-st}dx = \frac{M}{s-k}
\end{aligned} \tag{6.11}
$$

According to Eq. (6.11), $\mathscr{L}(f)$ exists if $\Re(s) > k$. Equation (6.10) can be easily verified; for example, $\cosh x < e^x$ and $x^n < n!e^x$ (since $x^n/n!$ is a single term in the Maclaurin series), and similar comparisons. A function that does not satisfy equation (6.10) for any constants M and k is e^{x^2} (this can be seen by taking logarithms) (Fig. 6.3). The conditions in Eq. (6.11) are sufficient but not necessary.

If f is of exponential order, there exist constants γ, $M_1 > 0$, and $X > 0$ so that $|f(x)| \leq M_1 e^{\gamma x}$ for $x > X$. Also, since f is piecewise continuous for $0 \leq x \leq X$, it is necessarily bounded on the interval; that is, $|f(x)| \leq M_2 = M_2 e^{0x}$. If M denotes the maximum of the set $\{M_1, M_2\}$ and k denotes the maximum of $\{0, \gamma\}$, then

$$
|F(s)| \leq \int_0^{\infty} e^{-sx}|f(x)|dx \leq M\int_0^{\infty} e^{-sx}e^{kx}dx = M\int_0^{\infty} e^{-(s-k)x}dx = \frac{M}{s-k} \tag{6.12}
$$

For $s > k$. As $s \to \infty$, we have $|F(s)| \to 0$, and so $F(s) = \mathscr{L}\{f(t)\} \to 0$.

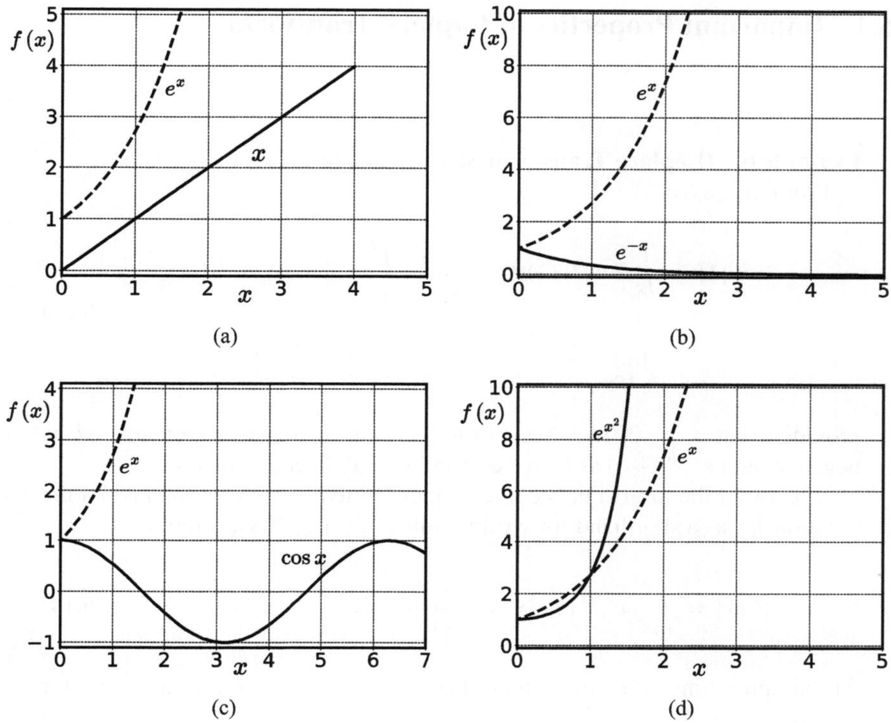

Fig. 6.3 Functions of exponential order. (**a**) Linear function (**b**) Exponential decaying function (**c**) Trigonometric function (**d**) Exponential of a quadratic function

The functions of s such as $F_1(s) = 1$ and $F_2(s) = s/(s+1)$ are not the Laplace transforms of piecewise continuous functions of exponential order, since $F_1(s) \nrightarrow 0$ and $F_2(s) \nrightarrow 0$ as $s \to \infty$. But this should not be concluded that $F_1(s)$ and $F_2(s)$ are not Laplace transforms. There are other kinds of functions.

Uniqueness. If a function has a Laplace transform, that transform is uniquely determined. Conversely, if two functions, both defined on the positive real axis, share the same Laplace transform, they cannot differ over any interval of positive length. However, they may differ at isolated points. Thus, the inverse of a given transform is essentially unique. Specifically, if two continuous functions have the same Laplace transform, they are entirely identical.

6.4 Important Properties of Laplace Transform

Example 6.1 (Laplace Transform of 1**)** Evaluate $\mathscr{L}\{1\}$.

From Eq. (6.6)

$$\begin{aligned}\mathscr{L}\{1\} &= \int_0^{\infty} e^{-sx}(1)dx = \lim_{b\to\infty}\int_0^b e^{-sx}dx \\ &= \lim_{b\to\infty}\left.\frac{-e^{-sx}}{s}\right|_0^b = \lim_{b\to\infty}\frac{-e^{-sb}+1}{s} = \frac{1}{s}\end{aligned} \tag{6.13}$$

provided that $s > 0$. In other words, when $s > 0$, the exponent $-sb$ is negative, and $e^{-sb} \to 0$ as $b \to \infty$. The integral diverges for $s < 0$.

The use of the limit sign becomes somewhat tedious, so we shall adopt the notation $|_0^{\infty}$ as a shorthand for writing $\lim_{b\to\infty}(\ )\,|_0^b$. For example,

$$\mathscr{L}\{1\} = \int_0^{\infty} e^{-sx}(1)dx = \left.\frac{-e^{-sx}}{s}\right|_0^{\infty} = \frac{1}{s}, \quad s > 0 \tag{6.14}$$

At the upper limit, it is understood that we mean $e^{-sx} \to 0$ as $x \to \infty$ for $s > 0$.

Example 6.2 (Laplace Transform of x**)** Evaluate $\mathscr{L}\{x\}$.

From definition in Eq. (6.6), we have

$$\mathscr{L}\{x\} = \int_0^{\infty} e^{-sx}x\,dx \tag{6.15}$$

Integrating by parts and using $\lim_{x\to\infty} te^{-sx} = 0, s > 0$, along with the result from Example 6.1, results in

$$\mathscr{L}\{x\} = \left.\frac{-xe^{-sx}}{s}\right|_0^{\infty} + \frac{1}{s}\int_0^{\infty} e^{-sx}dx = \frac{1}{s}\mathscr{L}\{1\} = \frac{1}{s}\left(\frac{1}{s}\right) = \frac{1}{s^2} \tag{6.16}$$

Example 6.3 (Laplace Transform of $e^{\pm ax}$) Evaluate $\mathscr{L}\left\{e^{\pm ax}\right\}$

From definition in Eq. (6.6), we have

$$\begin{aligned}\mathscr{L}\{e^{\pm ax}\} &= \int_0^\infty e^{-sx}e^{\pm ax}dx = \int_0^\infty e^{-(s\mp a)x}dx \\ &= \left.\frac{-e^{-(s\mp a)x}}{s\mp a}\right|_0^\infty = \frac{1}{s\mp a}\end{aligned} \tag{6.17}$$

The above result is valid for $s > \pm a$ because in order to have $\lim_{x\to\infty} e^{-(s\mp a)x} = 0$, we must require that $s\mp a > 0$ or $s > \pm a$.

6.4.1 *Linearity of the Laplace Transform*

The Laplace transform is a linear operation, meaning that if $f(x)$ and $g(x)$ are functions with existing transforms and a and b are constants, then the transform of $af(x)+bg(x)$ also exists, and

$$\begin{aligned}\mathscr{L}\{af(x)+bg(x)\} &= \int_0^\infty e^{-sx}[af(x)+bg(x)]dx \\ &= a\int_0^\infty e^{-sx}f(x)dx + b\int_0^\infty e^{-sx}g(x)dx \\ &= a\mathscr{L}\{f(x)\} + b\mathscr{L}\{g(x)\}\end{aligned} \tag{6.18}$$

Integration is a linear operation.

Example 6.4 (Laplace Transform of Hyperbolic Functions) Evaluate $\mathscr{L}\{\sinh x\}$ and $\mathscr{L}\{\cosh x\}$

Since $\cosh ax = \frac{1}{2}\left(e^{ax}+e^{-ax}\right)$ and $\sinh ax = \frac{1}{2}\left(e^{ax}-e^{-ax}\right)$, using Example 6.3 and principle of linearity,

$$\mathscr{L}\{\cosh ax\} = \frac{1}{2}\left(\mathscr{L}\left\{e^{ax}\right\}+\mathscr{L}\left\{e^{-ax}\right\}\right) = \frac{1}{2}\left(\frac{1}{s-a}+\frac{1}{s+a}\right) = \frac{s}{s^2-a^2} \tag{6.19}$$

$$\mathscr{L}\{\sinh ax\} = \frac{1}{2}\left(\mathscr{L}\left\{e^{ax}\right\}-\mathscr{L}\left\{e^{-ax}\right\}\right) = \frac{1}{2}\left(\frac{1}{s-a}-\frac{1}{s+a}\right) = \frac{a}{s^2-a^2} \tag{6.20}$$

Example 6.5 (Laplace Transform of $\sin \omega x$ and $\cos \omega x$) Evaluate $\mathscr{L}\{\sin \omega x\}$ and $\mathscr{L}\{\cos \omega x\}$

Let $L_c = \mathscr{L}\{\cos \omega x\}$ and $L_s = \mathscr{L}\{\sin \omega x\}$. Integrating by parts and noting that the integral free parts give no contribution from the upper limit, ∞

$$\begin{aligned} L_c &= \int_0^\infty e^{-sx} \cos \omega x dx = \frac{e^{-sx}}{-s} \cos \omega x \Big|_0^\infty - \frac{\omega}{s} \int_0^\infty e^{-sx} \sin \omega x dx \\ &= \frac{1}{s} - \frac{\omega}{s} L_s \end{aligned} \tag{6.21}$$

$$L_s = \int_0^\infty e^{-sx} \sin \omega x dx = \frac{e^{-sx}}{-s} \sin \omega x \Big|_0^\infty + \frac{\omega}{s} \int_0^\infty e^{-sx} \cos \omega x dx = \frac{\omega}{s} L_c \tag{6.22}$$

By substituting L_s into the formula for L_c on the right and then by substituting L_c into the formula for L_s on the right, we obtain

$$\begin{aligned} L_c &= \frac{1}{s} - \frac{\omega}{s} \left(\frac{\omega}{s} L_c \right) \\ L_c \left(1 + \frac{\omega^2}{s^2} \right) &= \frac{1}{s} \\ \mathscr{L}\{\cos \omega x\} &= \frac{s}{s^2 + \omega^2} \end{aligned} \tag{6.23}$$

$$\begin{aligned} L_s &= \frac{\omega}{s} \left(\frac{1}{s} - \frac{\omega}{s} L_s \right) \\ L_s \left(1 + \frac{\omega^2}{s^2} \right) &= \frac{\omega}{s^2} \\ \mathscr{L}\{\sin \omega x\} &= \frac{\omega}{s^2 + \omega^2} \end{aligned} \tag{6.24}$$

Example 6.6 (Laplace Transform of x^{n+1} for Any Integer $n \geq 0$) Evaluate $\mathscr{L}\{x^{n+1}\}$

$$\mathscr{L}\left\{x^{n+1}\right\} = \int_0^\infty e^{-sx} x^{n+1} dx = -\frac{1}{s} e^{-sx} x^{n+1} \Big|_0^\infty + \frac{n+1}{s} \int_0^\infty e^{-sx} x^n dx \tag{6.25}$$

(continued)

Example 6.6 (continued)
Now the integral-free part is zero and the last part is $(n+1)/s$ times $\mathscr{L}\{x^n\}$. From this and the induction hypothesis,

$$\mathscr{L}\left\{x^{n+1}\right\} = \frac{n+1}{s}\mathscr{L}\left\{x^n\right\} = \frac{n+1}{s} \cdot \frac{n!}{s^{n+1}} = \frac{(n+1)!}{s^{n+2}} \tag{6.26}$$

Example 6.7 (Laplace Transform of x^a for Any Real Number $a > 0$)
Evaluate $\mathscr{L}\{x^a\}$ for any real number $a > 0$

$$\mathscr{L}\left\{x^a\right\} = \int_0^\infty e^{-sx} x^a dx \tag{6.27}$$

setting $sx = y$

$$\int_0^\infty e^{-sx} x^a dx = \int_0^\infty e^{-y} \left(\frac{y}{s}\right)^a \frac{dy}{s} = \frac{1}{s^{a+1}} \int_0^\infty e^{-y} y^a dy = \frac{\Gamma(a+1)}{s^{a+1}} \tag{6.28}$$

$\Gamma(a) = (a-1)!$ represents the gamma function:

$$\Gamma(a) = \int_0^\infty e^{-x} x^{a-1} dx \quad a > 0 \tag{6.29}$$

$\Gamma(a+1) = a\Gamma(a)$,

$$\mathscr{L}\left\{x^a\right\} = \frac{\Gamma(a+1)}{s^{a+1}} \tag{6.30}$$

Example 6.8 (Laplace Transform of the Error Function) The error function (also called the Gauss error function) is defined as

$$\operatorname{erf} x = \frac{2}{\sqrt{\pi}} \int_0^x e^{-y^2}\, dy \tag{6.31}$$

The Laplace transform of the $\operatorname{erf}\left(\frac{a}{2\sqrt{x}}\right)$ is defined as

$$\mathscr{L}\left\{\operatorname{erf}\left(\frac{a}{2\sqrt{x}}\right)\right\} = \int_0^\infty e^{-sx} \left[\frac{2}{\sqrt{\pi}} \int_0^{a/2\sqrt{x}} e^{-y^2} dy\right] dx \tag{6.32}$$

(continued)

Example 6.8 (continued)
By putting $z = \dfrac{a}{2\sqrt{x}}$ or $x = \dfrac{a^2}{4z^2}$ and interchanging the order of integration,

$$\mathscr{L}\{\mathrm{erf}(z)\} = \frac{2}{\sqrt{\pi}} \int_0^\infty e^{-z^2} \left[\int_0^{a^2/4z^2} e^{-sx} dx \right] dz$$

$$= \frac{2}{\sqrt{\pi}} \int_0^\infty e^{-z^2} \frac{1}{s} \left\{ 1 - \exp\left(-\frac{a^2 s}{4z^2} \right) \right\} dz$$

$$= \frac{1}{s} \frac{2}{\sqrt{\pi}} \left[\int_0^\infty e^{-z^2} dz - \int_0^\infty \exp\left\{ -\left(z^2 + \frac{sa^2}{4z^2} \right) \right\} dz \right] \tag{6.33}$$

Taking $\alpha = \dfrac{a\sqrt{s}}{2}$, the second integral becomes

$$\int_0^\infty \exp\left\{ -\left(z^2 + \frac{\alpha^2}{z^2} \right) \right\} dz$$

$$= \frac{1}{2} \left[\int_0^\infty \left(1 - \frac{\alpha}{z^2} \right) \exp\left[-\left(z + \frac{\alpha}{z} \right)^2 + 2\alpha \right] \right. \tag{6.34}$$

$$\left. + \int_0^\infty \left(1 + \frac{\alpha}{z^2} \right) \exp\left[-\left(z - \frac{\alpha}{z} \right)^2 - 2\alpha \right] \right] dz$$

By putting $x = \left(z \pm \dfrac{a}{z} \right)$, $dx = \left(1 \mp \dfrac{a}{z^2} \right) dz$, and

$$\int_0^\infty \exp\left\{ -\left(z^2 + \frac{\alpha^2}{z^2} \right) \right\} dz = \frac{1}{2} e^{-2\alpha} \int_{-\infty}^\infty e^{-z^2} dz = \frac{\sqrt{\pi}}{2} e^{-2\alpha} \tag{6.35}$$

Consequently,

$$\mathscr{L}\left\{ \mathrm{erf}\left(\frac{a}{2\sqrt{2}} \right) \right\} = \frac{1}{s} \frac{2}{\sqrt{\pi}} \left[\frac{\sqrt{\pi}}{2} - \frac{\sqrt{\pi}}{2} e^{-a\sqrt{s}} \right] = \frac{1}{s} \left[1 - e^{-a\sqrt{s}} \right] \tag{6.36}$$

The Laplace transform of the complementary error function is

$$\mathscr{L}\left\{ \mathrm{erfc}\left(\frac{a}{2\sqrt{x}} \right) \right\} = \frac{1}{s} e^{-a\sqrt{x}} \tag{6.37}$$

Example 6.9 (Laplace Transform of the Bessel Function of Order Zero) Using the series representation of $J_0(at)$, the Laplace transform becomes

$$\begin{aligned}\mathscr{L}\{J_0(ax)\} &= \mathscr{L}\left[1 - \frac{a^2x^2}{2^2} + \frac{a^4x^4}{2^2\cdot 4^2} - \frac{a^6x^6}{2^2\cdot 4^2\cdot 6^2} + \cdots\right] \\ &= \frac{1}{s} - \frac{a^2}{2^2}\frac{2!}{s^3} + \frac{a^4}{2^2\cdot 4^2}\cdot\frac{4!}{s^5} - \frac{a^6}{2^2\cdot 4^2\cdot 6^2}\cdot\frac{6!}{s^7} + \cdots \\ &= \frac{1}{s}\left[1 - \frac{1}{2}\left(\frac{a^2}{s^2}\right) + \frac{1\cdot 3}{2\cdot 4}\left(\frac{a^4}{s^4}\right) - \frac{1\cdot 3\cdot 5}{2\cdot 4\cdot 6}\left(\frac{a^6}{s^6}\right) + \cdots\right] \\ &= \frac{1}{s}\left[\left(1 + \frac{a^2}{s^2}\right)^{-\frac{1}{2}}\right] = \frac{1}{\sqrt{a^2+s^2}}\end{aligned} \tag{6.38}$$

Similarly, using $J_0'(x) = -J_1(x)$ gives

$$\mathscr{L}\{J_1(ax)\} = \frac{1}{a}\left[\frac{s}{\sqrt{s^2+a^2}} - 1\right] \tag{6.39}$$

Example 6.10 (Laplace Transform of the Gaussian Function) From definition, the Laplace transform of the Gaussian function is

$$\begin{aligned}\mathscr{L}\left\{e^{-a^2x^2}\right\} &= \int_0^\infty e^{-a^2x^2 - xs}ds = e^{\frac{s^2}{4a^2}}\int_0^\infty e^{-a^2\left(1+\frac{s}{2a^2}\right)^2}dx \\ &= e^{\frac{s^2}{4a^2}}\int_{\frac{s^2}{2a^2}}^\infty e^{-a^2u^2}du = \frac{1}{a}e^{\frac{s^2}{4a^2}}\int_{\frac{s}{2a}}^\infty e^{-a^2u^2}du \\ &= \frac{\sqrt{\pi}}{2a}e^{\frac{a^2}{4a^2}}\,\mathrm{erfc}\left(\frac{s}{2a}\right)\end{aligned} \tag{6.40}$$

6.4.2 *Heaviside's First Shifting Theorem*

If $\mathscr{L}\{f(x)\} = F(s)$, then

$$\mathscr{L}\left\{e^{\pm ax}f(x)\right\} = \int_0^\infty e^{-(s\pm a)x}f(x)dx = F(s\mp a) \tag{6.41}$$

Taking the inverse of both sides,

$$e^{\pm ax} f(x) = \mathscr{L}^{-1}\{F(s \mp a)\} \tag{6.42}$$

Example 6.11 (Application of the First Shifting Theorem)

$$\mathscr{L}\left\{x^n e^{-ax}\right\} = \frac{n!}{(s+a)^{n+1}} \tag{6.43}$$

$$\mathscr{L}\left\{e^{-ax} \sin bx\right\} = \frac{b}{(s+a)^2 + b^2} \tag{6.44}$$

$$\mathscr{L}\left\{e^{-ax} \cos bx\right\} = \frac{s+a}{(s+a)^2 + b^2} \tag{6.45}$$

6.4.3 Time Shifting (t-Shifting): Second Shifting Theorem

If $\mathscr{L}\{f(x)\} = F(s)$, then for $a > 0$, the "shifted function" is defined as

$$\tilde{f}(x) = f(x-a)\mathcal{H}(x-a) = \begin{cases} 0 & \text{if } x < a \\ f(x-a) & \text{if } x > a \end{cases} \tag{6.46}$$

The Laplace transform of $\tilde{f}(x)$ is defined as

$$\begin{aligned}
\mathscr{L}\left\{\tilde{f}(x)\right\} &= \mathscr{L}\{f(x-a)\mathcal{H}(x-a)\} \\
&= \int_0^a e^{-sx} f(x-a) \underbrace{\mathcal{H}(x-a)}_{\substack{\text{zero for} \\ 0 \le t < a}} dx + \int_a^\infty e^{-sx} f(x-a) \underbrace{\mathcal{H}(x-a)}_{\substack{\text{one for} \\ t \ge a}} dx \\
&= \int_a^\infty e^{-sx} f(x-a) dx
\end{aligned} \tag{6.47}$$

Substituting $y = x - a$, $dy = dx$ in the last integral,

$$\begin{aligned}
\mathscr{L}\{f(x-a)\mathcal{H}(x-a)\} &= \int_0^\infty e^{-s(y+a)} f(y) dy \\
&= e^{-as} \int_0^\infty e^{-sx} f(x) dx \\
&= e^{-as} \mathscr{L}\{f(x)\} = e^{-as} F(s)
\end{aligned} \tag{6.48}$$

Taking the inverse Laplace transform of Eq. (6.48),

$$f(x-a)\mathcal{H}(x-a) = \mathscr{L}^{-1}\{e^{-as}F(s)\} \tag{6.49}$$

Putting $f(x) = 1$ in Eq. (6.48), the Laplace transform of a Heaviside function is

$$\mathscr{L}\{\mathcal{H}(x-a)\} = \frac{e^{-as}}{s} \tag{6.50}$$

6.4.4 *Scaling Property of the Laplace Transform*

For $a > 0$

$$\mathscr{L}\{f(ax)\} = \int_0^\infty e^{-sx} f(ax)dx \tag{6.51}$$

Substituting $y = ax$ and $dx = \frac{1}{a}dy$, Eq. (6.51)

$$\mathscr{L}\{f(ax)\} = \frac{1}{a}\int_0^\infty e^{-\frac{s}{a}y} f(y)dy = \frac{1}{a}F\left(\frac{s}{a}\right) \tag{6.52}$$

Combining both shifting and scaling properties together,

$$\mathscr{L}\{f(ax+b)\} = \frac{1}{a}e^{\frac{sb}{a}}F\left(\frac{s}{a}\right) \tag{6.53}$$

Example 6.12 (Application of Shifting Theorem) The Laplace transform of

$$f(x) = \begin{cases} 1 & 0 < x < 1 \\ -1 & 1 < x < 2 \\ 0 & x > 2 \end{cases} \tag{6.54}$$

Equation (6.54) is expressed as

$$f(x) = 1 - 2\mathcal{H}(x-1) + \mathcal{H}(x-2) \tag{6.55}$$

The $\mathscr{L}\{f(x)\}$ become

(continued)

Example 6.12 (continued)

$$F(s) = \mathscr{L}\{f(x)\} = \mathscr{L}\{1\} - 2\mathscr{L}\{\mathcal{H}(x-1)\} + \mathscr{L}\{\mathcal{H}(x-2)\}$$
$$= \frac{1}{s} - \frac{2e^{-s}}{s} + \frac{e^{-2s}}{s} \tag{6.56}$$

The Laplace transform of $g(x) = \sin x\mathcal{H}(x-\pi)$

$$G(s) = \mathscr{L}\{\sin x\mathcal{H}(x-\pi)\} = -e^{-\pi s}\mathscr{L}\{\cos x\} = -\frac{se^{-\pi s}}{s^2+1} \tag{6.57}$$

Example 6.13 (Laplace Transform of Square Wave Function) A square wave function of width a is defined as

$$f(x) = \mathcal{H}(x) - 2\mathcal{H}(x-a) + 2\mathcal{H}(x-2a) - 2\mathcal{H}(x-3a) + \cdots \tag{6.58}$$

Taking the Laplace transform of (6.58),

$$\mathscr{L}\{f(x)\} = \frac{1}{s} - 2\cdot\frac{e^{-as}}{s} + 2\cdot\frac{e^{-2as}}{s} - 2\cdot\frac{e^{-3as}}{s} + \cdots \tag{6.59}$$

Taking $r = e^{-as}$,

$$\mathscr{L}\{f(x)\} = \frac{1}{s}\left[1 - 2r\left(1 - r + r^2 - \cdots\right)\right]$$
$$= \frac{1}{s}\left[1 - \frac{2r}{1+r}\right] = \frac{1}{s}\left[1 - \frac{2e^{-as}}{1+e^{-as}}\right]$$
$$= \frac{1}{s}\left(\frac{1-e^{-as}}{1+e^{-as}}\right) = \frac{1}{s}\left(\frac{e^{\frac{sa}{2}} - e^{-\frac{as}{2}}}{e^{\frac{sa}{2}} + e^{-\frac{as}{2}}}\right) = \frac{1}{s}\tanh\left(\frac{as}{2}\right) \tag{6.60}$$

Example 6.14 (Laplace Transform of a Periodic Function) If $f(x)$ is a periodic function of period a, then its Laplace transform is defined as

$$\mathscr{L}\{f(x)\} = \int_0^\infty e^{-sx} f(x)dx = \int_0^a e^{-sx} f(x)dx + \int_a^\infty e^{-sx} f(x)dx \tag{6.61}$$

Substituting $x = y + a$ and $dy = dx$, in the second integral,

$$\mathscr{L}\{f(x)\} = \int_0^a e^{-sx} f(x)dx + e^{-sa} \int_0^\infty e^{-sy} f(y+a)dy \tag{6.62}$$

Due to periodicity $f(x + a) = f(x)$, and replacing the dummy variable y with x,

$$\begin{aligned} \mathscr{L}\{f(x)\} &= \int_0^a e^{-sx} f(x)dx + e^{-sa} \int_0^\infty e^{-sx} f(x)dx \\ &= \int_0^a e^{-sx} f(x)dx + e^{-sa} \mathscr{L}\{f(x)\} \end{aligned} \tag{6.63}$$

Solving equation (6.63),

$$\mathscr{L}\{f(x)\} = \frac{1}{e^{-sa}} \int_0^a e^{-sx} f(x)dx \tag{6.64}$$

Example 6.15 (Laplace Transform of Rectified Sine Wave) A rectified sine wave is defined as

$$f(x) = |\sin ax| \tag{6.65}$$

The rectified sine function has a period $\dfrac{\pi}{2}$

$$\begin{aligned} \int_0^{\frac{\pi}{a}} e^{-sx} \sin ax dx &= \left[\frac{e^{-sx}(-a\cos ax - s\sin ax)}{(s^2 + a^2)} \right]_0^{\frac{\pi}{a}} \\ &= \frac{a\left\{1 + \exp\left(-\frac{s\pi}{a}\right)\right\}}{(s^2 + a^2)} \end{aligned} \tag{6.66}$$

(continued)

Example 6.15 (continued)
The Laplace transform of the rectified sine wave becomes

$$\begin{aligned}\mathscr{L}\{f(x)\} &= \frac{a}{(s^2+a^2)} \frac{1+\exp\left(-\frac{s\pi}{a}\right)}{1-\exp\left(-\frac{s\pi}{a}\right)} \\ &= \frac{a}{(s^2+a^2)}\left[\frac{\exp\left(\frac{s\pi}{2a}\right)+\exp\left(-\frac{s\pi}{2a}\right)}{\exp\left(\frac{s\pi}{2a}\right)-\exp\left(-\frac{s\pi}{2a}\right)}\right] \\ &= \frac{a}{s^2+a^2}\coth\left(\frac{\pi s}{2a}\right)\end{aligned} \tag{6.67}$$

6.4.5 Laplace Transform of Impulse Response

Impulse response is a fundamental concept in signal processing and control systems. It refers to the output of a system when it is subjected to an impulse function, often represented by Dirac's delta function $\delta(x)$. The previous chapter discusses the impulse function and its Fourier transform in detail. Let us consider a model function to evaluate the Laplace transform of Dirac's delta function:

$$f_k(x-a) = \begin{cases} 1/k & \text{if } a \leqq x \leqq a+k \\ 0 & \text{otherwise} \end{cases} \tag{6.68}$$

This function represents a force of magnitude $\frac{1}{k}$ acting from $x = a$ to $x = a + k$, where k is positive and small:

$$I_k = \int_0^\infty f_k(x-a)dx = \int_a^{a+k} \frac{1}{k}dx = 1 \tag{6.69}$$

In mechanics, the integral of a force acting over an interval $a \leqq x \leqq a + k$ is called the impulse of the force.

Dirac's delta function represents the impulse function in the limiting situation $k \to 0$ for $k > 0$:

$$\delta(x-a) = \lim_{k\to 0} f_k(x-a) \tag{6.70}$$

$\delta(x-a)$ is Dirac's delta function or the unit impulse function at a:

$$\delta(x-a) = \begin{cases} \infty & \text{if } x = a \\ 0 & \text{otherwise} \end{cases} \tag{6.71}$$

$$\int_0^{\infty} \delta(x-a)dx = 1 \tag{6.72}$$

Operating on $\delta(x-a)$ is convenient as though it were an ordinary function. In particular, for a continuous function $g(x)$ using the property,

$$\int_0^{\infty} g(x)\delta(x-a)dx = g(a) \tag{6.73}$$

To evaluate the Laplace transform of $\delta(x-a)$, let us write

$$f_k(x-a) = \frac{1}{k}[\mathcal{H}(x-a) - \mathcal{H}(x-(a+k))] \tag{6.74}$$

Taking the Laplace transform of Eq. (6.74),

$$\mathcal{L}\{f_k(x-a)\} = \frac{1}{ks}\left[e^{-as} - e^{-(a+k)s}\right] = e^{-as}\frac{1-e^{-ks}}{ks} \tag{6.75}$$

Taking the limit as $k \to 0$. By l'Hôpital's rule, the quotient on the right has the limit 1 (differentiate the numerator and the denominator separately with respect to k, obtaining se^{-ks} and s, respectively, and use $se^{-ks}/s \to 1$ as $k \to 0$). Hence, the right side has the limit e^{-as}. This suggests defining the transform of $\delta(t-a)$ by this limit, that is,

$$\mathcal{L}\{\delta(x-a)\} = e^{-as} \tag{6.76}$$

The impulse response at the center can be obtained by taking the Laplace transform of Dirac's delta function δ_0:

$$\begin{aligned} \mathcal{L}\{\delta(0)\} &= \int_0^{\infty} e^{-sx}\delta(0)\,dx \\ &= e^{-s\cdot 0} = e^0 = 1 \end{aligned} \tag{6.77}$$

6.5 The Inverse Laplace Transform

The Laplace transform $F(s)$ of a given function $f(x)$ is calculated by direct integration. What about the inverse transform? A given Laplace transform $F(s)$ of an unknown function $f(x)$ is essentially concerned with the solution of the integral equation:

Table 6.1 Some functions $f(x)$ and their Laplace transforms $\mathscr{L}\{f(x)\}$

$f(x)$	$\mathscr{L}\{f(x)\}$	$f(x)$	$\mathscr{L}\{f(x)\}$
1	$1/s$	$\cos \omega x$	$\dfrac{s}{s^2+\omega^2}$
x	$1/s^2$	$\sin \omega x$	$\dfrac{\omega}{s^2+\omega^2}$
x^2	$2!/s^3$	$\cosh ax$	$\dfrac{s}{s^2-a^2}$
x^n for $(n = 0, 1, \cdots)$	$\dfrac{n!}{s^{n+1}}$	$\sinh ax$	$\dfrac{a}{s^2-a^2}$
x^a for $a > 0$	$\dfrac{\Gamma(a+1)}{s^{a+1}}$	$e^{ax} \cos \omega x$	$\dfrac{s-a}{(s-a)^2+\omega^2}$
e^{ax}	$\dfrac{1}{s-a}$	$e^{ax} \sin \omega x$	$\dfrac{\omega}{(s-a)^2+\omega^2}$

$$\int_0^{\infty} e^{-sx} f(x)dx = F(s) \tag{6.78}$$

It is quite difficult to handle this integral equation to determine $f(x)$. For simple cases, the inverse transform can be computed from the known forward Laplace transform given in Table 6.1. For example,

$$\mathscr{L}^{-1}\left\{\frac{1}{s^5}\right\} = \frac{1}{4!}\mathscr{L}^{-1}\left\{\frac{4!}{s^5}\right\} = \frac{1}{24}x^4 \tag{6.79}$$

$$\mathscr{L}^{-1}\left\{\frac{1}{s^2+7}\right\} = \frac{1}{\sqrt{7}}\mathscr{L}^{-1}\left\{\frac{\sqrt{7}}{s^2+7}\right\} = \frac{1}{\sqrt{7}}\sin\sqrt{7}x \tag{6.80}$$

However for complex case, the inverse Laplace transform is determined by using four methods: (i) partial fraction decomposition, (ii) convolution theorem, (iii) contour integration of the Laplace inversion integral, and (iv) Heaviside's expansion theorem.

6.5.1 Partial Fraction Decomposition Method

If

$$F(s) = \frac{P(s)}{Q(s)} \tag{6.81}$$

where $P(s)$ and $Q(s)$ are polynomials in s, and the degree of $P(s)$ is less than that of $Q(s)$, the method of partial fractions is used to express $F(s)$ as the sum of terms which can be inverted by using a table of Laplace transforms. For example,

$$F(s) = \frac{P(s)}{Q(s)} \tag{6.82}$$

Example 6.16 (Inverse Laplace Transform) Find the $\mathscr{L}^{-1}\left\{\frac{1}{s(s-a)}\right\}$

Using partial fraction,

$$\left\{\frac{1}{s(s-a)}\right\} = \frac{1}{a}\left\{\frac{1}{s-a} - \frac{1}{s}\right\} \tag{6.83}$$

So, the inverse transform is

$$\begin{aligned}\mathscr{L}^{-1}\left\{\frac{1}{s(s-a)}\right\} &= \mathscr{L}^{-1}\left[\frac{1}{a}\left\{\frac{1}{s-a} - \frac{1}{s}\right\}\right] \\ &= \frac{1}{a}\left[\mathscr{L}^{-1}\left\{\frac{1}{s-a}\right\} - \mathscr{L}^{-1}\left\{\frac{1}{s}\right\}\right] \\ &= \frac{1}{a}\left(e^{ax} - 1\right)\end{aligned} \tag{6.84}$$

Example 6.17 (Inverse Laplace Transform) Find the $\mathscr{L}^{-1}\left\{\frac{1}{\left(s^2+a^2\right)\left(s^2+b^2\right)}\right\}$.

Using partial fraction,

$$\begin{aligned}\mathscr{L}^{-1}\left\{\frac{1}{\left(s^2+a^2\right)\left(s^2+b^2\right)}\right\} &= \frac{1}{b^2-a^2}\left[\mathscr{L}^{-1}\left\{\frac{1}{s^2+a^2} - \frac{1}{s^2+b^2}\right\}\right] \\ &= \frac{1}{\left(b^2-a^2\right)}\left(\frac{\sin ax}{a} - \frac{\sin bx}{b}\right)\end{aligned} \tag{6.85}$$

Example 6.18 (Inverse Laplace Transform) Find the $\mathscr{L}^{-1}\left\{\dfrac{s+7}{s^2+2s+5}\right\}$.

Using partial fraction,

$$\begin{aligned}\mathscr{L}^{-1}\left\{\frac{s+7}{(s+1)^2+4}\right\} &= \mathscr{L}^{-1}\left\{\frac{s+1+6}{(s+1)^2+2^2}\right\}\\ &= \mathscr{L}^{-1}\left\{\frac{s+1}{(s+1)^2+2^2}\right\}+3\mathscr{L}^{-1}\left\{\frac{2}{(s+1)^2+2^2}\right\}\\ &= e^{-x}\cos 2x+3e^{-x}\sin 2x\end{aligned} \tag{6.86}$$

Example 6.19 (Inverse Laplace Transform) Find the $\mathscr{L}^{-1}\left\{\dfrac{2s^2+5s+7}{(s-2)\left(s^2+4s+13\right)}\right\}$.

Using partial fraction,

$$\begin{aligned}&\mathscr{L}^{-1}\left\{\frac{2s^2+5s+7}{(s-2)\left(s^2+4s+13\right)}\right\}\\ &= \mathscr{L}^{-1}\left\{\frac{1}{s-2}+\frac{s+2}{(s+2)^2+3^2}+\frac{1}{(s+2)^2+3^2}\right\}\\ &= \mathscr{L}^{-1}\left\{\frac{1}{s-2}\right\}+\mathscr{L}^{-1}\left\{\frac{s+2}{(s+2)^2+3^2}\right\}\\ &\quad+\frac{1}{3}\mathscr{L}^{-1}\left\{\frac{3}{(s+2)^2+3^2}\right\}\\ &= e^{2x}+e^{-2x}\cos 3x+\frac{1}{3}e^{-2x}\sin 3x.\end{aligned} \tag{6.87}$$

6.5.2 *Contour Integration of the Laplace Inversion Integral*

The Bromwich contour is named after English mathematician Thomas John I'Anson Bromwich (1875–1929). Bromwich made significant contributions to the theory of complex functions and integrals. Unlike real integrals, complex integrals depend on the integration path. The Bromwich contour is a path in the complex plane used in

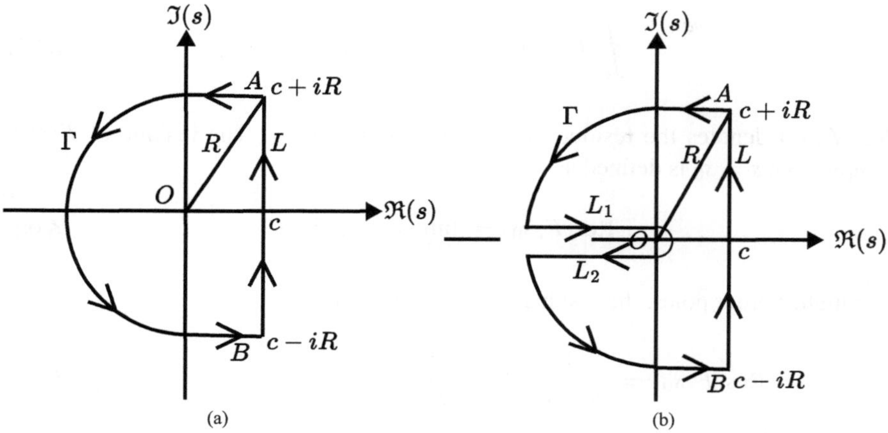

Fig. 6.4 The Bromwich contour and the contour of integration

the inverse Laplace transform to evaluate integrals. The complex integral formula defines the inverse Laplace transform:

$$\mathscr{L}^{-1}\{F(s)\} = f(x) = \frac{1}{2\pi i}\int_{c-i\infty}^{c+i\infty} e^{sx} F(s) ds \tag{6.88}$$

c is a suitable real constant chosen so that the contour path is to the right of all singularities (poles) of $F(s)$. $F(s)$ is an analytic function of the complex variable s in the right half-plane $\Re(s) > a$. Poles of a complex function are specific types of singularities or points where a function is not defined or does not behave normally. The evaluation of Eq. (6.88) depends on the nature of the singularities of $F(s)$. Typically, $F(s)$ is a single-valued function with either a finite or countably infinite number of pole singularities and often has branch points. The integration path is a straight line L (Fig. 6.4a) in the complex s-plane, given by the equation $s = c + iR$, $-\infty < R < \infty$, $\Re(s) = c$, chosen so that all the singularities of the integrand in Eq. (6.88) are to the left of line L. This line is known as the Bromwich contour. In practice, the Bromwich contour is closed by a circular arc of radius R as shown in Fig. 6.4a. The limit as $R \to \infty$ is then taken to expand the contour to infinity, ensuring all singularities of $F(s)$ lie within the integration contour. When $F(s)$ has a branch point at the origin, the integration contour is modified by making a cut along the negative real axis and drawing a small semicircle γ around the origin, as illustrated in Fig. 6.4b.

The Cauchy residue theorem is a powerful result in complex analysis that provides a method for evaluating complex integrals. Let $F(s)$ be a complex analytic function on a simple closed contour C, except for a finite number of isolated singularities $s_1, s_2, \ldots, s_n$ inside C. Then, the integral of $F(s)$ around C is given by

$$\oint_C F(s)\, ds = 2\pi i \sum_{k=1}^{n} \text{Res}(F, s_k) \tag{6.89}$$

$\text{Res}(F, s_k)$ denotes the residue of F at the singularity s_k. The residue of F at a simple pole $s = s_0$ is defined as

$$\text{Res}(F, s_0) = \lim_{s \to s_0} (s - s_0) F(s) \tag{6.90}$$

For higher-order poles, the residue can be found using

$$\text{Res}(F, s_0) = \frac{1}{(n-1)!} \lim_{s \to s_0} \frac{d^{n-1}}{ds^{n-1}} \left[(s - s_0)^n F(s) \right] \tag{6.91}$$

n is the order of the pole.

The Cauchy residue theorem is used to evaluate the Bromwich contour integral:

$$\int_L e^{st} F(s) ds + \int_\Gamma e^{st} F(s) ds = \int_C e^{st} F(s) ds = 2\pi i \sum_{k=1}^{n} \text{Res}(e^{sx} F(s), s_k) \tag{6.92}$$

The summation represents the sum of all the residues at the poles inside C. Letting $R \to \infty$, the integral over Γ tends to zero, which is true in most problems of interest. The complex integration is carried out along the line L in Fig. 6.4a. Simple examples illustrate the above method.

Example 6.20 (Inverse Laplace Transform Through Contour Integration) If $F(s) = \dfrac{s}{s^2 + a^2}$, then the integrand has two simple poles at $s = \pm ia$ and the residues at these poles are

$$\begin{aligned} R_1 &= \text{Residue of } e^{sx} F(s) \text{ at } s = ia \\ &= \lim_{s \to ia} (s - ia) \frac{s e^{sx}}{\left(s^2 + a^2\right)} = \frac{1}{2} e^{iax} \\ R_2 &= \text{Residue of } e^{sx} F(s) \text{ at } s = -ia \\ &= \lim_{s \to -ia} (s + ia) \frac{s e^{sx}}{\left(s^2 + a^2\right)} = \frac{1}{2} e^{-iax} \end{aligned} \tag{6.93}$$

Hence,

(continued)

Example 6.20 (continued)

$$f(x) = \frac{1}{2\pi i}\int_{c-i\infty}^{c+i\infty} e^{sx}F(s)ds = R_1 + R_2 = \frac{1}{2}\left(e^{iax} + e^{-iax}\right) = \cos ax \tag{6.94}$$

as obtained earlier.

Example 6.21 (Inverse Laplace Transform Through Contour Integration) If $F(s) = \dfrac{s}{\left(s^2+a^2\right)^2}$, then the integrand has double poles at $s = \pm ia$ and the residues at these poles are

$$\begin{aligned} R_1 &= \lim_{s\to ia}\frac{d}{ds}\left[(s-ia)^2\frac{se^{sx}}{\left(s^2+a^2\right)^2}\right] \\ &= \lim_{s\to ia}\frac{d}{ds}\left[\frac{se^{sx}}{(s+ia)^2}\right] = \frac{xe^{iax}}{4ia} \end{aligned} \tag{6.95}$$

Similarly, the residue at the double pole at $s = -ia$ is $\left(-xe^{-iax}\right)/4ia$. Thus,

$$f(x) = \text{ Sum of the residues } = \frac{x}{4ia}\left(e^{iax} - e^{-iax}\right) = \frac{x}{2a}\sin ax \tag{6.96}$$

Example 6.22 (Inverse Laplace Transform Through Contour Integration) If $F(s) = \dfrac{\cosh(\alpha y)}{s\cosh(\alpha\ell)}$ for $\alpha = \sqrt{\dfrac{s}{a}}$, then the integrand has double poles at $s = 0$ and $s = s_n = -(2n+1)^2\dfrac{a\pi^2}{4\ell^2}$, where $n = 0, 1, 2, \ldots$.

R_1 = Residue at the pole $s = 0$ is 1, and R_n = Residue at the pole $s = s_n$ is

$$\begin{aligned} &\frac{\exp(-s_n x)\cosh\left\{i(2n+1)\dfrac{\pi y}{2\ell}\right\}}{\left[s\dfrac{d}{ds}\left\{\cosh l\sqrt{\dfrac{s}{a}}\right\}\right]_{s=s_n}} \\ &= \frac{4(-1)^{n+1}}{(2n+1)\pi}\exp\left[-\left\{\frac{(2n+1)\pi}{2\ell}\right\}^2 ax\right]\cos\left\{(2n+1)\frac{\pi y}{2\ell}\right\} \end{aligned} \tag{6.97}$$

(continued)

Example 6.22 (continued)
Thus,

$$
\begin{aligned}
f(x) &= \text{Sum of the residues at the pole} \\
&= 1 + \frac{4}{\pi}\sum_{n=0}^{\infty}\frac{(-1)^{n+1}}{(2n+1)}\exp\left[-(2n+1)^2\frac{\pi^2 ax}{4\ell^2}\right]\times\cos\left\{(2n+1)\frac{\pi y}{2\ell}\right\}
\end{aligned}
\tag{6.98}
$$

Example 6.23 (Inverse Laplace Transform Through Contour Integration) If $F(s) = \dfrac{e^{-a\sqrt{s}}}{s}$, then the integrand has a branch point at $s = 0$. The contour of integration as shown in Fig. 6.4b needs to be performed, which excludes the branch point at $s = 0$. Thus, the Cauchy fundamental theorem gives

$$
\frac{1}{2\pi i}\left[\int_L + \int_\Gamma + \int_{L_1} + \int_{L_2} + \int_\gamma\right]\exp(sx - a\sqrt{s})\frac{ds}{s} = 0 \tag{6.99}
$$

The integral on Γ tends to zero as $R \to \infty$ and that on L gives the Bromwich integral. The remaining three integrals in Eq. (6.99) need to be evaluated. On L_1, $s = re^{i\pi} = -r$ and

$$
\begin{aligned}
\int_{L_1}\exp(sx - a\sqrt{s})\frac{ds}{s} &= \int_{-\infty}^{0}\exp(sx - a\sqrt{s})\frac{ds}{s} \\
&= -\int_0^{\infty}\exp\{-(rx + ia\sqrt{r})\}\frac{dr}{r}
\end{aligned}
\tag{6.100}
$$

On L_2, $s = re^{-i\pi} = -r$

$$
\begin{aligned}
\int_{L_2}\exp(sx - a\sqrt{s})\frac{ds}{s} &= \int_0^{-\infty}\exp(sx - a\sqrt{s})\frac{ds}{s} \\
&= \int_0^{\infty}\exp\{-rx + ia\sqrt{r}\}\frac{dr}{r}
\end{aligned}
\tag{6.101}
$$

Thus, the integrals along L_1 and L_2 combined yield

(continued)

Example 6.23 (continued)

$$-2i\int_0^{\infty} e^{-rx}\sin(a\sqrt{r})\frac{dr}{r} = -4i\int_0^{-\infty} e^{-y^2x}\frac{\sin ay}{y}dy, \quad (\sqrt{r}=y) \tag{6.102}$$

Integrating the following standard integral with respect to β

$$\int_0^{\infty} e^{-y^2\alpha^2}\cos(2\beta y)dy = \frac{\sqrt{\pi}}{2\alpha}\exp\left(-\frac{\beta^2}{\alpha^2}\right) \tag{6.103}$$

This results in

$$\begin{aligned}\frac{1}{2}\int_0^{\infty} e^{-y^2\alpha^2}\frac{\sin 2\beta y}{y}dy &= \frac{\sqrt{\pi}}{2\alpha}\int_0^{\beta}\exp\left(-\frac{\beta^2}{\alpha^2}\right)d\beta \\ &= \frac{\sqrt{\pi}}{2}\int_0^{\beta/\alpha} e^{-u^2}du, \quad (\beta=\alpha u) \\ &= \frac{\pi}{4}\operatorname{erf}\left(\frac{\beta}{\alpha}\right)\end{aligned} \tag{6.104}$$

From results in Eq. (6.104), the integral in Eq. (6.102) becomes

$$-4i\int_0^{\infty}\exp\left(-xy^2\right)\frac{\sin ay}{y}dy = -2\pi i\operatorname{erf}\left(\frac{a}{2\sqrt{x}}\right) \tag{6.105}$$

Finally, on γ, $s = re^{i\theta}$, $ds = ire^{i\theta}d\theta$, and

$$\begin{aligned}\int_{\gamma} |\exp(sx - a\sqrt{s})|\frac{ds}{s} &= i\int_{\pi}^{-\pi}\exp\left(rx\cos\theta - a\sqrt{r}\cos\frac{\theta}{2}\right)d\theta \\ &= i\int_{-\pi}^{\pi} d\theta = 2\pi i\end{aligned} \tag{6.106}$$

in which the limit as $r \to 0$ is used and integration from π to $-\pi$ is interchanged to make γ in the counterclockwise direction. Thus, the final result is as follows:

$$\begin{aligned}\mathscr{L}^{-1}\left\{\frac{e^{-a\sqrt{s}}}{s}\right\} &= \frac{1}{2\pi i}\int_{c-i\infty}^{c+i\infty}\exp(sx - a\sqrt{s})\frac{ds}{s} \\ &= \left[1 - \operatorname{erf}\left(\frac{a}{2\sqrt{x}}\right)\right] = \operatorname{erfc}\left(\frac{a}{2\sqrt{x}}\right)\end{aligned} \tag{6.107}$$

6.5.3 *Heaviside's Expansion Theorem for Inverse Laplace Transform*

Suppose $F(s)$ is the Laplace transform of $f(x)$, which has a Maclaurin power series expansion in the form

$$f(x) = \sum_{r=0}^{\infty} a_r \frac{x^r}{r!} \tag{6.108}$$

Taking the Laplace transform, it is possible to write formally

$$F(s) = \sum_{r=0}^{\infty} \frac{a_r}{s^{r+1}} \tag{6.109}$$

Heaviside's expansion theorem, also known as the partial fraction expansion method, is a technique used for finding the inverse Laplace transform of a rational function. It involves expressing the given Laplace transform as a sum of simpler fractions, each of which can be easily inverted. Here is the step-by-step process for the same.

Express the function $F(s)$ as a ratio of polynomials, $F(s) = \frac{P(s)}{Q(s)}$, where $P(s)$ and $Q(s)$ are polynomials, and the degree of $P(s)$ is less than the degree of $Q(s)$. Factor the denominator $Q(s)$ into linear factors (if possible) or irreducible quadratic factors $Q(s) = (s - a_1)(s - a_2) \cdots (s - a_n)$. Decompose $F(s)$ as a sum of partial fractions, $F(s) = \frac{A_1}{s - a_1} + \frac{A_2}{s - a_2} + \cdots + \frac{A_n}{s - a_n} = \sum_{k=1}^{n} \frac{A_k}{(s-a_k)}$, where $A_1, A_2, \ldots, A_n$ are constants to be determined. Determine the constants A_i by multiplying both sides of the partial fraction expansion by $(s - a_i)$ and then evaluate at $s = a_i$, i.e., $A_i = \lim_{s \to a_i} (s - a_i)F(s)$. Then use of the known inverse Laplace transforms for each partial fraction term results in

$$\mathscr{L}^{-1}\left\{\frac{A_i}{s - a_i}\right\} = A_i e^{a_i x} \tag{6.110}$$

The final solution is the combination of all inverse transforms:

$$f(x) = \sum_{i=1}^{n} A_i e^{a_i x} \tag{6.111}$$

Example 6.24 (Inverse Laplace Transform Through Heaviside's Expansion) The inverse Laplace transform of

$$F(s) = \frac{3s+2}{(s+1)(s-2)} \tag{6.112}$$

Decomposing equation (6.110) into partial fractions,

$$F(s) = \frac{A_1}{s+1} + \frac{A_2}{s-2} \tag{6.113}$$

The constants A_1 and A_2 are

$$A_1 = \lim_{s\to -1}(s+1)F(s) = \lim_{s\to -1}\frac{3s+2}{s-2} = \frac{3(-1)+2}{-1-2} = \frac{-3+2}{-3} = \frac{-1}{-3} = \frac{1}{3} \tag{6.114}$$

$$A_2 = \lim_{s\to 2}(s-2)F(s) = \lim_{s\to 2}\frac{3s+2}{s+1} = \frac{3(2)+2}{2+1} = \frac{6+2}{3} = \frac{8}{3} \tag{6.115}$$

So

$$F(s) = \frac{1/3}{s+1} + \frac{8/3}{s-2} \tag{6.116}$$

Inverse Laplace transform each term:

$$\begin{aligned} \mathscr{L}^{-1}\left\{\frac{1/3}{s+1}\right\} &= \frac{1}{3}e^{-x} \\ \mathscr{L}^{-1}\left\{\frac{8/3}{s-2}\right\} &= \frac{8}{3}e^{2x} \end{aligned} \tag{6.117}$$

The sum of the inverse Laplace transform is

$$\mathscr{L}^{-1}\left\{\frac{3s+2}{(s+1)(s-2)}\right\} = f(x) = \frac{1}{3}e^{-x} + \frac{8}{3}e^{2x} \tag{6.118}$$

6.6 Derivatives and Integration of Laplace Transforms

As the Laplace transform gained prominence, its derivative and integral properties emerged as central tools for solving complex dynamic systems during the twentieth century. With the rise of electrical engineering and control theory, engineers began to work on complex electrical circuits, communication systems, and automatic control systems that often need derivative and integral properties of the Laplace transform. These properties provide a means to transform the differential operations in the time domain into simpler algebraic manipulations in the Laplace domain, making the Laplace transform a powerful method for solving differential equations.

6.6.1 Laplace Transform of Derivatives of a Function

If a function $f(x)$ is differentiable, the Laplace transform of its derivatives $f'(x)$ follows specific rules. For example, if $f'(x)$ is continuous for $x \geq 0$, then integration by parts gives

$$\begin{aligned}\mathscr{L}\left\{f'(x)\right\} &= \int_0^\infty e^{-sx} f'(x)dx = e^{-sx} f(x)\big|_0^\infty + s\int_0^\infty e^{-sx} f(x)dx \\ &= -f(0) + s\mathscr{L}\{f(x)\} \\ &= sF(s) - f(0)\end{aligned} \tag{6.119}$$

Here, it is assumed that $e^{-sx} f(x) \to 0$ as $x \to \infty$. Similarly,

$$\begin{aligned}\mathscr{L}\left\{f''(x)\right\} &= \int_0^\infty e^{-sx} f''(x)dx = e^{-sx} f'(x)\big|_0^\infty + s\int_0^\infty e^{-sx} f'(x)dx \\ &= -f'(0) + s\mathscr{L}\left\{f'(x)\right\} \\ &= s[sF(s) - f(0)] - f'(0) \\ &= s^2F(s) - sf(0) - f'(0)\end{aligned} \tag{6.120}$$

In the similar manner, it can be shown that

$$\mathscr{L}\left\{f'''(x)\right\} = s^3F(s) - s^2f(0) - sf'(0) - f''(0) \tag{6.121}$$

The recursive nature of the Laplace transform of the derivatives of a function $f(x)$ is apparent. The Laplace transform of the nth derivative of $f(x)$ becomes

$$\mathscr{L}\left\{f^{(n)}(x)\right\} = s^nF(s) - s^{n-1}f(0) - s^{n-2}f'(0) - \cdots - f^{(n-1)}(0) \tag{6.122}$$

where $F(s) = \mathscr{L}\{f(x)\}$. This property is particularly useful for solving higher-order differential equations using Laplace transforms.

Example 6.25 (First-Order ODE)

$$f'(x) + f(x) = 0, \quad f(0) = 2$$

Taking the Laplace transform of both sides,

$$sF(s) - f(0) + F(s) = 0$$

Substitute $f(0) = 2$

$$sF(s) - 2 + F(s) = 0 \quad \Rightarrow \quad (s+1)F(s) = 2$$

$$F(s) = \frac{2}{s+1}$$

Taking the inverse Laplace transform,

$$f(x) = 2e^{-x}$$

Example 6.26 (Second-Order Homogeneous ODE)

$$f''(x) + 4f(x) = 0, \quad f(0) = 0, \quad f'(0) = 1$$

Taking the Laplace transform of both sides,

$$s^2F(s) - sf(0) - f'(0) + 4F(s) = 0$$

Substituting $f(0) = 0$ and $f'(0) = 1$:

$$s^2F(s) - 1 + 4F(s) = 0 \quad \Rightarrow \quad (s^2+4)F(s) = 1$$

$$F(s) = \frac{1}{s^2+4}$$

Taking the inverse Laplace transform,

$$f(x) = \frac{1}{2}\sin(2x)$$

Example 6.27 (Second-Order ODE with Damping)

$$f''(x) + 3f'(x) + 2f(x) = 0, \quad f(0) = 1, \quad f'(0) = 0$$

Taking the Laplace transform of both sides,

$$s^2F(s) - sf(0) - f'(0) + 3(sF(s) - f(0)) + 2F(s) = 0$$

Substitute $f(0) = 1$ and $f'(0) = 0$:

$$s^2F(s) - s + 3(sF(s) - 1) + 2F(s) = 0$$

On simplifying

$$(s^2 + 3s + 2)F(s) = s - 3$$

$$F(s) = \frac{s-3}{s^2 + 3s + 2} = \frac{(s-3)}{(s+1)(s+2)}$$

Partial fraction decomposition:

$$F(s) = \frac{A}{(s+1)} + \frac{B}{(s+2)}$$

Solving for A and B,

$$F(s) = \frac{-2}{s+1} + \frac{3}{s+2}$$

Taking the inverse Laplace transform,

$$f(x) = -2e^{-t} + 3e^{-2x}$$

Example 6.28 (First-Order Nonhomogeneous ODE)

$$f'(x) + f(x) = e^{2x}, \quad f(0) = 1$$

Taking the Laplace transform of both sides,

(continued)

Example 6.28 (continued)

$$sF(s) - f(0) + F(s) = \frac{1}{s-2}$$

Substituting $y(0) = 1$,

$$sF(s) - 1 + F(s) = \frac{1}{s-2} \quad \Rightarrow \quad (s+1)F(s) = 1 + \frac{1}{s-2}$$

On simplifying,

$$F(s) = \frac{1}{s+1} + \frac{1}{(s+1)(s-2)}$$

Using partial fraction decomposition,

$$F(s) = \frac{1}{s+1} + \frac{A}{s+1} + \frac{B}{s-2}$$

Solving for A and B,

$$F(s) = \frac{1}{s+1} + \frac{1}{3}\left(\frac{1}{s-2} - \frac{1}{s+1}\right)$$

Taking the inverse Laplace transform,

$$f(x) = e^{-x} + \frac{1}{3}(e^{2x} - e^{-x})$$

On simplifying,

$$f(x) = \frac{2}{3}e^{-x} + \frac{1}{3}e^{2x}$$

Example 6.29 (Second-Order Nonhomogeneous ODE)

$$f''(x) + f(x) = \cos(x), \quad y(0) = 0, \quad f'(0) = 0$$

Taking the Laplace transform of both sides,

$$s^2F(s) - sf(0) - f'(0) + F(s) = \frac{s}{s^2+1}$$

(continued)

Example 6.29 (continued)
Substituting $f(0) = 0$ and $f'(0) = 0$,

$$s^2F(s) + F(s) = \frac{s}{s^2+1} \quad \Rightarrow \quad (s^2+1)F(s) = \frac{s}{s^2+1}$$

Solving for $F(s)$,

$$F(s) = \frac{s}{(s^2+1)^2}$$

Taking the inverse Laplace transform,

$$f(x) = \frac{1}{2}x\sin(x)$$

Example 6.30 (Second-Order Nonhomogeneous ODE)

$$\frac{d^2f(x)}{dx^2} + 3\frac{df(x)}{dx} + 2f(x) = \cos(2x)$$

With initial conditions $f(0) = 0$ and $f'(0) = 0$. Applying the Laplace transform to both sides,

$$\mathscr{L}\left\{\frac{d^2f(x)}{dx^2}\right\} + 3\mathscr{L}\left\{\frac{df(x)}{dx}\right\} + 2\mathscr{L}\{f(x)\} = \mathscr{L}\{\cos(2x)\}$$

Using the differentiation properties of the Laplace transform,

$$s^2F(s) - sf(0) - f'(0) + 3[sF(s) - f(0)] + 2F(s) = \frac{s}{s^2+4}$$

Substituting $f(0) = 0$ and $f'(0) = 0$,

$$s^2F(s) + 3sF(s) + 2F(s) = \frac{s}{s^2+4}$$

Simplifying it,

$$F(s) = \frac{s}{(s^2+4)(s^2+3s+2)}$$

(continued)

Example 6.30 (continued)
Now performing partial fraction decomposition to the right side and taking the inverse Laplace transform,

$$f(x) = \mathscr{L}^{-1}\left\{\frac{-s}{20\left(s^2+4\right)} + \frac{6}{20\left(s^2+4\right)} - \frac{1}{5\left(s+1\right)} + \frac{1}{4\left(s+2\right)}\right\}$$

$$= -\frac{1}{20}\mathscr{L}^{-1}\left\{\frac{s}{s^2+4}\right\} + \frac{6}{20}\mathscr{L}^{-1}\left\{\frac{1}{s^2+4}\right\} - \frac{1}{5}\mathscr{L}^{-1}\left\{\frac{1}{s+1}\right\}$$

$$+ \frac{1}{4}\mathscr{L}^{-1}\left\{\frac{1}{s+2}\right\}$$

$$= -\frac{1}{20}\cos\left(2x\right) + \frac{3\sin\left(2x\right)}{20} - \frac{1}{5}e^{-x} + \frac{1}{4}e^{-2x}$$

Example 6.31 (First-Order ODE with Discontinuous Forcing Function)

$$f'(x) + 2f(x) = \mathcal{H}(x-1), \quad f(0) = 0$$

$\mathcal{H}(x)$ represents the Heaviside function. Taking the Laplace transform of both sides,

$$sF(s) - f(0) + 2F(s) = \frac{e^{-s}}{s}$$

Substituting $f(0) = 0$,

$$(s+2)F(s) = \frac{e^{-s}}{s}$$

Solving for $F(s)$,

$$F(s) = \frac{e^{-s}}{s(s+2)}$$

Using partial fraction decomposition,

$$F(s) = \frac{A}{s} + \frac{B}{s+2}$$

(continued)

Example 6.31 (continued)
Solving for A and B, and taking the inverse Laplace transform,

$$f(x) = \frac{1}{2}\left(1 - e^{-2(x-1)}\right)\mathcal{H}(x-1)$$

Example 6.32 (Second-Order ODE with Resonance)

$$f''(x) + 4f(x) = \cos(2x), \quad y(0) = 0, \quad y'(0) = 0$$

Taking the Laplace transform,

$$s^2F(s) + 4F(s) = \frac{s}{s^2+4}$$

Solving for $F(s)$,

$$F(s) = \frac{s}{(s^2+4)^2}$$

Using the inverse Laplace transform,

$$f(x) = \frac{x}{4}\sin(2x)$$

Example 6.33 (First-Order Linear ODE)

$$f'(x) + 3f(x) = 2, \quad f(0) = 0$$

Taking the Laplace transform,

$$sF(s) + 3F(s) = \frac{2}{s}$$

Solving for $F(s)$,

$$F(s) = \frac{2}{s(s+3)}$$

(continued)

Example 6.33 (continued)
Using partial fractions,

$$F(s) = \frac{2}{3}\left(\frac{1}{s} - \frac{1}{s+3}\right)$$

Taking the inverse Laplace transform,

$$f(x) = \frac{2}{3}(\mathcal{H}(x) - e^{-3x})$$

Example 6.34 (Second-Order ODE with Delta Function)

$$f''(x) + 2f'(x) + f(x) = \delta(x-1), \quad y(0) = 0, \quad y'(0) = 0$$

Taking the Laplace transform,

$$s^2F(s) + 2sF(s) + F(s) = e^{-s}$$

Solving for $F(s)$,

$$F(s) = \frac{e^{-s}}{(s+1)^2}$$

Taking the inverse Laplace transform,

$$f(x) = (x-1)e^{-(x-1)}\mathcal{H}(x-1)$$

Example 6.35 (First-Order Nonhomogeneous ODE with Exponential Forcing)

$$f'(x) + 2f(x) = e^{-x}, \quad f(0) = 0$$

Taking the Laplace transform,

$$sF(s) + 2F(s) = \frac{1}{s+1}$$

(continued)

Example 6.35 (continued)
Solving for $F(s)$,

$$F(s) = \frac{1}{(s+1)(s+2)}$$

Using partial fractions,

$$F(s) = \frac{1}{s+1} - \frac{1}{s+2}$$

Taking the inverse Laplace transform,

$$f(x) = e^{-x} - e^{-2x}$$

6.6.2 Laplace Transform of the Integral of a Function

Differentiation and integration are inverse operations and so are multiplication and division. Since differentiation of a function $f(x)$ corresponds to the multiplication of its transform $\mathscr{L}\{f(x)\}$ by s, it is expected that the integration of $f(x)$ to correspond to the division of $\mathscr{L}\{f(x)\}$ by s. Let $F(s)$ denote the transform of a function $f(x)$, which is piecewise continuous for $x \geqq 0$ and satisfies a growth restriction. $g(x)$ is continuous, as $f(x)$ is piecewise continuous. Then, for $s > 0, s > k$, and $x > 0$,

$$|g(x)| = \left| \int_0^x f(\tau)d\tau \right| \leqq \int_0^x |f(\tau)|d\tau \leqq M \int_0^x e^{k\tau}d\tau = \frac{M}{k}\left(e^{kt} - 1\right) \leqq \frac{M}{k}e^{kt} \tag{6.123}$$

for $(k > 0)$. This shows that $g(x)$ also satisfies a growth restriction. Also, $g'(x) = f(x)$, except at points at which $f(x)$ is discontinuous. Hence, $g'(x)$ is piecewise continuous on each finite interval. Since $g(0) = 0$ (the integral from 0 to 0 is zero),

$$\mathscr{L}\{f(x)\} = \mathscr{L}\left\{g'(x)\right\} = s\mathscr{L}\{g(x)\} - g(0) = s\mathscr{L}\{g(x)\} \tag{6.124}$$

Dividing by s and interchanging of the left and right sides gives

$$\mathscr{L}\left\{\int_0^x f(\tau)d\tau\right\} = \frac{1}{s}F(s) \tag{6.125}$$

By taking the inverse Laplace transform on both sides,

$$\int_0^x f(\tau)d\tau = \mathscr{L}^{-1}\left\{\frac{1}{s}F(s)\right\} \tag{6.126}$$

Fredholm integral equations, named after Swedish mathematician Erik Ivar Fredholm (1866–1927), are often encountered in physics and engineering, particularly in problems involving boundary conditions or when describing systems with a fixed interaction range. For Fredholm integral equations of the first kind, the unknown function appears only inside the integral equation:

$$f(x) = \int_a^b K(x,t)\phi(t)\,dt \tag{6.127}$$

The unknown function appears both inside and outside the integral for the Fredholm integral equation of the second kind:

$$\phi(x) = \lambda \int_a^b K(x,t)\phi(t)\,dt + g(x) \tag{6.128}$$

$f(x)$ and $g(x)$ are the given functions and $\phi(t)$ is the unknown function to be solved for, and $K(x,t)$ is the kernel of the integral equation. The parameter λ is a constant, and the limits a and b are fixed.

Volterra integral equation is named after Italian mathematician Vito Volterra (1860–1940). It differs from the Fredholm type in that the upper limit of integration is a variable, usually dependent on the independent variable x. This makes the Volterra equation more suited to modeling processes where the interaction or evolution depends on time or another evolving parameter. Similar to Fredholm equations, Volterra equations are classified into two types. Volterra integral equation of the first kind is

$$f(x) = \int_a^x K(x,t)\phi(t)\,dt \tag{6.129}$$

Volterra integral equation of the second kind is

$$\phi(x) = \int_a^x K(x,t)\phi(t)\,dt + g(x) \tag{6.130}$$

Volterra equations are widely used to model processes in biology, physics, and economics, particularly when the system has memory or past influences, such as population dynamics or viscoelasticity.

Example 6.36 (Integral Equation)

$$f(x) = \int_0^x e^{-2\tau}\, d\tau$$

Applying the Laplace transform to both sides,

$$\mathscr{L}\{f(x)\} = \mathscr{L}\left\{\int_0^x e^{-2\tau}\, d\tau\right\}$$

Using the integral property of the Laplace transform,

$$F(s) = \frac{\mathscr{L}\{e^{-2x}\}}{s} = \frac{1}{s(s+2)}$$

Taking the inverse Laplace transform through partial fraction,

$$f(x) = 1 - e^{-2x}$$

Example 6.37 (Fredholm Integral Equation of the Second Kind)

$$f(x) = 2 + \int_0^x f(\tau)\, d\tau$$

Taking the Laplace transform of both sides,

$$\mathscr{L}\{f(x)\} = \mathscr{L}\{2\} + \mathscr{L}\left\{\int_0^x f(\tau)\, d\tau\right\}$$

Using properties of the Laplace transform,

$$F(s) = \frac{2}{s} + \frac{F(s)}{s}$$

Rearranging the equation,

$$F(s)\left(1 - \frac{1}{s}\right) = \frac{2}{s}$$

(continued)

Example 6.37 (continued)
Simplifying,

$$F(s) = \frac{2}{(s-1)}$$

Taking the inverse Laplace transform,

$$f(x) = 2e^x$$

Example 6.38 (Volterra Integral Equation of the First Kind)

$$\int_0^x f(\tau)\,d\tau = \sin(x)$$

Taking the Laplace transform of both sides,

$$\mathcal{L}\left\{\int_0^x f(\tau)\,d\tau\right\} = \mathcal{L}\{\sin(x)\}$$

$$\frac{F(s)}{s} = \frac{1}{s^2+1}$$

Multiplying both sides by s,

$$F(s) = \frac{s}{s^2+1}$$

Taking the inverse Laplace transform,

$$f(x) = \cos(x)$$

Example 6.39 (Fredholm Integral Equation of the First Kind)

$$\int_0^x (x-\tau)f(\tau)\,d\tau = x^2$$

Taking the Laplace transform of both sides,

(continued)

Example 6.39 (continued)

$$\mathscr{L}\left\{\int_0^t (t-\tau)f(\tau)\,d\tau\right\} = \mathscr{L}\{x^2\}$$

Using the convolution property,

$$\mathscr{L}\{x * f(x)\} = \frac{F(s)}{s^2}$$

So the equation becomes

$$\frac{F(s)}{s^2} = \frac{2}{s^3}$$

Multiplying both sides by s^2,

$$F(s) = \frac{2}{s}$$

Taking the inverse Laplace transform,

$$f(x) = 2$$

Example 6.40 (Volterra Integral Equation of the Second Kind)

$$f(x) = 1 + \int_0^x e^{(x-\tau)} f(\tau)\,d\tau$$

Taking the Laplace transform of both sides,

$$F(s) = \frac{1}{s} + \mathscr{L}\left\{\int_0^x e^{(x-\tau)} f(\tau)\,d\tau\right\}$$

Using the convolution property,

$$\mathscr{L}\left\{e^{(x)} * f(x)\right\} = \frac{1}{s-1}F(s)$$

So the equation becomes

(continued)

Example 6.40 (continued)

$$F(s) = \frac{1}{s} + \frac{F(s)}{s-1}$$

Rearranging,

$$F(s)\left(1 - \frac{1}{s-1}\right) = \frac{1}{s}$$

Simplifying,

$$F(s) = \frac{(s-1)}{s(s-2)}$$

Taking the inverse Laplace transform using the partial fraction,

$$f(x) = \frac{1}{2}(1 + e^{2x})$$

Example 6.41 (Fredholm Integral Equation with Kernel)

$$f(x) = x + \int_0^x \tau f(\tau)\, d\tau$$

Taking the Laplace transform of both sides,

$$F(s) = \frac{1}{s^2} + \mathscr{L}\left\{\int_0^x \tau f(\tau)\, d\tau\right\}$$

Using the convolution property,

$$\mathscr{L}\{\tau f(\tau)\} = \frac{F(s)}{s^2}$$

So the equation becomes

$$F(s) = \frac{1}{s^2} + \frac{F(s)}{s^2}$$

(continued)

Example 6.41 (continued)
Rearranging,

$$F(s)\left(1-\frac{1}{s^2}\right)=\frac{1}{s^2}$$

Solving for $F(s)$,

$$F(s)=\frac{1}{s^2-1}=\frac{1}{2}\left(\frac{1}{s-1}-\frac{1}{s+1}\right)$$

Taking the inverse Laplace transform,

$$f(x)=\frac{1}{2}(e^t-e^{-t})$$

6.6.3 Derivative of the Laplace Transform of a Function

If $f(x)$ and $F(s)$ are Laplace transform pairs, and $F(s)$ sufficiently smooth and of exponential order, then differentiating $F(s)$ with respect to s,

$$\frac{d}{ds}F(s)=\frac{d}{ds}\left(\int_0^\infty e^{-sx}f(x)\,dx\right) \tag{6.131}$$

Since the differentiation is with respect to s, the derivative can move inside the integral:

$$\begin{aligned}\frac{d}{ds}F(s)&=\int_0^\infty \frac{d}{ds}\left(e^{-sx}f(x)\right)\,dx\\&=\int_0^\infty -xf(x)e^{-sx}\,dx\\&=-\mathscr{L}\{xf(x)\}\end{aligned} \tag{6.132}$$

This result can be extended to higher-order derivatives of the Laplace transform with respect to s. For the n-th derivative of $F(s)$, the same differentiation process is applied iteratively:

$$\frac{d^n}{ds^n}F(s)=(-1)^n\mathscr{L}\{x^nf(x)\} \tag{6.133}$$

So, the n-th derivative of the Laplace transform corresponds to the Laplace transform of $(-1)^n x^n f(x)$. Taking the inverse Laplace of both sides,

$$\mathcal{L}^{-1}\{\frac{d^n}{ds^n}F(s)\} = (-1)^n x^n f(x) \tag{6.134}$$

For $n = 1$,

$$\mathcal{L}^{-1}\{F'(s)\} = -xf(x) \tag{6.135}$$

In this way, the differentiation of the transform of a function $f(x)$ corresponds to the multiplication of the function by x. This is useful in control theory.

Example 6.42 (Derivative of Laplace Transform)

$$F(s) = \frac{1}{s+1}$$

Differentiating $F(s)$,

$$\frac{d}{ds}\left(\frac{1}{s+1}\right) = -\frac{1}{(s+1)^2}$$

Applying the inverse Laplace transform using the known transform pair,

$$\mathcal{L}^{-1}\left\{-\frac{1}{(s+1)^2}\right\} = xe^{-x}$$

Thus,

$$f(x) = e^{-x}$$

6.6.4 *Integral of the Laplace Transform of a Function*

If $f(x)$ and $F(s)$ are Laplace transform pairs, and the limit of $\dfrac{f(x)}{x}$ exists and as x approaches 0 from the right, then for $s > k$, from the definition it follows that

$$\begin{aligned}\int_s^\infty F(\tilde{s})d\tilde{s} &= \int_s^\infty \left[\int_0^\infty e^{-\tilde{s}x} f(x)dx\right] d\tilde{s} \\ &= \int_0^\infty \left[\int_s^\infty e^{-\tilde{s}x} f(x)d\tilde{s}\right] dx \\ &= \int_0^\infty f(x) \left[\int_s^\infty e^{-\tilde{s}x} d\tilde{s}\right] dx\end{aligned} \tag{6.136}$$

Integration of $e^{-\tilde{s}x}$ with respect to $\tilde{s}$ gives $-\dfrac{e^{-\tilde{s}x}}{x}$. Here the integral over $\tilde{s}$ on the right equals $\dfrac{e^{-sx}}{x}$. Therefore,

$$\int_s^\infty F(\tilde{s})d\tilde{s} = \int_0^\infty e^{-sx} \frac{f(x)}{x} dx = \mathscr{L}\left\{\frac{f(x)}{x}\right\} \quad (s > k) \tag{6.137}$$

Taking the inverse Laplace transform of both sides of Eq. (6.133),

$$\mathscr{L}^{-1}\left\{\int_s^\infty F(\tilde{s})d\tilde{s}\right\} = \frac{f(x)}{x} \tag{6.138}$$

The integration of the Laplace transform with respect to s leads to a complex expression, often involving terms proportional to $\dfrac{f(x)}{x}$. For specific functions $f(x)$, it is possible to compute the exact result, but in general, this expression cannot be reduced to a simple formula without knowing $f(x)$ explicitly. Thus, the integration of $F(s)$ with respect to s involves time-domain behavior in terms of the original function $f(x)$ and its integral.

6.7 Initial and Final Value Theorems of the Laplace Transform

The initial value theorem and the final value theorem of the Laplace transform are two important theorems that provide information about the behavior of a time-domain function $f(x)$ at $x = 0$ (initial time) and as $x \to \infty$ (final time). These theorems are particularly useful in control systems and signal processing to predict system behavior without performing the full inverse Laplace transform.

6.7.1 Initial Value Theorem (IVT)

The initial value theorem allows us to determine the value of a function $f(t)$ at $t = 0^+$ (just after $t = 0$) directly from its Laplace transform $F(s)$, without performing the inverse Laplace transform. If $f(x)$ and $F(s)$ are two Laplace pair, then multiplying both sides with s,

$$sF(s) = s\int_0^\infty e^{-sx} f(x)\,dx \tag{6.139}$$

As $s \to \infty$, the exponential term e^{-sx} decays very rapidly for any $x > 0$, so most of the contribution to the integral comes from the behavior of $f(x)$ very close to $x = 0$. Thus, for large s, the integral approximates $f(0)$, giving

$$\lim_{s\to\infty} sF(s) = \lim_{t\to 0^+} f(x) \tag{6.140}$$

The limit $\lim_{x\to 0^+} f(x)$ must exist and the function $f(x)$ must not have an impulse (Dirac delta function) or a discontinuity at $x = 0$. For example, let $f(x) = 5e^{-3x}$. The Laplace transform is

$$F(s) = \frac{5}{s+3}$$

Using the initial value theorem,

$$\lim_{x\to 0^+} f(x) = \lim_{s\to\infty} sF(s) = \lim_{s\to\infty} s\cdot\frac{5}{s+3} = \lim_{s\to\infty}\frac{5s}{s+3} = 5$$

So, $f(0^+) = 5$, which is consistent with the time-domain expression $f(t) = 5e^{-3t}$, where $f(0) = 5$.

6.7.2 Final Value Theorem (FVT)

The final value theorem allows us to determine the steady-state (long-term) behavior of $f(x)$ as $x \to \infty$ directly from its Laplace transform $F(s)$. In Eq. (6.139), as $s \to 0$, $e^{-sx} \to 1$, so the integral approaches $\int_0^\infty f(x)\,dx$, which corresponds to the steady-state value of $f(x)$ as $x \to \infty$). Thus,

$$\lim_{s\to 0} sF(s) = \lim_{t\to\infty} f(t) \tag{6.141}$$

$f(x)$ must tend to a finite value as $x \to \infty$. If $f(x)$ oscillates or grows without bound, the final value theorem is not applicable. All poles of $F(s)$ must be in the left half of the complex plane (i.e., they must have negative real parts). If there are poles on the right half-plane or on the imaginary axis (except at $s = 0$), the final value theorem does not apply.

For example, let $f(x) = 5e^{-3x}$. The Laplace transform is

$$F(s) = \frac{5}{s+3}$$

Using the final value theorem,

$$\lim_{x \to \infty} f(x) = \lim_{s \to 0} sF(s) = \lim_{s \to 0} s \cdot \frac{5}{s+3} = \lim_{s \to 0} \frac{5s}{s+3} = 0$$

Thus, $f(x) \to 0$ as $x \to \infty$, which makes sense because $5e^{-3x} \to 0$ as $x \to \infty$.

6.8 Laplace Transform for System Modeling

Figure 6.5 shows the block diagram, which gives the graphical representation of the system. A system can be an open-loop or a closed-loop system. An open-loop

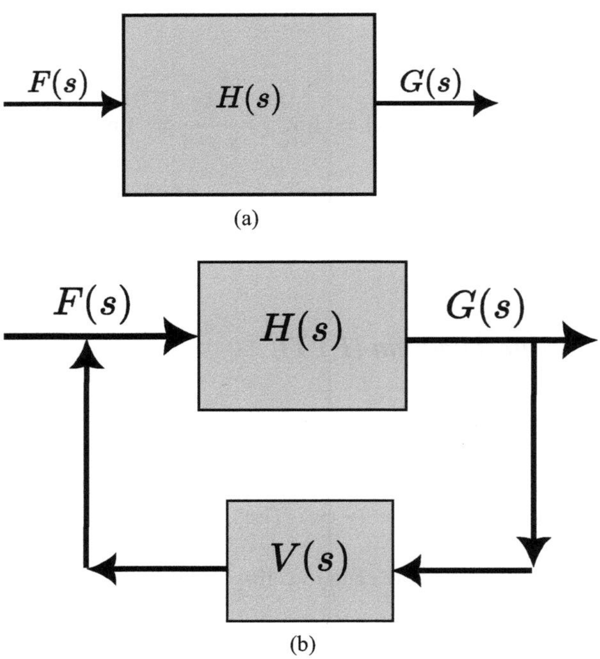

Fig. 6.5 Block diagram of a system. (**a**) Open-loop system. (**b**) Closed-loop system

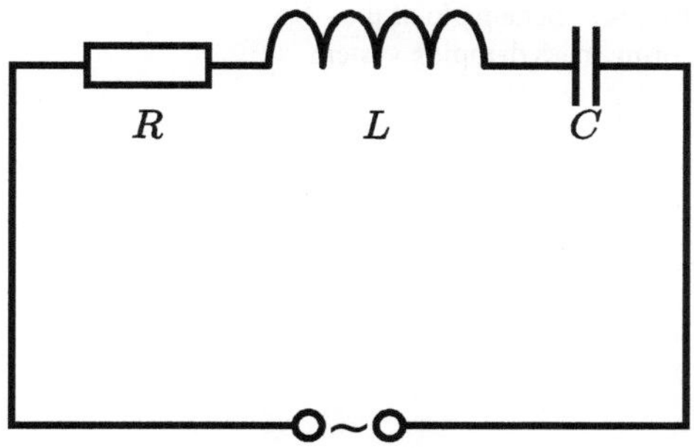

Fig. 6.6 Schematic of the RLC circuit

system is a control system where the output is not fed back to the input for correction or adjustment (Fig. 6.5a). In such systems, the control action is independent of the desired output. This means the system operates based on a set input, and there is no mechanism to adjust for disturbances or errors. On the contrary, a closed-loop system is a control system in which the output is continuously monitored, and a part of it is fed back to the input (Fig. 6.5b). This feedback allows the system to adjust and correct its operation based on the difference between actual and desired outputs. These systems are designed to reduce errors and improve accuracy and stability. The system is fed with an input signal $f(x)$ and provides the output signal $g(x)$. A transfer function is a mathematical representation of mapping the output to the input signal in the frequency domain. It is typically used to analyze electrical, mechanical, and control systems to determine how the system responds to different inputs, especially in frequency. The transfer function $H(s)$ is defined as the ratio of the Laplace transform of the output $G(s)$ to the Laplace transform of the input $F(s)$, assuming all initial conditions are zero:

$$H(s) = \frac{G(s)}{F(s)} \tag{6.142}$$

s is the complex frequency variable in the Laplace domain, which is used to analyze the system's behavior in terms of both frequency ($i\omega$) and exponential growth or decay (σ). The transfer function provides insight into how a system reacts to different input frequencies. For example, it shows whether the system amplifies, attenuates, or phase shifts certain frequencies. Zeros are the values of s that make the transfer function $H(s) = 0$. Poles are the values of s that make the transfer function blow up to infinity.

Figure 6.6 shows the schematic diagram for a series RLC circuit. For a series RLC circuit, the differential equation relating the input voltage $f(x)$ to the output voltage $g(x)$ across the capacitor is

$$L\frac{d^2g(x)}{dx^2} + R\frac{dg(x)}{dx} + \frac{1}{C}g(x) = f(x) \tag{6.143}$$

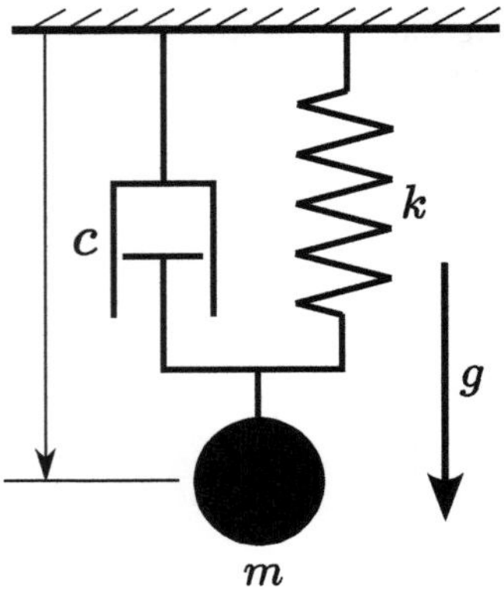

Fig. 6.7 Schematic of the spring mass damping system

L is the inductance, R is the resistance, and C is the capacitance. Applying the Laplace transform to Eq. (6.143),

$$Ls^2G(s) + RsG(s) + \frac{1}{C}G(s) = F(s) \tag{6.144}$$

The transform function for this system is

$$H(s) = \frac{G(s)}{F(s)} = \frac{1}{Ls^2 + Rs + \dfrac{1}{C}} \tag{6.145}$$

Figure 6.7 shows the schematic diagram for a spring mass damping system. The differential equation of a mass-spring-damper system is expressed as

$$m\frac{d^2g(x)}{dx^2} + c\frac{dg(x)}{dx} + kg(x) = f(x) \tag{6.146}$$

m is the mass, c is the damping coefficient, and k is the spring constant. $f(x)$ is the input force and $g(x)$ is the displacement. Taking Laplace transform of Eq. (6.146), assuming zero initial conditions,

$$ms^2G(s) + csG(s) + kG(s) = F(s) \tag{6.147}$$

The transfer function for this system is

$$H(s) = \frac{G(s)}{F(s)} = \frac{1}{ms^2 + cs + k} \tag{6.148}$$

The transfer functions are used to study the system's different features and help model the system. System stability refers to the ability of a system to return to equilibrium after being disturbed. In control systems, a stable system is one where the output remains bounded for a bounded input. Conversely, if small perturbations or inputs cause the output to diverge, the system is considered unstable. The locations of poles and zeros in the complex plane are critical for determining system stability and behavior. The location of the transfer function poles in the complex plane determines the system's stability. If real parts of all the system poles are negative, i.e., located on the left half of the complex plane, then any disturbance given to the system exponentially damps out, and the system remains stable. But if any of the poles of the system has a positive real part, then even infinitesimal small disturbance exponentially grows up and makes the system unstable. If some poles of the system have zero real parts, then the system is marginally stable. The system exhibits sustained oscillations but does not grow unbounded over time.

The step response of a system is the output when the input is a unit step function (Heaviside function). The output of a system to a step response is expressed as

$$G(s) = H(s)\frac{1}{s} \tag{6.149}$$

If transfer function for a given system is $\frac{1}{(s^2 + 3s + 2)}$, then the corresponding step response is

$$G(s) = \frac{1}{(s^2 + 3s + 2)}\frac{1}{s} \tag{6.150}$$

The inverse Laplace transform determines the system's output to this step response in the physical domain:

$$g(x) = \frac{1}{2} - e^{-x} + \frac{1}{2}e^{-2x} \tag{6.151}$$

Figure 6.8 shows the graphical representation of the step response of the system. This helps to visualize the behavior of the system over time.

The frequency response of a system describes how the system reacts to sinusoidal inputs of varying frequencies. The frequency response of the system is obtained by evaluating its transfer function $H(s)$on the imaginary axis. For the given transfer function, the frequency response is

$$G(i\omega) = \frac{1}{(-\omega^2 + 3i\omega + 2)} \tag{6.152}$$

The magnitude of the frequency response is

$$|H(i\omega)| = \frac{1}{\sqrt{(-\omega^2 + 2)^2 + (3\omega)^2}} \tag{6.153}$$

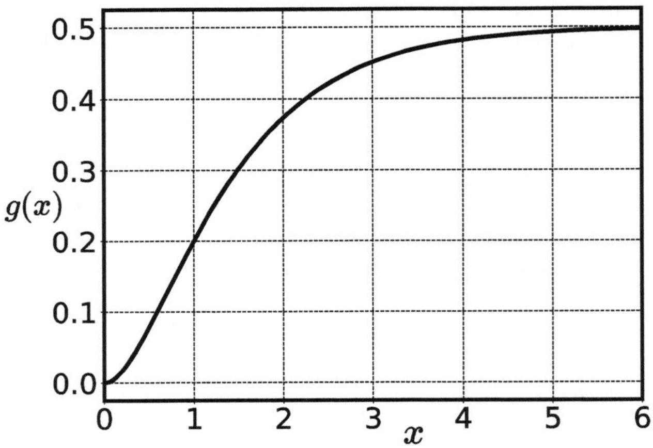

Fig. 6.8 Step response of a system in time domain

Phase for the frequency response is

$$\angle H(i\omega) = \tan^{-1}\left(\frac{3\omega}{2-\omega^2}\right) \tag{6.154}$$

As $\omega \rightarrow 0$, $|H(i\omega)| \approx \frac{1}{2}$, and $\angle H(j\omega) \approx 0°$. The system behaves like passing low-frequency signals with a gain and no phase shift. When $\omega \rightarrow \infty$, the system attenuates high-frequency signals and introduces a phase shift of $-180°$, meaning the output is inverted compared to the input.

Swedish-American physicist Harry Nyquist (1889–1976) proposed a chart for visualization of the transfer function in the complex domain, known as the Nyquist chart. This plot shows how the frequency response (magnitude and phase) of the open-loop transfer function $H(i\omega)$ behaves in the complex plane as the frequency ω varies from $-\infty$ to ∞. The magnitude and phase of the transfer function are computed from Eqs. (6.153) and (6.154), respectively. For each value of frequency $\infty < \omega < -\infty$, the real part $\Re\{H(i\omega)\}$ is plotted in the real axis and the imaginary part $\Im\{H(i\omega)\}$ along the imaginary axis in the complex plane. Figure 6.9 shows the Nyquist chart for the transfer function $\frac{1}{(s^2+3s+2)}$. Any crossing of the negative real axis $-1+0i$ is marked in the above plot. This point is crucial for determining the stability of the system using the Nyquist stability criterion:

$$Z = P + N \tag{6.155}$$

P is the number of unstable poles of the open-loop system. N is the number of clockwise encirclements of the point $-1+0i$ in the Nyquist plot. Z is the number of poles of the closed-loop transfer function in the right half-plane. According to

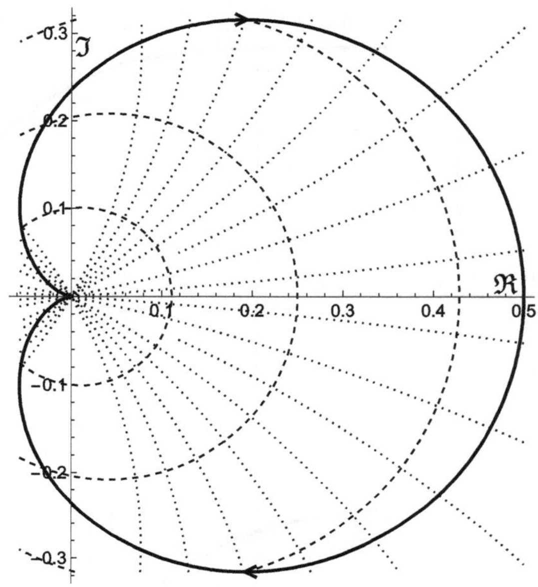

Fig. 6.9 Nyquist plot for the transfer function $\dfrac{1}{(s^2+3s+2)}$

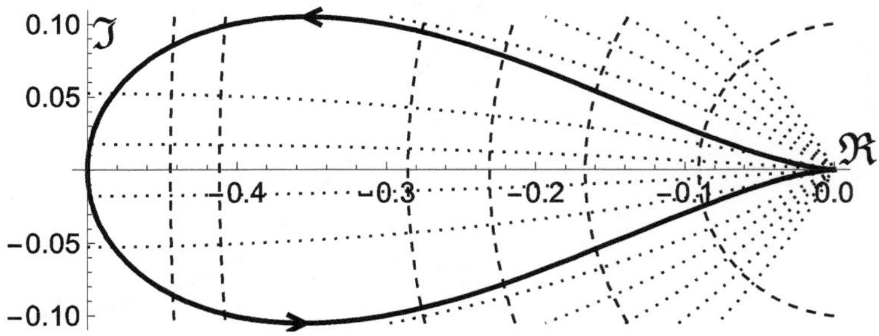

Fig. 6.10 Nyquist plot for the transfer function $\dfrac{1}{(s-1)(s+2)}$

the Nyquist stability criterion, the closed-loop system is stable if the number of encirclements of $-1 + 0i$ in the clockwise direction equals the number of unstable poles in the open-loop transfer function ($Z = 0$). The closed-loop system is unstable for $Z > 0$. Let us consider another open-loop transfer function $\dfrac{1}{(s-1)(s+2)}$, which has an unstable pole at $s = 1$ and gives the impression that the system is unstable. Figure 6.10 shows the Nyquist chart for the above transfer function. An anticlockwise encirclement can be seen in the plot along the negative real axis. For this system, $Z = 0$, so there are no closed-loop poles in the right half-plane, and the closed-loop system is stable despite the open-loop system having an unstable pole.

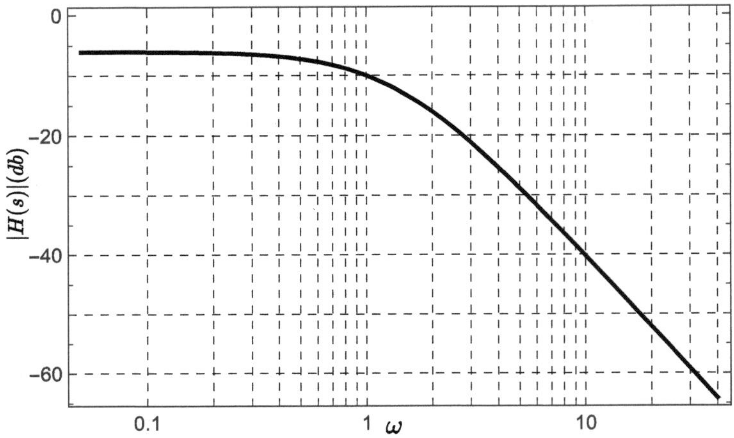

(a)

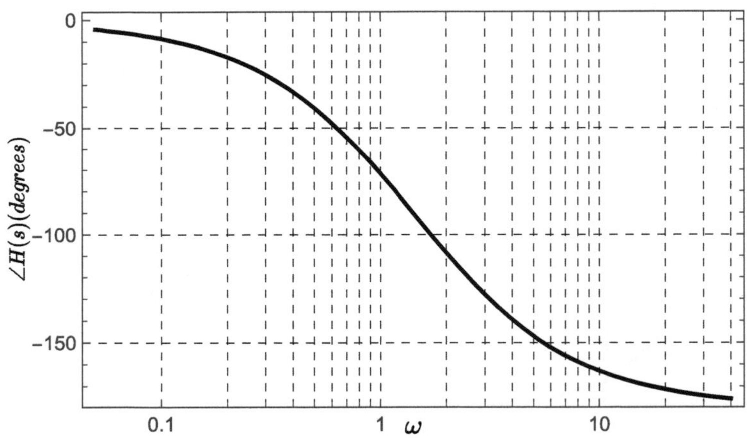

(b)

Fig. 6.11 Bode plot for the transfer function $\dfrac{1}{(s^2 + 3s + 2)}$. (**a**) Magnitude. (**b**) Phase

The Bode diagram is another graphical representation of the transfer function and is named after Dutch ancestry American engineer Hendrik Wade Bode (1905–1982). The Bode diagram consists of two superimposed graphs in which the amplitude and the phase shift computed from Eqs. (6.153) and (6.154) are plotted as a function of frequency ω. The magnitude plot typically shows the gain in decibels (dB) versus the logarithm of the frequency, while the phase plot shows the phase shift in degrees versus the logarithm of the frequency. Figure 6.11 shows the Bode plot for $\dfrac{1}{(s^2 + 3s + 2)}$.

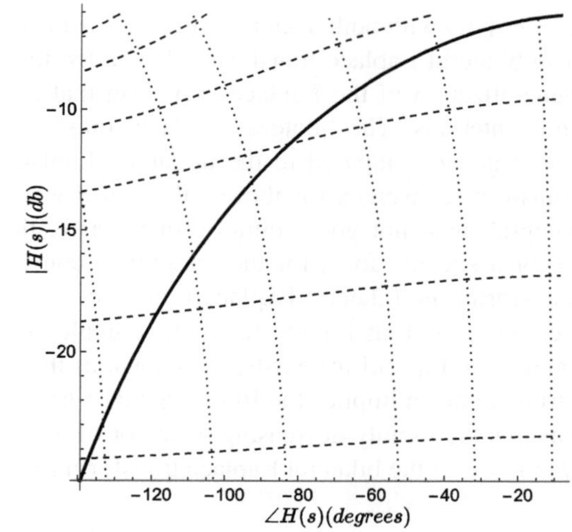

Fig. 6.12 Nichols plot for the transfer function $\dfrac{1}{(s^2+3s+2)}$

American control system engineer Nathaniel B. Nichols (1914–1997) introduced a plot with the transfer function's phase in abscissa and magnitude inordinate. This plot is known as the Nichols plot. It combines the gain (magnitude) and phase information of a system's open-loop transfer function into a single plot, making it useful for stability and performance analysis, especially in feedback control systems. Nichols plot represents the variation of both gain and phase as the frequency ω varies, but unlike Bode plots, the frequency itself is not directly plotted. Instead, it is implicit in the curve's shape. The gain margin is the vertical distance between the curve and the critical $-180°$ phase line when the gain crosses 0 dB. The phase margin is the horizontal distance between the curve and the 0 dB gain line when the phase crosses $-180°$. The Nichols plot provides a visual way to assess the impact of phase and gain changes on system stability and performance. Figure 6.12 shows the Nichols plot for the transfer function $\dfrac{1}{(s^2+3s+2)}$.

6.9 The Bilateral Laplace Transform

Van der Pol and Bremmer (1950) derived the bilateral Laplace transform for a function $f(x)$, defined for $-\infty < x < \infty$ as

$$\mathscr{L}\{f(x)\} = \int_{-\infty}^{\infty} f(x)e^{-sx}\,dx = F(s) \tag{6.156}$$

$s = \sigma + i\omega$ with both σ and ω are real. x is the real variable in the physical domain. The bilateral Laplace transform, also called the two-sided Laplace transform, is a generalization of the Laplace transform that considers both positive and negative time intervals. The bilateral Laplace transform requires stringent conditions for convergence compared to the unilateral Laplace transform. The bilateral Laplace transform converges for those values of s where the integral exists, meaning the integral must not go to infinity. In practice, this requires that the function $f(x)$ be of a specific form for the integral to converge. This implies, like the Fourier transform, the bilateral Laplace transform exists for rapidly decreasing functions, i.e., $f(x) \to 0$ as $|x| \to \infty$. In the complex plane, the values of s, for which the integral in Eq. (6.156) exists, are known as the region of convergence (ROC). This stringent norm implies the ROC has that s having positive σ. The positive real part ensures the rapidly decreasing behavior of $f(x)e^{-sx}$. For example, let us consider $f(x) = e^{ax}$; the bilateral Laplace transform for this function is

$$\begin{aligned} \mathscr{L}\{f(x)\} &= \int_{-\infty}^{\infty} e^{ax} e^{-sx}\, dx \\ &= \int_{-\infty}^{\infty} e^{(a-s)x}\, dx \end{aligned} \tag{6.157}$$

Integral in Eq. (6.157) converges only when $\Re(a - s) < 0$, i.e., $\Re(s) > \Re(a)$. The function grows up for $\Re(a - s) > 0$, and the integral diverges. The bilateral transform for e^{ax} is

$$F(s) = \frac{1}{s - a}, \qquad \text{for } \Re(s) > \Re(a) \tag{6.158}$$

Bilateral Laplace transforms satisfy most of the properties of one-sided Laplace transforms, but due to the stringent condition for convergence, they are rarely used. The bilateral Laplace transform is used when signals or systems involving noncausal systems need to be analyzed.

6.10 Laplace Transform of Some Standard Functions

	$f(x)$	$F(s) = \int_0^\infty f(x)\, e^{-sx} dx$
1.	$x^n \quad (n = 0, 1, 2, 3, \ldots)$	$\frac{n!}{s^{n+1}}$
2.	e^{ax}	$\frac{1}{s-a}$
3.	$\cos ax$	$\frac{s}{s^2+a^2}$
4.	$\sin ax$	$\frac{a}{s^2+a^2}$
5.	$\cosh ax$	$\frac{s}{s^2-a^2}$
6.	$\sinh ax$	$\frac{a}{s^2-a^2}$
7.	$x^n e^{-ax}$	$\frac{\Gamma(n+1)}{(s+a)^{n+1}}$
8.	$x^a \quad (a > -1)$	$\frac{\Gamma(a+1)}{s^{a+1}}$
9.	$e^{ax} \cos bx$	$\frac{s-a}{(s-a)^2+b^2}$
10.	$e^{ax} \sin bx$	$\frac{b}{(s-a)^2+b^2}$
11.	$\left(e^{ax} - e^{bx}\right)$	$\frac{a-b}{(s-a)(s-b)}$
12.	$\frac{1}{(a-b)}\left(ae^{ax} - be^{bx}\right)$	$\frac{s}{(s-a)(s-b)}$
13.	$x \sin ax$	$\frac{2as}{\left(s^2+a^2\right)^2}$
14.	$x \cos ax$	$\frac{s^2-a^2}{\left(s^2+a^2\right)^2}$
15.	$\sin ax \sinh ax$	$\frac{2sa^2}{\left(s^4+4a^4\right)}$
16.	$(\sinh ax - \sin ax)$	$\frac{2a^3}{\left(s^4-a^4\right)}$
17.	$(\cosh ax - \cos ax)$	$\frac{2a^2 s}{\left(s^4-a^4\right)}$

	$f(x)$	$F(s) = \int_0^\infty f(x)\, e^{-sx} dx$
18.	$\frac{\cos ax - \cos bx}{\left(b^2 - a^2\right)} \left(a^2 \neq b^2\right)$	$\frac{s}{\left(s^2 + a^2\right)\left(s^2 + b^2\right)}$
19.	$\frac{1}{\sqrt{x}}$	$\sqrt{\frac{\pi}{s}}$
20.	$2\sqrt{x}$	$\frac{1}{s}\sqrt{\frac{\pi}{s}}$
21.	$x \cosh ax$	$\frac{\left(s^2 + a^2\right)}{\left(s^2 - a^2\right)^2}$
22.	$x \sinh ax$	$\frac{2as}{\left(s^2 - a^2\right)^2}$
23.	$\frac{\sin(ax)}{x}$	$\tan^{-1}\left(\frac{a}{s}\right)$
24.	$x^{\frac{-1}{2}} e^{-\frac{a}{x}}$	$\sqrt{\frac{\pi}{s}} e^{-2\sqrt{as}}$
25.	$x^{\frac{-3}{2}} e^{-\frac{a}{x}}$	$\sqrt{\frac{\pi}{a}} e^{-2\sqrt{as}}$
26.	$\frac{1}{\sqrt{\pi x}}(1 + 2ax)e^{ax}$	$\frac{s}{(s-a)\sqrt{s-a}}$
27.	$(1 + ax)e^{ax}$	$\frac{s}{(s-a)^2}$
28.	$\frac{1}{2\sqrt{\pi x^3}}\left(e^{bx} - e^{ax}\right)$	$\sqrt{s-a} - \sqrt{s-b}$
29.	$\delta(x - a)$	$e^{-as}, \quad a \geq 0$
30.	$\mathcal{H}(x - a)$	$\frac{1}{s}e^{-as}, \quad a \geq 0$
31.	$\delta'(x - a)$	$se^{-as}, \quad a \geq 0$
32.	$\delta^{(n)}(x - a)$	$s^n e^{-as}$
33.	$\lvert \sin ax \rvert, \quad (a > 0)$	$\frac{a}{\left(s^2 + a^2\right)} \coth\left(\frac{\pi s}{2a}\right)$
34.	$\frac{1}{\sqrt{\pi x}} \cos(2\sqrt{ax})$	$\frac{1}{\sqrt{s}} e^{-\frac{a}{s}}$
35.	$\sqrt{\pi x} \sin(2\sqrt{ax})$	$\frac{1}{s\sqrt{s}} e^{-\frac{a}{s}}$
36.	$\frac{1}{\sqrt{\pi a}} \cosh(2\sqrt{ax})$	$\frac{1}{\sqrt{s}} e^{\frac{a}{s}}$

	$f(x)$	$F(s) = \int_0^\infty f(x)e^{-sx}dx$
37.	$\frac{1}{\sqrt{\pi a}}\sinh(2\sqrt{ax})$	$\frac{1}{s\sqrt{s}}e^{\frac{a}{s}}$
38.	$\frac{1}{x}\left(e^{bx} - e^{ax}\right)$	$\log\left(\frac{s-a}{s-b}\right)$
39.	$\frac{1+2ax}{\sqrt{\pi x}}$	$\frac{s+a}{s\sqrt{s}}$
40.	$\frac{1}{\sqrt{x}}e^{-\frac{a}{x}}$	$2\sqrt{\frac{\pi}{a}}e^{-2\sqrt{as}}$
41.	$\sin(2\sqrt{ax})$	$\frac{\sqrt{\pi}}{2}\left(\frac{a}{s}\right)^{\frac{3}{2}}e^{-\frac{a}{4s}}$
42.	$\frac{\cos(2\sqrt{ax})}{\sqrt{x}}$	$\sqrt{\frac{\pi}{s}}e^{-\frac{a}{s}}$
43.	$\frac{e^{-bx} - x^{-ax}}{(b-a)\sqrt{4\pi x^3}}$	$\frac{1}{\sqrt{s+a}+\sqrt{s+b}}$
44.	$Si(x) = \int_0^x \frac{\sin x}{x}dx$	$\frac{1}{s}\cot^{-1}(s)$
45.	$Ci(x) = -\int_x^\infty \frac{\cos x}{x}dx$	$-\frac{1}{2s}\log\left(1+s^2\right)$
46.	$-Ei(-x) = \int_x^\infty \frac{e^{-x}}{x}dx$	$\frac{1}{s}\log(1+s)$

6.11 Closure

This chapter explored the fundamental concepts of the Laplace transform and its key properties and methods for finding inverse Laplace transforms. It began by defining the Laplace transform and highlighting its linearity, which simplifies the transformation of complex functions by breaking them down into manageable components. Essential properties like the first and second shifting theorems were thoroughly discussed, showcasing how they facilitate the transformation process and extend the range of functions that can be effectively analyzed. The chapter also covered the inverse Laplace transform, a vital tool for converting solutions back into the time domain. Techniques such as partial fraction decomposition, Heaviside's expansion theorem, known transform pairs, and Bromwich contour integrals were examined, providing a comprehensive approach to recovering time-domain representations from their transformed forms. Mastering the Laplace and inverse Laplace transforms is crucial for solving differential equations and analyzing system dynamics. The differentiation and integration properties of the Laplace transform enable the conversion of linear ODEs into algebraic equations, simplifying their solution. Detailed methods for solving ODEs and integral equations using the Laplace transform were presented, along with applications in system modeling, stability analysis, and system response. Finally, the concept of the bilateral Laplace transform was introduced.

In conclusion, the Laplace transform is an invaluable analytical tool that simplifies the solution of differential equations, system modeling, and signal analysis by converting time-domain problems into the frequency domain. It also allows for a comprehensive analysis of signals and systems across the entire time axis, making it a powerful resource for engineers and scientists working in diverse fields such as control theory and physics.

Reference

Van der Pol, B., & Bremmer, H. (1950). *Operational calculus: Based on the two-sided Laplace integral* (1st ed.). Cambridge University Press.

Chapter 7
Discrete Fourier Transform

Life is good for only two things, discovering mathematics and teaching mathematics.

— Siméon Denis Poisson

Abstract This chapter provides a comprehensive look at the discrete Fourier transform (DFT) and its broad applications, from signal processing and communications to medical imaging and economic analysis. It starts by introducing the transition from continuous to discrete data, presenting the DFT with a focus on its properties, including linearity, circular shifts, and symmetry, which underpin digital signal analysis. The chapter then explores the fast Fourier transform (FFT), highlighting the efficiency of the Cooley-Tukey algorithm and mixed radix methods for faster DFT computation. Spectrum analysis, sampling, and interpolation techniques are also discussed, along with windowing functions to reduce spectral leakage in applied contexts. A section on noise characterization explains various noise types, clarifying the challenges of maintaining signal fidelity. The chapter wraps up with insights into quantization and other practical considerations in DFT, equipping readers to handle complex digital systems in real-world, noisy environments.

Keywords Discrete Fourier transform · DFT · Circular shift · Fast Fourier transform · FFT · Cooley-Tukey algorithm · Mixed radix fast transforms · Spectrum analysis · Sampling and interpolation · Windowing · Noise characterization · Quantization

7.1 Introduction

In earlier chapters, we explored the Fourier series and Fourier transform of various functions. The Fourier series takes a function $f(x)$ and produces a sequence of coefficients, while the Fourier transform maps $f(x)$ to a new function $F(\omega)$. Calculating either the Fourier series or the Fourier transform requires evaluating an integral, meaning it is only feasible for functions that can be described analytically and are

S. K. Jena, *Fourier, Laplace, and the Tangled Love Affair with Transforms*,
https://doi.org/10.1007/978-3-031-80165-5_7

relatively simple. However, such neatly defined functions are uncommon in the real world, so we rely on digital computers. Unlike mathematical analysis, computers process sequences of numbers that represent functions, not the functions themselves. This process involves digitizing a function to create a numerical sequence that a computer can handle. Any time-varying quantity, such as sound wave pressure, radio signals, or daily temperature readings, can be sampled over a finite period for digital signal processing through a window function. In image processing, these samples correspond to pixel values along a row or column of a raster image. The discrete Fourier transform (DFT) converts a finite sequence of equally spaced samples of a function into a sequence of equally spaced samples of the discrete-time Fourier transform (DTFT), which is a complex-valued function of frequency. The sampling interval of the DTFT is the inverse of the input sequence's duration. The inverse DFT produces a Fourier series using the DTFT samples as coefficients of complex sinusoids at the corresponding DTFT frequencies, reproducing the original input sequence. The DFT provides a frequency-domain representation of the original sequence. If the original sequence contains all nonzero values of a function, its DTFT is continuous and periodic, with the DFT providing discrete samples of one cycle. On the other hand, if the original sequence represents one cycle of a periodic function, the DFT captures all nonzero values within a single DTFT cycle. The discrete Fourier transform is also known as the finite Fourier transform and is a crucial tool for various applications.

7.1.1 Signal Processing

Signal transformation into the frequency domain allows for efficient filtering, compression, and analysis in signal processing. For instance, in audio processing, the DFT enables noise reduction, echo cancellation, and enhancement of audio quality. Similarly, these transforms are used for image compression, enhancement, and feature extraction in image processing.

7.1.2 Communications

The DFT is crucial for modulating and demodulating signals in communications. It enables efficient data transmission by converting signals from the time domain to the frequency domain, allowing for techniques such as orthogonal frequency division multiplexing (OFDM), which is widely used in modern communication systems like Wi-Fi and Long Term Evolution (LTE).

7.1.3 Medical Imaging

In medical imaging, the DFT is essential for analyzing signals from magnetic resonance imaging (MRI) and computed tomography (CT). These transforms help reconstruct images from raw data, allowing for detailed visualization of internal body structures, essential for diagnosis and treatment planning.

7.1.4 Scientific Research

In scientific research, DFT analyzes experimental data and models physical phenomena. It is indispensable in fields such as quantum mechanics, seismology, and astrophysics, where analyzing the frequency components of signals and data is critical for understanding underlying processes.

7.1.5 Economic and Financial Analysis

In economics and finance, the DFT is applied to analyze time-series data, detect cycles, and identify trends. It is used in stock market analysis, economic forecasting, and algorithmic trading, where understanding the frequency components of financial data can lead to better decision-making and investment strategies.

7.2 Discrete Representation of Data

A finite sequence of N terms, or Nth-order sequence, is defined as a function whose domain is the set of integers $\{0, 1, 2, \ldots, N-1\}$ and whose range is the set of terms $\{f[0], f[1], f[2], \ldots, f[N-1]\}$. Formally, there is an Nth-order sequence in the set of ordered pairs:

$$\langle(0, f[0]), (1, f[1]), (2, f[2]), \ldots, (N-1, f[N-1])\rangle \tag{7.1}$$

In this chapter, a shorter notation is followed for denoting the sequence as $\{f[k]\}$ and the kth term of the sequence as $f[k]$. Let us consider the sequence $\{f[k]\}$ whose terms are defined as

$$f[k] = \frac{1}{k+1}, \quad k = 0, 1, \ldots, N-1 \tag{7.2}$$

Thus,

$$\{f[k]\} = \left\{1, \frac{1}{2}, \frac{1}{3}, \ldots, \frac{1}{N}\right\} \tag{7.3}$$

This is misleading to assume, that $f[k]$ has a formula or equation in terms of k to define a sequence. For example, the sequence $\{f[k]\} = \{1, -3, 4, 7.8, -3, \pi, 41\}$ is an eighth-order sequence with no specific formula that maps k to $f[k]$ for $k \in [0, 7]$.

A sequence is considered real if all of its terms have real values. Similarly, a sequence is termed imaginary (or purely imaginary) if every term within it has imaginary values. For a given sequence $\{f[k]\}$, the complex conjugate sequence $\{f^*[k]\}$ is defined as the sequence where each term is the complex conjugate of the corresponding term in $\{f[k]\}$. Some basic algebraic operation for the sequences are given below.

The sum of two Nth-order sequences $\{f[k]\}$ and $\{g[k]\}$ is denoted as $\{f[k]\} + \{g[k]\rangle$ and is defined as the sequence whose terms are given by $f[k] + g[k]$ for $k \in [0, N-1]$. In other words, we add the individual term, for example,

$$\{0, 1, 2\} + \{4, 2, 1\} = \{4, 3, 3\} \tag{7.4}$$

Subtraction is the inverse of addition. That is, to subtract two sequences, the respective individual terms are subtracted. Note that addition and subtraction are only defined for sequences with the same order. The product of two sequences $\{f[k]\}$ and $\{g[k]\}$, denoted as $\{f[k]\}\{g[k]\}$, is defined as the sequence whose terms are given by $f[k]g[k]$ for $k \in [0, N-1]$. In other words, we multiply the individual terms, for example,

$$\{0, 1, 2\}\{4, 2, 1\} = \{0, 2, 2\} \tag{7.5}$$

Division is the inverse of multiplication, and thus to divide the sequence $\{f[k]\}$ by $\{g[k]\}$, the individual terms are divided:

$$\frac{f[k]}{g[k]}, \quad g[k] \neq 0, \quad k \in [0, N-1] \tag{7.6}$$

When the terms of the sequences are complex, the algebraic rules for multiplication (and division) of complex numbers are followed. To multiply a sequence by a constant, each term of the sequence is multiplied by the constant. For example,

$$3\{1, 2, 3\} = \{3, 6, 9\} \tag{7.7}$$

An Nth-order sequence $\{f[k]\}$ is even if and only if

$$f[N-k] = f[-k] = f[k] \quad k \in [0, N-1] \tag{7.8}$$

Similarly, an Nth-order sequence $\{f[k]\}$ is odd if and only if

$$f[N-k] = f[-k] = -f[k] \quad k \in [0, N-1] \tag{7.9}$$

The sequence $\{6, 1, 3, 3, 1\}$ is even and the sequence $\{1, 3, -3, 1\}$ is odd.

7.3 From Continuous to Discrete Fourier Transform

The transition from continuous (analog) to discrete (digital) Fourier transform became necessary with the advent of digital computers in the mid-twentieth century. Discrete Fourier transform (DFT) was developed to analyze signals and data in a digital format at discrete data points using effective numerical algorithms on digital computers. German mathematician Carl Friedrich Gauss (1777–1855) developed efficient methods for calculating trigonometric sums while working on the problem of interpolating periodic data to compute asteroid orbits. However, Gauss's contributions to the fast computation of the DFT were not widely recognized. In the 1940s and 1950s, the Nyquist-Shannon sampling theorem, formulated by Swedish-American physicist Harry Nyquist (1889–1976) and American mathematician Claude Elwood Shannon (1916–2001), provided the theoretical foundation for converting continuous signals into discrete sequences. Later, American computer scientist Norbert Wiener (1894–1964) and American mathematician and statistician John Wilder Tukey (1915–2000) made significant contributions that further developed the field. Some books on image processing and digital signal processing directly introduce the DFT. But in this chapter, we introduced the DFT from the knowledge of continuous Fourier transform studied in the previous chapters.

If $f(x)$ is a rapidly decreasing time-limited signal $0 \le x \le L$, then assume its Fourier transform $\mathscr{F}\{f(x)\} = F(s)$ is band-limited $0 \le s \le 2B$. This is an abuse of the sampling theorem, as a signal cannot be time- and band-limited simultaneously. But no offence in this approximation as long as time interval $\Delta x = \dfrac{1}{2B}$ is used for sampling to identify the discrete variable pairs from the continuous variables. The continuous $0 \le x \le L$ is sampled at $N+1$ discrete value as $x = \left\{0, \dfrac{1}{2B}, \dfrac{2}{2B}, \ldots, \dfrac{N}{2B}\right\}$ with $\dfrac{N}{2B} = L$ or $N = 2BL$. The sample points can be written as

$$x_0 = \frac{0}{2B}, x_1 = \frac{1}{2B}, x_2 = \frac{2}{2B}, \ldots, x_{N-1} = \frac{N-1}{2B} \tag{7.10}$$

The sampled form of the function $f(x)$ can be written using the delta function as

$$f(x) = \sum_{k=0}^{N-1} \delta(x - x_k) f(x_k) \tag{7.11}$$

The Fourier transform of Eq. (7.11) becomes

$$\mathscr{F}\left\{f(x)_{\text{sampled}}\right\} = \sum_{k=0}^{N-1} f(x_k)e^{-2\pi i s x_k} = F(s) \tag{7.12}$$

The Fourier transform is continuous function $F(s)$ and not yet discrete. In the frequency domain, the signal is band-limited $0 \leq s \leq 2B$. s is sampled with an interval $\Delta s = \frac{1}{L}$ (both domains are in reciprocal relationship). Suppose M discrete samples are taken in the frequency domain:

$$\frac{M}{L} = 2B \tag{7.13}$$

or

$$M = 2BL = N \tag{7.14}$$

The number of sampling points in both domains is equal. The sampled frequencies are expressed as

$$s_0 = \frac{0}{L}, s_1 = \frac{1}{L}, s_2 = \frac{2}{L}, \ldots, s_{N-1} = \frac{N-1}{L} \tag{7.15}$$

The sampled form of the Fourier transform is expressed using the delta function as

$$\begin{aligned} \mathscr{F}\{f(x)\}_{\text{sampled}} = F\left(s_{\text{sampled}}\right) &= \sum_{k=0}^{N-1} f(x_k)\, e^{-2\pi i s x_k} \sum_{m=0}^{N-1} \delta(s - s_m) \\ &= \sum_{k,m=0}^{N-1} f(x_k)\, e^{-2\pi i s x_k} \delta(s - s_m) \end{aligned} \tag{7.16}$$

The sampled form of the Fourier transform is expressed as

$$\begin{aligned} F(s_0) &= \sum_{k=0}^{N-1} f(x_k)e^{-2\pi i s_0 x_k} \\ F(s_1) &= \sum_{k=0}^{N-1} f(x_k)e^{-2\pi i s_1 x_k} \\ &\vdots \\ F(s_{N-1}) &= \sum_{k=0}^{N-1} f(x_k)e^{-2\pi i s_{N-1} x_k} \end{aligned} \tag{7.17}$$

The following substitution helps to eliminate the continuous variable signature:

$$\begin{aligned} x_k &= \frac{k}{2B} \\ s_m &= \frac{m}{L} \end{aligned} \tag{7.18}$$

$$x_k s_m = \frac{km}{2BL} = \frac{km}{N} \tag{7.19}$$

So

$$e^{-2\pi i s_m x_k} = e^{-\frac{2\pi ikm}{N}} \tag{7.20}$$

k and m are the indices. With these indices, the sampled function f and its Fourier transform F are expressed as follows:

$$f(x_k) = f[k] \tag{7.21}$$

$$F(s_m) = F[m] \tag{7.22}$$

After removing the traces of all continuous variables, the discrete Fourier transform (DFT) is expressed as

$$F[m] = \sum_{n=0}^{N-1} f[n]\, e^{-\frac{2\pi imn}{N}} = \sum_{n=0}^{N-1} f[n]\, w^{mn} \tag{7.23}$$

$w = e^{-\frac{2\pi i}{N}}$ is the weighting kernel. DFT in Eq. (7.23) can be expressed in vector form as

$$\begin{bmatrix} F[0] \\ F[1] \\ \vdots \\ F[N-1] \end{bmatrix} = \begin{bmatrix} w^{00} & w^{01} & \cdots & w^{0(N-1)} \\ w^{10} & w^{11} & \cdots & w^{1(N-1)} \\ \vdots & \vdots & \vdots & \vdots \\ w^{(N-1)0} & w^{(N-1)1} & \cdots & w^{(N-1)(N-1)} \end{bmatrix} \begin{bmatrix} f[0] \\ f[1] \\ \vdots \\ f[N-1] \end{bmatrix} \tag{7.24}$$

Expression for DFT in Eq. (7.24) can be interpreted as mapping of a Nth-order sampled vector $\{f[m]\}$ from one domain to Nth-order sequence $\{F[n]\}$ in other domain, through the mapping $N \times N$ DFT matrix W_N. W_N is a complex and symmetric matrix. Equation (7.24) is also known as direct discrete Fourier transform.

From the time- and band-limited approximation,

$$\begin{aligned} N\Delta x &= L \\ N\Delta s &= 2B \end{aligned} \tag{7.25}$$

So

$$\Delta x \Delta s = \frac{2BL}{N^2} = \frac{1}{N} \tag{7.26}$$

Once the sampling rate Δx and the number of samples N are determined, the value of Δs gets fixed by default. Often, the sampling rate of any device Δx is fixed; if the number of samples N is not sufficient to produce the desired frequency resolution Δs, then a new sequence of zeros is added at the end of the signal vector $\{f[m]\}$. This process is called zero padding, which increases the number of points in the frequency domain, providing a finer frequency resolution. Zero padding does not add new information to the signal but makes the frequency spectrum appear smoother, which helps identify the precise frequencies present.

From the definition of inner product,

$$W_N^{km} W_N^{*kn} = \sum_{k=0}^{N-1} e^{\frac{2\pi ik(m-n)}{N}} \tag{7.27}$$

The above sequence represents a geometric series for $m \neq n$, and the sum is

$$W_N^{km} W_N^{*kn} = \frac{1 - e^{2\pi ik(m-n)}}{1 - e^{\frac{2\pi ik(m-n)}{N}}} = 0 \tag{7.28}$$

For $m = n$, Eq. (7.27) becomes

$$W_N^{km} W_N^{*kn} = N \tag{7.29}$$

So, each column vector of the DFT matrix is orthogonal:

$$W_N^{km} W_N^{*kn} = \begin{cases} 0\ , & m \neq n \\ N\ , & m = n \end{cases} = NI \tag{7.30}$$

I is the identity matrix. It is a crucial point to notice that the DFT matrix is orthogonal, not orthonormal; this results in a multiplicative factor $\frac{1}{N}$ in the inverse discrete Fourier transform (IDFT):

$$f[m] = \frac{1}{N} \sum_{n=0}^{N-1} F[n] e^{\frac{2\pi imn}{N}} = \frac{1}{N} \sum_{n=0}^{N-1} F[n] \left(w^*\right)^{mn} \tag{7.31}$$

$w^* = e^{\frac{2\pi i}{N}}$. Equation (7.31) can be verified by using the relation $\mathscr{F}^{-1}\{\mathscr{F}\{f[k]\}\} = f[k]$. The inverse discrete Fourier transform (IDFT) allows us to reconstruct the original discrete sequence from its frequency components.

Fig. 7.1 Input sequence

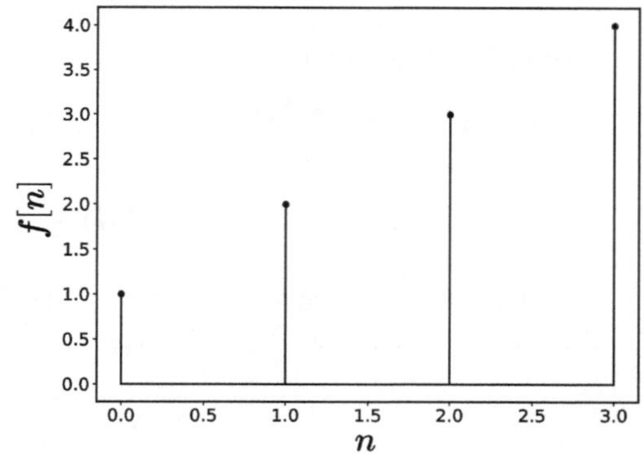

Example 7.1 (Example of DFT) Find the DFT for a given sequence $f[n] = [1, 2, 3, 4]$. The above sequence is presented in Fig. 7.1. The DFT of the sequence is expressed as

For $k = 0$

$$F[0] = \sum_{n=0}^{3} f[n]e^{-2\pi i 0n/4} = f[0] + f[1] + f[2] + f[3]$$
$$= 1 + 2 + 3 + 4 = 10$$

For $k = 1$

$$F[1] = \sum_{n=0}^{3} f[n]e^{-2\pi i 1n/4} = f[0] + f[1]e^{-\pi i/2} + f[2]e^{-\pi i} + f[3]e^{-3\pi i/2}$$
$$= 1 + 2(-i) + 3(-1) + 4(i) = -2 + 2i$$

For $k = 2$

$$F[2] = \sum_{n=0}^{3} f[n]e^{-2\pi i 2n/4}$$
$$= f[0] + f[1]e^{-\pi i} + f[2]e^{-2\pi i} + f[3]e^{-3\pi i}$$
$$= 1 + 2(-1) + 3(1) + 4(-1) = -2$$

For $k = 3$

(continued)

Example 7.1 (continued)

$$F[3] = \sum_{n=0}^{3} f[n]e^{-2\pi i 3n/4} = f[0] + f[1]e^{-3\pi i/2} + f[2]e^{-3\pi i} + f[3]e^{-9\pi i/2}$$

$$= 1 + 2(i) + 3(-1) + 4(-i) = -2 - 2i$$

The DFT of the sequence $f[n] = [1, 2, 3, 4]$ is $F[k] = [10, -2 + 2i, -2, -2 - 2i]$. The magnitudes and phases of the DFT coefficients are

k	$\lvert X(k)\rvert$	$\angle X(k)$
0	10	0
1	$2\sqrt{2}$	$-\frac{\pi}{4}$
2	2	π
3	$2\sqrt{2}$	$\frac{\pi}{4}$

The DFT magnitude (Fig. 7.2) and phase (Fig. 7.3) give the insights into the frequency components of the given sequence. The reconstructed sequence through IDFT is presented in Fig. 7.4.

So far the sequences $\{f[m]\}$ and $\{F[n]\}$ are considered over the Nth-order domain $[0, N-1]$. Let us look at the behavior of $\{f[m]\}$ and $\{F[n]\}$ for values of m and n outside this domain. Some basic properties of the weighting kernel W_N:

$$\begin{aligned} W_N^{nN} &= e^{2\pi i nN/N} = e^{2\pi i n} = 1 \quad \text{for any integer} \quad n \\ W_N^{(m+n)} &= e^{2\pi i(m+n)/N} = e^{2\pi i m/N} e^{2\pi i n/N} = W_N^m W_N^n \end{aligned} \tag{7.32}$$

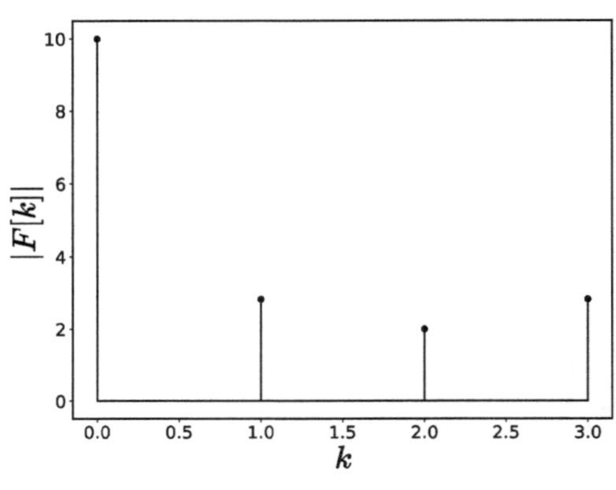

Fig. 7.2 Magnitude $|F[k]|$

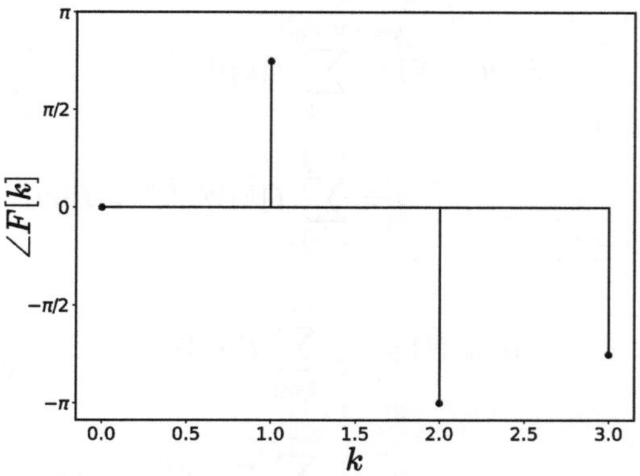

Fig. 7.3 Phase $\angle F[k]$

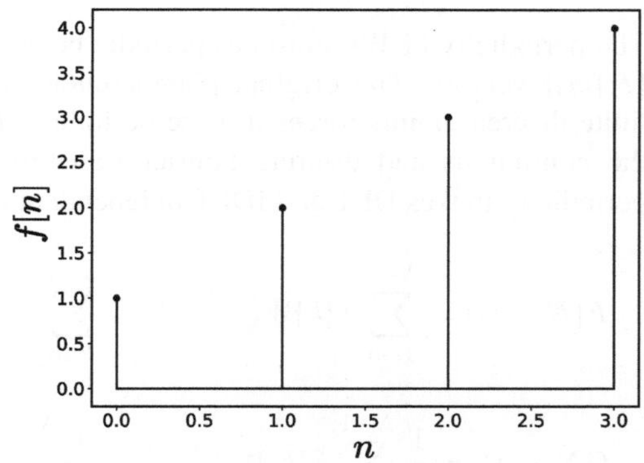

Fig. 7.4 Reconstructed sequence $f[n]$

w^{mn} is periodic with a period N. Using the above two properties, it is a trivial task to show that W_N is periodic in both m and n with period N and DFT is the linear combination of these periodic signals:

$$\begin{aligned}
W_N^{(m+N)n} &= W_N^{mn} W_N^{Nn} = W_N^{mn} \\
W_N^{m(n+N)} &= W_N^{mn} W_N^{mN} = W_N^{mn} \\
W_N^{m(N-n)} &= W_N^{mN} W_N^{-mn} = W_N^{-mn} \\
W_N^{(N-m)n} &= W_N^{Nn} W_N^{-mn} = W_N^{-mn}
\end{aligned} \tag{7.33}$$

The discrete Fourier transform $\{F[m]\}$ and the discrete inverse Fourier transform $\{f[n]\}$ are both periodic with periodicity N; that is, $F[m+N] = F[m]$ and $f[n+N] = f[n]$:

$$F[m+N] = \sum_{k=0}^{N-1} f[k] W_N^{-(m+N)k} = \sum_{k=0}^{N-1} f[k] W_N^{-Nk} W_N^{-km}$$
$$= \sum_{k=0}^{N-1} f[k] W_N^{-km} = F[m] \tag{7.34}$$

$$f[n+N] = \frac{1}{N}\sum_{k=0}^{N-1} F[k] W_N^{(n+N)k} = \frac{1}{N}\sum_{k=0}^{N-1} F[k] W_N^{nk} W_N^{Nn}$$
$$= \frac{1}{N}\sum_{k=0}^{N-1} F[k] W_N^{nk} = f[n] \tag{7.35}$$

The periodicity of W_N enforces periodic behavior of both input $\{f\,[n]\}$ and output $\{F\,[m]\}$ vectors. The original phenomenon may not be periodic, but taking the finite discrete points forces it to be periodic. This is the major difference between the continuous and discrete Fourier transforms. The consequence of this force periodicity makes DFT and IDFT independent from indexing:

$$F[N-m] = \sum_{k=0}^{N-1} f[k] W_N^{-k(N-m)} = \sum_{k=0}^{N-1} f[k] W_N^{-k(-m)} = F[-m] \tag{7.36}$$

$$f[N-n] = \frac{1}{N}\sum_{k=0}^{N-1} F[k] W_N^{k(N-n)} = \frac{1}{N}\sum_{k=0}^{N-1} F[k] W_N^{k(-n)} = f[-n] \tag{7.37}$$

The negative indices now have a meaning and considering $f[-n]$ and $F[-m]$ are justified:

$$f[n] = \{1, -3, 2, 4\} \Longleftrightarrow F[m] = \{4+0i, -1+7i, 2+0i, -1-7i\}$$

$f[-4] = f[0] = 1$ and $F[5] = F[1] = -1 + 7i$ For even N, some author prefers to indexing from $-\dfrac{N}{2} + 1$ to $\dfrac{N}{2}$ instead of 0 to $N - 1$.

When the signal is forced to be periodic, then the reverse signal formulates duality. The DFT and IDFT operations are almost alike. The signs of the exponents in the definitions of forward and inverse kernel are opposite, and either one of the transform needs to accommodate the multiplying factor $\dfrac{1}{N}$. If $f\{[n]\}$ and $\{F[m]\}$ and Nth-order transform pairs, the independent variable from either domain can be interchanged due to the dual nature of the transform:

$$\{F[m]\} \Longleftrightarrow N\{f[N-n]\} \tag{7.38}$$

For example, the DFT of $\{1, 2, 3, 4\}$ is $\{10 + 0i, -2 + 2i, -2 + 0i, -2 - 2i\}$. The DFT of $\{10 + 0i, -2 + 2i, -2 + 0i, -2 - 2i\}$ is $4\{1, 4, 3, 2\}$.

For any given bounded sequence $\{f(n)\}$, the DFT is calculated using Eq. (7.24). This calculation requires $O\left(N^2\right)$ mathematical multiplications and additions, and for large values of N, this can become unwieldy. Several properties of the DFT often simplifies its calculation.

7.3.1 Linearity of DFT

Like the continuous transform, the DFT is a linear transformation. Mathematically, for two sequences $f[n]$ and $g[n]$, and constants a and b, the linear property of DFT is expressed as

$$\mathscr{F}\{af[n] + bg[n]\} = a\mathscr{F}\{f[n]\} + b\mathscr{F}\{g[n]\} \tag{7.39}$$

The linearity is fundamental in signal processing, as it allows for the superposition of signals and simplifies the analysis of systems that respond linearly to multiple inputs. This property is utilized extensively in various applications, such as filtering, modulation, and system analysis.

Two sixth-order sequences are given below:

$$\{\delta_0[k]\} = \{1, 0, 0, 0, 0, 0\}$$
$$\{f[k]\} = \{1, 2, 3, 3, 2, 1\}$$

The DFT of the above two sequence is

$$\mathscr{F}\{\delta_0[k]\} = \{1, 1, 1, 1, 1, 1\}$$
$$\mathscr{F}\{f[k]\} = \{12 + 0i, -3 - 1.732i, 0 + 0i, 0 + 0i, 0 + 0i, -3 + 1.732i\}$$

Thus, if $\{g[k]\} = 6\{\delta_0[k]\} + \{f[k]\} = \{7, 2, 3, 3, 2, 1\}$, then

$$\mathscr{F}\{g[k]\} = \{18 + 0i, 3 - 1.732i, 6 + 0i, 6 + 0i, 6 + 0i, 3 + 1.732i\}$$

Linearity is defined and applicable only when the sequences have the same order.

7.3.2 Circular Shift of a Sequence

Since a periodic sequence is fully defined by its elements within one period, a shifted version of such a sequence can be generated by circularly shifting its elements within that period. Both the time-domain sequence $\{f[n]\}$ and its DFT $\{F[m]\}$ are periodic, so any shift in these sequences is referred to as a circular shift. For instance, the delayed sequence $\{f[n-1]\}$ is created by moving the last sample of $\{f[n]\}$ to the start of the sequence. Similarly, the advanced sequence $\{f[n+2]\}$ is formed by moving the first two samples of $\{f[n]\}$ to the sequence's end. For a sequence with N samples, only $(N-1)$ unique shifts are possible:

$$\{f[n \pm k]\}\{F[m]W_N^{\mp mk}\} \tag{7.40}$$

This theorem tells that if the discrete Fourier transform of the sequence $\{f[n]\}$ is known, then the discrete Fourier transform of be shifted sequence $\{f[n-k]\}$ can be computed with only N additional multiplications.

For a given sequence, $\{f[n]\} = \{1, 2, 3, 3, 2, 1\}$.

The shifted sequence $\{f[n-3]\} = \{3, 2, 1, 1, 2, 3\}$.

The DFT of this shifted sequence is computed as

$$\begin{aligned}\mathscr{F}\{f[n-3]\} &= \{F[m]W_6^{-3m}\} = \{F[m]e^{-2\pi i 3m/6}\} = \{F[m](-1)^m\} \\ &= \{12+0i, 3+1.732i, 0+0i, 0+0i, 0+0i, 3-1.732i\}\end{aligned}$$

7.3.3 *Circular Shift of a Spectrum*

The spectrum $\{F[m]\}$, of a Nth-order signal $\{f[n]\}$ can be shifted by multiplying the signal by a complex exponential, $e^{\mp\frac{2\pi ink}{N}}$, where k is an integer. The new spectrum is $\{F[m \pm k]\}$

$$e^{\pm\frac{2\pi ink}{N}}\{f[n]\} = \{W_N^{\pm kn} f[n]\} \Longleftrightarrow \{F[m \mp k]\} \tag{7.41}$$

The spectrum is circularly shifted by k sample intervals. For example, if $k = 1$ or $k = N + 1$, then the DC spectral value of the original signal appears at $k = 1$. With $k = -1$ or $k = N - 1$, it appears at $k = N - 1$.

For example, $\{f[n]\} = \{1, 2, 3, 4\} \Longleftrightarrow \{F[m]\} = \{10+0i, -2+2i, -2+0i, -2-2i\}$:

$$\begin{aligned}e^{\frac{2\pi in}{4}}\{f[n]\} &= \{1+0i, 0+2i, -3+0i, 0-4i\} \\ \Longleftrightarrow \{F[m]\} &= \{-2-2i, 10+0i, -2+2i, -2+0i\} \\ e^{\frac{2\pi i2n}{4}}\{f[n]\} &= \{1+0i, -2+0i, 3+0i, -4+0i\} \\ \Longleftrightarrow \{F[m]\} &= \{-2+0i, -2-2i, 10+0i, -2+2i\}\end{aligned}$$

7.3.4 Circular Time Reversal

If $\{f[n]\}$ and $\{F[m]\}$ and the DFT pair, then the reversal pair $\{f[N-n]\}$ and $\{F[N-m]\}$

$$\{f[N-n]\} \Longleftrightarrow \{F[N-m]\} \tag{7.42}$$

$$\begin{bmatrix} F[0] \\ F[3] \\ F[2] \\ F[1] \end{bmatrix} = \begin{bmatrix} 1 & 1 & 1 & 1 \\ 1 & -j & -1 & j \\ 1 & -1 & 1 & -1 \\ 1 & j & -1 & -j \end{bmatrix} \begin{bmatrix} f[0] \\ f[3] \\ f[2] \\ f[1] \end{bmatrix} = F(N-m) \tag{7.43}$$

For example,

$$\{f[n]\} = \{2, 1, 3, 4\} \Longleftrightarrow \{F[m]\} = \{10+0i, -1+3i, 0+0i, -1-3i\}$$
$$\{f[4-n]\} = \{2, 4, 3, 1\} \Longleftrightarrow \{F[4-m]\} = \{10+0i, -1-3i, 0+0i, -1+3i\}$$

7.3.5 Symmetry Relations of DFT

$\{F[m]\}$ is the discrete Fourier transform of the Nth-order sequence $\{f[n]\}$; then it follows the following symmetric relationship (Table 7.1).

Table 7.1 Symmetry relations of DFT

$\{f[n]\}$	$\{F[m]\}$
Even sequence	Even sequence
Odd sequence	Odd sequence
Real sequence	$\{F[m]\} = \{F^*[-m]\}$
$\{f[n]\} = \{f^*[-n]\}$	Real sequence
Imaginary sequence	$\{F[m]\} = -\{F^*[-m]\}$
$\{f[n]\} = -\{f^*[-n]\}$	Imaginary sequence
Real and even sequence	Real and even sequence
Real and odd sequence	Imaginary and odd sequence
Imaginary and even sequence	Imaginary and even sequence
Imaginary and odd sequence	Real and odd sequence

7.3.6 Difference of Sequence and Its DFT

The difference of a sequence is the sequence of forward difference:

$$\{\Delta f[n]\} = \{f[n+1] - f[n]\} \tag{7.44}$$

$\{F[m]\}$ is the discrete Fourier transform of the Nth-order sequence $\{f[n]\}$. DFT of the difference of a sequence is

$$\mathscr{F}\{\Delta f[n]\} = \{F[m]\left(W_N^m - 1\right)\} = \{W_N^m - 1\}\{F[m]\} \tag{7.45}$$

Difference of the transform is

$$\{\Delta F[m]\} = \mathscr{F}\left[\{\left(W_N^{-n} - 1\right) f[n]\}\right] \tag{7.46}$$

The DFT of the sixth-order sequence is $\{g[n]\} = \{1, 1, 0, -1, -1, 0\}$.
From closer examination of the sequence,

$$\{g[n]\} = \{\Delta f[n]\}$$

where $\{f[n]\} = \{1, 2, 3, 3, 2, 1\}$ so $\{F[m]\} = \{12 + 0i, -3 - 1.732i, 0 + 0i, 0 + 0i, 0 + 0i, -3 + 1.732i\}$ Using Eq. (7.40),

$$\begin{aligned}\{G[m]\} &= \{12 + 0i, -3 - 1.732i, 0 + 0i, 0 + 0i, 0 + 0i, -3 + 1.732i\}\\ &\qquad \{0 + 0i, -0.5 + 0.866i, -1.5 + 0.866i, -2 + 0i, -1.5 - 0.866i,\\ &\qquad -0.5 - 0.866i\}\\ &= \{0 + 0i, 3 - 1.732i, 0 + 0i, 0 + 0i, 0 + 0i, 3 + 1.732i\}\end{aligned}$$

7.3.7 Simultaneous Calculation of Real Transforms

The linearity and symmetry properties of the DFT can be combined for efficient simultaneous calculation of the DFT of two real sequences. $\{f_1[n]\}$ and $\{f_2[n]\}$ are two real sequences with DFT $\{F_1[m]\}$ and $\{F_2[m]\}$. These two sequences can be combined to obtain a complex sequence:

$$\{f[n]\} = \{f_1[n]\} + i\,\{f_2[n]\} \tag{7.47}$$

From the principle of linearity,

$$\mathscr{F}\{f[n]\} = \{F[m]\} = \{F_1[m]\} + i\{F_2[m]\} \tag{7.48}$$

Taking the complex conjugate of both sides of Eq. (7.48),

$$\{F^*[m]\} = \{F_1^*[m]\} - i\{F_2^*[m]\} \tag{7.49}$$

Substituting $N - m$ for m in Eq. (7.49),

$$\{F^*[N - m]\} = \{F_1^*[N - m]\} - i\{F_2^*[N - m]\} \tag{7.50}$$

From symmetry, for a real sequence,

$$\begin{aligned} \{F_1[m]\} &= \{F_1^*[-m]\} = \{F_1^*[N - m]\} \\ \{F_2[m]\} &= \{F_2^*[-m]\} = \{F_2^*[N - m]\} \end{aligned} \tag{7.51}$$

Substituting relations in Eq. (7.51) in Eq. (7.50),

$$\{F^*[N - m]\} = \{F_1[m]\} - i\{F_2[m]\} \tag{7.52}$$

From Eqs. (7.48) and (7.52),

$$\begin{aligned} \{F_1[m]\} &= \frac{\{F^*[m]\} + \{F^*[N - m]\}}{2} \\ \{F_2[m]\} &= \frac{\{F^*[m]\} - \{F^*[N - m]\}}{2i} \end{aligned} \tag{7.53}$$

The Fourier transform of two real sequences can be computed just once by combining it.

7.3.8 Upsampling of a Sequence

For a sequence and its DFT,

$$\{f[n]\} = \{1 + 1i, 2 - 3i\} \Longleftrightarrow \{F[m]\} = \{3 - 2i, -1 + 4i\}$$

$\{f[n]\}$ is upsampled by a factor of 2:

$$\begin{aligned} \{f_u[n]\} &= \{1 + 1i, 0 + 0i, 2 - 3i, 0 + 0i\} \\ &\Longleftrightarrow \{F_u[m]\} = \{3 - 2i, -1 + 4i, 3 - 2i, -1 + 4i\} \end{aligned}$$

The spectrum is repeated. In general, with

$$\{f[n]\} \Longleftrightarrow \{F[m]\}, n, m = 0, 1, \ldots, N - 1$$

and a positive integer upsampling factor L,

$$\{f_u[n]\} = \begin{cases} \{f[\frac{n}{L}]\} \text{ for } n = 0, L, 2L, \ldots, L(N-1) \\ 0 \quad \text{otherwise} \end{cases}$$
$$\Longleftrightarrow \{F[m]\} = \{F[m \bmod N]\}, m = 0, 1, \ldots, LN - 1 \tag{7.54}$$

The spectrum $\{F[m]\}$ is repeated L times. The same thing happens in the upsampling of a spectrum, except for a constant factor in the amplitude of the time-domain signal. Consider the sequence and its DFT:

$$\{f[n]\} = \{1 + 1i, 2 - 3i\} \Longleftrightarrow \{F[m]\} = \{3 - 2i, -1 + 4i\}$$

$\{F[m]\}$ is upsampled by a factor of 2 to form

$$\{F_u[m]\} = \{3 - 2i, 0 + 0i, -1 + 4i, 0 + 0i\} \Longleftrightarrow \{1 + 1i, 2 - 3i, 1 + 1i, 2 - 3i\}/2$$

The time-domain sequence is repeated.

7.4 Fast Fourier Transform

The naive computation of the DFT involves N summations, each requiring N multiplications. Therefore, the computational complexity of the naive DFT is $O(N^2)$. For large N, direction computation of DFT is extremely expensive. During the Cold War, the United States and the Soviet Union were locked in an intense arms race, particularly focused on developing nuclear weapons. Monitoring and detecting nuclear tests, especially underground ones, became a critical national security concern. This required analyzing seismic data to identify the unique signatures of nuclear explosions, necessitating the processing of large datasets to distinguish between natural seismic events and man-made explosions. In 1965, Cooley and Tukey (1965) published a groundbreaking paper on the fast Fourier transform (FFT) algorithm, marking a breakthrough in the field. The FFT algorithm leverages the symmetry and periodicity properties of the discrete Fourier transform (DFT), breaking down the DFT computation into smaller DFTs that can be solved recursively. Using a divide-and-conquer approach, the FFT reduces the computational complexity of the DFT from $O(N^2)$ to $O(N \log N)$ for N sampled data points. The United States deployed an extensive network of seismic sensors, and the FFT algorithm was quickly adopted for seismic monitoring to detect nuclear tests conducted by the Soviet Union. Identifying nuclear explosions amid seismic noise possible through the FFT enabled rapid and accurate seismic data analysis. The dramatic improvement in speed and efficiency brought by the FFT algorithm revolutionized seismic monitoring and had a transformative impact on various other fields, including telecommunications, audio processing, medical imaging, and more. The FFT algorithm is considered one of the most significant contributions to numerical analysis of this century.

7.4.1 Cooley-Tukey Algorithm

Cooley and Tukey (1965) algorithm tremendously reduces the number of computations required to compute the discrete Fourier transform of a sequence. A concise understanding of the DFT matrix W_N helps to understand this efficient algorithm. W_N stores the entry $e^{\frac{2\pi imn}{N}}$ for mapping $\{f[n]\}$ to $\{F[m]\}$. The entries of this matrix represent N complex numbers such that each raised to the power N is equal to 1.

$$\left(w^{mn}\right)^N = e^{2\pi imn} = 1 \tag{7.55}$$

These N complex numbers are the Nth roots of unity, located on the unit circle equidistantly. These complex exponentials are located on a unit circle and exhibit periodic behavior. When multiplied with a vector, the $N \times N$ complex DFT matrix (often called twiddle factors) changes its angle, not the magnitude. The vector is rotated on the unit circle with its own magnitude. In rectangular coordinates, these complex numbers are expressed as

$$1^{\frac{1}{N}} = \cos\left(\frac{2k\pi}{N}\right) + i\sin\left(\frac{2k\pi}{N}\right), \quad k = 0, 1, 2, \ldots, N-1 \tag{7.56}$$

The solution of Eq. (7.56) forms N basis. For $N = 8$, the basis on the unit circle is shown in Fig. 7.5. These N basis are used to form the entries of the $N \times N$ DFT matrix W_N:

$$W_8^{nm} = \begin{bmatrix} 1 & 1 & 1 & 1 & 1 & 1 & 1 & 1 \\ 1 & \frac{\sqrt{2}}{2} - i\frac{\sqrt{2}}{2} & -i & -\frac{\sqrt{2}}{2} - i\frac{\sqrt{2}}{2} & -1 & -\frac{\sqrt{2}}{2} + i\frac{\sqrt{2}}{2} & i & \frac{\sqrt{2}}{2} + i\frac{\sqrt{2}}{2} \\ 1 & -i & -1 & i & 1 & -i & -1 & i \\ 1 & -\frac{\sqrt{2}}{2} - i\frac{\sqrt{2}}{2} & i & \frac{\sqrt{2}}{2} - i\frac{\sqrt{2}}{2} & -1 & \frac{\sqrt{2}}{2} + i\frac{\sqrt{2}}{2} & -i & -\frac{\sqrt{2}}{2} + i\frac{\sqrt{2}}{2} \\ 1 & -1 & 1 & -1 & 1 & -1 & 1 & -1 \\ 1 & -\frac{\sqrt{2}}{2} + i\frac{\sqrt{2}}{2} & -i & \frac{\sqrt{2}}{2} + i\frac{\sqrt{2}}{2} & -1 & \frac{\sqrt{2}}{2} - i\frac{\sqrt{2}}{2} & i & -\frac{\sqrt{2}}{2} - i\frac{\sqrt{2}}{2} \\ 1 & i & -1 & -i & 1 & i & -1 & -i \\ 1 & \frac{\sqrt{2}}{2} + i\frac{\sqrt{2}}{2} & i & -\frac{\sqrt{2}}{2} + i\frac{\sqrt{2}}{2} & -1 & -\frac{\sqrt{2}}{2} - i\frac{\sqrt{2}}{2} & -i & \frac{\sqrt{2}}{2} - i\frac{\sqrt{2}}{2} \end{bmatrix} \tag{7.57}$$

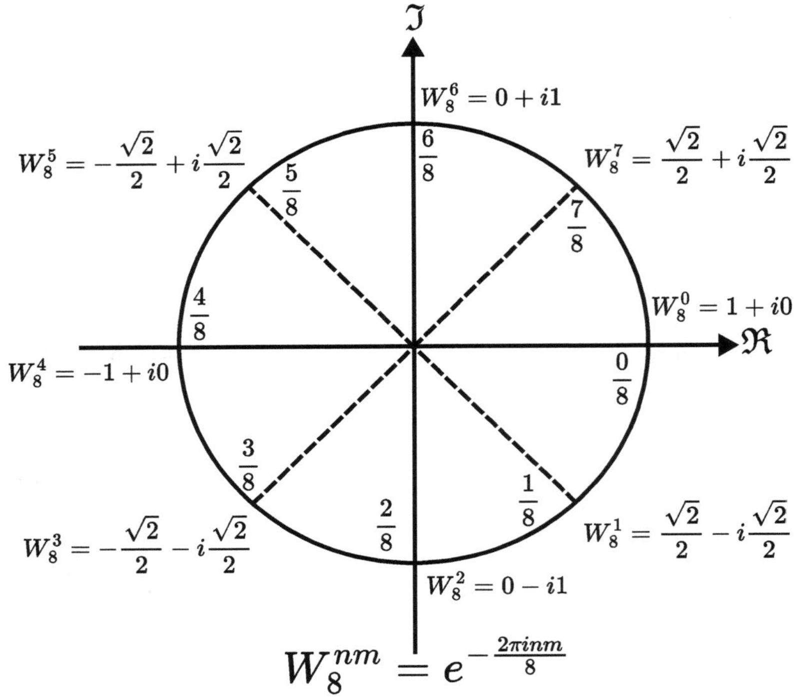

Fig. 7.5 Periodicity of the twiddle factors with $N = 8$ and the discrete frequencies at which the DFT coefficients are computed

Entries of the W_N follow some pattern. Cooley and Tukey (1965) algorithm exploits the symmetries and periodicities of the DFT to break down the computation into smaller, more manageable parts. The algorithm is based on the divide-and-conquer strategy and recursively divides the DFT computation into smaller DFTs.

Any periodic function can be uniquely decomposed into even half-wave and odd half-wave symmetric components. The even half-wave symmetric component is composed of the even-indexed frequency components, and the odd half-wave symmetric component is composed of the odd-indexed frequency components. Therefore, if an arbitrary function is decomposed into its even half-wave and odd half-wave symmetric components, then the original problem of finding the N frequency coefficients is divided into two problems, each of them being the determination of $N/2$ frequency coefficients. This decomposition is continued until each frequency component is isolated.

Let us consider an Nth-order sequence $\{f[n]\}$ with discrete Fourier transform $\{F[m]\}$. N is assumed to be an even integer and forms two new subsequences:

$$\begin{aligned}\{f_1[n]\} &= \{f[2n]\} \\ \{f_2[n]\} &= \{f[2n+1]\}, \quad n = 0, \ldots, M-1, \text{ where } M = N/2\end{aligned} \tag{7.58}$$

For example, if $\{f[n]\} = \{0, 1, 2, 14, 16, 20\}$, then $\{f_1[n]\} = \{0, 2, 16\}$ and $\{f_2[n]\} = \{1, 14, 20\}$.

$$\begin{aligned}
\{f_1[n+M]\} &= \{f[2(n+M)]\} = \{f[2n+N]\} = \{f[2n]\} = \{f_1[n]\} \\
\{f_2[n+M]\} &= \{f[2(n+M)+1]\} = \{f[2n+N+1]\} = \{f[2n+1]\} = \{f_2[n]\}
\end{aligned} \tag{7.59}$$

Both $\{f_1[n]\}$ and $\{f_2[n]\}$ are periodic sequences with periodicity M. DFTs of these two sequences are

$$\begin{aligned}
\{F_1[m]\} &= \sum_{n=0}^{M-1} f_1[n] W_M^{mn} \\
\{F_2[m]\} &= \sum_{n=0}^{M-1} f_2[n] W_M^{mn} \quad m = 0, \ldots, M-1,
\end{aligned} \tag{7.60}$$

Both $\{F_1[m]\}$ and $\{F_2[m]\}$ are periodic with periodicity M. DFT of the Nth-order sequence is $\{f[n]\}$

$$\{F[m]\} = \sum_{n=0}^{N-1} \{f[n]\} W_N^{mn} \tag{7.61}$$

The summation in Eq. (7.61) can be split as

$$\{F[m]\} = \sum_{n=0}^{M-1} \{f[2n]\} W_N^{2nm} + \sum_{n=0}^{M-1} \{f[2n+1]\} W_N^{(2n+1)m} \tag{7.62}$$

However,

$$\begin{aligned}
W_N^{2nm} &= e^{\frac{-2\pi i nm}{N/2}} = W_M^{nm} \\
W_N^{(2n+1)m} &= e^{\frac{-2\pi i (2n+1)m}{N}} = W_M^{nm} W_N^{m}
\end{aligned} \tag{7.63}$$

So Eq. (7.62) becomes

$$\begin{aligned}
\{F[m]\} &= \sum_{n=0}^{M-1} \{f_1[n]\} W_M^{nm} + W_N^{m} \sum_{n=0}^{M-1} \{f_2[n]\} W_M^{nm} \\
& \qquad m = 0, \ldots, N-1,
\end{aligned} \tag{7.64}$$

Using the relationship in Eq. (7.60),

$$\{F[m]\} = \{F_1[m]\} + W_N^m\{F_2[m]\} \qquad m = 0, \ldots, N-1 \tag{7.65}$$

As $\{F_1[m]\}$ and $\{F_2[m]\}$ are periodic with period M,

$$\begin{aligned} \{F[m]\} &= \{F_1[m]\} + W_N^m\{F_2[m]\} \\ \{F[M+m]\} &= \{F_1[m]\} - W_N^m\{F_2[m]\} \\ m &= 0, \ldots, M-1 \end{aligned} \tag{7.66}$$

Calculation of the DFT of $\{f[n]\}$ is of $O\left(N^2\right)$, whereas DFTs of $\{f_1[n]\}$ and $\{f_2[n]\}$ are of $O\left(M^2 = N^2/4\right)$. Computation of $\{F_1[m]\}$ and $\{F_2[m]\}$ is of $O\left(N^2/4\right)$ followed by N additional operations from Eq. (7.66). Finally, the $\{F[m]\}$ computation from Eq. (7.66) is of $O\left(N + 2N^2/4\right)$. The computation is reduced from $O\left(N^2\right)$ to $O\left(N + N^2/2\right)$.

For $N = 2$, the second-order sequence $\{f[n]\}$ is divided into two first-order sequence $\{f_1[0]\}$ and $\{f_2[1]\}$. A first-order sequence is its own transform, i.e., $\{f_1[0]\} = \{F_1[0]\}$ and $\{f_2[0]\} = \{F_2[0]\}$, and does not require any complex multiplication or addition to obtain these transforms. Using Eq. (7.66), this sequence requires only $N = 2$ operations to obtain $\{F[m]\}$. For example, if $\{f[n]\} = \{1, 2\}$, then $\{f_1[n]\} = \{1\}$ and $\{f_2[n]\} = \{2\}$. Also $\{F_1[n]\} = \{1\}$ and $\{F_2[n]\} = \{2\}$. Now using Eq. (7.66),

$$\begin{aligned} \{F[0]\} &= \{F_1[0]\} + \{F_2[0]\}W_2^0 = 1 + 2 = 3 \\ \{F[1]\} &= \{F_1[0]\} - \{F_2[0]\}W_2^0 = 1 - 2 = -1 \end{aligned} \tag{7.67}$$

$\{F[n]\} = \{3, -1\}$ requires only two operations instead of four operations required by direct DFT using Eq. (7.24).

If N is divisible by 4 or $M = N/2$ is divisible by 2, then $\{f_1[n]\}$ and $\{f_2[n]\}$ can be further subdivided into four $M/2$:

$$\begin{aligned} \{g_1[n]\} &= \{f_1[2n]\} \\ \{g_2[n]\} &= \{f_1[2n+1]\}, \\ \{h_1[n]\} &= \{f_2[2n]\} \\ \{h_2[n]\} &= \{f_2[2n+1]\} \end{aligned} \tag{7.68}$$

This requires $N + 2\left(M + M^2/2\right) = 2N + N^2/4$ operations. When the sequence is divided twice, the number of operations decreases from N^2 to $2N + N^2/4$. $N = 2^p$ is used for complete reduction of FFT, where p is a positive integer. If the number of samples $N < 2^p$, then zero padding is done with the sequence to obtain $N = 2^p$. Keep on continuing the reduction process recursively; the computation reduces to a factor

$$\frac{pN}{N^2} = \frac{p}{N} = \frac{\log_2 N}{N} \tag{7.69}$$

For $p = 10$, $N = 1024$ and direct DFT using Eq. (7.24) requires $\approx 10^6$ computation, while FFT using Cooley and Tukey (1965) algorithm requires $\approx 10^4$ computations. For practical purposes, $p \approx 18 - 22$.

7.4.2 Mixed Radix Fast Transforms

Cooley and Tukey (1965) algorithm for computing DFT of Nth-order sequence requires N to be even. Let $\{f[n]\}$ be an Nth-order sequence having DFT {F[m]} and $L = \frac{N}{3}$ be an integer. Three new sequences can be formed from the original sequence:

$$\begin{aligned}
\{f_1[n]\} &= \{f[3n]\} \\
\{f_2[n]\} &= \{f[3n+1]\} \\
\{f_3[n]\} &= \{f[3n+2]\} \quad n = 0, \ldots, L-1, \text{ where } L = N/3
\end{aligned} \tag{7.70}$$

A 9th-order sequence $\{f[n] = \{1, 3, 7, 5, 9, 2, 4, 8, 6\}$ can be subdivided into three sequences:

$$\begin{aligned}
\{f_1[n]\} &= \{f[1, 5, 4]\} \\
\{f_2[n]\} &= \{f[3, 9, 8]\} \\
\{f_3[n]\} &= \{f[7, 2, 6]\}
\end{aligned}$$

$\{f_1[n]\}$, $\{f_2[n]\}$ and $\{f_3[n]\}$ are periodic with periodicity $L = N/3$. As L is an integer, direct DFT using Eq. (7.24) can be used for these sequences:

$$\begin{aligned}
\{F_1[m]\} &= \sum_{n=0}^{L-1} \{f_1[n]\} W_L^{nm} \\
\{F_2[m]\} &= \sum_{n=0}^{L-1} \{f_2[n]\} W_L^{nm} \\
\{F_3[m]\} &= \sum_{n=0}^{L-1} \{f_3[n]\} W_L^{nm} \quad m = 0, \ldots, L-1
\end{aligned} \tag{7.71}$$

$\{F_1[m]\}$, $\{F_2[m]\}$, and $\{F_3[m]\}$ are periodic with periodicity L. The Fourier transform of the Nth-order sequence $\{f[n]\}$ is

$$\{F[m]\} = \sum_{n=0}^{N-1} \{f[n]\} W_N^{nm} \tag{7.72}$$
$$m = 0, \ldots, N-1$$

The above summation can be split as follows:

$$\begin{aligned}\{F[m]\} &= \sum_{n=0}^{L-1} \{f[3n]\} W_L^{3nm} + \sum_{n=0}^{L-1} \{f[3n+1]\} W_L^{(3n+1)m} \\ &+ \sum_{n=0}^{L-1} \{f[3n+2]\} W_L^{(3n+2)m}\end{aligned} \tag{7.73}$$

However,

$$\begin{aligned}W_N^{3mn} &= W_L^{mn} \\ W_N^{(3n+1)m} &= W_L^{mn} W_N^m \\ W_N^{(3n+2)m} &= W_L^{mn} W_N^{2m}\end{aligned} \tag{7.74}$$

Using the relationship in Eqs. (7.74), (7.72) becomes

$$\{F[m]\} = \{F_1[m]\} + \{F_2[m]\} W_N^m + \{F_3[m]\} W_N^{2m} \tag{7.75}$$

$\{F_1[m]\}$,$\{F_2[m]\}W_N^m$ and $\{F_3[m]\}W_N^{2m}$ are periodic with periodicity L:

$$\begin{aligned}&\{F[m]\} = \{F_1[m]\} + \{F_2[m]\} W_N^m + \{F_3[m]\} W_N^{2m} \\ &\{F[m+L]\} = \{F_1[m]\} + e^{\frac{-2\pi i}{3}} \{F_2[m]\} W_N^m + e^{\frac{-4\pi i}{3}} \{F_3[m]\} W_N^{2m} \\ &\{F[m+2L]\} = \{F_1[m]\} + e^{\frac{-4\pi i}{3}} \{F_2[m]\} W_N^m + e^{\frac{-2\pi i}{3}} \{F_3[m]\} W_N^{2m} \\ &m = 0, \ldots, L-1\end{aligned} \tag{7.76}$$

To calculate the DFT of $\{f[n]\}$ requires N^2 complex operations, whereas to calculate the DFT of the subsequences $\{f_1[n]\}$, $\{f_2[n]\}$, or $\{f_3[n]\}$ requires only L^2 or $N^2/9$ complex operations. Equation (7.75) requires $3N + N^2/3$ complex operations to compute $\{F[m]\}$ using $\{F_1[m]\}$, $\{F_2[m]\}$, and $\{F_3[m]\}$. Initially, $3\left(N^2/9\right)$ operations are required to compute $\{F_1[m]\}$, $\{F_2[m]\}$, and $\{F_3[m]\}$and then $3N$ additional operations required for Eq. (7.75). Radix-3 fast algorithm reduces the number of operations from N^2 to $3N + N^2/3$. If N is divisible by 9, then the sequence can be subdivided twice, and the number of operations becomes $2(3N) + N^2/9$. In the general $q(3N) + N^2/3^q$, the numbers of complex operations are required to subdivide the sequence $q$times (N divisible by 3^q). This results in a reduction factor:

$$\frac{q(3N)}{N^2} = \frac{3q}{N} = \frac{3\log_3 N}{N} \tag{7.77}$$

It's interesting to note that if a sequence isn't reduced to its smallest form, the radix-3 method offers greater computational savings compared to the radix-2 method. However, the radix-2 method becomes more efficient when complete reduction is applied. For example, with $N = 36$, the sequence can be reduced twice using either the radix-2 or radix-3 method, as N is divisible by both 9 and 4. The radix-2 method requires $2N + \frac{N^2}{4} = 396$ operations, while the radix-3 method only needs $2\left(3N + \frac{N^2}{9}\right) = 360$ operations. Now, consider determining the discrete Fourier transform (DFT) of a sequence with 700 terms. Since 700 is neither a power of 2 nor 3, zeros must be artificially added. For radix-3, the next largest power of 3 is $729 = 3^6$, requiring the addition of 29 zeros. The complete reduction would need $6 \times 3 \times 729 = 13{,}122$ operations. For the radix-2 method, 324 zeros are added to reach $N = 1024 = 2^{10}$, requiring $10 \times 1024 = 10{,}240$ operations, which is 22% fewer than the radix-3 method. Analogously, generating radix-5, radix-7, and higher radix methods is possible. It is also possible to combine these various methods into one algorithm. That is, suppose

$$N = 2^q 3^q 5^r \tag{7.78}$$

For the computation of the discrete Fourier transform of this Nth-order sequence digitally, the sequence would first be subdivided r times using the radix-5 method, followed by four subdivisions using the radix-3 method, and finally, p times using the radix-2 method. These methods can be easily implemented on a digital computer. More sophisticated version of algorithm like Rader (1968) and Bluestein (1970) are available for efficient computation of DFT. Python script to compute FFT using Cooley and Tukey (1965) is provided below.

```
import os
import pandas as pd
import numpy as np
import matplotlib.pyplot as plt
from scipy import fftpack
import math
import mplcursors

dlt = 0.001                        #Time interval of recording
Fs = 1/dlt;                        #Sampling frequency

Time_series = pd.read_csv("read_time_series.csv",sep=r'\s*,\s*'
                          , header=0, encoding='ascii',
                          usecols=[id], engine='python',
                          low_memory = True)
```

```
N = len(Time_series)
nfft = 2**23                    #Length of FFT\% padding with zeros
Time = np.zeros(nfft)
Signal = np.zeros(nfft)

for i in range(0,nfft-1):
    Time[i+1] = Time[i]+dlt

Mean_value = np.mean(Time_series[id])
print(Mean_value)

for i in range(N):
    Signal[i] = Time_series[id][i] - Mean_value

FT = fftpack.fft(Signal)
Amplitude  = np.abs(FT)
Power = Amplitude**2
Angle = np.angle(FT)
sample_frequency = fftpack.fftfreq(Signal.size, d = dlt)
Amp_frequency  = np.array([Amplitude,sample_frequency])
Amp_position = Amp_frequency[0,:].argmax()
Peak_frequency = Amp_frequency[1,Amp_position_1]
Half_frequency = sample_frequency[0:nfft//2]
Half_power = (2.0/nfft)*Power[0:nfft//2]


fig = plt.figure(figsize=(8,5))
ax = fig.add_subplot(1, 1, 1)
plt.plot(Time,Signal,"-k", linewidth = 2)
plt.xlabel("t", fontsize='40')
plt.ylabel("$f$'", fontsize='40')
eps = 0.03
plt.xlim([0, N*dlt])
ytick_loc = [-0.15, 0.000, 0.15]
ax.set_yticks(ytick_loc)
plt.xticks(fontsize=30)
plt.yticks(fontsize=30)
ratio = 0.25
x_left, x_right = ax.get_xlim()
y_low, y_high = ax.get_ylim()
ax.set_aspect(abs((x_right-x_left)/(y_low-y_high))*ratio)
ax = fig.gca()
for axis in ['top','bottom','left','right']:
    ax.spines[axis].set_linewidth(3.5)
plt.show()
plt.close()


fig = plt.figure(figsize=(8,5))
ax = fig.add_subplot(1, 1, 1)
lines = ax.plot(Half_frequency, Half_power,"-k", label = "",
                                     linewidth = 2)
ax.set_xscale('log')
#ax.set_yscale('log')
```

```
plt.xlim([10**-3, 10**1])
#plt.ylim([10**-9, 10**-4])
plt.xlabel("$f$", fontsize='40')
plt.ylabel("psd \TagInMath{$\EID{IEq262}(u_2$}')", fontsize='40
                                        ')
plt.tight_layout()
plt.grid(True, which="both", axis='both', ls="--")
plt.xticks(fontsize=20)
plt.yticks(fontsize=20)
ratio = 0.8
x_left, x_right = ax.get_xlim()
y_low, y_high = ax.get_ylim()
ax.set_aspect(abs((x_right-x_left)/(y_low-y_high))*ratio)
ax = fig.gca()
for axis in ['top','bottom','left','right']:
    ax.spines[axis].set_linewidth(3.5)
mplcursors.cursor(multiple=True).connect(
    "add", lambda sel: sel.annotation.draggable(False))
plt.show()
```

7.5 Spectrum Analysis

Physical signals are primarily recorded in the time domain but are often transformed into the frequency domain for more efficient analysis and easier interpretation of operations like filtering. The frequency-domain representation offers a clearer understanding of the signal's characteristics. For example, operations such as convolution become much simpler, reducing to basic multiplication in the frequency domain. Classifying signals into different frequency bands, such as low and high frequencies, is crucial for designing signal processing systems. In the frequency domain, signals can be compressed more effectively since the frequency content of practical signals diminishes at higher frequencies. Spectrum analysis, which involves examining the frequency content of signals, is essential in many applications. The fast Fourier transform (FFT) is a powerful tool for this purpose, providing a detailed view of the signal's frequency components, which is vital in fields like radar, sonar, and communication systems. A signal's power spectral density (PSD) measures its power distribution across frequencies and is calculated using the FFT, revealing key signal characteristics like dominant frequencies and bandwidth. For a discrete signal $\{f[n]\}$, the PSD is estimated as follows:

$$psd[f] = \frac{1}{N}\,|\mathrm{FFT}(\{f[n]\})|^2 \tag{7.79}$$

(NB: in LHS, psd[f] represents that the *psd* is a function of frequency f; this should not be confused with the time-domain function $\{f[n]\}$). Figure 7.6 represents the turbulent velocity time series from natural convection and their respective *psd*.

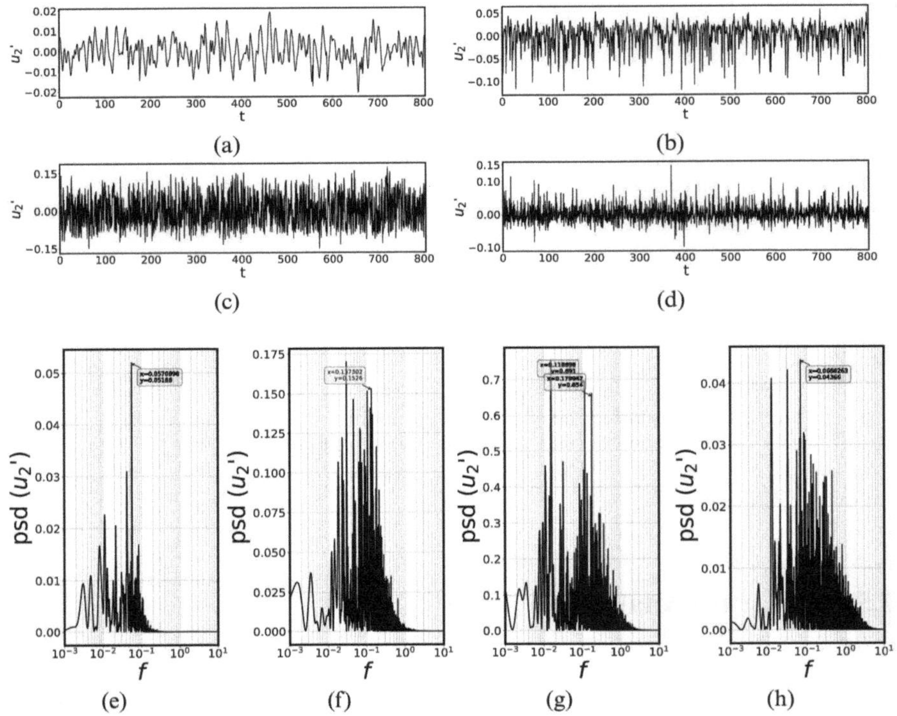

Fig. 7.6 Time series and power spectral density (PSD) from turbulent velocity signal (Jena & Manceau, 2024)

The power spectrum provides information about the dominating frequency in the process.

In the process of transferring the signal from time to the spectral domain or vice versa, the total energy content of the signal remains unaltered. The same is represented through Parseval's theorem (named after French mathematician Marc-Antoine Parseval des Chênes (1755–1836)). Let $\{f[n]\} \Longleftrightarrow \{F[m]\}$ be the N-order sequence pair. Parseval's theorem expresses the power of a signal in terms of its DFT spectrum. These samples occur on the unit circle. The sum of the squared magnitude of the samples of a complex exponential with amplitude one over the period N is N. The DFT decomposes a signal in terms of complex exponentials with coefficients $\{F[m]\}/N$. The power of the signal is the sum of the powers of its constituent complex exponentials and is given as

$$\sum_{n=0}^{N-1} |\{f[n]\}|^2 = \frac{1}{N} \sum_{m=0}^{N-1} |\{F[m]\}|^2 \tag{7.80}$$

The total energy content of the system is conserved by transferring the signal from one domain to another domain. Consider the sequence and its transform

$$\{2, 1, 3, 3\} \Longleftrightarrow \{9 + 0i, -1 + 2i, 1 + 0i, -1 - 2i\}$$

The sum of the squared magnitude of the data sequence is 23 and that of the DFT coefficients divided by 4 is also 23.

7.6 Sampling and Interpolation

In the real world, recording and analyzing signals from various phenomena or physical events are often of interest. Although these signals can rarely, if ever, be described by an analytical expression, they are typically well-behaved enough to allow for the existence of their Fourier transforms. The analytical determination of an integral equation is required for both the Fourier series and Fourier transform. These integrals depend on the complexity of the involved functions and are extremely difficult (if not impossible) to evaluate for functions not having a straightforward antiderivative. However, the discrete Fourier transform, which deals with bounded sequences, requires only straightforward addition and multiplication. All real-world signals are analog (continuous in time and space). The measurement of these real-world signals is discrete. Since these functions lack a mathematical or analytical expression, they must first be digitized and then analyzed using numerical algorithms on a digital computer. In the previous section, the FFT algorithm was presented for the efficient computation of the DFT. The relatively simple DFT can be used to obtain a function's Fourier transform and Fourier series coefficients. Digital signal processing is preferred for its numerous advantages. A signal $f([k]) = \{f(x[0]), f(x[1]), f(x[2]), f(x[3]), \ldots, f(x[k])\}$ is measured at k discrete points $x[k] = \{x[1], x[2], x[3], x[4], \ldots\}$ (instantaneous value at the desired points) at specific interval. Each measurement is referred to as a sample, and the continuous signal is reduced to a discrete sequence through signal sampling. These discrete sample locations are evenly spaced in x, with the distance between any two samples being Δx. The distance of a sample point at k location is

$$\{x[k]\} = x[0] + k\Delta x \tag{7.81}$$

$x[0]$ is the location of the first sample point. After sampling, the signal forms a sequence {f[k]}. The goal of sampling is to capture the essential information of the continuous signal in a discrete form, allowing it to be processed, stored, and analyzed using digital systems. The continuous function is supposed to be recovered from the discrete sequence through mathematical operations, like functional value estimation between two discrete points $f[n]$ and $f[n+1]$. Discrete sequences are used to reconstruct a continuous signal, and the continuous signal can be adequately processed through its samples when the sampling size (or sampling rate) Δx must

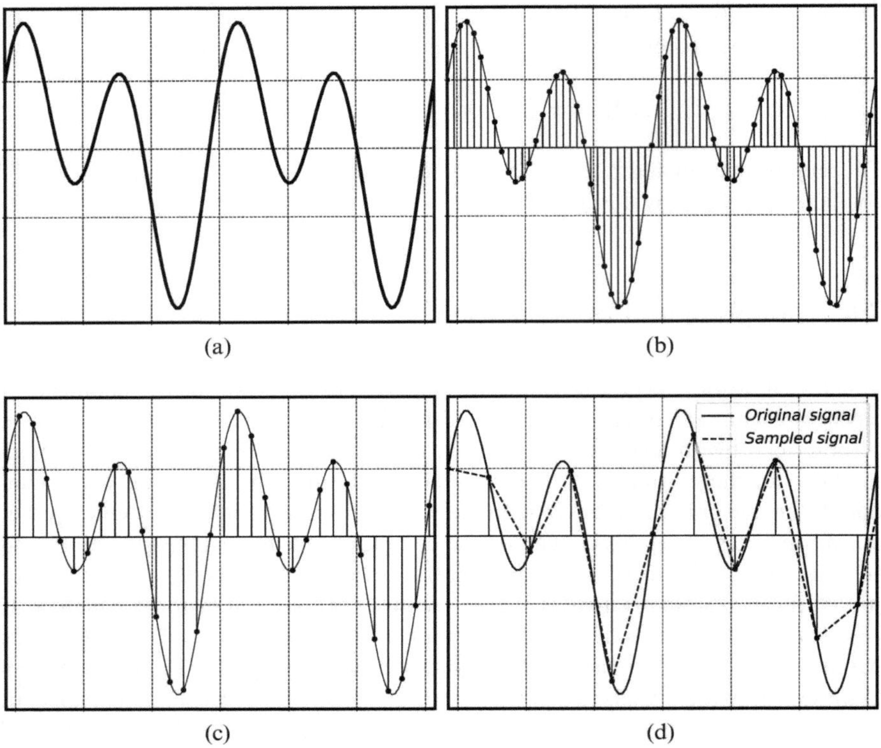

Fig. 7.7 Sampling of a continuous function

be small. A small Δx accurately represents the signal. However, an excessively large sample size increases the computation cost. The number of samples must be optimized to reduce processing costs. A continuous signal is heuristically sampled with a finite number of points; however, some problems are encountered during the sampling. Figure 7.7a represents an arbitrary continuous periodic signal. Figure 7.7b and c represent the discrete sampling of function in Fig. 7.7a with different values of Δx. As Δx increases, the discrete signal shows a different pattern from the original signal (Fig. 7.7d). What is the optimum value of the Δx for which the sampled sequence properly represents the original signal?

A signal consists of infinite frequency components; some information content is lost during sampling. When the continuous analog signal is sampled, the resulting discrete signal has more frequency components than the analog signal. The high-frequency signals make the data noisy, and it is difficult to interpolate any functional value between two points. Often, the magnitude of the higher-frequency components of practical signals gradually weakens. Therefore, the frequency components above a certain threshold can be considered insignificant, and the signal can be regarded as band-limited for practical purposes. The discrete-time signal is made band-limited by eliminating all frequencies beyond the threshold ($F(\omega) = 0, \quad |\omega| \geq \frac{\omega_b}{2}$). The smallest value of ω_b is known as bandwidth. Impulse train or Dirac's comb discussed in Chap. 5 is used for sampling. Dirac's comb of period p or Ш function is represented as

$$Ш(x) = \sum_{n=-\infty}^{\infty} \delta(x - np) \tag{7.82}$$

This represents periodically distributed Dirac delta function p apart. The Fourier transform of Dirac's comb results in another Dirac's comb in the frequency domain:

$$\mathscr{F}\{Ш(x)\} = \frac{1}{2\pi}\frac{1}{p}Ш\left(\frac{n}{p}\right) \tag{7.83}$$

Let $f(x)$ be a continuous function over the entire domain; its functional values at a discrete location can be pulled by multiplying the function with Dirac's comb:

$$\begin{aligned} f_s(x) &= f(x)Ш(x) = f(x) \sum_{n=-\infty}^{\infty} \delta(x - np) \\ &= \sum_{n=-\infty}^{\infty} f(np)\delta(x - np) \end{aligned} \tag{7.84}$$

The right side of Eq. (7.84) represents the sampling of a continuous function, and the left side shows the sampling value of the function at some discrete locations. Taking the Fourier transfer of Eq. (7.84),

$$\begin{aligned} \mathscr{F}\{f_s(x)\} &= F(n) * \frac{1}{2\pi p}Ш\left(\frac{n}{p}\right) \\ &= \frac{1}{p}\sum_{n=-\infty}^{\infty} F\left(\frac{n}{p}\right) \end{aligned} \tag{7.85}$$

Multiplication in the physical domain becomes convolution in the spectral domain. Taking the inverse Fourier transform of (7.85),

$$f_s(x) = \sum_{n=-\infty}^{\infty} f(np) \tag{7.86}$$

$f(x) \rightleftharpoons F(n)$ are Fourier pairs. The continuous signal $f(x)$ is sampled on a regular grid at intervals p. The left side of Eq. (7.86) represents the sampled value of the signal at discrete points p apart. This formula provides a relationship between periodic functions' spatial and frequency representations. But the puzzle remains with p. What should be the correct value of p? Swedish-American physicist and electronic engineer Harry Nyquist (1889–1976) and American mathematician and electrical engineer Claude Elwood Shannon (1916–2001) provided the Nyquist-Shannon sampling theorem for selecting appropriate values of p:

$$p \leq \frac{1}{2f_{\max}} \tag{7.87}$$

Alternatively, the sampling frequencies f_s, which are the reciprocals of the sampling intervals, must satisfy

$$f_s \geq 2f_{\max} \tag{7.88}$$

$f_{\max}$ is the maximum frequency present in the signal. Shannon is known as the "Father of Information Theory" and performed research at Bell labs during the World War on communication, encryption, encoding, and code-breaking. Two classic scholar articles Nyquist (1928) and Shannon (1949) give information on how much information can be compressed to communicate over a long distance and decompression downstream.

Nyquist-Shannon sampling theorem dictates that if more than two samples of the highest frequency content of a signal are taken, then the signal can be reconstructed from its samples exactly. In other words, resolving all frequencies in a function must be sampled at twice the highest frequency present. A function containing no frequency over ω Hz is completely determined by sampling at $2\omega\ Hz$ (Nyquist sampling frequency). The human ear can sense audio signals up to 22 kHz, so the $.mp3$ audio signals are encoded at 44 kHz. In practice, a higher number of samples are taken due to the limitation of physical device response. If the sampling theorem is not satisfied, the representation of the signal by samples is distorted. If a signal is slowly varying, the sampling interval can be longer. For rapidly varying signals, the sampling interval should be sufficiently short. Suppose a signal is sampled with a longer sampling interval than necessary. In that case, that component loses its identity and impersonates a lower-frequency component, a phenomenon called aliasing (sampled below the Nyquist sampling frequency). Figure 7.8 represents the sampling of simple sinusoids below the Nyquist sampling frequency, which results in aliased sample data (dotted line) with a period different from the original signal (solid line). Aliasing is also known as frequency folding, and to avoid aliasing, either a sufficient number of samples must be taken or, with a given number of samples, the number of frequency components in the signal by pre-filtering. In practice, the aliasing effect cannot be eliminated but is limited so that the accuracy of representation is adequate.

British mathematician Sir Edmund Taylor Whittaker (1873–1956) discovered the Nyquist-Shannon sampling theorem (Whittaker, 1915) earlier and Shannon (1949) cited Whittaker's work. In honor of all three contributors, this theorem is also known as the Whittaker-Nyquist-Shannon or cardinal theorem of interpolation. The Whittaker-Nyquist-Shannon or cardinal interpolation theorem offers a method to

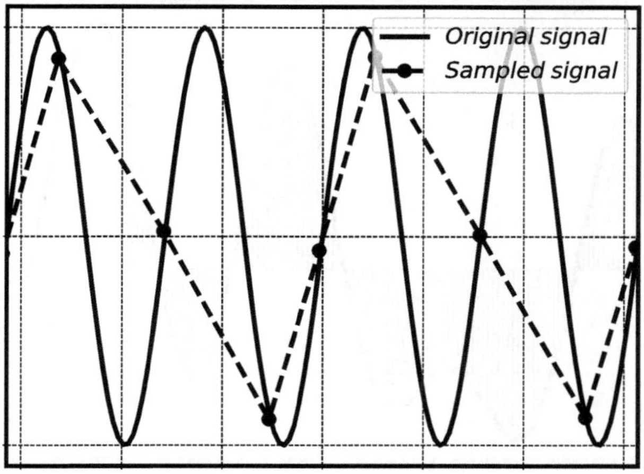

Fig. 7.8 Sampling of a function below Nyquist sampling frequency

reconstruct a continuous function from its Nyquist samples. This method involves creating a *sinc* function centered at each sampled point and then summing all these weighted *sinc* functions at each point. The resulting sum exactly reproduces the original continuous function.

A band-limited function defined over a single independent variable can be accurately reconstructed from samples taken at a uniform rate exceeding a critical minimum. Interpolation is the process of reconstructing a continuous signal from its discrete samples. The goal is to recover the original signal (or a close approximation) using the sampled points. The Poisson summation formula provides insight into how sampled data can be reconstructed back into a continuous signal. A *sinc* interpolation for band-limited signals is done for the reconstruction:

$$f(x) = \sum_{n=-\infty}^{\infty} f(np)\, sinc\left(\frac{x - np}{p}\right) \tag{7.89}$$

7.7 Windowing

Windowing is a common challenge in function sampling. Signal sampling with an adequate sampling rate Δx is illustrated in Fig. 7.9. However, instead of capturing the entire signal, only a portion is sampled, like placing a window over the signal. Data is ideally collected at uniform intervals in practical measurements, such as temperature and pressure recordings. Unfortunately, due to unavoidable circumstances, signals are often not captured continuously but rather from different instances of experiments. These signals are then sequentially combined to create a larger sequence for the discrete Fourier transform (DFT). The combined sequence may

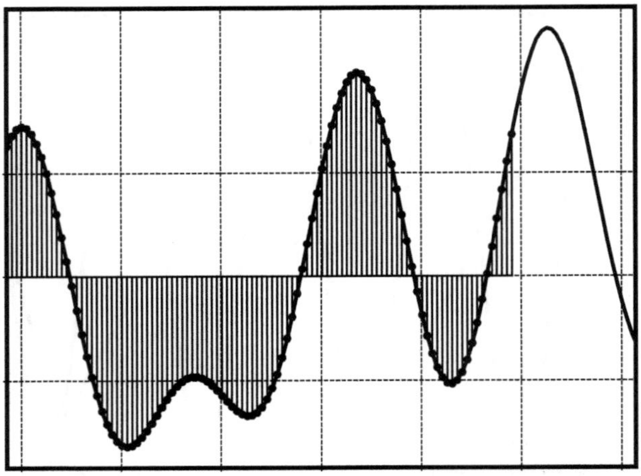

Fig. 7.9 Signal windowing

not be perfectly periodic, often exhibiting sharp discontinuities at the endpoints. The DFT algorithm interprets these sharp discontinuities as impulse responses, leading to a phenomenon known as spectral leakage. Spectral leakage results in the spreading of signal energy across multiple frequency bins in the spectrum, potentially obscuring or distorting the true frequency components of the original signal.

The DFT of a discrete-time sequence $\{f[n]\}$ of length N is defined as

$$F[k] = \sum_{n=0}^{N-1} f[n]e^{-i\frac{2\pi}{N}kn}, \quad k = 0, 1, 2, \ldots, N-1 \tag{7.90}$$

In this formulation, $\{f[n]\}$ is assumed to be periodic with period N. However, in practical applications, a signal is often sampled for a finite portion that is not periodic. This leads to spectral leakage, and energy from one frequency component spreads into neighboring frequency bins. The windowing technique is used to mitigate the effects of spectral leakage. It involves multiplying the signal by a window function that smoothly tapers it to zero at the boundaries. The goal is to reduce the sharp discontinuities at the edges of the sampled signal. The window function $\{w[n]\}$ with the same length N as the original function is multiplied with $\{f[n]\}$. The windowed signal becomes

$$f_w[n] = f[n] \cdot w[n] \tag{7.91}$$

The DFT of the windowed signal is then

$$F_w[k] = \sum_{n=0}^{N-1} f_w[n]e^{-i\frac{2\pi}{N}kn} = \sum_{n=0}^{N-1} f[n]w[n]e^{-i\frac{2\pi}{N}kn} \tag{7.92}$$

The window function $w[n]$ modifies the original signal in the physical domain and its spectrum in the frequency domain. These functions gradually taper the signal, smoothing abrupt discontinuities and minimizing spectral leakage. This process, known as windowing or gating, restricts the signal to a finite interval. However, this truncation introduces artifacts in the frequency domain. Several window functions are commonly used to mitigate spectral leakage. The main lobe refers to the central peak in the frequency spectrum of a windowed signal. The main lobe width is the range of frequencies over which most of the signal's energy is concentrated. Narrow main lobes provide better frequency resolution, i.e., closely spaced frequency components are more easily distinguished. A wider main lobe reduces frequency resolution and makes differentiating between closely spaced frequencies harder. The sidelobes are the smaller peaks or ripples that appear around the main lobe in the frequency spectrum. These sidelobes are undesirable because they represent spectral leakage, energy from one frequency leaking into other frequency bins. Sidelobe suppression refers to reducing the amplitude of these sidelobes to minimize the effects of spectral leakage. Effective sidelobe suppression means that the sidelobes are much smaller in amplitude compared to the main lobe, reducing the interference caused by leakage from other frequencies. Each has different properties in terms of main lobe width and sidelobe suppression.

7.7.1 Rectangular Window

The rectangular window corresponds to no windowing at all, which means

$$w[n] = 1, \quad 0 \leq n \leq N-1 \tag{7.93}$$

This is the simplest window, but it has high spectral leakage due to the sharp discontinuities at the boundaries of the signal.

7.7.2 Hann (or Hanning) Window

The Hanning window smoothly tapers the signal to zero at the boundaries, reducing the discontinuities:

$$w[n] = 0.5\left(1 - \cos\left(\frac{2\pi n}{N-1}\right)\right), \quad 0 \leq n \leq N-1 \tag{7.94}$$

The Hanning window reduces sidelobe levels significantly but at the cost of widening the main lobe in the frequency domain.

7.7.3 Hamming Window

The Hamming window is similar to the Hanning window but provides better sidelobe suppression at the expense of slightly higher main lobe width:

$$w[n] = 0.54 - 0.46\cos\left(\frac{2\pi n}{N-1}\right), \quad 0 \leq n \leq N-1 \tag{7.95}$$

7.7.4 Blackman Window

The Blackman window offers even more sidelobe suppression:

$$w[n] = 0.42 - 0.5\cos\left(\frac{2\pi n}{N-1}\right) + 0.08\cos\left(\frac{4\pi n}{N-1}\right), \quad 0 \leq n \leq N-1 \tag{7.96}$$

This window has higher sidelobe attenuation but the widest main lobe among these windows.

7.7.5 Gaussian Window

The Gaussian window applies a normal distribution shape to the data before performing DFT. It is popular because it provides reasonable control over the main lobe width and sidelobe suppression, as well as smooth transitions that minimize spectral leakage:

$$w[n] = e^{-\frac{1}{2}\left(\frac{n-N/2}{\sigma(N/2)}\right)^2} \tag{7.97}$$

σ is the standard deviation, which controls the width of the Gaussian curve. The main lobe width is controlled by the standard deviation σ. A smaller σ results in a narrower main lobe and a better frequency resolution, while a larger σ widens the main lobe. The Gaussian window is known for excellent sidelobe suppression. The sidelobes decay very quickly, which reduces spectral leakage significantly.

Multiplying a signal by a window in the time domain corresponds to convolution in the frequency domain (formally introduced in Chap. 8). If $F(f)$ is the Fourier transform of the original signal and $W(f)$ is the Fourier transform of the window function, then the Fourier transform of the windowed signal $f_w[n]$ is

$$F_w(f) = F(f) * W(f) \tag{7.98}$$

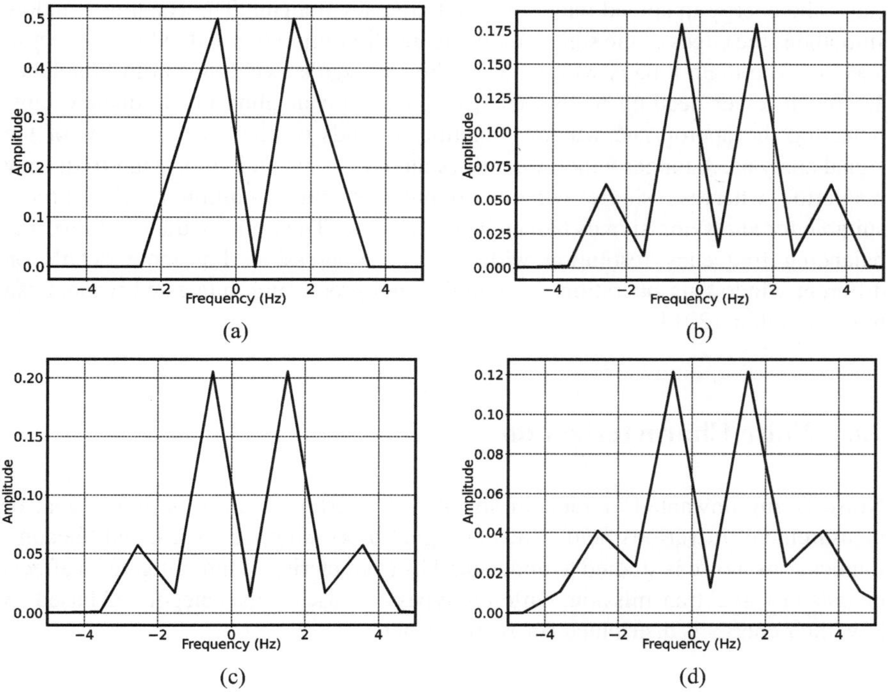

Fig. 7.10 Spectrum of the signal with different windowing. (**a**) Rectangular window. (**b**) Hanning window. (**c**) Hamming window. (**d**) Blackman window

Thus, windowing the signal smoothens its spectrum by convolving it with the spectrum of the window function. This reduces spectral leakage by attenuating the sidelobes (which are responsible for spreading the signal's energy into neighboring frequency bins). For example, let us consider a signal composed of two sinusoidal components with different frequencies:

$$f[n] = \sin(2\pi f_1 n) + 0.5\sin(2\pi f_2 n)$$

with $f_1 = 1Hz$ and $f_2 = 2\ Hz$. The signal is sampled with a total of 50 points. If DFT to this signal is applied using the rectangular window (i.e., without windowing), the signal does not complete an integer number of periods within the observation window (Fig. 7.10a). As a result, spectral leakage occurs, and the energy from the sinusoids spreads across multiple frequency bins. Applying the Hanning window to the signal reduces the spectral leakage. It shows two distinct peaks corresponding to f_1 and f_2 with less energy spread into neighboring bins than the rectangular window (Fig. 7.10b).

The rectangular window has a narrow main lobe but high sidelobes, which leads to significant spectral leakage. The sharp discontinuities at the edges of the signal cause the energy to spread widely in the frequency domain. The Hanning window smoothens the edges of the signal, reducing the discontinuities and reducing spectral leakage. The main lobe is wider than in the rectangular window, but the sidelobes are much lower, leading to less energy spread into neighboring frequency bins. Selecting the appropriate window function depends on the specific needs of the signal analysis. A longer window captures more of the signal, improving frequency resolution, whereas a shorter window offers better time resolution, making it more suitable for analyzing signals that change over time. The primary trade-off involves balancing frequency resolution with leakage suppression. For more details on different windowing operations in digital signal processing, interested readers can refer to Prabhu (2014).

7.8 Noise Characterization

Noise is any unwanted or random signal that interferes with a desired signal or measurement. In various fields, such as signal processing, acoustics, and telecommunications, noise is studied to understand its characteristics and mitigate its effects on data or signal transmission. Different types of noise can be categorized based on how their energy is distributed across frequencies.

7.8.1 White Noise

White noise is a random signal with equal intensity at all frequencies within a given bandwidth, meaning its power spectral density (PSD) is constant across the entire frequency spectrum:

$$\text{PSD}(f) = \text{Constant} \tag{7.99}$$

In physical space, the white noise is represented as a sequence of uncorrelated random variables with zero mean and a constant variance σ^2.

7.8.2 Pink Noise

Pink noise, called $1/f$ noise, has a PSD that decreases with frequency. Specifically, its power decreases inversely with frequency. This gives pink noise a richer sound at lower frequencies than white noise:

$$\text{PSD}(f) \propto \frac{1}{f} \tag{7.100}$$

Lower frequencies dominate pink noise, which falls off at a rate of 3 dB per octave. In the time domain, pink noise is generated by applying a low-pass filter to white noise to modify its frequency characteristics.

7.8.3 Brown Noise (Red Noise)

Brown noise, also called Brownian or red noise, has a power spectral density that decreases more steeply with frequency than pink noise. It is named after the Brownian motion, which describes the random movement of particle suspension in a fluid. The power spectral density of brown noise is inversely proportional to the square of the frequency:

$$\mathrm{PSD}(f) \propto \frac{1}{f^2} \tag{7.101}$$

Brown noise contains even more low-frequency components than pink noise. Its energy decreases by 6 dB per octave.

7.8.4 Blue Noise (Azure Noise)

Blue noise, also known as azure noise, has a power spectral density that increases with frequency. It is the opposite of brown noise and emphasizes higher frequencies:

$$\mathrm{PSD}(f) \propto f \tag{7.102}$$

This results in more energy at higher frequencies, and the noise sounds sharper and more high-pitched.

7.8.5 Violet Noise (Purple Noise)

Violet noise, also known as purple noise, has a power spectral density that increases with the square of the frequency. This results in even more emphasis on high-frequency components than blue noise:

$$\mathrm{PSD}(f) \propto f^2 \tag{7.103}$$

This gives violet noise a very sharp, hiss-like sound.

7.8.6 Gray Noise

Gray noise contains all frequencies with equal loudness, as opposed to white noise, which contains all frequencies with equal energy. The difference between the two is the result of psychoacoustics. More specifically, human hearing is more sensitive to some frequencies than others. The gray noise represents the equal loudness curve, which reflects the human ear's sensitivity to different frequencies.

7.8.7 Black Noise

Black noise refers to silence or near-silence with occasional random spikes. Its power spectral density is concentrated at very low or zero frequencies. Black noise has a spectrum corresponding to blackbody radiation (thermal noise).

Figure 7.11 presents the power spectral density (PSD) plots for various types of noise, including white, pink, brown, blue, purple, gray, and black noise. These PSD plots reveal how each noise interacts with frequency, highlighting their importance in specific real-world applications. Each noise type exhibits distinct frequency characteristics critical for various applications in signal processing and communication systems.

7.9 Quantization

Quantization is a fundamental aspect of almost all digital signal processing, as converting a signal to digital form typically involves rounding. The actual value of the function at the sampled points is rounded to the nearest digital level. Figure 7.12 illustrates a magnified visualization of quantization error. The smooth wave represents the original continuous signal. The quantized signal appears as a staircase function due to the limited number of discrete levels. Round-off and truncation errors are the primary sources of quantization errors. When a continuous value is mapped to the nearest discrete level, the difference between the true and quantized values introduces a round-off error. For some systems, quantization truncates values (e.g., rounding down instead of rounding to the nearest value), leading to a systematic bias in the error. Additionally, quantization is central to most lossy compression algorithms. The difference between an input value $f(x)$ and its quantized value $\hat{f}(x)$, i.e., the value mapped to the closest discrete level, is known as quantization error $e_q(x)$:

$$e_q(x) = f(x) - \hat{f}(x) \tag{7.104}$$

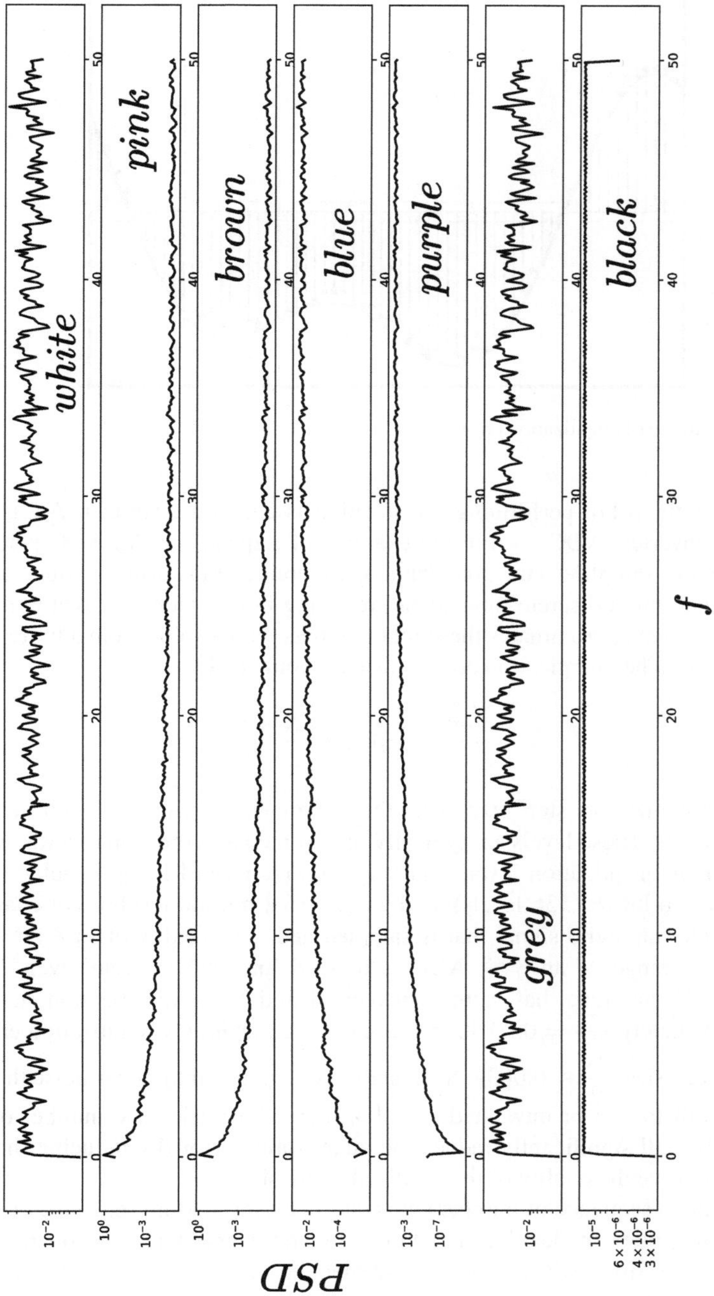

Fig. 7.11 Noise characterization based on power spectrum

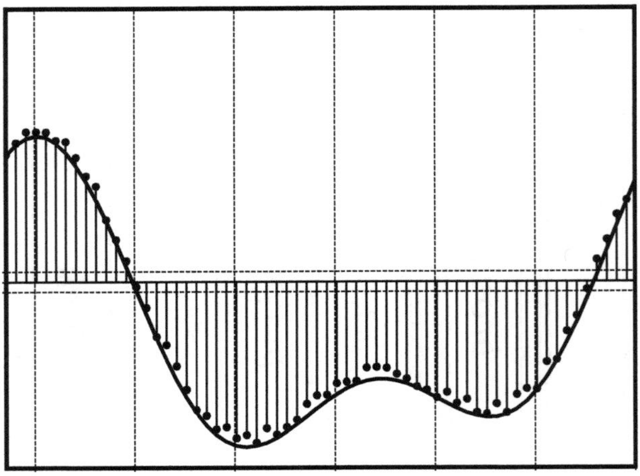

Fig. 7.12 Example of quantization error

A device or algorithm performing quantization is called a quantizer. An analog-to-digital converter (ADC) is a prime example of a quantizer. An ADC performs two processes: sampling and quantization. Sampling transforms a time-varying voltage signal into a discrete-time signal, resulting in a sequence of real numbers. Quantization then approximates these real numbers with a value from a finite set of discrete levels. The quantization error is usually bounded by

$$-\frac{\Delta}{2} \leq e_q(x) \leq \frac{\Delta}{2} \tag{7.105}$$

Δ is the quantization step size, i.e., the difference between two consecutive quantized levels. These levels are typically represented as fixed-point words. While the number of quantization levels can vary, common word lengths include 8-bit (256 levels), 16-bit (65,536 levels), and 24-bit (16.8 million levels). For example, let us consider an audio signal that is sampled and quantized with an 8-bit ADC. The dynamic range of an 8-bit ADC is divided into 256 discrete levels (since $2^8 = 256$). If the signal has values between -1 and 1, the quantization step size Δ is approximately $\frac{2}{256} = 0.0078$. In this case, the maximum quantization error is half of the step size, $\dfrac{\Delta}{2} \approx 0.0039$. In an audio system, this introduces noise that can manifest as distortion or unwanted sound artifacts. Increasing the number of bits from 8 to 16 will significantly reduce the quantization error (by reducing the step size) and improve the quality of the digitized audio signal.

Quantizing a sequence of numbers generates a corresponding sequence of quantization errors, often modeled as an additive random signal known as quantization noise due to its stochastic nature. The greater the number of quantization levels used, the lower the quantization noise power. The signal-to-quantization noise ratio

(SQNR) quantifies the relationship between the power of the original signal and the power of the quantization error. It is a key performance metric in digital signal processing. For a uniform quantizer with N bit quantization, the SQNR can be approximated as

$$\text{SQNR (dB)} = 6.02N + 1.76\ \text{dB} \tag{7.106}$$

This equation indicates that for every additional bit of resolution, the SQNR improves by approximately 6 dB, equating to a fourfold improvement in the signal-to-noise ratio.

The quantization error can be either deterministic or stochastic. For simple input signals like a sine wave or other periodic function, the uniform quantizer exhibits a pattern of the quantization error. This error is predicted and may introduce harmonic distortions in the signal. For complex or random input signals (such as speech or noise), the quantization error is often treated as random and behaves like additive white noise with a uniform distribution. The quantization error is modeled as additive noise for many systems and is usually called quantization noise. The quantization error is often assumed to be uniformly distributed between $-\frac{\Delta}{2}$ and $\frac{\Delta}{2}$, with a zero mean for signals with complex waveforms, such as speech, audio, or video signals. The power of the quantization noise is expressed as

$$P_q = \frac{\Delta^2}{12} \tag{7.107}$$

White noise approximation for quantization noise holds reasonably well for a high dynamic range signal. Dithering is done to reduce the effects of deterministic quantization error. Dithering involves adding a small amount of random noise to the input signal before quantization, which helps to randomize the quantization error and reduce patterns in the error. This technique is instrumental in audio and image processing, where quantization artifacts can be noticeable. Quantization error is a trade-off between accuracy and efficiency in digital systems. It can be minimized but not completely eliminated in practical applications.

7.10 Practicality in DFT Analysis

Let's shift from the good news to the less favorable side. The sampling theorem applies under the ideal condition where we can sample a function with perfect precision and no errors from quantization or windowing. It assumes that samples extend from $-\infty$ to ∞, meaning the sampled sequence contains an infinite number of terms, which is impractical in real-world applications. Thus, the sampling theorem should be viewed as an idealized model intended to provide insight into the challenges of function sampling. It serves as a guideline rather than an inflexible rule. One needs to know the function's highest frequency to set the appropriate sampling rate, which requires evaluating its Fourier transform. However, the primary reason for digitizing a function is often to compute its Fourier transform using a computer. Practical judgment and intuition are used to resolve this challenge. When the function's bandwidth is unknown, a sampling rate that smoothly draws the curve through the samples and closely approximates the original function is acceptable for DFT.

For practical applications, a common guideline is to sample the signal at relatively fine intervals. Capturing 10–20 samples per cycle of the highest frequency is generally adequate for discrete Fourier transform purposes. The interpolation method suggested by the sampling theorem is often complex and computationally demanding. The sampling rate used with actual data values usually reproduces the original function with adequate precision. However, it is essential to note that the statement "adequate precision" is nebulous. Each case should be evaluated individually, balancing the level of accuracy required against practical considerations.

7.11 Closure

This chapter has thoroughly explored fundamental concepts in the discrete Fourier transform (DFT), fast Fourier transform (FFT), signal sampling, windowing, and quantization. The DFT is introduced with a clear definition and an overview of its properties. The chapter also details the efficient computation of DFT using the FFT algorithm, including Python code examples for calculating power spectral density (PSD). The Nyquist-Shannon sampling theorem is discussed to ensure that continuous signals are accurately represented in their discrete forms. The chapter addresses the challenges associated with signal sampling, windowing, and quantization and presents practical guidelines for applying DFT in real-world scenarios. Mastery of these concepts will facilitate effective signal analysis and manipulation, opening doors to advanced applications across various fields of technology and science.

References

Bluestein, L. (1970). A linear filtering approach to the computation of discrete fourier transform. *IEEE Transactions on Audio and Electroacoustics, 18*(4), 451–455.

Cooley, J. W., & Tukey, J. W. (1965). An algorithm for the machine calculation of complex fourier series. *Mathematics of Computation, 19*(90), 297–301.

Jena, S. K., & Manceau, R. (2024). Dynamics of turbulent natural convection in a cubic cavity with centrally placed partially heated inner obstacle. *Physics of Fluids, 36*(8), 085123 .

Nyquist, H. (1928). Certain topics in telegraph transmission theory. *Transactions of the American Institute of Electrical Engineers, 47*(2), 617–644.

Prabhu, K. M. (2014). *Window functions and their applications in signal processing*. Taylor & Francis.

Rader, C. M. (1968). Discrete Fourier transforms when the number of data samples is prime. *Proceedings of the IEEE, 56*(6), 1107–1108.

Shannon, C. E. (1949). Communication in the presence of noise. *Proceedings of the IRE, 37*(1), 10–21.

Whittaker, E. T. (1915). XVIII.—On the functions which are represented by the expansions of the interpolation-theory. *Proceedings of the Royal Society of Edinburgh, 35*, 181–194.

Chapter 8
Convolution, Cross-correlation, and Stochastic Analysis

The mathematician's patterns, like the painter's or the poet's must be beautiful; the ideas, like the colors or the words, must fit together in a harmonious way. Beauty is the first test: there is no permanent place in this world for ugly mathematics.
— Godfrey H. Hardy

Abstract This chapter explores convolution, cross-correlation, and stochastic analysis for signal processing and random signal analysis. It begins by examining convolution in both continuous and discrete forms, highlighting applications such as determining inverse Laplace transforms, Volterra integral solutions, signal filtering, time series analysis, and image processing. The section on circular convolution addresses techniques for efficiently processing periodic data. The chapter then introduces probability theory, covering its historical roots, fundamental principles, joint probability distributions, and applications in random signal analysis. It discusses the role of convolution in probability distributions, especially in relation to the central limit theorem and other statistical properties. Cross-correlation is introduced as a method to measure similarities between signals or datasets, with detailed coverage of its mathematical properties and applications in stochastic system analysis. This chapter integrates these tools, making it a valuable resource for theoretical and practical applications across diverse fields.

Keywords Convolution · Cross-correlation · Stochastic analysis · Volterra integral · Signal filtering · Time series analysis · Image processing · Image filtering · Circular convolution · Random variable · Ensembles · Central limit theorem

8.1 Introduction

Convolution and cross-correlation are fundamental mathematical operations with profound applications in signal processing, image analysis, and various fields of science and engineering. At their core, these operations combine two sequences or

S. K. Jena, *Fourier, Laplace, and the Tangled Love Affair with Transforms*,
https://doi.org/10.1007/978-3-031-80165-5_8

functions to produce a third sequence or function that reveals vital characteristics of the original inputs. Although they share some similarities, convolution and cross-correlation serve distinct purposes and are employed in different contexts. The historical development of these operations reflects their broad utility and versatility. From their mathematical foundations in the works of Pierre-Simon Laplace and Joseph Fourier to their practical applications in modern technologies, convolution and cross-correlation have become indispensable tools in both theoretical analysis and practical implementation. Understanding these concepts provides a gateway to mastering various techniques in signal processing, control systems, and beyond.

8.2 Convolution

What is convolution? There's not a single answer to that question. The fact is that convolution is used in many ways and for many reasons, and it can be a mistake to try to attach to it one particular meaning or interpretation. This multitude of interpretations and applications is somewhat making convolution a special mathematical entity. From abstract mathematical prospective, it is safe to state that convolution is a mathematical operation that combines two functions or sequences to generate a third, often representing how one function's shape is altered by the other. As the application of convolution for particular application will be explored, it will be interpreted in a unique way. The mathematical concept of convolution's roots can be traced back to the study of integrals and differential equations in the eighteenth and nineteenth centuries. D'Alembert (1756) employed the convolution integral during the derivation of Taylor's theorem. In its integral form, convolution first appeared in probability theory and in solving differential equations. French mathematician Sylvestre François Lacroix (1765–1843) utilized the convolution integral in his book Lacroix and Courcier (1819). The convolution process involves integrating the product of two functions, with one function being reversed and shifted. This idea naturally arose when tackling problems related to the distribution of random variables and system responses to various inputs. Pierre-Simon Laplace (1749–1827) significantly contributed to the early development of convolution through his work on the Laplace transform, although he did not explicitly define convolution. His work laid the groundwork for its later formalization. Convolution operations also appeared in the work of Siméon Denis Poisson (1781–1840). A significant advancement followed with Joseph Fourier (1768–1830), whose research on Fourier series and Fourier transforms in the early nineteenth century introduced tools closely related to convolution. Convolution was formalized and became widely used in the early twentieth century. American transmission theorist John Renshaw Carson (1886–1940) applied this superposition integral in early communication systems, and Italian mathematician and physicist Vito Volterra (1860–1940) derived the integral for specific composition products. The term "convolution" itself did not gain widespread use until the 1950s or 1960s. It was previously known by its German name, "faltung," meaning folding. American mathematician Norbert Wiener (1894–1964) relied heavily on convolution in his work on filtering and prediction in noisy environments.

Convolution became a cornerstone in the analysis and design of filters, the modeling of physical systems, and the study of probability distributions. Convolution is used in image processing to detect edges, blur images, and enhance features. In machine learning, convolutional neural networks (CNNs) leverage convolution to automatically and efficiently identify patterns within data, driving advances in computer vision and other AI applications. The concept of convolution has also expanded beyond its traditional boundaries, influencing areas such as graph theory, quantum mechanics, and financial modelling. Its utility in both theoretical and practical applications continues to grow, making it a critical operation in understanding and manipulating complex systems and signals.

Convolution is fundamental to almost every field within the physical sciences and engineering. In mechanics, it is known as the superposition or Duhamel integral and is used to calculate the response of linear systems and structures to arbitrary, time-varying external disturbances. In systems theory, it serves as the impulse response integral, while in optics, it is recognized as the point spread or smearing function. Before diving into the mathematical details of this integral, understanding its physical interpretation through visual representation can greatly enhance comprehension of the convolution process. For instance, Fig. 8.1a depicts uniformly placed delta functions on a two-dimensional lattice. When this lattice is convolved with the image in Fig. 8.1b, it produces the lattice structure shown in Fig. 8.1c. The German term "faltung," meaning folding, hints at how convolution involves the multiplication of transformed functions.

Now, a distinct and innovative method will be employed to introduce convolution mathematically. It is often advantageous to work with functions in the spectral domain. When two functions are multiplied in the spectral domain, what is the relationship between their corresponding functions in the spatial domain? If $f(x) \leftrightarrow F(\omega)$ and $g(x) \leftrightarrow G(\omega)$ are two separate Fourier pairs, and their product in the spectral domain is $G(\omega)F(\omega)$, we must uncover the relationship between $f(x)$ and $g(x)$. To reveal this connection, we will use the principle of ab initio:

$$\begin{aligned} G(\omega) F(\omega) &= \mathscr{F}\{g(x)\}\mathscr{F}\{f(x)\} \\ &= \frac{1}{2\pi}\int_{-\infty}^{\infty} g(x) e^{-i\omega x} dx \frac{1}{2\pi}\int_{-\infty}^{\infty} f(x) e^{-i\omega x} dx \end{aligned} \tag{8.1}$$

Equation (8.1) can be interpreted as $F(\omega)$ is scaled by an amount $G(\omega)$. For definite integral, variables under the integral vanished at the end of integration. Substitution of the variable x with y is permitted without any violation of the principle:

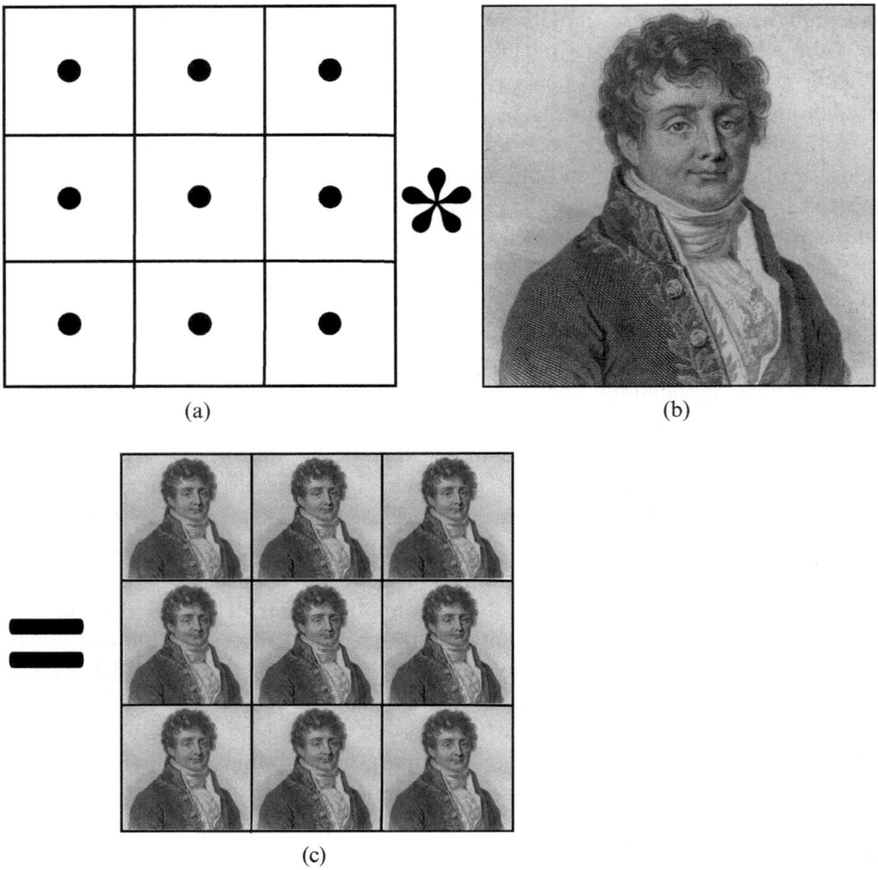

Fig. 8.1 Graphical representation of convolution

$$
\begin{aligned}
G(\omega)\, F(\omega) &= \frac{1}{2\pi} \int_{-\infty}^{\infty} g(y)\, e^{-i\omega y} dy \frac{1}{2\pi} \int_{-\infty}^{\infty} f(x)\, e^{-i\omega x} dx \\
&= \frac{1}{2\pi} \frac{1}{2\pi} \int_{-\infty}^{\infty} \int_{-\infty}^{\infty} g(y)\, f(x)\, e^{-i\omega(y+x)} dx dy \\
&= \frac{1}{2\pi} \int_{-\infty}^{\infty} \left(\frac{1}{2\pi} \int_{-\infty}^{\infty} g(y)\, e^{-i\omega(y+x)} dy \right) f(x)\, dx
\end{aligned}
\tag{8.2}
$$

With the substitution $z = x + y$, $y = z - x$ and $dy = dx$:

$$
\begin{aligned}
G(\omega) F(\omega) &= \frac{1}{2\pi} \int_{-\infty}^{\infty} \left(\frac{1}{2\pi} \int_{-\infty}^{\infty} g(z-x) e^{-i\omega z} dz \right) f(x)\, dx \\
&= \frac{1}{2\pi} \int_{-\infty}^{\infty} \left(\frac{1}{2\pi} \int_{-\infty}^{\infty} g(z-x) f(x)\, dx \right) e^{-i\omega z} dz \\
&= \frac{1}{2\pi} \int_{-\infty}^{\infty} h(z) e^{-i\omega z} dz = H(\omega)
\end{aligned}
\tag{8.3}
$$

$h(x) \leftrightarrow H(\omega)$ is the new Fourier pair. The new function $h(x)$ represents the desired functional relationship between $f(x)$ and $g(x)$ in the time/spatial domain and is known as a convolution operation:

$$
h(x) = (g * f)(x) = \int_{-\infty}^{\infty} g(x-y) f(y)\, dy \tag{8.4}
$$

The factor $\frac{1}{2\pi}$ is a matter of choice. In the study of convolution, this factor is often omitted as it does not affect the properties of convolution. Convolution is a mathematical operation on two functions $f(x)$ and $g(x)$ that produce third function $h(x) = (g * f)(x)$. Convolution is defined as the integral of the product of the two functions after one is reflected and shifted. Figure 8.2a shows one of the convolution functions $g(y)$ and its reflection $g(-y)$ is shown in Fig. 8.2b. Figure 8.2c shows the x shift in the function after reflection $g(x-y)$. Then $g(x-y)$ is multiplied with $f(y)$ and the integral is calculated across all possible shifts x from $-\infty$ to ∞. At each value of x, the convolution formula in Eq. (8.4) can be interpreted as the area under the curve of the function $f(y)$, weighted by the function $g(x-y)$ after x shift. Different regions of the input function $f(y)$ highlighted as $g(x-y)$ slide on x toward ∞. When x is positive, $g(y-x)$ shifts to the right (toward $+\infty$) by x. Conversely, when x is negative, $g(y-x)$ shifts to the left (toward $-\infty$) by $|x|$. The result of the integral remains unchanged regardless of which function is reflected and shifted before integration:

$$
h(x) = (g * f)(x) = \int_{-\infty}^{\infty} g(y) f(x-y)\, dy \tag{8.5}
$$

The convolution is commutative. A graphical illustration of the convolution operation is represented in Fig. 8.3. For computation simplicity, a rectangular pulse function is considered for both $f(y)$ and $g(y)$. The value of these functions is one within the pulse width and zero outside. Convolution operation starts as one of the

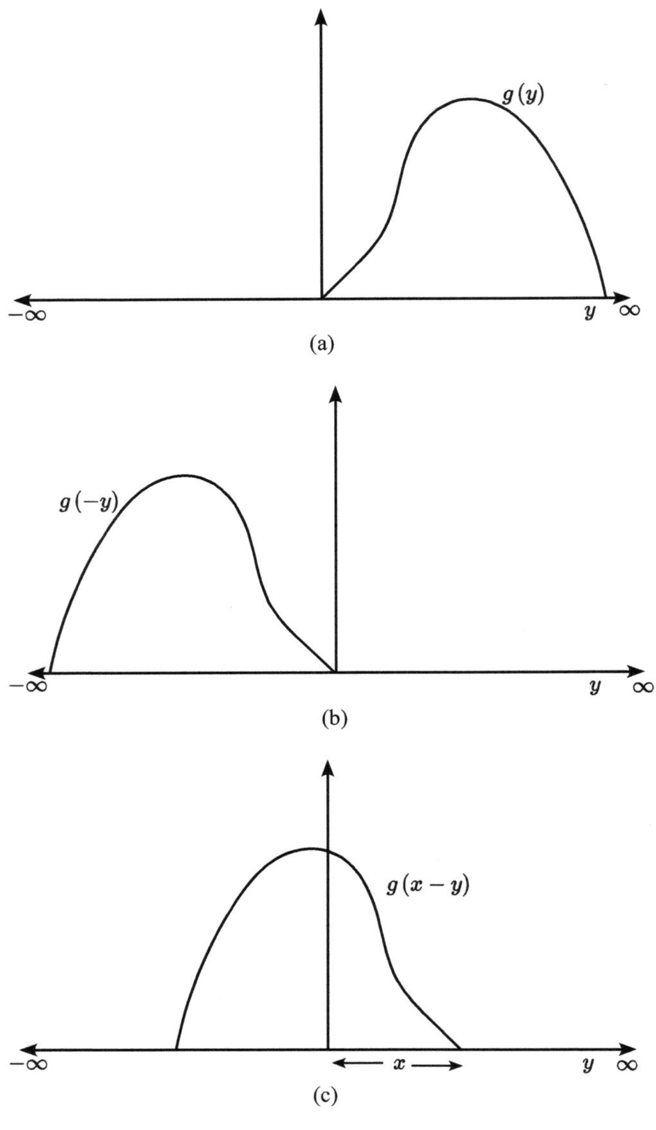

Fig. 8.2 Graphical representation of reflection and shift in $g(y)$. (**a**) $g(y)$. (**b**) $g(-y)$. (**c**) $g(x-y)$

functions starts sliding in x. In the sliding process, both $f(y)$ and $g(y-x)$ intersect for some range of x. The integral of the product of $f(y)$ and $g(y-x)$ is represented by the intersection area between the two functions (shown as a hatched mark in Fig. 8.3). This product is stored within the third function $h(x)$. $h(x)$ is a continuous function of x as the integral is evaluated for all possible shifts. Convolution between

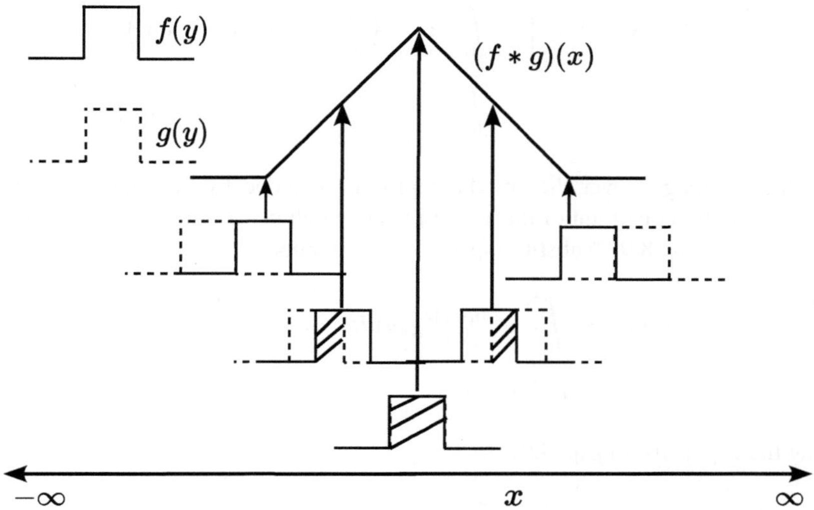

Fig. 8.3 Convolution of two rectangular pulses

two rectangular pulses results in a triangular pulse in Fig. 8.3. Multiplication in the spectral domain results in a convolution relationship in the time domain as in Eq. (8.5). The symmetry between Fourier and inverse Fourier transformation results in

$$\mathscr{F}^{-1}\left\{(G*F)(\omega)\right\} = \mathscr{F}^{-1}\left\{G(\omega)\right\}\mathscr{F}^{-1}\left\{F(\omega)\right\} = g(x)\,f(x) \tag{8.6}$$

Equation (8.6) represents convolution in the frequency domain, resulting in the multiplication of the two functions in the time domain. The same can be derived from the principle of duality:

$$\mathscr{F}\left\{g(x)\right\} * \mathscr{F}\left\{f(x)\right\} = G(\omega) * F(\omega) = \mathscr{F}\left\{g(x)f(x)\right\} \tag{8.7}$$

When $g(x)$ and $f(x)$ are both causal functions, i.e., defined only on $[0, \infty)$, Eq. (8.4) becomes

$$(g*f)(x) = \int_0^x f(y)g(x-y)dy \quad \text{for } f, g : [0,\infty) \to \mathbb{R} \tag{8.8}$$

The upper limit of the integral in Eq. (8.8) is x instead of ∞ because the causal functions $g(y) = 0$ for $y > x$. If the Laplace transform of the two functions $\mathscr{L}\{f(x)\} = F(s)$ and $\mathscr{L}\{g(x)\} = G(s)$, then the product of their transform $F(s)G(s)$ is the Laplace transform of the convolution of original function $(g*f)(x)$:

$$\mathscr{L}\left\{\int_0^x f(y)g(x-y)dy\right\} = \int_0^\infty e^{-sx}\left(\int_0^x f(y)g(x-y)dy\right)dx$$
$$= \int_0^\infty \left(f(y)\int_y^\infty e^{-sx}g(x-y)dx\right)dy \tag{8.9}$$

The double integral over the angular region is bounded by the lines $y = 0$ and $y = x$ in the first quadrant of the $x - y$-plane, which permits changing the order of integration in Eq. (8.9). Substituting $x - y = u$ results in

$$\int_y^\infty e^{-sx}g(x-y)dy = \int_0^\infty e^{-(u+y)s}g(u)du = e^{-ys}\int_0^\infty e^{-su}g(u)du$$
$$= e^{-ys}G(s) \tag{8.10}$$

Using the Eq. (8.10) in Eq. (8.9),

$$\mathscr{L}\left\{\int_0^x f(y)g(x-y)dy\right\} = \int_0^\infty f(y)e^{-ys}G(s)dy = G(s)\int_0^\infty f(y)e^{-ys}dy$$
$$= F(s)G(s) \tag{8.11}$$

The convolution of two functions in physical space is associated with multiplying transform pairs in Laplace space. The mathematical properties of the convolution integral are provided in Table 8.1.

Table 8.1 Properties of convolution

$f * g = g * f$	(commutative)
$(f * g) * h = f * (g * h)$	(associative)
$f * (g + h) = f * g + f * h$	(distributive over addition)
$f * \delta = \delta * f = f$	(identity)
$a(f * g) = (af) * g$	(associativity with scalar multiplication)
$\overline{f * g} = \overline{f} * \overline{g}$	(complex conjugation)
If $h(x) = f(x) * g(x)$ then $h(-x) = f(-x) * g(-x)$	(reversal)
$\frac{d}{dx}(f * g) = \frac{df}{dx} * g = f * \frac{dg}{dx}$	(differentiation)
$\int_{\mathbf{R}^d}(f * g)(x)\,dx = \left(\int_{\mathbf{R}^d} f(x)\,dx\right)\left(\int_{\mathbf{R}^d} g(x)\,dx\right)$	(integration)

Example 8.1 (Convolution of Two Functions) Convolution of $f(x) = \cos x$ and $g(x) = \exp(-a|x|), \quad a > 0$.

From the definition,

$$\begin{aligned}(f * g)(x) &= \int_{-\infty}^{\infty} f(x-y)g(y)dy = \int_{-\infty}^{\infty} \cos(x-y)e^{-a|y|}dy \\ &= \int_{-\infty}^{0} \cos(x-y)e^{ay}dy + \int_{0}^{\infty} \cos(x-y)e^{-ay}dy \\ &= \int_{0}^{\infty} \cos(x+y)e^{-ay}dy + \int_{0}^{\infty} \cos(x-y)e^{-ay}dy \\ &= 2\cos x \int_{0}^{\infty} \cos y e^{-ay}dy = \frac{2a\cos x}{(1+a^2)}\end{aligned}$$

If $a = 1$, then $(f * g)(x) = f(x)$ and g becomes an identity element of convolution.

Example 8.2 (Convolution of Two Functions) Convolution of $f(x) = \chi_{[a,b]}(x)$ and $g(x) = x^2$, where $\chi_{[a,b]}(x)$ is the characteristic function of the interval $[a, b] \subseteq \mathbb{R}$ defined by

$$\chi_{[a,b]}(x) = \begin{cases} 1, & a \leq x \leq b \\ 0, & \text{otherwise} \end{cases}$$

From the definition,

$$\begin{aligned}(f * g)(x) &= \int_{-\infty}^{\infty} f(x-y)g(y)dy = \int_{-\infty}^{\infty} \chi_{[a,b]}(x-y)g(y)dy \\ &= \int_{-\infty}^{\infty} \chi_{[a,b]}(y)g(x-y)dy = \int_{a}^{b} g(x-y)dy = \int_{a}^{b} (x-y)^2dy \\ &= \frac{1}{3}\left\{(x-a)^3 - (x-b)^3\right\}\end{aligned}$$

Example 8.3 (Convolution of Two Causal Functions) Convolution of two causal functions $f(x) = x$ and $g(x) = x^2$ is

$$(f * g)(x) = \int_0^x (y - x)y^2 dx = \left[\frac{y^3}{3}x - \frac{y^4}{4}\right]_0^x$$

$$= \frac{x^4}{3} - \frac{x^4}{4} = \frac{x^4}{12}$$

Example 8.4 (Convolution of Two Causal Functions) Convolution of two causal functions $f(x) = x$ and $g(x) = \sin x$ is

$$\begin{aligned}(f * g)(x) &= \int_0^x (x - y) \sin y dy \\ &= [-(x - y) \cos y]_0^x - \int_0^x (-1)(- \cos y)dy \\ &= x - \int_0^x \cos y dy \\ &= x - [\sin y]_0^x = x - \sin x\end{aligned}$$

8.3 Discrete Convolution

Before delving into the formal mathematical formula for the discrete convolution, let us physically interpret the convolution operation once again. The possible sequence from the outcome of rolling a ludo dice once is expressed as

$$\{f[n]\} = \{1, 2, 3, 4, 5, 6\} \tag{8.12}$$

If two ludo dice are rolled simultaneously, all possible sequences are graphically expressed in Fig. 8.4. The secondary or counter diagonal of the matrix represents the sequence of all the pairs having a similar sum from the two dice. Another unique way to get this sequence is to have an identical sum from the simultaneous running of two ludo dice. Keep one copy of the outcome sequence from the rolling dice fixed, flip the other copy, and shift it to the extreme left. Then, slide the second sequence over the first sequence with one index movement at a time. Figure 8.5 provides the graphical representation for such a process. During the sliding process, take all overlapping pairs from the two sequences and keep them in a new sequence,

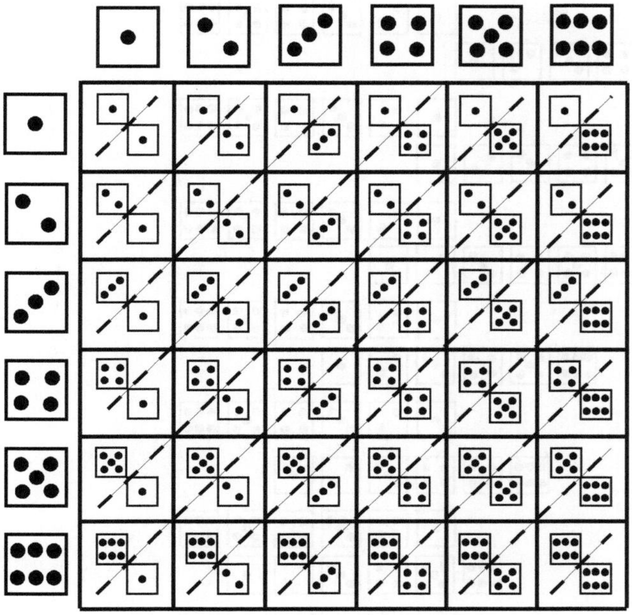

Fig. 8.4 Possible sequence from rolling two ludo dice simultaneously

as shown in Fig. 8.6. As shown in Fig. 8.6, a new sequence is at the end of the sliding process formed. The elements of this new sequence represent all possible combinations of two dice that give a particular sum. Dividing the total number of possible combining pairs (i.e., 36) with the elements of this sequence provides the probability of getting a particular sum from the rolling of two dice. This process of obtaining this sequence, as shown in Fig. 8.5, is mathematically expressed as

$$h[n] = \sum_{m=-\infty}^{\infty} f[m] \cdot f[n-m] \tag{8.13}$$

Equation (8.13) is nothing but discrete convolution of the sequence $\{f[n]\}$ with itself. In a more generic way, the discrete convolution between two sequences $\{f[n]\}$ and $\{g[n]\}$ are expressed as

$$\begin{aligned} h[n] = f[n] * g[n] &= \sum_{m=-\infty}^{\infty} f[m] \cdot g[n-m] \\ &= \sum_{m=-\infty}^{\infty} f[n-m] \cdot g[m] \end{aligned} \tag{8.14}$$

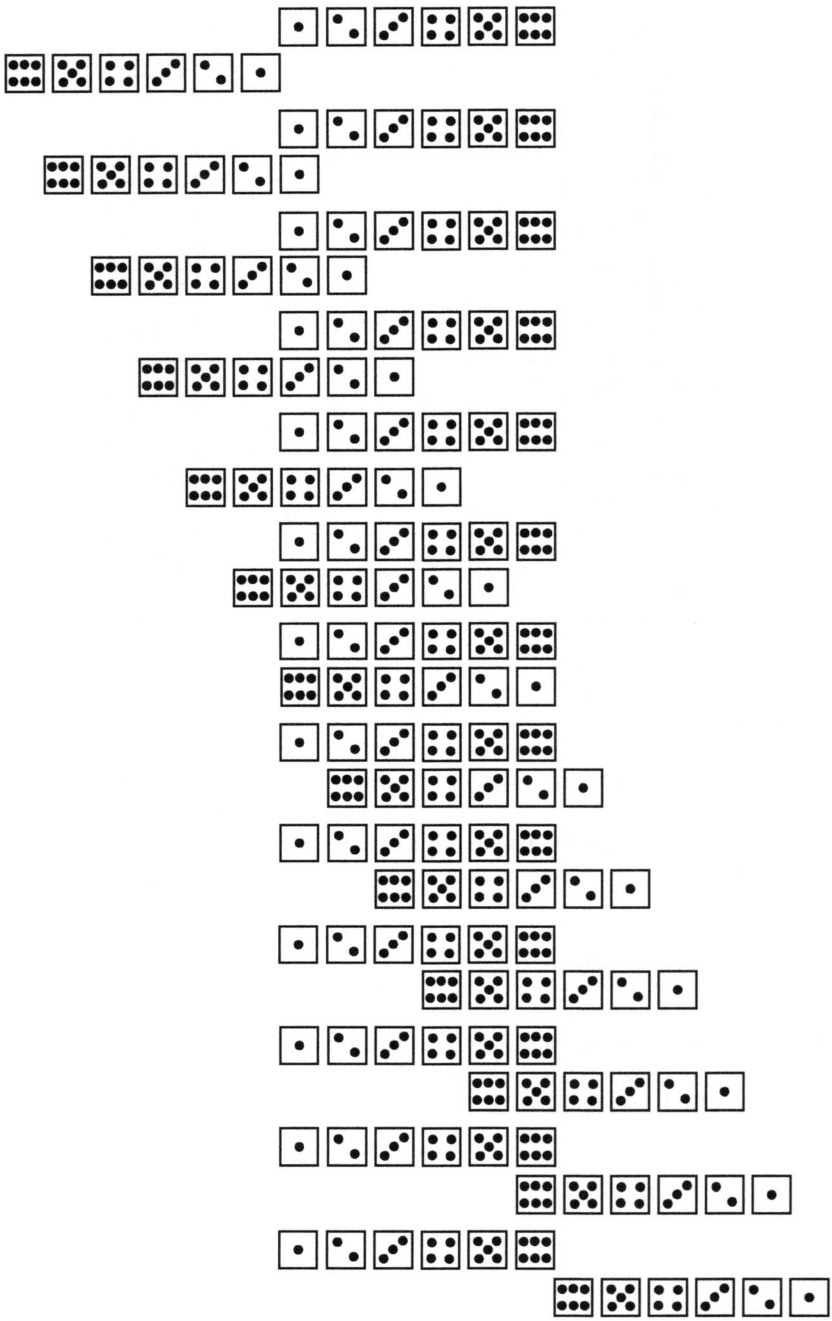

Fig. 8.5 Graphical representation of convolution of sequences from rolling two ludo dice simultaneously

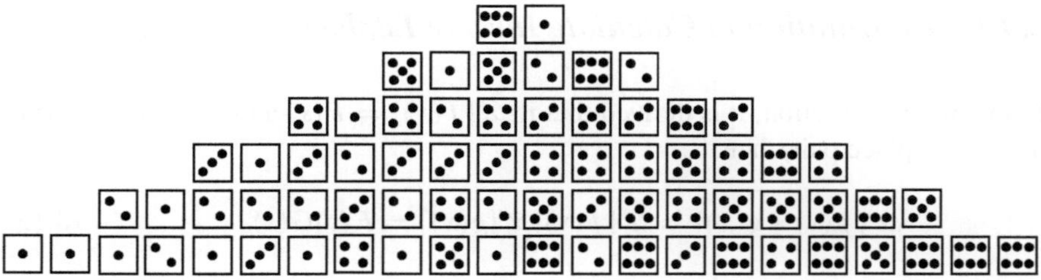

Fig. 8.6 Convolution of two sequence from rolling two ludo dice simultaneously

The "n" dependency of $h[n]$ is computed by summing over the dummy variable m for each n.

Example 8.5 (Discrete Convolution) $f = \begin{bmatrix} 1 \\ 0 \\ -1 \end{bmatrix}$ and $g = \begin{bmatrix} 0 \\ 1 \\ 2 \end{bmatrix}$

The discrete convolution becomes

$$f * g = \begin{bmatrix} f[1]g[1] \\ f[1]g[2] + f[2]g[1] \\ f[1]g[3] + f[2]g[2] + f[3]g[1] \\ f[2]g[3] + f[3]g[2] \\ f[3]g[3] \end{bmatrix} = \begin{bmatrix} 1 \cdot 0 \\ 1 \cdot 1 + 0 \cdot 0 \\ 1 \cdot 2 + 0 \cdot 1 + (-1)0 \\ 0 \cdot 2 + (-1)1 \\ (-1)2 \end{bmatrix}$$

$$= \begin{bmatrix} 0 \\ 1 \\ 2 \\ -1 \\ -2 \end{bmatrix}$$

8.4 Application of Convolution

Convolution is a versatile mathematical operation with broad applications across many modern technologies. It plays a crucial role in areas such as probability theory, time series analysis, signal filtering, and image processing. It is also fundamental to advanced AI applications, including deep learning with convolutional neural networks (CNNs) and natural language processing (NLP). While a comprehensive exploration of all its uses is beyond the scope of this book, this section will highlight some of the most relevant applications of the convolution operator.

8.4.1 Convolution to Calculate Inverse Laplace Transform

In the previous section, it is derived that if $\mathscr{L}\{f(x)\} = F(s)$ and $\mathscr{L}\{g(x)\} = G(s)$ are two Laplace pair, then

$$\mathscr{L}\{f(x) * g(x)\} = \mathscr{L}\{f(x)\}\mathscr{L}\{g(x)\} = F(s)G(s) \tag{8.15}$$

Taking the Laplace transform of both sides of Eq. (8.15),

$$\mathscr{L}^{-1}\{F(s)G(s)\} = f(x) * g(x) \tag{8.16}$$

Equation (8.16) is used to calculate the inverse Laplace transfer.

Example 8.6 (Inverse Laplace Transform) Inverse Laplace transform $H(s) = \dfrac{1}{(s-a)s}$:

$$H(s) = \frac{1}{(s-a)s} = F(s)G(s)$$

where

$$F(s) = \frac{1}{(s-a)}$$

$$G(s) = \frac{1}{s}$$

So,

$$f(x) = e^{ax}$$

$$g(x) = 1$$

From convolution of $f(x)$ and $g(x)$,

$$h(x) = f(x) * g(x) = \int_0^x e^{ay} \cdot 1dy = \frac{1}{a}\left(e^{ax} - 1\right)$$

To cross-check it,

(continued)

Example 8.6 (continued)

$$H(s) = \mathscr{L}\{h(x)\} = \frac{1}{a}\mathscr{L}\{(e^{ax} - 1)\} = \frac{1}{a}\left(\frac{1}{s-a} - \frac{1}{s}\right) = \frac{1}{a}\cdot\frac{a}{s^2 - as}$$

$$= \frac{1}{s-a}\cdot\frac{1}{s} = \mathscr{L}\left(e^{ax}\right)\mathscr{L}(1)$$

Example 8.7 (Inverse Laplace Transform) Inverse Laplace transform $H(s) = \frac{1}{\left(s^2+\omega^2\right)^2}$:

$$H(s) = \frac{1}{\left(s^2+\omega^2\right)}\frac{1}{\left(s^2+\omega^2\right)} = F(s)G(s)$$

where

$$F(s) = G(s) = \frac{1}{\left(s^2+\omega^2\right)}$$

So,

$$f(x) = g(x) = \frac{\sin\omega x}{\omega}$$

From convolution,

$$h(x) = f(x) * f(x) = \frac{\sin\omega x}{\omega} * \frac{\sin\omega x}{\omega} = \frac{1}{\omega^2}\int_0^x \sin\omega y \sin\omega(x-y)dy$$

$$= \frac{1}{2\omega^2}\left[-y\cos\omega x + \frac{\sin\omega y}{\omega}\right]_{y=0}^{t}$$

$$= \frac{1}{2\omega^2}\left[-x\cos\omega x + \frac{\sin\omega x}{\omega}\right] = \frac{1}{2\omega^3}\left[\sin\omega x - \omega x\cos\omega x\right]$$

Example 8.8 (Inverse Laplace Transform) Inverse Laplace transform $H(s) = \frac{1}{s^2\left(s^2 + a^2\right)}$:

$$H(s) = \frac{1}{s^2\left(s^2 + a^2\right)} = F(s)G(s)$$

where

$$F(s) = \frac{1}{s^2}$$

$$G(s) = \frac{1}{\left(s^2 + a^2\right)}$$

So,

$$f(x) = x$$

$$g(x) = \frac{\sin ax}{a}$$

From convolution of $f(x)$ and $g(x)$,

$$\begin{aligned} h(x) = f(x) * g(x) &= \frac{1}{a}\int_0^x (x - y) \cdot \sin aydy \\ &= \frac{x}{a}\int_0^x \sin aydy - \frac{1}{a}\int_0^x y \sin aydy \\ &= \frac{1}{a^2}\left(x - \frac{1}{a}\sin ax\right) \end{aligned}$$

Example 8.9 (Inverse Laplace Transform) Inverse Laplace transform $H(s) = \frac{1}{\sqrt{s}(s - a)}$:

$$H(s) = \frac{1}{\sqrt{s}(s - a)} = F(s)G(s)$$

where

(continued)

Example 8.9 (continued)

$$F(s) = \frac{1}{\sqrt{s}}$$

$$G(s) = \frac{1}{(s-a)}$$

So,

$$f(x) = \frac{1}{\sqrt{\pi x}}$$

$$g(x) = e^{ax}$$

for $a \geq 0$. From convolution of $f(x)$ and $g(x)$,

$$\begin{aligned} h(x) = f(x) * g(x) &= \frac{1}{\sqrt{\pi}} \int_0^x \frac{1}{\sqrt{y}} e^{a(x-y)} dy \\ &= \frac{2e^{ax}}{\sqrt{\pi a}} \int_0^{\sqrt{ax}} e^{-z^2} dz, \quad (\text{putting } \sqrt{ay} = z) \\ &= \frac{e^{ax}}{\sqrt{a}} \operatorname{erf}(\sqrt{ax}) \end{aligned}$$

Example 8.10 (Inverse Laplace Transform) Inverse Laplace transform $H(s) = \frac{1}{s} e^{-a\sqrt{s}}$:

$$H(s) = \frac{1}{s} e^{-a\sqrt{s}} = F(s)G(s)$$

where

$$F(s) = \frac{1}{s}$$

$$G(s) = e^{-a\sqrt{s}}$$

So,

(continued)

Example 8.10 (continued)

$$f(x) = 1$$

$$g(x) = \frac{a}{2} \frac{e^{-a^2/4x}}{\sqrt{\pi x^3}}$$

for $a \geq 0$. From convolution of $f(x)$ and $g(x)$,

$$h(x) = f(x) * g(x) = \frac{a}{2\sqrt{\pi}} \int_0^x \frac{e^{-a^2/4y}}{y^{3/2}} dy, \quad \text{(putting } \frac{a}{2\sqrt{y}} = z\text{)}$$

$$= \frac{2}{\sqrt{\pi}} \int_{\frac{a}{2\sqrt{x}}}^{\infty} e^{-z^2} dz = \text{erfc}\left(\frac{a}{2\sqrt{x}}\right)$$

8.4.2 Convolution for Solution of Volterra Integral

If the upper limit of integration is a variable, then the integral is known as Volterra integral. This is named after Italian mathematician Vito Volterra (1860–1940). Convolution helps solve a few Volterra integral equations.

Example 8.11 (Volterra Integral) A Volterra integral equation of the second kind is expressed as

$$f(x) - \int_0^x f(y) \sin(x - y) dy = x$$

The given Volterra integral equation can be written as

$$f(x) - f(x) * \sin x = x$$

Let $F(s)$ be the Laplace transform of $f(x)$. Taking Laplace transform of the above equation,

$$F(s) - F(s) \frac{s^2}{s^2 + 1} = \frac{1}{s^2}$$

$$F(s) = \frac{s^2 + 1}{s^4} = \frac{1}{s^2} + \frac{1}{s^4}$$

(continued)

Example 8.11 (continued)
Now taking the inverse Laplace transform gives

$$f(x) = x + \frac{x^3}{6}$$

Example 8.12 (Volterra Integral) A Volterra integral equation of the second kind is expressed as

$$f(x) - \int_0^x (1 + y) f(x - y) dy = 1 - \sinh x$$

The given Volterra integral equation can be written as

$$f(x) - f(x) * (1 + x) = 1 - \sinh x$$

Let $F(s)$ be the Laplace transform of $f(x)$. Taking Laplace transform of the above equation,

$$F(s) - F(s) \left(\frac{1}{s} + \frac{1}{s^2} \right) = \frac{1}{s} - \frac{1}{s^2 - 1}$$

Simplifying the above equation algebraically,

$$F(s) \cdot \frac{s^2 - s - 1}{s^2} = \frac{s^2 - 1 - s}{s\left(s^2 - 1\right)}$$

$$F(s) = \frac{s}{s^2 - 1}$$

Now taking the inverse Laplace transform gives

$$f(x) = \cosh x$$

8.4.3 Convolution for Signal Filtering

Instrumental measurement errors are inherent in any measured data, often leading to significant noise that can obscure meaningful information, as illustrated in Fig. 8.7a. This high-frequency noise complicates data interpretation, necessitating additional steps to filter out these unwanted disturbances. Convolution plays a crucial role in the design of these filters. The process involves taking the noisy input signal $f(x)$ and convolving it with an impulse or transfer function $g(x)$ in the time or spatial domain. This convolution corresponds to the multiplication of their Fourier transforms in the spectral domain.

To illustrate, consider the design of a low-pass filter, which allows signals with lower frequencies to pass through while blocking higher frequencies. In this case, a rectangular function $\Pi(\omega)$ is used as the impulse, defined by a cut-off frequency range from $-\omega_c$ to ω_c. The desired outcome in the spectral space is obtained by multiplying $\Pi(\omega)$ with the Fourier transform of the input signal $F(\omega)$. This product is zero outside the specified frequency range $(-\omega_c, \omega_c)$, effectively removing high-frequency components.

The inverse Fourier transform of this product, $F(\omega)\Pi(\omega)$, yields the filtered signal, as shown in Fig. 8.7b. This inverse transform is equivalent to the convolution of the impulse function with the noisy input signal, producing a cleaner version of the data. Figure 8.7 visually represents this entire process in the spectral domain, demonstrating how convolution and Fourier transform work together to filter out the noise and enhance data quality.

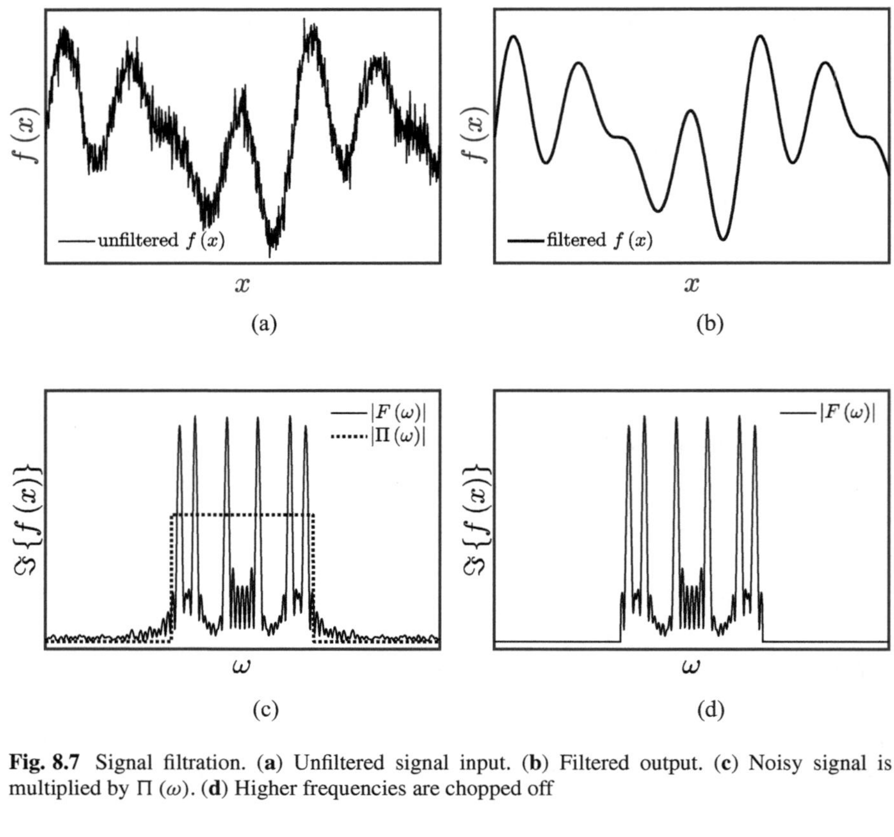

Fig. 8.7 Signal filtration. (**a**) Unfiltered signal input. (**b**) Filtered output. (**c**) Noisy signal is multiplied by $\Pi(\omega)$. (**d**) Higher frequencies are chopped off

8.4.4 Convolution for Time Series Analysis

A moving average is a widely used technique in time series analysis to smooth data and reveal underlying trends by reducing noise. Convolution is an effective way to compute a moving average. The noisy input data is considered as input sequence $\{f[n]\}$, while the moving averaging kernel $\{g[n]\}$ is the second sequence. For example, a simple moving average filter might assign equal weights to a specified number of adjacent data points. By convolving this averaging kernel with the input data, the output is a smoothed version of the original data:

$$h[n] = (f * g)[n] = \sum_{k=-K}^{K} f[n-k] \cdot g[k] \tag{8.17}$$

$h[n]$ is the resulting smoothed data point after applying the moving average. Inputs of an Nth-order uniform kernel is expressed as

$$\{g[n]\} = \{1/N, 1/N, \ldots, 1/N\} \tag{8.18}$$

Figure 8.8 shows the United States dollar (USD) to Indian national rupee (INR) currency conversion price for each day, starting from January 1, 2020. The original conversion rate varies too frequently (shown in the dotted mark). However, convolution of the original sequence with a uniform kernel produces a smooth moving average for the currency exchange rate. Python script for moving the average of a time series data is provided below.

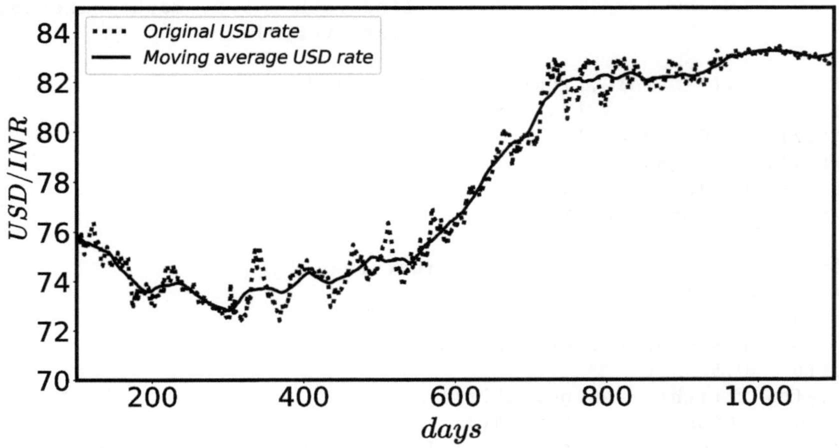

Fig. 8.8 Moving average of USD/INR from January 1, 2020

```
# Moving average of time series
import matplotlib.pyplot as plt
import numpy as np
import pandas as pd

#import the data
series = pd.read_csv("USD_INR_Historical_Data.csv.csv",sep=r'\s
                                  *,\s*', header=0, encoding='
                                  ascii', engine='python',usecols
                                  =["Date","Price"])
USD = np.flip(series["Price"])

#Define window size
w=51
#Uniform kernel
Uniform_kernel=np.ones((1,w))/w
Uniform_kernel=Uniform_kernel[0,:]

#Convolve the data for moving average
convolved_USD=np.convolve(USD,Uniform_kernel,'same')

fig = plt.figure()
fig.set_figheight(2.5)
fig.set_figwidth(4)
ax = fig.add_subplot(1, 1, 1)

plt.plot(series["Date"], USD,":k", label = "$Original \; USD \;
                                   rate$", linewidth = 4.0,
                                  markersize=18, markevery=1)
plt.plot(series["Date"], convolved_USD,"-k", label = "$Moving
                                  \; average \; USD \; rate$",
                                  linewidth = 3.0, markersize=18,
                                   markevery=1)

plt.xlabel("$days$", fontsize='40', labelpad=-5)
plt.ylabel("$USD/INR$", fontsize='40', labelpad=-5)
plt.xlim([100, 1100])
plt.ylim([70, 85])
plt.xticks(fontsize=30)
plt.yticks(fontsize=30)

plt.legend(loc = "upper left", fontsize='20')
ax = fig.gca()
for axis in ['top','bottom','left','right']:
    ax.spines[axis].set_linewidth(3.5)
ratio = 0.5
x_left, x_right = ax.get_xlim()
y_low, y_high = ax.get_ylim()
ax.set_aspect(abs((x_right-x_left)/(y_low-y_high))*ratio)
plt.show()
plt.close()
```

More sophisticated kernels can be applied to achieve different types of smoothing or emphasize certain data features. This versatility makes convolution a vital tool in time series analysis, enabling analysts to extract meaningful patterns from noisy datasets.

8.4.5 Image Processing

In the age of Instagram reels and YouTube shorts, image filters have become an essential tool, particularly cherished by millennials and Gen Z. These filters add a touch of magic to every captured moment, turning the ordinary into something extraordinary. Take, for instance, the timeless beauty of Marilyn Monroe, captured in high resolution in Fig. 8.9a. Her image, filled with intricate details, is like a memory brimming with vivid emotions. When this image is gently convolved with a Gaussian blurring function or point spread function, as shown in Fig. 8.9b, it softens the noise and smooths out the excess detail, much like nostalgia gently blurring the sharp edges of the past. The result, displayed in Fig. 8.9a, is a dreamier, more romantic version of the original, an image that feels like it's been wrapped in the soft glow of reminiscence. To truly grasp this image filtering process, one must delve

(a) (b)

Fig. 8.9 Image smearing through convolution

into the enchanting world of higher-dimensional convolution operations, beautifully encapsulated in the two-dimensional convolution equation (8.19):

$$\begin{aligned} f(x, y) * g(x, y) &= \int_{-\infty}^{\infty} \int_{-\infty}^{\infty} f(x-\alpha, y-\beta) g(\alpha, \beta) d\alpha d\beta \\ &= \int_{-\infty}^{\infty} \int_{-\infty}^{\infty} f(\alpha, \beta) g(x-\alpha, y-\beta) d\alpha d\beta \end{aligned} \tag{8.19}$$

α and β are the dummy variables. The discrete version of the two-dimensional convolution formula is expressed as

$$\begin{aligned} h[r, c] = f[r, c] * g[r, c] &= \sum_{m_r=-\infty}^{\infty} \sum_{m_c=-\infty}^{\infty} f\left[m_r, m_c\right] g\left[r-m_r, c-m_c\right] \\ &= \sum_{m_r=-\infty}^{\infty} \sum_{m_c=-\infty}^{\infty} g\left[m_r, m_c\right] f\left[r-m_r, c-m_c\right] \end{aligned} \tag{8.20}$$

These two-dimensional convolutions are like the hidden brushstrokes in a masterpiece, bringing life to digital images through edge detection, blurring, sharpening, and embossing. They add depth and emotion, transforming ordinary pixels into captivating visuals that evoke feelings and tell stories. Their role in digital image processing is vital, driving a billion-dollar industry that turns technology into art. This section delves into the mathematical intricacies behind these enchanting image transformations, exploring how these operations breathe life into every digital canvas.

Digital images are stored as matrices of pixels, where each pixel represents a small square of color that makes up the entire image. In the most common format, images are represented using the RGB color model, which stands for red, green, and blue. Each pixel in an image is associated with three values corresponding to the intensity of these three primary colors. The image is stored as a three-dimensional matrix, where the dimensions correspond to the height and width of the image and the color channels (red, green, and blue). For example, if an image is 1000 pixels wide and 800 pixels tall, it would be stored as a matrix with dimensions $800 \times 1000 \times 3$. The third dimension holds the color information: one matrix for red, one for green, and one for blue. Figure 8.10 shows the pixel space of a digital image. Each element within these matrices contains an integer value ranging from 0 to 255, which indicates the intensity of the respective color in that pixel. A value of 0 means that the color is absent, while 255 indicates the maximum intensity. When these color channels are combined, they create a full spectrum of colors, allowing the image to display everything from subtle shades to vibrant hues. This matrix format is what enables computers and digital devices to store, manipulate, and display images with incredible precision and detail. A small matrix called the filter matrix is convoluted with the original image matrix. This convolution operation

Fig. 8.10 Pixel space of a digital image

122	180	195	223	032	067	212	137
182	128	163	156	194	023	043	231
123	122	098	142	134	102	043	129
152	222	147	132	171	111	127	162
009	062	087	102	125	123	117	106
102	212	063	034	085	074	129	142
159	174	137	185	197	206	232	016
042	065	152	129	139	172	222	082
134	157	189	164	127	153	132	172

0.0625	0.125	0.0625
0.125	0.25	0.125
0.0625	0.125	0.0625

Blur

0.0	-1.0	0.0
-1.0	5.0	-1.0
0.0	-1.0	0.0

Sharpen

-2.0	-1.0	0.0
-1.0	1.0	1.0
0.0	1.0	2.0

Emboss

-1.0	-1.0	-1.0
-1.0	8.0	-1.0
-1.0	-1.0	-1.0

Outline

1.0	2.0	1.0
0.0	0.0	0.0
-1.0	-2.0	-1.0

Top sobel

-1.0	-2.0	-1.0
0.0	0.0	0.0
1.0	2.0	1.0

Bottom sobel

1.0	0.0	-1.0
2.0	0.0	-2.0
1.0	0.0	-1.0

Left sobel

-1.0	0.0	1.0
-2.0	0.0	2.0
-1.0	0.0	1.0

Right sobel

Fig. 8.11 3×3 image filter matrix

adjusts various image processing techniques, such as blurring, edge detection, and sharpening, for visual effects and corrections. Figure 8.11 shows various 3×3 filter matrices. To understand the $2D$ convolution process, let us consider a 9×9 image matrix:

$$\text{Image} = \begin{bmatrix} 1 & 2 & 3 & 4 & 5 & 6 & 7 & 8 & 9 \\ 10 & 11 & 12 & 13 & 14 & 15 & 16 & 17 & 18 \\ 19 & 20 & 21 & 22 & 23 & 24 & 25 & 26 & 27 \\ 28 & 29 & 30 & 31 & 32 & 33 & 34 & 35 & 36 \\ 37 & 38 & 39 & 40 & 41 & 42 & 43 & 44 & 45 \\ 46 & 47 & 48 & 49 & 50 & 51 & 52 & 53 & 54 \\ 55 & 56 & 57 & 58 & 59 & 60 & 61 & 62 & 63 \\ 64 & 65 & 66 & 67 & 68 & 69 & 70 & 71 & 72 \\ 73 & 74 & 75 & 76 & 77 & 78 & 79 & 80 & 81 \end{bmatrix}$$

Now, the convolution of the above matrix will be done with the emboss filter:

$$\text{Emboss filter} = \begin{bmatrix} -2 & -1 & 0 \\ -1 & 1 & 1 \\ 0 & 1 & 2 \end{bmatrix}$$

The first 3×3 region from the top-left corner of the image matrix is selected:

$$\begin{bmatrix} 1 & 2 & 3 \\ 10 & 11 & 12 \\ 19 & 20 & 21 \end{bmatrix}$$

Then, the filtration operation is done by first combining the patch of the image matrix with an emboss filter element-wise:

$$\begin{bmatrix} 1 \cdot (-2) & 2 \cdot (-1) & 3 \cdot 0 \\ 10 \cdot (-1) & 11 \cdot 1 & 12 \cdot 1 \\ 19 \cdot 0 & 20 \cdot 1 & 21 \cdot 2 \end{bmatrix} = \begin{bmatrix} -2 & -2 & 0 \\ -10 & 11 & 12 \\ 0 & 20 & 42 \end{bmatrix}$$

Then, the sum of all values is

$$-2 + (-2) + 0 + (-10) + 11 + 12 + 0 + 20 + 42 = 71$$

So, the first value of the filtered image matrix is 70. No moving one element further in the horizontal direction, a new 3 patch is extracted from the image matrix:

$$\begin{bmatrix} 2 & 3 & 4 \\ 11 & 12 & 13 \\ 20 & 21 & 22 \end{bmatrix}$$

Again, the same process is repeated with this patch of the matrix as

$$\begin{bmatrix} 2\cdot(-2) & 3\cdot(-1) & 4\cdot 0 \\ 11\cdot(-1) & 12\cdot 1 & 13\cdot 1 \\ 20\cdot 0 & 21\cdot 1 & 22\cdot 2 \end{bmatrix} = \begin{bmatrix} -4 & -3 & 0 \\ -11 & 12 & 13 \\ 0 & 21 & 44 \end{bmatrix}$$

Sum of all values:

$$-4 + (-3) + 0 + (-11) + 12 + 13 + 0 + 21 + 44 = 72$$

So, the second value of the output matrix is 72. After applying the convolution operation to the entire image, the result is

$$\text{Output} = \begin{bmatrix} 71 & 72 & 73 & 74 & 75 & 76 & 77 \\ 80 & 81 & 82 & 83 & 84 & 85 & 86 \\ 89 & 90 & 91 & 92 & 93 & 94 & 95 \\ 98 & 99 & 100 & 101 & 102 & 103 & 104 \\ 107 & 108 & 109 & 110 & 111 & 112 & 113 \\ 116 & 117 & 118 & 119 & 120 & 121 & 122 \\ 125 & 126 & 127 & 128 & 129 & 130 & 131 \end{bmatrix}$$

Figure 8.12 graphically represents this $2D$ convolution process. The Python code used for the convolution is given below.

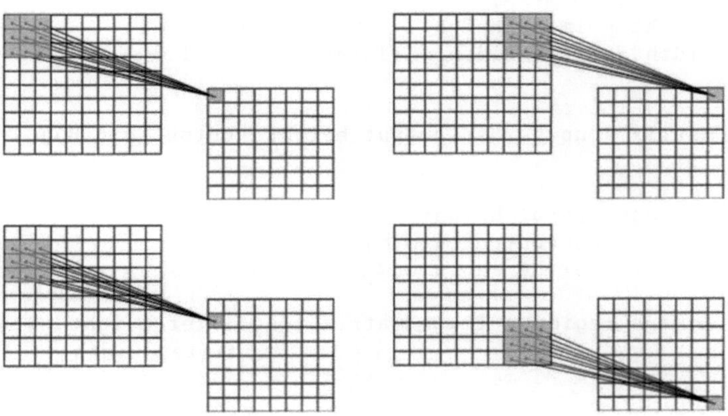

Fig. 8.12 Visualization of the 2D convolution operation

```
#Matrix convolution
import numpy as np

# Define the 9x9 image matrix
image_matrix = np.array([
    [1, 2, 3, 4, 5, 6, 7, 8, 9],
    [10, 11, 12, 13, 14, 15, 16, 17, 18],
    [19, 20, 21, 22, 23, 24, 25, 26, 27],
    [28, 29, 30, 31, 32, 33, 34, 35, 36],
    [37, 38, 39, 40, 41, 42, 43, 44, 45],
    [46, 47, 48, 49, 50, 51, 52, 53, 54],
    [55, 56, 57, 58, 59, 60, 61, 62, 63],
    [64, 65, 66, 67, 68, 69, 70, 71, 72],
    [73, 74, 75, 76, 77, 78, 79, 80, 81]
])

# Define the 3x3 emboss filter
emboss_filter = np.array([
    [-2, -1, 0],
    [-1,  1, 1],
    [ 0,  1, 2]
])

# Get dimensions
image_height, image_width = image_matrix.shape
filter_height, filter_width = emboss_filter.shape

# Output matrix dimensions (valid convolution)
output_height = image_height - filter_height + 1
output_width = image_width - filter_width + 1

# Initialize the output matrix
output_matrix = np.zeros((output_height, output_width))

# Perform convolution
for i in range(output_height):
    for j in range(output_width):
        # Extract the region of the image that matches the
                                filter size
        image_region = image_matrix[i:i+filter_height, j:j+
                                filter_width]
        # Perform element-wise multiplication and sum the
                                result
        output_matrix[i, j] = np.sum(image_region *
                                emboss_filter)

# Output the result
print("Convolved Matrix:\n", output_matrix)
```

In the previous example, the filtered image from a 9×9 image shrinks to 7×7 after convolution with a 3×3 filter. Often, to keep the original dimension of the image unchanged, special treatment is done with the boundary of the original matrix.

000	000	000	000	000	000	000	000
000	128	163	156	194	023	043	000
000	122	098	142	134	102	043	000
000	222	147	132	171	111	127	000
000	062	087	102	125	123	117	000
000	212	063	034	085	074	129	000
000	174	137	185	197	206	232	000
000	065	152	129	139	172	222	000
000	000	000	000	000	000	000	000

(a)

222	065	152	129	139	172	222	065
043	128	163	156	194	023	043	128
043	122	098	142	134	102	043	122
127	222	147	132	171	111	127	222
117	062	087	102	125	123	117	062
129	212	063	034	085	074	129	212
232	174	137	185	197	206	232	174
222	065	152	129	139	172	222	065
043	128	163	156	194	023	043	128

(b)

Fig. 8.13 Handling pixel close to boundary. (**a**) Pad with zeros. (**b**) Warp around

In the first approach, zero padding is done with the image matrix by adding a border of zeros around the input matrix so that the filter can still be applied to the edges (Fig. 8.13a). This results in an output matrix with the same dimensions as the input. In the second approach, wraparound padding (or circular padding) is done with the original image matrix. Wraparound padding takes the values from the opposite side of the image when reaching the borders, essentially "wrapping" the image around (Fig. 8.13b).

Through the lens of image processing, the iconic sunset scene of Jack and Rose from Titanic takes on a new depth. As the hues of the sunset blend into each other and the details soften, the essence of their love remains sharp and vivid. "In a sea of memories, even time blurs, but love stays in focus"—this sentiment captures how, even as the world around them fades into a gentle blur, their love stands out in clear, unwavering focus, much like the enduring power of true love amidst the passage of time. Python code used for image processing in Fig. 8.14 is given below.

```
#Image processing through convolution
import cv2
import numpy as np
import matplotlib.pyplot as plt

def rgb2gray(rgb):
    return np.dot(rgb[...,:3], [0.2989, 0.5870, 0.1140])

# Reading the image from the disk using cv2.imread() function
img = cv2.imread('Titanic.jpg')[:,:,::-1]

img = rgb2gray(img)

# Different filters
```

```
# Edge Detection1
kernel1 = np.array([[0, -1, 0],
                    [-1, 4, -1],
                    [0, -1, 0]])
# Edge Detection2
kernel2 = np.array([[-1, -1, -1],
                    [-1, 8, -1],
                    [-1, -1, -1]])
# Bottom Sobel Filter
kernel3 = np.array([[-1, -2, -1],
                    [0, 0, 0],
                    [1, 2, 1]])
# Top Sobel Filter
kernel4 = np.array([[1, 2, 1],
                    [0, 0, 0],
                    [-1, -2, -1]])
# Left Sobel Filter
kernel5 = np.array([[1, 0, -1],
                    [2, 0, -2],
                    [1, 0, -1]])
# Right Sobel Filter
kernel6 = np.array([[-1, 0, 1],
                    [-2, 0, 2],
                    [-1, 0, 1]])
# Sharpen
kernel7 = np.array([[0, -1, 0],
                    [-1, 5, -1],
                    [0, -1, 0]])
# Emboss
kernel8 =  np.array([[-2, -1, 0],
                     [-1,  1, 1],
                     [ 0,  1, 2]])
# Box Blur
kernel9 = (1 / 9.0) * np.array([[1, 1, 1],
                                [1, 1, 1],
                                [1, 1, 1]])
# Gaussian Blur 3x3
kernel10 = (1 / 16.0) * np.array([[1, 2, 1],
                                  [2, 4, 2],
                                  [1, 2, 1]])
# Gaussian Blur 5x5
kernel11 = (1 / 256.0) * np.array([[1, 4, 6, 4, 1],
                                   [4, 16, 24, 16, 4],
                                   [6, 24, 36, 24, 6],
                                   [4, 16, 24, 16, 4],
                                   [1, 4, 6, 4, 1]])
# Unsharp masking 5x5
kernel12 = -(1 / 256.0) * np.array([[1, 4, 6, 4, 1],
                                    [4, 16, 24, 16, 4],
                                    [6, 24, -476, 24, 6],
                                    [4, 16, 24, 16, 4],
                                    [1, 4, 6, 4, 1]])
```

```
# Sharpeneded image is obtained using the variable sharp_img
# cv2.fliter2D() is the function used
# src is the source of image(here, img)
# ddepth is destination depth. -1 will mean output image will
                                        have same depth as input image
# kernel is used for specifying the kernel operation (here,
                                        sharp_kernel)
img_kernel = cv2.filter2D(src=img, ddepth=-1, kernel=kernel8)

#showing the original image
plt.imshow(img, cmap="gray")
plt.show()
#showing the processed image
plt.imshow(img_kernel, cmap="gray")
plt.show()
```

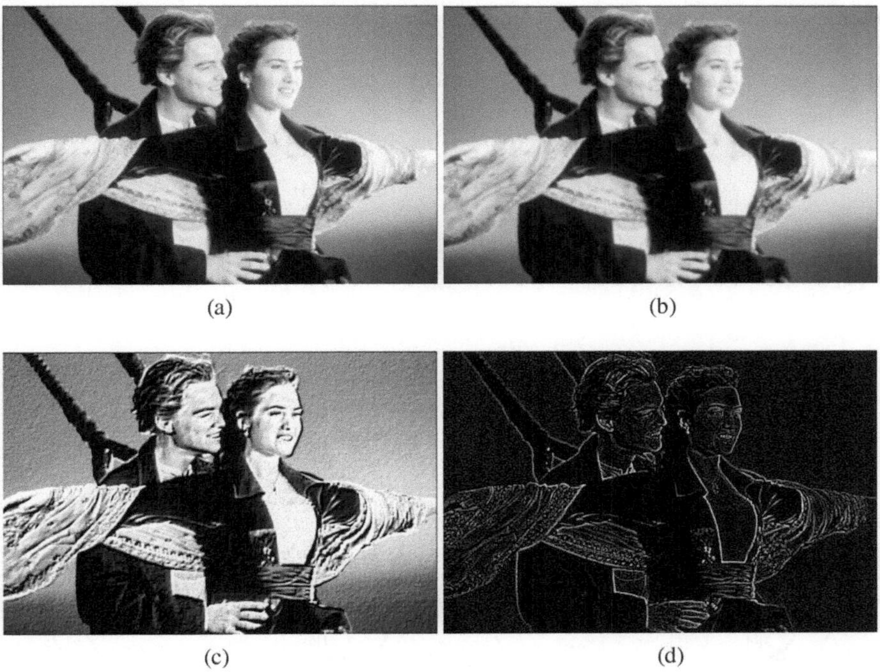

Fig. 8.14 Image processing of the iconic sunset scene from Titanic. (**a**) Original. (**b**) Blurred. (**c**) Embossed. (**d**) Outline

8.5 Circular Convolution

In linear convolution, either of the signals can be extended beyond their sequence through zero padding. However, this flexibility is not possible when the signals are periodic. If $f\{[n]\}$ and $g\{[n]\}$ are two periodic time-domain sequences of the same period N, then the circular convolution of the sequences is defined as

$$h[n] = (f \circledast g)[n] = \sum_{m=0}^{N-1} f[m]g[n-m] = \sum_{m=0}^{N-1} f[n-m]g[m] \,, \quad n = 0, 1, \ldots, N-1 \tag{8.21}$$

The principal difference between cyclic or circular convolution and linear convolution is that the summation range is restricted to a single period. Figure 8.15 shows the position of two sequences $f\{[n]\}$ and $g\{[n]\}$, of length 8 for finding their convolution output. One sequence is placed clockwise and the other counterclockwise. Figure 8.15a shows the convolution for $h[0]$. The sum of products of the corresponding elements of the sequences is the convolution output $h[0]$:

$$\begin{aligned} h[0] = {} & f[0]g[0] + f[1]g[7] + f[2]g[6] + f[3]g[5] \\ & + f[4]g[4] + f[5]g[3] + f[6]g[2] + f[7]g[1] \end{aligned} \tag{8.22}$$

The inner sequence is kept fixed, and the outer sequence is rotated 3 units to compute $h[3]$ as shown in Fig. 8.15b:

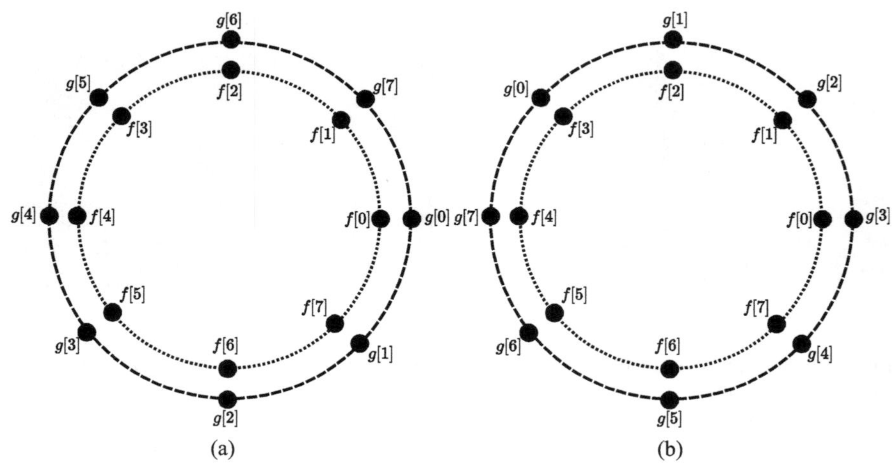

Fig. 8.15 Circular convolution. (**a**) $n = 0$. (**b**) $n = 3$

$$\begin{aligned} h[3] &= f[0]g[3] + f[1]g[2] + f[2]g[1] + f[3]g[0] \\ &\quad + f[4]g[7] + f[5]g[6] + f[6]g[5] + f[7]g[4] \end{aligned} \tag{8.23}$$

The process is similar to linear convolution, except that the sequences are placed in a circle. For circular convolution, both sequences must be of the same length. For N-point sequences, the output is also periodic of period N.

The DFT of a sequence $f[n]$ of length N is

$$F[k] = \sum_{n=0}^{N-1} f[n]e^{-i\frac{2\pi}{N}kn} \tag{8.24}$$

Similarly, the DFT of sequence $g[n]$ of length N is

$$G[k] = \sum_{n=0}^{N-1} g[n]e^{-i\frac{2\pi}{N}kn} \tag{8.25}$$

If $H[k]$ is the point-wise multiplication of the two sequences in the frequency domain,

$$H[k] = F[k] \cdot G[k] \tag{8.26}$$

The inverse DFT of $H[k]$ is

$$\begin{aligned} h[n] &= \frac{1}{N}\sum_{k=0}^{N-1} H[k]e^{i\frac{2\pi}{N}kn} \\ &= \frac{1}{N}\sum_{k=0}^{N-1} (F[k] \cdot G[k])\, e^{i\frac{2\pi}{N}kn} \end{aligned} \tag{8.27}$$

This inverse DFT is the circular convolution of $f[n]$ and $g[n]$:

$$h[n] = (f \circledast g)[n] \tag{8.28}$$

The circular convolution of two sequences in the time domain is equivalent to multiplying their DFTs in the frequency domain. So, the circular convolution can be used to compute DFT efficiently. For example, let's take two sequences:

$$f[n] = \{1, 2, 3, 4\}$$

$$g[n] = \{1, 1, 1, 1\}$$

The length of both sequences is $N = 4$. Using the DFT formula, DFT of both sequences is expressed as

$$F[k] = [10, -2+2j, -2, -2-2j]$$

$$G[k] = [4, 0, 0, 0]$$

The point-wise multiplication of DFTs gives

$$\begin{aligned} H[k] = F[k] \cdot G[k] &= \{10 \cdot 4, (-2+2j) \cdot 0, -2 \cdot 0, (-2-2j) \cdot 0\} \\ &= \{40, 0, 0, 0\} \end{aligned}$$

The circular convolution of $f[n] = \{1, 2, 3, 4\}$ and $g[n] = \{1, 1, 1, 1\}$ is

$$h[n] = \{10, 10, 10, 10\}$$

So the inverse DFT of $H[k]$ can be obtained from the circular convolution result:

$$h[n] = \text{IDFT}(H[k]) = [10, 10, 10, 10]$$

Let $F[k]$ and $G[k]$ be two periodic frequency-domain sequences of the same period N. Then, the circular convolution of the sequences, divided by N, is given as

$$f[n]g[n] \Longleftrightarrow \frac{1}{N}\sum_{m=0}^{N-1} F[m]F[k-m] = \frac{1}{N}\sum_{m=0}^{N-1} F[k-m]F[m] \tag{8.29}$$

$f[n]$ and $g[n]$ are the IDFT $F[k]$ and $G[k]$, respectively. For example, two sequences in the time domain is $F[k] = \{9, -1+j2, 1, -1-j2\}$ and $G[k] = \{7, -1+j4, -1, -1-j4\}$. Their IDFT are $f[n] = \{2, 1, 3, 3\}$ and $g[n] = \{1, 0, 2, 4\}$. The circular convolution of $F[k]$ and $G[k]$ is

$$4f[n]g[n] = 4\{2, 0, 6, 12\} = \{8, 0, 24, 48\}$$

The steps involved in calculating the circular convolution using the DFT are the following:

1. Compute the DFT of both sequences.
2. Multiply the DFTs element-wise in the frequency domain.
3. Apply the IDFT to return to the time domain, obtaining the circular convolution.

This method leveraged the efficiency of the fast Fourier transform (FFT) for larger sequences, making it computationally attractive compared to directly computing the convolution.

8.6 Introduction to Probability Theory

We exist in a realm characterized by uncertainty. Every measurement within this universe carries an element of unpredictability. Let's examine a vibrating structure while undergoing significant deflection to grasp the concept of uncertainty. The equation that governs this system can be written as

$$\ddot{x} + 0.05\dot{x} + x^3 = 7.5\cos t \tag{8.30}$$

Imagine two neighboring particles that start off with an infinitesimally small separation and the same initial velocity. The first particle is positioned at $x_1 = 1.0$ and the second at $x_2 = 1.001$. Both particles begin their motion with the same speed ($\dot{x}_1 = 4.1$). By using the fourth-order Runge-Kutta (RK4) method to numerically solve their trajectories (as illustrated in Fig. 8.16), we observe that even a minor difference in their initial positions leads to significant divergence in their paths over time. This divergence is a result of the nonlinear nature of Eq. (8.30), which causes both position and velocity to become chaotic (see Fig. 8.16). Nonlinear equations, such as the Navier-Stokes equation for fluid dynamics, the Stefan-Boltzmann law for radiation, the Black-Scholes equation in finance, and Lotka-Volterra equations in biology, describe many natural phenomena. However, our measurement instruments and computational algorithms have limited precision and accuracy. Any rounding error in measurements can lead to chaotic results over time. Our measurements are inherently uncertain due to various unidentifiable factors and are often treated as random variables. There isn't a universally agreed definition for random variables, but the essence of randomness is unpredictability. In modern mathematics, a random variable is a measurable function from a probability space to a measurable state space. In the late nineteenth century, observations of natural phenomena such as turbulence, molecular motion in gases, and Brownian motion highlighted the complexities of randomness, leading to the development of stochastic processes by scientists like Einstein, Smoluchowski, and Maxwell. The work of Russian mathematician Andrei Kolmogorov (1903–1987) in the twentieth century advanced the study of random and pseudo-random processes. Essentially, a random variable represents a quantity whose long-term behavior cannot be predicted with certainty, as demonstrated in Fig. 8.17, where the velocity and acceleration of the two adjacent particles become unpredictable over time.

Fortunately, despite the randomness of the universe, many systems operate in a deterministic manner. For instance, an aeroplane, engineered through various experiments, can accurately fly between two cities. Similarly, precise temperature and pressure measurements in a boiler keep power plants running smoothly. The stock market, trade, and the economy also function in an orderly manner. Probability plays a crucial role in the smooth operation of numerous random processes around us. It involves assessing the likelihood or chance of a particular event or outcome within the uncertain realm of all possible values. However, this explanation of probability is somewhat circular. Probability has evolved over time and, in

Fig. 8.16 Chaotic response of a time series. (**a**) $x_1 = 1.0$. (**b**) $x_2 = 1.001$. (**c**) $x_{x_1} - x_{x_2}$

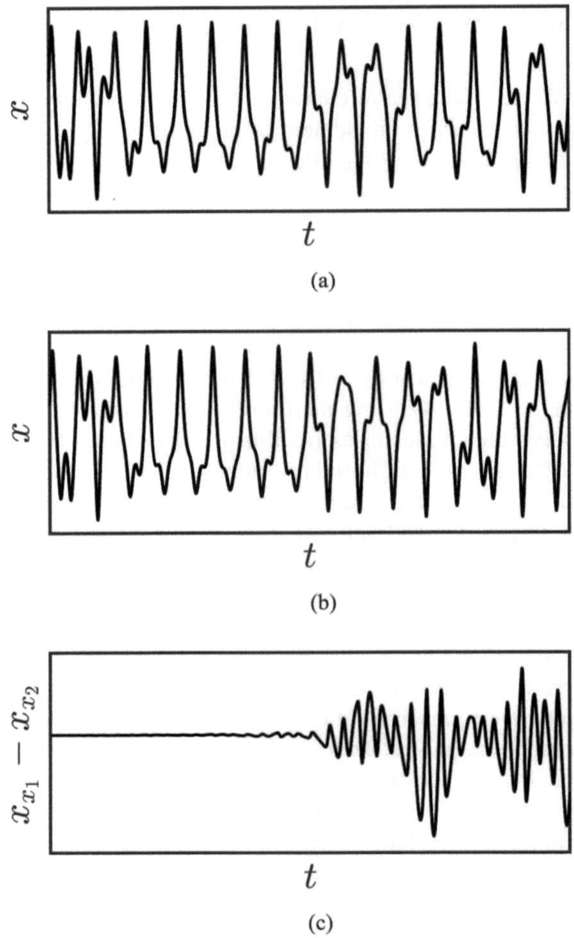

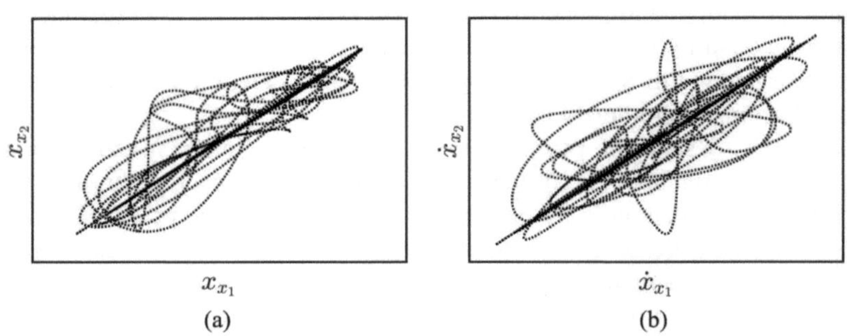

Fig. 8.17 Phase plane representation for position and velocity of two particles. (**a**) Trajectory of two particles. (**b**) Velocity of two particles

modern mathematics, lacks a concrete definition akin to points, lines, or planes in Euclidean geometry. Probability theory is based on axioms and doesn't have a direct measurement tool like a ruler or thermometer. Instead, probability is a theoretical construct expressed as a value between zero and one, derived through mathematical analysis of extensive datasets rather than physical measurements.

8.6.1 History of Probability Theory

The word probability is derived from the Latin word "probo" and the English words "probe" and "probable." The Oxford dictionary explains the word probability as "The quality or state of being probable; the extent to which something is likely to happen or be the case." The mathematical sense had a meaning that was more or less like plausibility. In the words of Laplace, "Probability has reference partly to our ignorance, partly to our knowledge.... The Theory of chances consists in reducing all events of the same kind to a certain number of cases equally possible, that is, such that we are equally undecided as to their existence; and determining the number of these cases which are favorable to the event sought. The ratio of that number to the number of all the possible cases is the measure of the Probability...."

The application of probability theory to address universal uncertainty is a relatively recent development in human history. Probability was practiced long before it was formalized mathematically. In ancient Egypt, people played Astragalus, dice made from the bones of sheep, deer, and dogs, which could land on any of its four sides after being tossed. This game of chance had entirely random outcomes. Approximately 4500 years ago, the "Royal Game of Ur" emerged in Mesopotamia (present-day Iraq), one of the world's earliest literate civilizations. This board game, played between two individuals, combined elements of luck and strategy. Players rolled dice to generate random numbers and moved their pieces on a board, like modern Ludo. The game later spread to India, where it remained popular for centuries. As the Mesopotamian civilization declined around 2500 BC, the Roman Empire rose. The Romans, known for gambling, combined skill and chance to determine outcomes. Gambling was regulated and allowed during certain holidays to mitigate its negative social effects. Roman emperors like Augustus and Vitellius were avid gamblers who relished the randomness of dice throws. More accurate than ancient Astragalus, modern dice are uniformly weighted cubes that produce unbiased results. The origins of the card game rummy are uncertain, with suggestions of either Italian or Eastern roots. Card printing, which began in the fourteenth century with the advent of paper production and block printing, eventually made rummy popular. Initially condemned by the church, rummy became well-known for its inherent randomness and association with gambling over time. Despite dealing with randomness for millennia, no effort was made by any ancient civilization to develop any mathematical form to quantify these random outcomes formally. The ancient emperor used to make many decisions based on the outcome of dice or Astragali. A list of possible actions was assigned to the sides of the dice, and the final decision was based on dice output. Greek civilization contributed significantly to many fields of science and mathematics, but surprisingly, Greeks did not pay any attention to developing any mathematical theory for these games of chance.

In ancient Rome, many emperors allowed lotteries as an essential means of government funding. Over time, Romans developed a fondness for lotteries and games like rummy and dice for entertainment and financial gain. This growing interest led to an increased curiosity about analyzing these games and calculating the odds of winning. It was in the sixteenth century that the Italian Renaissance mathematician Girolamo Cardano (1501–1576) challenged the notion of divine influence in games of chance. Cardano applied mathematics, logic, and reasoning to evaluate the odds of winning these games, establishing himself as a pioneer in probability theory. After experiencing substantial losses in gambling, Cardano authored a book on games of chance, "*Liber de Ludo Aleae*" (*Games of Chance*, 1663), which shifted the focus from divine intervention to mathematical analysis. In this book, Cardano explored various card games, including Primero (an early version of poker), and introduced the concept of probability, denoted as a value between 0 and 1, to represent the likelihood of random events. He also formulated the law of large numbers, stating that with a large number of trials, the probability of an event approximates the expected value. Cardano further contributed by identifying combinatorial numbers, representing the number of ways to choose k items from n items (David, 1962).

Another notable Italian Renaissance figure, Galileo Galilei (1565–1642), explored randomness in his article "Thoughts about Dice-Games." He examined why certain numbers, like 10 and 11, appeared more frequently than others, such as 9 and 12, when rolling three dice. Galileo discovered that there were 16 possible sums from rolling three dice, ranging from a minimum of 3 to a maximum of 18. He found that the sums of 3 and 18 could be achieved in only one way each, while other sums had multiple possible outcomes. This led Galileo to the realization that the sum of 10 appeared in 27 different ways, compared to the sum of 9, which appeared in only 25 ways. Galileo laid the groundwork for probability theory by quantifying random outcomes through counting principles. Interestingly, a Latin poem, *De Vetula,* notes that Bishop Wibold of Cambrai had correctly enumerated 216 possible outcomes for three dice, with 56 different results, as early as 960 AD, long before the formal study of permutations and probability.

In the seventeenth century, France and Italy were central to scientific advancements. Numerous formal and informal groups of like-minded individuals frequently gathered to discuss the latest developments in science and mathematics. During this period, the foundation of modern probability theory was laid in France by two dedicated mathematicians, Pierre de Fermat (1601–1665) and Blaise Pascal (1623–1662). Young Pascal often visited Marin Mersenne's (1588–1648) home to engage in various mathematical and scientific discussions. On one occasion, the renowned French gambler Chevalier de Méré asked Pascal to determine the better strategy between rolling a fair die four times to get a six and rolling two fair dice 24 times to get double sixes. Pascal reached out to Fermat for assistance, leading to a series of letters exchanged between July and October 1654. Pascal and Fermat addressed numerous gambling problems using mathematical methods in these correspondences. They applied the principle of enumerating equally possible outcomes to determine the most advantageous scenarios for each player, much like Cardano and Galileo had previously done. Initially, they utilized combinatorial theory to solve these problems. For example, when a coin is tossed twice, there are four possible outcomes (HH, HT, TH, TT), and the probability of each event is linked to their binomial coefficients:

$$(H + T)^2 = H^2 + 2HT + T^2 \tag{8.31}$$

The sample space for tossing a coin three times is $HHH, HHT, HTH, THH, HTT, THT, TTH, TTT$.

$$(H + T)^3 = H^3 + 3H^2T + 3HT^2 + T^3 \tag{8.32}$$

The ratio of the individual binomial coefficient of a particular event to the sum of binomial coefficients of all plausible events gives the probability of that particular event.

Pascal and Fermat used a triangle of numbers, now famous as the Pascal triangle. The first row and first column of the triangle have value unity. Each entry of the subsequent row is constructed by adding the value from the left entry and top entry of the concerned position. Binomial coefficients are placed among the diagonal of the triangle for respective order of expansion. Pascal's arithmetic triangle appeared in "Traite du triangle arithmétique":

$$\begin{array}{llllll} 1 & 1 & 1 & 1 & 1 & 1 \\ 1 & 2 & 3 & 4 & 5 & \\ 1 & 3 & 6 & 10 & & \\ 1 & 4 & 10 & & & \\ 1 & 5 & & & & \\ 1 & & & & & \end{array} \tag{8.33}$$

The number of ways r objects can be selected from a total of n distinct objects is known as a combination:

$$\binom{n}{r} = {}^{n}C_r = \frac{n!}{r!\,(n-r)!} \tag{8.34}$$

The early history of combinatorics is obscure. Indian mathematician Bhaskara knew the general formula for permutation and combination of objects about 1150 AC. From India, this knowledge spread to Europe through Arabs. The binomial expansion and the corresponding arithmetical triangle of coefficients also originated among Hindu and Arab mathematicians (al-Tusi in 1265 AC). The Pascal triangle was found earlier in the work of Chinese mathematician Chu Shih-Chieh in 1303 AC. But it was not found in European works before the sixteenth century. Before Pascal and Fermat, Italian mathematician Niccolo Tartaglia (1499–1557) formulated a similar arithmetical triangle for binomial coefficients up to $n = 12$ through the addition rule. Cardano used such a triangle for his derivations. In 1636, Mersenne extended the Cardano-Tartagila table to get a combination of $n = 36$ to arrange musical notes. It was quite cumbersome to formulate such a table for a large number. In 1654, Pascal published a unified theory for combinatorial numbers and binomial coefficients. Coefficient of the term $a^r b^{n-r}$ in the expansion of $(a+b)^n$ is expressed by ${}^nC_r$, for $n \geq 0$ and $0 \leq r \leq n$. Unfortunately, in the later phase of life, Pascal became a religious Pasteur and gave up mathematics.

Young Dutch mathematician Christiaan Huygens (1629–1695) was fascinated with the theory of dice games during his Paris visit in 1655. Huygens started solving several complex game problems on his arrival in Holland, including those earlier solved by Pascal and Fermat. During this time, Huygen and Fermat communicated through letters from mutual friends. Huygens quickly solved five problems posted by Fermat and published a book entitled "*De Rationiis in Ludo Aleae (On Reasoning in Games of Dice)*" in 1657. This was the first mathematical book on probability, and for the next 50 years, it helped probability theory to reach a broad spectrum of audiences. In this treatise, Huygens introduced the definition of the probability of an event as a quotient of favorable cases to all possible cases. He also reintroduced the concept of mathematical expectation (mean) and elaborated on Cardano's idea of the expected value of random variables. Between 1679 and 1688, he solved several problems for popular card games.

Calculus was discovered during the middle of the seventeenth century by English mathematician Sir Issac Newton (1643–1727) and German mathematician Gottfried Leibniz (1646–1716). Mathematical problems that remained unsolved for centuries were solved using calculus. Probability theory benefited from the new ideas and techniques of calculus. Swiss mathematician Jacob Bernoulli (1654–1705) was the first to recognize the importance of calculus in probability theory and the role of probability theory beyond the study of games. Bernoulli was impressed by Huygens's book and had several correspondences with Leibniz regarding the mathematical development of probability theory. In contemporary times, "*Ars Conjectandi (Arts of Prediction)*, 1713," by Jacob Bernoulli, was the major work in probability. Ars Conjectandi extended the purview of the subject beyond dice and card games to theory for the prediction of various uncertainty in life, like criminal justice and computation of human mortality, along with various civil, moral, social, and economic problems. Although the initial success was not significant, it stands as an inceptor for applications of probability in physics, engineering, and astronomy. Ars Conjectandi was published by Nicolas Bernoulli (nephew of Jacob Bernoulli) after 8 years of the death of Jacob Bernoulli. In Ars Conjectandi, Bernoulli discussed a very famous theorem known as the "law of large numbers." To derive that theorem, Bernoulli himself took 20 years. This theorem itself has been debated for centuries among mathematicians and philosophers. If two events do not influence each other, they are called independent events in probability theory. Two successive flips of a coin or roll of the dice are independent events. Bernoulli used the binomial expansion of an event p and its complement $q = (1-p)$ for a large exponent n. This is also known as Bernoulli's binomial distribution of the random variable, having r successes in n trials. Bernoulli repeated what Cardano, Fermat, and Pascal tried (computing binomial coefficient for large numbers n was difficult earlier). Bernoulli trials are binomial trials, like tossing a coin where outcome 1 is a success and 0 is a failure. If the Bernoulli trials are conducted for n time, having a probability of success p in each trial. The sum of likelihood between i and j successes trials in n trials is expressed as

$$f(r) = \sum_{r=i}^{j} {}^{n}C_r p^r (1-p)^{n-r} \tag{8.35}$$

Consider tossing a ludo die ten times and calculate the probability of getting the index 3 each time. Probability of getting digit 3 from one roll is $p = \frac{1}{6}$. By this law of large numbers, Bernoulli made an important and explicit relationship between what we observe and compute for a special class of random processes. Bernoulli was also interested in the converse law of probability, which was a tough job.

Another great French mathematician, Abraham de Moivre (1667–1754), was inspired by Huygens and published a book entitled "*Doctrine of Chances*" in 1718. de Moivre used the latest calculus and some algebra discoveries to derive his own conclusions. In the eighteenth century, it was an excellent book for understanding the probability theory. de Moivre started the book by presenting solutions to simple gambling problems and gradually derived solutions to complex problems. Toward the end of the book, de Moivre approximated the binomial coefficient with a relatively easy-to-compute normalized function using Stirling's formula and gamma function:

$${}^{n}C_r p^r (1-p)^{n-k} = \frac{1}{\sqrt{2\pi np(1-p)}} e^{-\frac{(k-np)^2}{2np(1-p)}} \tag{8.36}$$

This approximation facilitated de Moivre for easy computation of a tedious task involving determining the binomial coefficient for a large number n, which Cardano and Bernoulli discussed earlier. For a large number of trials, de Moivre got an exciting profile for the probability distribution, which looked like a bell. Later, this profile became famous as a bell curve or normal curve. Unlike a contentious modern bell-shaped curve, de Moivre's distribution curve was from discrete variables. Like Bernoulli, de Moivre showed the application of this curve in various other fields apart from the game of chance. Cardano originally postulated the idea of the law of large numbers, but subsequently, it got a legitimate mathematical form from Jacob Bernoulli. de Moivre showed Bell shape distribution for the law of large numbers. But interestingly, the name "law of large numbers" was coined by French mathematician Siméon Poisson. Although de Moivre realized that the histogram of the frequency distribution is very close to the bell-shaped normal curve, he could not write down its equation. Gauss derived the formal equation for normal distribution during the derivation of the error curve. Subsequently, it is recognized that when independent measurements are averaged, they tend to look like the bell-shaped normal curve, which is known as the central limit theorem (the core of probability theory).

8.6.2 Grammar of Probability Theory

Just as mastering a new language such as French, English, Spanish, or German involves understanding basic vocabulary and grammar rules to construct meaningful sentences, learning probability theory for signal processing requires familiarizing oneself with key terminology and mathematical definitions to apply the concepts effectively.

The sample space Ω is the set of all possible outcomes of an experiment or random process. For example, the sample space for tossing a coin is $\Omega = \{H, T\}$ and rolling a die is $\Omega = \{1, 2, 3, 4, 5, 6\}$. An event is any subset of the sample space. An event can consist of one outcome or multiple outcomes. For example, rolling die to get an even number is $E_1 = \{2, 4, 6\}$, and to get a number less than 3 is $E_2 = \{1, 2\}$. The probability of an event is a measure of how likely that event is to occur. Mathematically, the probability of an event E is defined as

$$P(E) = \frac{E}{\Omega} \tag{8.37}$$

E is the number of outcomes in the event, and Ω is the number of outcomes in the sample space. For example, in a fair die roll, the probability of rolling an even number (event $E_1 = \{2, 4, 6\}$) is $P(E_1) = \dfrac{3}{6} = 0.5$.

Complementary probability refers to the probability of the event not happening when the probability of it happening is known. If the probability of an event E happening is $P(E)$, the probability of the event E not happening, which is called the complement of E, is denoted as

$$P(E^c) = 1 - P(E) \tag{8.38}$$

The probability of both E_1 and E_2 is the probability that both events A and B occur. It is called the joint probability or the intersection of E_1 and E_2 and denoted as $P(E_1 \cap E_2)$. Its calculation depends on whether the events are independent or dependent. If the probability of an event occurring depends on another event that has already occurred, then it is called conditional probability. Conditional probability is denoted as $P(E_1|E_2)$, meaning the probability of event E_1 happening given that E_2 has occurred. Mathematically,

$$P(E_1|E_2) = \frac{P(E_1 \cap E_2)}{P(E_2)} \tag{8.39}$$

For example, from rolling the die, if the result is an even number, then the conditional probability of rolling a 2 is $P(\text{Roll a 2} \mid \text{Even number}) = \frac{P(\{2\})}{P(\{2, 4, 6\})} = \frac{1/6}{1/2} = \frac{1}{3}$.

Events E_1 and E_2 are said to be independent if the occurrence of one event does not affect the occurrence of the other. For independent events, the probability of both E_1 and E_2 happening is the product of their individual probabilities.

$$P(E_1 \cap E_2) = P(E_1) \cdot P(E_2) \tag{8.40}$$

For example, tossing two coins. The result of the first coin toss does not affect the outcome of the second. Hence, these events are independent if E_1 is the event that the first coin lands head and E_2 is the event that the second coin lands head. $P(E_1 \cap E_2) = P(E_1) \cdot P(E_2) = \frac{1}{2} \cdot \frac{1}{2} = \frac{1}{4}$. If E_1 and E_2 are dependent events, then the occurrence of one event affects the probability of the other:

$$P(E_1 \cap E_2) = P(E_1) \cdot P(E_2|E_1) \tag{8.41}$$

$P(E_2|E_1)$ is the conditional probability of E_2 given that E_1 has already occurred. For example, the probability that a person owns a car is $P(\text{Car}) = 0.7$, and the probability that a person owns a house, given that they own a car, is $P(\text{House} \mid \text{Car}) = 0.8$. The probability that a person owns both a car and a house is $P(\text{Car} \cap \text{House}) = 0.7 \cdot 0.8 = 0.56$.

The probability that either event E_1 or event E_2 occurs is called the union of E_1 and E_2, denoted as $P(E_1 \cup E_2)$. It includes the probability of E_1 occurring, E_2 occurring, or both occurring. The following formula avoids double-counting the overlap between E_1 and E_2:

$$P(E_1 \cup E_2) = P(E_1) + P(E_2) - P(E_1 \cap E_2) \tag{8.42}$$

If events E_1 and E_2 are mutually exclusive (they cannot both occur at the same time), then $P(E_1 \cap E_2) = 0$, and the formula simplifies to

$$P(E_1 \cup E_2) = P(E_1) + P(E_2) \tag{8.43}$$

If the total number of random variables in the sample space is finite, it is called a discrete random variable. The probability mass function (PMF) is used to describe the probability distribution of a discrete random variable, which can only take on a countable number of distinct values (e.g., integers). The PMF provides the probability that the random variable X takes a specific value x. For example, for a fair six-sided die, the PMF is $P(X = x) = \frac{1}{6}$ for $x = 1, 2, 3, 4, 5, 6$. PMF is denoted as $P(x)$ and defined as

$$P(x) = P(X = x) \tag{8.44}$$

With the property,

$$P(x) \geq 0 \tag{8.45}$$

for all x

$$\sum_{x} P(x) = 1 \tag{8.46}$$

In the limiting case, the sample space is called a continuous random variable when the distance between two random variables decreases to zero. The notion of individual point probabilities changes from discrete to continuous distributions. In the continuous case, the probability mass spreads out over intervals, leading to the concept of a density instead of specific point probabilities. The probability density function (PDF) is used to describe the probability distribution of a continuous random variable, which can take an uncountably infinite number of values (e.g., real numbers). The PDF gives the relative likelihood of the random variable taking a value within a specific interval. For a continuous random variable x with a PDF $f(x)$, the probability that x lies within an interval $[a, b]$ is given by the area under the PDF curve over that interval:

$$P(a \leq X \leq b) = \int_{a}^{b} f(x)\, dx \tag{8.47}$$

To understand it more clearly, let us assume a random variable x represents temperature measured by a thermometer at a point in space. For precise temperature measurement, the thermometer's bulb tip should converge to a point (zero areas), which is not physically possible. Hence, the measured temperature at a point is the average temperature measured over the thermometer bulb area. The value of random variable x is uncertain at any point, and its product with probability density function $xf(x)$ represents the plausibility of existence. When it is integrated over the entire region, it gives the average or expected measure of the random variable. The probability density function is a positive quantity, having a sum equal to one in the entire sample space:

$$f(x) \geq 0 \tag{8.48}$$

$$\int_{-\infty}^{\infty} f(x)\, dx = 1 \tag{8.49}$$

Cumulative distribution function (CDF) $F(x)$ provides a way to describe the probability that a random variable takes on a value less than or equal to a given value $P(X \leq x)$. It provides the cumulative probability up to that point, allowing one to determine the likelihood of observing values within a certain range:

$$F(x) = \int_{-\infty}^{x} f(X)\, dX \tag{8.50}$$

X is the dummy variable for integration, post integration, F is the function of x. The probability density function of a random variable is expressed in the form of CDF as

$$f(x) = \frac{dF(x)}{dx} \tag{8.51}$$

In broad terms, statistics refers to the science of gathering, organizing, presenting, and interpreting numerical information. Michael J. Moroney (1918–1990), an English statistician, aptly described it as "A statistical analysis, properly conducted, is a delicate dissection of uncertainties, a surgery of suppositions." Random variables are frequently summarized using specific statistical measures when dealing with discrete or continuous data. These measures, often called averages, represent the central point of the data range and are also known as measures of central tendency. The mean, median, and mode are the primary types of averages.

- Mean is calculated by dividing the sum of all values by the total number of data points.
- Median is the value found in the middle when data points are arranged in order.
- Mode identifies the most frequently occurring value within the dataset.

There is no strict rule for choosing which measure of central tendency to use. When the dataset is skewed, the mean can be misleading. For example, the average income in areas where billionaires like Bill Gates or Warren Buffett reside would be disproportionately high, misrepresenting most people's earnings. The mode might emphasize outliers, such as the most common diameter of guitar strings. On the other hand, the median is particularly valuable for situations like exit polls or audience voting in reality shows, where understanding the most common response is critical.

In probability theory and statistics, the expected value of a random variable provides crucial insights into its distribution, central tendency, variability, and shape. The expectation operator $E(\{\})$ is defined as

$$E(\{\}) = \int_{-\infty}^{\infty} \{\} dx \tag{8.52}$$

which gives the integral over the sample space. The statistical moments provide important information about the shape and characteristics of the probability distribution of a random variable. Statistical moments are defined as the expected values of powers of the random variable. The n-th moment of a random variable x is given by

$$E\left(x^n\right) = \int_{-\infty}^{\infty} x^n\, f(x)\, dx \tag{8.53}$$

The first-order statistical moment for a continuous random variable is expressed as

$$E(x) = \mu = \int_{-\infty}^{\infty} x\, f(x)\, dx \tag{8.54}$$

μ provides the measure of the central tendency of the distribution and is commonly referred to as the average or mean. However, the mean is not enough to give a single summary of the set of random variables. Knowing the mean shoe size is not helpful for the tannery. Variability or spread is essential to estimate the range of the random sample. Any departure from the mean value gives $(x - \mu)$ and the fluctuation of the random variable, which is both positive and negative. By definition, these fluctuation damps out to zero by taking the average over the entire sample space:

$$E(x - \mu) = \int_{-\infty}^{\infty} (x - \mu)\, f(x)\, dx = 0 \tag{8.55}$$

Variance is the expected value of the random variable's second central statistical moment $(x - \mu)^2$:

$$\mathrm{var}(x) = \sigma^2(x) = \int_{-\infty}^{\infty} (x - \mu)^2 f(x)\, dx = \int_{-\infty}^{\infty} x^2 f(x)\, dx - \mu^2 \tag{8.56}$$

Variance measures the spread or dispersion of the random variable around the mean. A large variance indicates that the values of X are spread out, while a small variance indicates that the values are clustered closely around the mean. The positive square root of the variance is known as standard deviation (σ), which measures the spread. The nth central moment for a continuous random variable is expressed as

$$E\left((x - \mu)^n\right) = \mu_n = \int_{-\infty}^{\infty} (x - \mu)^n\, f(x)\, dx \tag{8.57}$$

Eventually, this give $\mu_0 = 1$, $\mu_1 = \mu$, $\mu_2 = \sigma$, and so on. The third central statistical moment of a random variable is known as skewness S, which measures the asymmetry of the distribution:

$$S = E\left[\frac{(x-\mu)^3}{\sigma^3}\right] \tag{8.58}$$

The distribution has a long right tail for a positive skewness $S > 0$, indicating more values on the right (higher than the mean). For a negative skewness $S < 0$, the distribution has a long left tail, indicating more values on the left (lower than the mean). The fourth central moment of a random variable is known as kurtosis (K), which is the descriptor of the shape of the random variable:

$$K = E\left[\frac{(x-\mu)^3}{\sigma^4}\right] \tag{8.59}$$

Kurtosis measures the tailedness of the probability distribution or how much probability mass is concentrated in the tails and the peak of the distribution. Leptokurtic (a high value of kurtosis $K > 3$) indicates heavy tails and a sharp peak. Platykurtic (a low value of kurtosis $K < 3$) indicates lighter tails and a flatter peak. Mesokurtic ($K < 3$) indicates the normal distribution.

It is often convenient to work in terms of standardized random variables, which, by definition, have zero mean and unit variance. The standardized random variable $\hat{X}$ corresponding to X is

$$\hat{x} \equiv \frac{(x-\mu)}{\sigma} \tag{8.60}$$

The characteristic function of the random variable X is defined by

$$\Psi(s) \equiv E\left(e^{ixs}\right) = \int_{-\infty}^{\infty} f(x)e^{ixs}\,dx \tag{8.61}$$

The integral in Eq. (8.61) is an inverse Fourier transform of the probability density function $f(x)$. $\Psi(s)$ and $f(x)$ form a Fourier transform pair and contain the same information. $\Psi(s)$ is a nondimensional complex function of the real variable s (s has the dimensions of x^{-1}). Like the PDF $f(x)$, the characteristic function $\Psi(s)$ fully characterizes the random variable x. The characteristic function is a mathematical device that facilitates some derivations. Because $f(x)$ is real, $\Psi(s)$ has conjugate symmetry: $\Psi(s) = \Psi^*(-s)$. The behavior at the origin can be obtained by setting s to zero:

$$\Psi(0) = 1 \tag{8.62}$$

The k th derivative of $\Psi(s)$ is

$$\Psi^{(k)}(s) \equiv \frac{d^k\Psi(s)}{ds^k} = i^k E\left(x^k e^{ixs}\right) \tag{8.63}$$

The kth moment of x (about the origin) is given by

$$E\left(x^k\right) = (-i)^k \Psi^{(k)}(0) \tag{8.64}$$

The linear transformation of a random variable

$$\widetilde{x} = a + bx \tag{8.65}$$

is also a random variable with a and b being constants. This transformation has the characteristic function

$$\begin{aligned}\widetilde{\Psi}(s) &= E\left(e^{i\tilde{x}s}\right) = E\left(e^{ias+ibxs}\right) \\ &= e^{ias}\Psi(bs)\end{aligned} \tag{8.66}$$

If X_1 and X_2 are independent random variables with characteristic functions $\Psi_1(s)$ and $\Psi_2(s)$, then their sum is also a random variable having characteristic function

$$\begin{aligned}\Psi(s) &= E\left(e^{i(x_1+x_2)s}\right) = E\left(e^{ix_1s}e^{ix_2s}\right) \\ &= E\left(e^{ix_1s}\right)E\left(e^{ix_2s}\right) = \Psi_1(s)\Psi_2(s)\end{aligned} \tag{8.67}$$

The characteristic function $\Psi(s)$ of the sum of two random variables X_1 and X_2 is the product of their individual characteristic functions $\Psi_1(s)$ and $\Psi_2(s)$.

Any function $Q(x)$ of a random variable x is also random, having a mean

$$E\left(Q(x)\right) = \int_{-\infty}^{\infty} Q(x)f(x)\,dx \tag{8.68}$$

The average of the random function $Q(x)$ exists only when the integral in Eq. (8.68) converges absolutely. If $Q(x)$ and $R(x)$ are two functions of x, with a and b being constants, then

$$E\left(aQ(U) + bR(U)\right) = a\left(Q(x)\right) + b\left(R(x)\right) \tag{8.69}$$

$E\left(\{\}\right)$ is a linear operator. x, $Q(x)$, and $R(x)$ are random variables, while $E\left(x\right)$, $E\left(Q(x)\right)$, and $E\left(R(x)\right)$ are not.

Suppose that X is a random variable that takes value a with probability p, and the value $b(b > a)$ with probability $1 - p$. It is straightforward to deduce the CDF of X as

$$F(x) = P\{X < x\} = \begin{cases} 0, & \text{for } x \le a \\ p, & \text{for } a < x \le b \\ 1, & \text{for } x > b \end{cases} \tag{8.70}$$

Equation (8.70) can be written in terms of Heaviside functions as

$$F(x) = p\mathcal{H}(x-a) + (1-p)\mathcal{H}(x-b) \tag{8.71}$$

The corresponding PDF is obtained by differentiating equation (8.71)

$$f(x) = p\delta(x-a) + (1-p)\delta(x-b) \tag{8.72}$$

A random variable that can take only a finite number of values is a discrete random variable (as opposed to a continuous random variable). With the aid of Heaviside and Dirac delta functions, the discrete random variables can also be treated. Furthermore, if x is a sure variable, with probability one of having the value a, its CDF and PDF are consistently given by

$$F(x) = \mathcal{H}(x-a) \tag{8.73}$$

$$f(x) = \delta(x-a) \tag{8.74}$$

The probability density function (PDF) for some common distribution and their cumulative distribution function (CDF) is provided in Fig. 8.18.

8.6.2.1 Uniform Distribution

PDF for a uniform random variable x on the interval $[a, b]$ is defined as

$$f(x) = \frac{1}{b-a}, \quad \text{for } a \le x \le b \tag{8.75}$$

The uniform distribution is often used in simulations or random sampling where each outcome in a specified range is equally likely. In Monte Carlo simulations, uniform distribution is used to generate random numbers over a specific interval. Fair games like rolling a fair die or flipping a fair coin are modeled using a discrete uniform distribution. Many random number generators are based on the uniform distribution to generate equally probable outcomes.

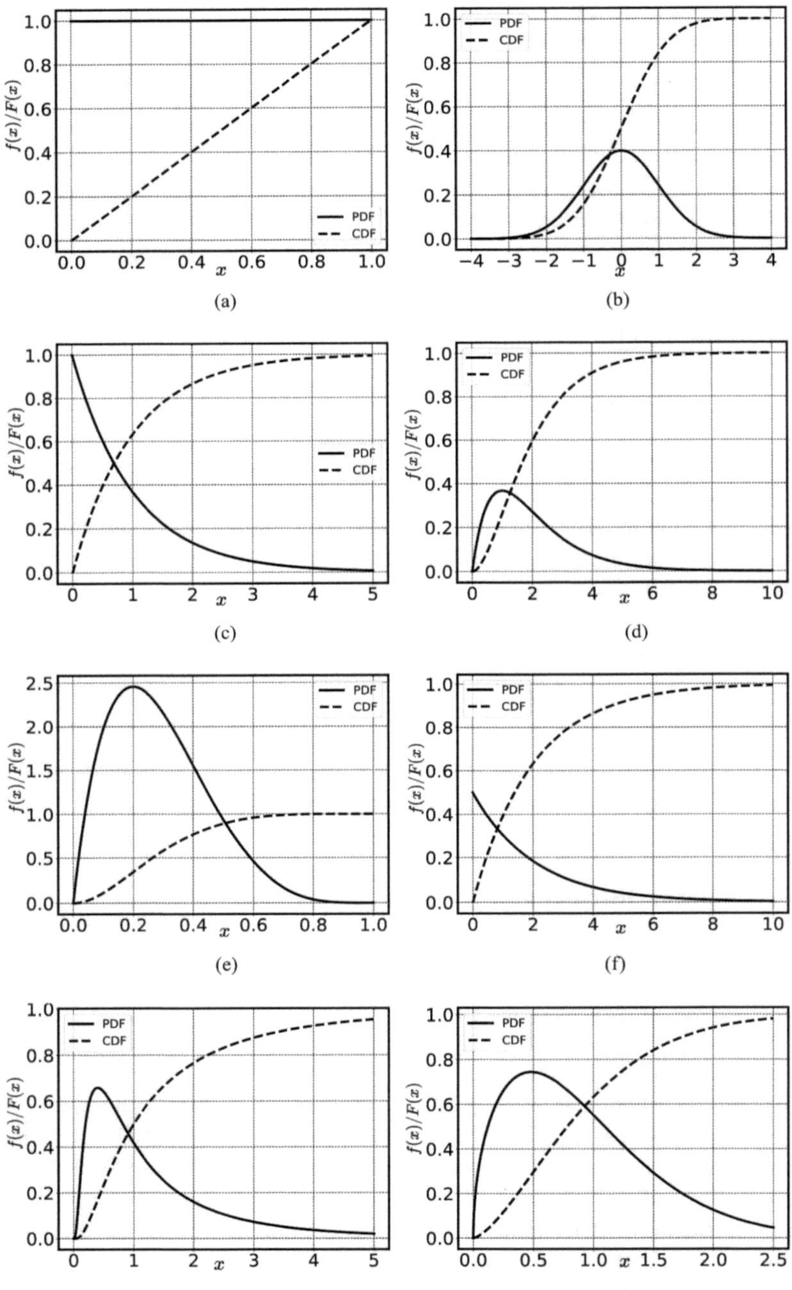

Fig. 8.18 PDF and CDF of some common distribution functions. (**a**) Uniform distribution. (**b**) Normal distribution. (**c**) Exponential distribution. (**d**) Gamma distribution. (**e**) Beta distribution. (**f**) Chi-square distribution. (**g**) Log-normal distribution. (**h**) Weibull distribution

8.6.2.2 Normal (Gaussian) Distribution

PDF for a normal random variable x with mean μ and variance σ^2 is expressed as

$$f(x) = \frac{1}{\sqrt{2\pi\sigma^2}} \exp\left(-\frac{(x-\mu)^2}{2\sigma^2}\right) \tag{8.76}$$

Many naturally occurring variables, such as height, weight, intelligence scores, and measurement errors, follow a normal distribution. The normal distribution is central to inferential statistics, including hypothesis testing, confidence intervals, and regression analysis. Noise in signals is often modeled as Gaussian noise in engineering and communication systems.

8.6.2.3 Exponential Distribution

PDF for an exponential random variable x with rate parameter λ is expressed as

$$f(x) = \lambda e^{-\lambda x}, \quad \text{for } x \geq 0 \tag{8.77}$$

The exponential distribution models the time between events in a Poisson process, such as the time between arrivals in a queue or the time between machine failures. It is also used in survival analysis to model the time until an event, like the lifespan of electronic components or the time between radioactive decay events.

8.6.2.4 Gamma Distribution

PDF for a gamma random variable x with shape parameter k and rate parameter λ is expressed as

$$f(x) = \frac{\lambda^k x^{k-1} e^{-\lambda x}}{\Gamma(k)}, \quad \text{for } x \geq 0 \tag{8.78}$$

The gamma distribution models waiting times in systems where the waiting time until the k-th event occurs is of interest (e.g., the number of phone calls received). In actuarial science, the gamma distribution models claim sizes and the waiting times between insurance claims. It is also used to model the distribution of lifetimes in biological organisms and for processes like the time it takes to complete biochemical reactions.

8.6.2.5 Beta Distribution

PDF for a beta random variable x with shape parameter α and β is expressed as

$$f(x) = \frac{x^{\alpha-1}(1-x)^{\beta-1}}{B(\alpha, \beta)}, \quad \text{for } 0 \leq x \leq 1 \tag{8.79}$$

In Bayesian statistics, the beta distribution is used as a prior distribution for parameters of binomial distributions due to its flexibility in modeling probabilities. The beta distribution is used to model random variables that represent proportions or probabilities (e.g., the success rate of a process). In PERT (Program Evaluation and Review Technique), the beta distribution models the uncertainty in task completion times. It is also used to model uncertainties in classification problems and reinforcement learning.

8.6.2.6 Chi-Square Distribution

PDF for a chi-square random variable x with k degrees of freedom is expressed as

$$f(x) = \frac{x^{\frac{k}{2}-1} e^{-\frac{x}{2}}}{2^{\frac{k}{2}} \Gamma\left(\frac{k}{2}\right)}, \quad \text{for } x \geq 0 \tag{8.80}$$

The chi-square distribution is used in statistical tests like the chi-square goodness-of-fit test and the chi-square test for independence. It is used to estimate the variance of a normally distributed population, particularly in confidence intervals and hypothesis tests for population variance (ANOVA: analysis of variance). Chi-square tests are used to analyze the distribution of genetic traits and verify if they follow expected Mendelian ratios.

8.6.2.7 Log-Normal Distribution

PDF for a log-normal random variable x where $\ln(x)$ is normally distributed with mean μ and variance σ^2 is expressed as

$$f(x) = \frac{1}{x\sigma\sqrt{2\pi}} \exp\left(-\frac{(\ln x - \mu)^2}{2\sigma^2}\right), \quad \text{for } x > 0 \tag{8.81}$$

The log-normal distribution is widely used in finance to model stock prices and asset returns because prices cannot go below zero, and small percentage changes are often multiplicative. In biology, the log-normal distribution models processes that involve growth rates, such as the distribution of body weights or organism sizes. The log-normal distribution is often used to model income or wealth distributions, where many small values and a few large values are common. Environmental variables such as concentrations of pollutants, wind speeds, and rainfall are often modeled with a log-normal distribution.

8.6.2.8 Weibull Distribution

PDF for a Weibull random variable x with scale parameter λ and shape parameter k is expressed as

$$f(x) = \frac{k}{\lambda}\left(\frac{x}{\lambda}\right)^{k-1} \exp\left(-\left(\frac{x}{\lambda}\right)^{k}\right), \quad \text{for } x \geq 0 \tag{8.82}$$

The Weibull distribution is extensively used in reliability engineering to model the time to failure of systems, especially in product manufacturing and life testing. It is also commonly used in renewable energy to model wind speed and assess wind energy potential. In materials science, the Weibull distribution models the breaking strength of materials, capturing the probability of material failure under stress. The Weibull distribution is also used to model the bubbles and droplet distribution in multiphase flow.

What is the average weight of a kangaroo in Australia? What is the average height of a kiwi in New Zealand? What is the average size of a mango produced in India? To determine the actual average, one would need to measure the weight of every kangaroo in Australia, the height of every kiwi in New Zealand, and the size of all mangoes produced in India over a year. This is impractical, so a limited number of samples are randomly selected from the entire population, and statistical analysis is performed. This process is known as statistical sampling. A sample is the subset of the population.

In statistical sampling, randomly selected samples are assumed to be independent and identically distributed (i.i.d.), and the selection process must be free from bias. Any bias in the sampling process would lead to inaccurate conclusions about the overall population. A "statistic" refers to any function calculated from the observations of a random variable. For instance, the mean and standard deviation derived from random samples are considered statistics, whereas the same measures computed for the entire population are referred to as "parameters." The probability distribution of these sample statistics is known as the "sampling distribution."

If $x_1, x_2, x_3, \ldots, x_n$ represent the heights of N randomly selected kiwi samples, then the sample mean $\mu_{\bar{x}}$ is calculated as follows:

$$\mu_{\bar{x}} = \frac{x_1 + x_2 + x_3 + \cdot + x_n}{N} = \frac{1}{N}\sum_{i=1}^{N} x_i = \frac{s_N}{N} \tag{8.83}$$

s_n is the sum of the cumulative height of the sample. Spread of the sample is expressed through the variance:

$$\sigma_{\bar{x}} = \frac{1}{(N-1)} \sum_{i=1}^{N} (x_i - \mu_{\bar{x}})^2 \tag{8.84}$$

The task is to get the true estimator of the population (mean μ and variance σ^2) from the sampling statistics $\mu_{\bar{x}}$ and $\sigma_{\bar{x}}$. This might stun for the time being, but an American pioneer of survey sampling, George Horace Gallup (1901–1984), explains it as "If you have cooked a large pan of soup, you do not need to eat it all to find out if it needs more seasoning. You can just taste a spoonful, provided you have given it a good stir." A good stirring is an analogy to randomize the entire sample.

Probability distribution functions are not merely abstract mathematical concepts; they serve as practical tools for signal analysis. Figure 8.19 presents an example of ECG signal analysis from the MIT-BIH long-term database (Goldberger et al., 2000). The discrete ECG data, recorded over time, is first processed by calculating the mean μ and standard deviation σ using the formulas from Eqs. (8.83)–(8.84). These μ and σ values are then applied to the Gaussian distribution formula in Eq. (8.76) to plot the distribution. The analysis in Fig. 8.19b illustrates a left-skewed distribution, while Fig. 8.19d shows a right-skewed distribution, and Fig. 8.19f displays a symmetric distribution. This demonstrates how statistical techniques can be used to examine the characteristics of signals.

8.6.3 *Joint Probability Distributions*

Multiple random variables often characterize a system; in such cases, the probability density function depends on all those variables. This leads to a joint probability distribution, which captures the likelihood of two or more random variables occurring together. It gives the probability of different combinations of values for the random variables. Consider a system influenced by just two random variables to keep things simple. For two discrete random variables X and Y on the sample space, the joint probability mass function (PMF) is represented as $P(x, y) = P(X = x, Y = y)$, and it has the following properties:

$$P(x, y) \geq 0 \text{ for all } x \text{ and } y \tag{8.85}$$

$$\sum_{x} \sum_{y} P(x, y) = 1 \tag{8.86}$$

The joint probability density function (JPDF) is denoted by $f(x, y)$ for continuous random variables:

$$f(x, y) = \frac{\partial^2}{\partial x \, \partial y} F(x, y) \tag{8.87}$$

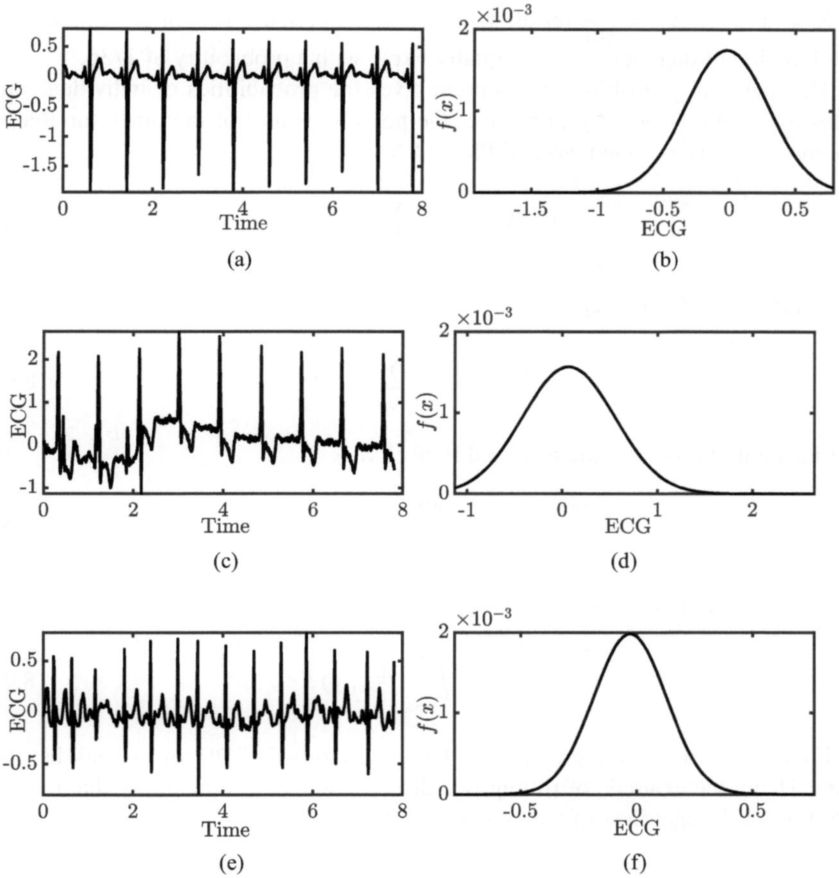

Fig. 8.19 ECG data and their distribution from MIT-BIH long-term database (Goldberger et al., 2000). (**a**) Left skew data. (**b**) Left skew distribution. (**c**) Right skew data. (**d**) Right skew distribution. (**e**) Symmetric data. (**f**) Symmetric distribution

with $F(x, y) = F(X \leq x, Y \leq y)$ being the joint cumulative distribution function (JCDF). The JPDF satisfies the following conditions:

$$f(x, y) \geq 0 \text{ for all } x \text{ and } y \tag{8.88}$$

$$\int_{-\infty}^{\infty} \int_{-\infty}^{\infty} f(x, y)\, dx\, dy = 1 \tag{8.89}$$

Rolling two ludo dice, let X be the outcome of the first die and Y the result of the second die. Each pair of outcomes $(X = x, Y = y)$ has a joint probability of

$P(X = x, Y = y) = \dfrac{1}{36}$ for all $x, y \in \{1, 2, 3, 4, 5, 6\}$. This indicates that each combination of dice outcomes is equally likely with a probability of $1/36$.

The marginal probability distribution gives the probabilities of individual variables by summing or integrating over the possible values of the other variable(s). For the discrete case, the marginal PMF of X is

$$P_x(X = x) = \sum_y P(x, y) \tag{8.90}$$

The marginal PMF of Y is

$$P_y(Y = y) = \sum_x P(x, y) \tag{8.91}$$

For the continuous case, the marginal PDF of X is

$$f_X(x) = \int_{-\infty}^{\infty} f(x, y)\, dy \tag{8.92}$$

The marginal PDF of Y is

$$f_Y(y) = \int_{-\infty}^{\infty} f(x, y)\, dx \tag{8.93}$$

Figure 8.20 shows the graph for JPDF and marginal JPDF for normal distributions. Using the example of rolling two dice, the marginal probability that the first die shows a 4, regardless of the outcome of the second die, is

Fig. 8.20 JPDF and marginal JPDF for normal distribution

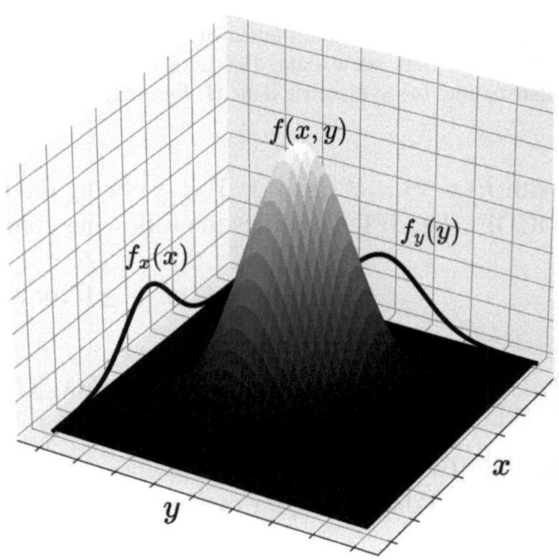

$$P(X = 4) = \sum_{y=1}^{6} P(X = 4, Y = y) = 6 \times \frac{1}{36} = \frac{1}{6}$$

Thus, the marginal probability of rolling a 4 on the first die is $1/6$.

Conditional probability describes the probability of one event occurring, given that another event has already occurred. It tells us how the probability of one event changes based on the knowledge of another event. The conditional probability of $X = x$ given that $Y = y$ is denoted by $P(X = x \mid Y = y)$, and it is defined as

$$P_{x|y}(X = x \mid Y = y) = \frac{P(x, y)}{P_y(y)} \tag{8.94}$$

This assumes that $P_y(y) > 0$. The conditional probability of $Y = y$ given $X = x$ is similarly

$$P_{y|x}(Y = y \mid X = x) = \frac{P(x, y)}{P_x(x)} \tag{8.95}$$

For continuous random variables, the conditional probability density function is

$$f_{x|y}(x \mid y) = \frac{f(x, y)}{f_y(y)} \tag{8.96}$$

where $f_y(y)$ is the marginal PDF of Y and $f(x, y)$ is the joint PDF. Similarly,

$$f_{y|x}(y \mid x) = \frac{f(x, y)}{f_x(x)} \tag{8.97}$$

In the case of rolling two dice, the conditional probability that the first die shows a 4, given that the second die shows a 3, is

$$P(X = 4 \mid Y = 3) = \frac{P(X = 4, Y = 3)}{P(Y = 3)} = \frac{\frac{1}{36}}{\frac{1}{6}} = \frac{1}{6}$$

Thus, the probability of rolling a 4 on the first die, given that the second die is a 3, is $1/6$. Two events, X and Y, are said to be independent if the occurrence of one event does not affect the occurrence of the other. Mathematically, X and Y are independent if

$$P(x, y) = P_x(x) \cdot P_y(y) \tag{8.98}$$

The JPDF is the product of individual marginal PDF. This can be extended to conditional probability as well. If X and Y are independent, the conditional probability simplifies to the marginal probability:

$$P_{x|y}(x \mid y) = P_x(x) \tag{8.99}$$

$$P_{y|x}(y \mid x) = P_y(y) \tag{8.100}$$

Similarly, for continuous random variables,

$$f(x, y) = f_x(x) \cdot f_y(y) \tag{8.101}$$

$$f_{x|y}(x \mid y) = f_x(x) \tag{8.102}$$

$$f_{y|x}(y \mid x) = f_x(y) \tag{8.103}$$

Bayes' theorem is a powerful formula in probability theory that relates conditional probabilities. It allows us to update the probability of a hypothesis based on new evidence:

$$P(A \mid B) = \frac{P(B \mid A) \cdot P(A)}{P(B)} \tag{8.104}$$

For example, consider a medical test for a rare disease. To know the probability that a person has the disease given that test is positive, let A be the event that the person has the disease and B the event that the test result is positive. Suppose the prior probability that the person has the disease is $P(A) = 0.01$ (1% of people have the disease). The likelihood of testing positive if the person has the disease is $P(B \mid A) = 0.95$ (the test correctly identifies the disease 95% of the time). The probability of testing positive without the disease is $P(B \mid \neg A) = 0.05$ (5% of healthy people still test positive, i.e., false positives). The marginal likelihood $P(B)$ is calculated using the law of total probability:

$$\begin{aligned} P(B) &= P(B \mid A)P(A) + P(B \mid \neg A)P(\neg A) \\ &= (0.95)(0.01) + (0.05)(0.99) = 0.059 \end{aligned}$$

Using Bayes' theorem,

$$P(A \mid B) = \frac{P(B \mid A) \cdot P(A)}{P(B)} = \frac{(0.95)(0.01)}{0.059} \approx 0.161$$

So, even though the person tested positive, the probability that they actually have the disease is only about 16.1%. This is because the disease is rare, and the test produces false positives.

In a joint discrete probability system, the mean μ_X or expected value $E[X]$ of random variable X is defined as

$$E(X) = \mu_X = \sum_x x\, P_x(x) = \sum_y \sum_x x\, P(x, y) \tag{8.105}$$

Similarly, the mean μ_Y or expected value $E[Y]$ for random variable Y is

$$E(Y) = \mu_Y = \sum_y y\, P_y(y) = \sum_x \sum_y y\, P(x, y) \tag{8.106}$$

For a contentious function, the same is expressed as

$$E(X) = \mu_X = \int_{-\infty}^{\infty} \int_{-\infty}^{\infty} x f(x, y)\, dxdy \tag{8.107}$$

$$E(Y) = \mu_Y = \int_{-\infty}^{\infty} \int_{-\infty}^{\infty} y f(x, y)\, dxdy \tag{8.108}$$

For single random variable X, the variance is already defined in Eq. (8.56). From Eq. (8.65), the linear transformation of random variable is a random variable too. It is interesting to see some more property of linear transform of random variables $\widetilde{x} = a + bx$:

$$\begin{aligned}
\text{Var}(a + bx) &= E\left((a + bx - E[a + bx])^2\right) \\
&= E\left((a + bx - a - b\mu)^2\right) \\
&= E\left((bx - b\mu)^2\right) \\
&= E\left(a^2(x - \mu)^2\right) \\
&= b^2 E\left((x - \mu)^2\right) \\
&= b^2 \text{Var}(x)
\end{aligned} \tag{8.109}$$

So, $\text{Var}(bx) = b^2\text{Var}(x)$. The expectation of a sum of random variables is equal to the sum of their expectations. The corresponding result for variances is, however, not generally valid. For example,

$$\begin{aligned}
\text{Var}(x + x) &= \text{Var}(2x) \\
&= 2^2\text{Var}(x) \\
&= 4\text{Var}(x) \\
&\neq \text{Var}(x) + \text{Var}(x)
\end{aligned} \tag{8.110}$$

There is, however, an important case in which the variance of a sum of random variables is equal to the sum of the variances; and this is when the random variables are independent. Before proving this, however, let us define the concept of the variance and covariance of joint distribution of random variables. The variance of X in a joint distribution is expressed as

$$\begin{aligned}\mathrm{Var}(X) = \sigma_X^2 &= \sum_x \sum_y (x-\mu_X)^2 P(x,y) \\ &= \left(\sum_x \sum_y x^2 P(x,y)\right) - \mu_X^2 \\ &= E\left(X^2\right) - (E(X))^2\end{aligned} \tag{8.111}$$

Similarly, the variance of Y is expressed as

$$\begin{aligned}\mathrm{Var}(Y) = \sigma_Y^2 &= \sum_x \sum_y (y-\mu_Y)^2 P(x,y) \\ &= \left(\sum_x \sum_y y^2 P(x,y)\right) - \mu_Y^2 \\ &= E\left(Y^2\right) - (E(Y))^2\end{aligned} \tag{8.112}$$

For continuous variable,

$$\begin{aligned}\mathrm{Var}(X) = \sigma_X^2 &= \int_{-\infty}^{\infty} (x-\mu_X)^2 f_x(x)\,dx \\ &= \int_{-\infty}^{\infty}\int_{-\infty}^{\infty} (x-\mu_X)^2 f(x,y)\,dxdy\end{aligned} \tag{8.113}$$

$$\begin{aligned}\mathrm{Var}(Y) = \sigma_Y^2 &= \int_{-\infty}^{\infty} (y-\mu_Y)^2 f_y(y)\,dy \\ &= \int_{-\infty}^{\infty}\int_{-\infty}^{\infty} (y-\mu_Y)^2 f(x,y)\,dxdy\end{aligned} \tag{8.114}$$

When two or more random variables are defined within a probability space, it is helpful to describe how they vary in relation to one another. In other words, it is useful to quantify the relationship between these variables. A widely used metric for assessing the relationship between two random variables is covariance, which measures the extent of their linear relationship. The covariance between X and Y is defined as

$$\begin{aligned}\mathrm{Cov}(XY) = \sigma_{XY} &= \sum_x \sum_y (x - \mu_X)(y - \mu_Y)\, P(x, y) \\ &= \left(\sum_x \sum_y xy\, P(x, y) \right) - \mu_X \mu_Y \\ &= E\,(XY) - (E\,(X)\, E\,(Y))\end{aligned} \tag{8.115}$$

From the definition, $\sigma_{XY} = \sigma_{YX}$ and $\sigma_{XX} = \sigma_X^2$. Suppose the relationship between the random variables is nonlinear. In that case, the covariance might not be sensitive to the relationship, which means it does not relate to the correlation between the two variables. The covariance of X and Y for a continuous is also known as the mixed second moment. For a continuous variable, it is expressed as

$$\sigma_{XY} = \int_{-\infty}^{\infty} \int_{-\infty}^{\infty} (x - \mu_X)\,(y - \mu_Y)\, f(x, y) dxdy \tag{8.116}$$

8.6.4 Correlations and Random Signals

A correlation analysis provides a clear understanding of the relationships between random variables. For example, consider three sets of ordered pairs of random variables x and y, visualized in the scatter plots in Fig. 8.21. A closer look at these plots shows that the ordered pairs in Fig. 8.21b and c can be described by a precise mathematical rule, while the pairs in Fig. 8.21a cannot. It's visually apparent that the data in Fig. 8.21b and c are strongly related since they follow a precise curve. In contrast, the data in Fig. 8.21a lack a discernible pattern, indicating a weak or low correlation. This visual approach works well when the relationships are as evident as those in Fig. 8.21a, b, and c. However, consider data like that in Fig. 8.21d. While there appears to be a trend where y increases as x increases, no precise curve fits all the data points. To more accurately assess the strength of the relationship, we introduce a definition: a set of data pairs (x, y) is said to be perfectly correlated if a mathematical equation defines a curve on which all data points lie. Let this equation be $y = f(x)$. It is rare to find an equation that fits every data point exactly. The challenge is determining the best-fitting curve or equation representing the data. This curve is typically evaluated by minimizing the least squares sum:

$$\varepsilon = \sum_{i=0}^{N-1} [f\,(x_i) - y_i]^2 \tag{8.117}$$

When the data are perfectly correlated, each pair (x_i, y_i) is positioned on the curve $y_i = f\,(x_i)$, resulting in $\varepsilon = 0$. As the data deviate from this curve, the value of ε increases. In this way, a quantitative measure of the strength of the relationship

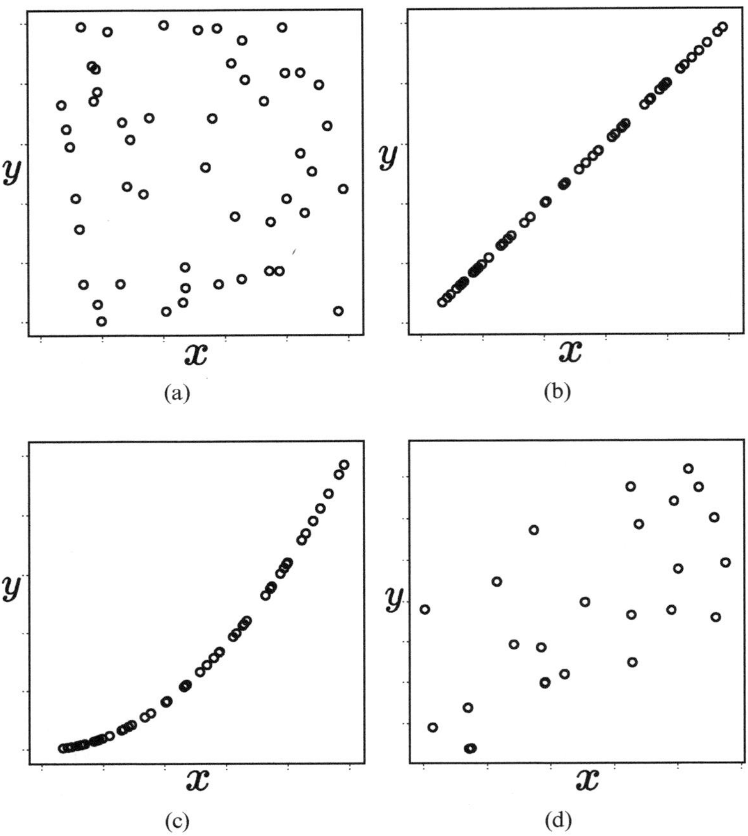

Fig. 8.21 Scatter diagram of two random variables x and y

is obtained. But what kind of relationship is being measured? It has been implicitly assumed that the form of the equation $y = f(x)$ is known, which could be a straight line, a parabola, a Gaussian, a Lorentzian, or another form. The types of curves that can be used are limited only by creativity; however, the nature of the data typically guides the appropriate choice, requiring sound judgment (or perhaps luck). Focus is now placed on the scenario where the curve is a straight line:

$$y = mx + b \tag{8.118}$$

This is known as linear correlation analysis, and it defines the strength of a relation in terms of the data dispersion of a straight line. In this case, the values of m and b must be optimized to best fit the data. The values of m and b that minimize the error measure given by

$$\varepsilon = \sum_{i=0}^{N-1} [y_i - (mx_i + b)]^2 \tag{8.119}$$

To determine the best-fit parameters m and b, Eq. (8.118) must be minimized for ε in terms of these parameters. After solving $\frac{\partial \varepsilon}{\partial m} = 0$ and $\frac{\partial \varepsilon}{\partial b} = 0$, the resulting expression becomes

$$m = \rho_{XY} \frac{\sigma_x}{\sigma_y} \tag{8.120}$$

and

$$b = \mu_y - m\mu_x \tag{8.121}$$

μ_x and σ_x are the mean and standard deviation of the distribution of x values and μ_y and σ_y are those for the y values. ρ_{XY} is known as Pearson's r coefficient and is defined in terms of the z-scores of the two distributions:

$$\rho_{X,Y} = \frac{E(XY) - E(X)E(Y)}{\sqrt{E\left(X^2\right) - (E(X))^2}\sqrt{E\left(Y^2\right) - (E(Y))^2}} \tag{8.122}$$

A little algebraic manipulation results in the following formula:

$$\rho_{XY} = \frac{\sigma_{XY}}{\sigma_X \sigma_Y} \tag{8.123}$$

Pearson's ρ_{XY} can be used as a measure of the strength of the (linear) relation between a set of data pairs $\{(x, y)\}$. A positive correlation coefficient arises when positive excursions from the mean for one random variable $(X - \mu_X) > 0$ are preferentially associated with positive excursions for the other $(Y - \mu_Y) > 0$. Conversely, if positive excursions for $(X - \mu_X) > 0$ are preferentially related to negative excursions of $(Y - \mu_Y) < 0$, then the correlation coefficient is negative. In general, we have the Cauchy-Schwarz inequality:

$$-1 \le \rho_{XY} \le 1 \tag{8.124}$$

The correlation coefficient $\rho_{XY} = 0$ implies uncorrelated X and Y. For perfectly correlated X and Y, $\rho_{XY} = 1$ and perfectly negative correlated X and Y, $\rho_{XY} = -1$. Values of $|\rho_{XY}|$ between these extremes indicate varying degrees of correlation ranging from weakly related to strongly related data.

This section provides a heuristic discussion on how correlation analysis tests a function's randomness. Attempts have been made to physically interpret autocorrelation when it is applied to random or stochastic signals. Subsequent sections present a more formal explanation of autocorrelation.

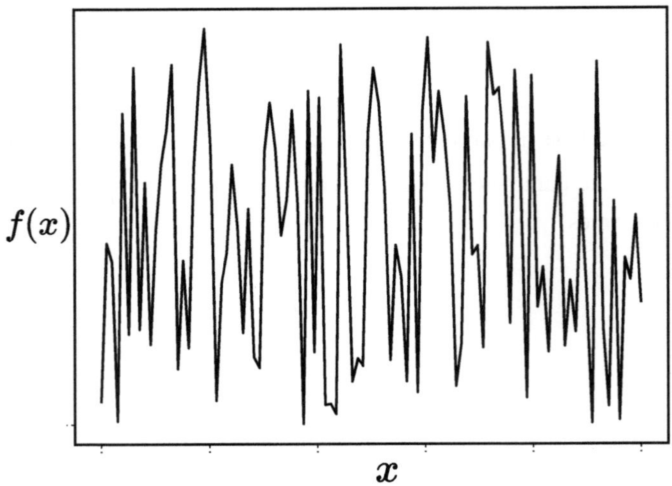

Fig. 8.22 Random signals

Let $f(x)$ represent a random function or signal similar to the one shown in Fig. 8.22. For any specific x, a mathematical equation cannot describe the value of $f(x)$. Instead, it must be expressed in terms of an expected value and standard deviation. Consequently, no analytical relationship is expected between any two functional values $f(x_1)$ and $f(x_2)$ for a random function $f(x)$. To clarify this further, a brief digression into deterministic or analytical functions is necessary. The issue to be addressed is determining the relationship between two values $f(x_1)$ and $f(x_2)$, or, when $x_1 = x$ and $x_2 = x + \tau$, the relationship between $f(x)$ and $f(x + \tau)$. For,

$$f(x) = e^{-ax} \tag{8.125}$$

then

$$f(x + \tau) = e^{-a(x+\tau)} = e^{-ax}e^{-a\tau} = e^{-a\tau} f(x) \tag{8.126}$$

For

$$f(x) = \sin x \tag{8.127}$$

then

$$\begin{aligned} f(x + \tau) &= \sin(x + \tau) = \sin x \cos \tau + \cos x \sin \tau \\ &= f(x) \cos \tau + \frac{df(x)}{dx} \sin \tau \end{aligned} \tag{8.128}$$

Clearly, $f(x+\tau)$ and $f(x)$ are related. However, for the random signals, the relationship is not as simple as in the previous example for deterministic signals. $\tau = 0$ represents the perfect correlation, i.e., $f(x+\tau) = f(x)$. Now, the original task of examining the relationship between two values of a random function, $f(x)$ and $f(x+\tau)$, will be revisited. When $\tau = 0$, the equation $f(x+\tau) = f(x)$ holds, indicating perfect correlation. However, no correlation seems to be present for any other value of τ. An autocorrelation function of $f(x)$ with itself is defined as

$$R_x(\tau) = \int_{-\infty}^{\infty} f(x) f(x+\tau) dx \tag{8.129}$$

It turns out that this assumption is quite reasonable when the signals are long in duration compared to the offset value τ. Additionally, the limit is considered as these sampled signals become continuous and infinite in duration. It can be argued that for a completely random continuous signal, its autocorrelation $R_x(\tau)$ will resemble an impulse or delta function (with a constant added). Therefore, it can be concluded that the autocorrelation integral is a test to determine whether a signal is entirely random. Conversely, if $f(x)$ is a completely deterministic signal, its autocorrelation will also be deterministic. For instance, if $f(x)$ is a pulse function, $R_x(\tau)$ will be a triangle function. Since real-world signals typically fall between these extremes, their autocorrelation lies somewhere between an impulse and a deterministic function. Admittedly, this is a somewhat vague statement, and assessing a signal's randomness through its autocorrelation must be developed through practice.

8.6.5 Ensembles and Expected Values of a Random Signal

Any specific mathematical formula cannot represent a random signal. Its pointwise properties cannot be defined, so it must be described using its overall or statistical characteristics. This statistical representation of a random signal can be achieved in one of two ways. One method involves defining the average value of the random signal through the following equation:

$$E\left(f(x)\right) = \lim_{T\to\infty} \frac{1}{2T} \int_{-T}^{T} f(x) dx \tag{8.130}$$

The alternative approach treats a random signal as a member of an ensemble of similar signals. For instance, if wind velocity were recorded daily at a specific location over a year, an ensemble of 365 recordings of the random signal is obtained. This ensemble's individual members are called sample signals or sample functions. In Fig. 8.23, four sample signals from the ensemble are shown. The position of a particular value of the independent variable x is marked in Fig. 8.23. For this value of x, the average value of the random signal across the ensemble is calculated as

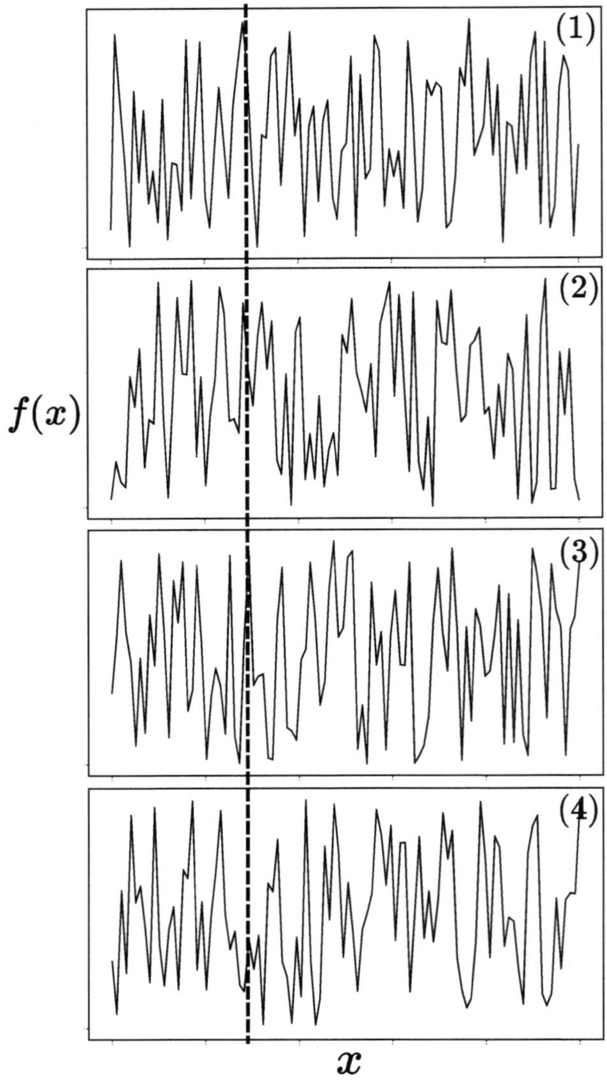

Fig. 8.23 Random signals

$$\mu(x) = \sum_{i=1}^{N} \frac{f_i(x)}{N} \tag{8.131}$$

This work focuses on the scenario where the ensemble is extremely large or, for practical purposes, infinite. Under such conditions, the expected value of the signal $f(x)$ (denoted as μ_{f_x} or $E(f(x))$) is defined as follows:

$$\mu_{f_x} = E\left(f(x)\right) = \lim_{N\to\infty} \frac{1}{N}\sum_{i=1}^{N} f_i(x) \tag{8.132}$$

Now, a few interesting properties of the expected value function are considered. Firstly, it maps any value x to μ_{f_x} using the Eq. (8.132), making it a deterministic function. Additionally, defining μ_{f_x} implied dealing with the random signals generated through the ensemble. However, if the ensemble is instead generated by a deterministic signal, then $f_i(x) = f(x)$ for all i, and $\mu_{f_x} = f(x)$. The expected value function is linear. Let $f(x)$, $g(x)$, and $h(x)$ represent three different signals generating three distinct ensembles, and let a, b, and c be arbitrary constants:

$$E\left(af(x) + bg(x) + ch(x)\right) = aE\left(f(x)\right) + bE\left(g(x)\right) + cE\left(h(x)\right) \tag{8.133}$$

This is a rather straightforward task:

$$\begin{aligned} E\left(af(x) + bg(x) + ch(x)\right) &= \lim_{N\to\infty} \frac{1}{N}\sum_{i=1}^{N} \left[af_i(x) + bg_i(x) + ch_i(x)\right] \\ &= \lim_{N\to\infty} \left[a\sum_{i=1}^{N}\frac{f_i(x)}{N} + b\sum_{i=1}^{N}\frac{g_i(x)}{N} + c\sum_{i=1}^{N}\frac{h_i(x)}{N}\right] \\ &= aE\left(f(x)\right) + bE\left(g(x)\right) + cE\left(h(x)\right) \end{aligned} \tag{8.134}$$

This relationship can be generalized to include an infinite number of random signals.

The cross-correlation for two signals $f(x_1)$ and $g(x_2)$ is denoted as $R_{f_{x_1}g_{x_2}}$ and defined in terms of ensembles as

$$\begin{aligned} R_{f_{x_1}g_{x_2}} &= E\left(f(x_1)\, g^*(x_2)\right) \\ &= \lim_{N\to\infty} \sum_{i=0}^{N-1} \frac{f_i(x_1)\, g_i^*(x_2)}{N} \end{aligned} \tag{8.135}$$

The product of the signal $f(x_1)$ for each x_1 and the complex conjugate of the signal $g(x_2)$ for each x_2 is first formed for each ensemble member to evaluate this expected value. Similarly, the autocorrelation of a signal $f(x)$ is defined as follows:

$$\begin{aligned} R_{f_{x_1}f_{x_2}} &= E\left(f(x_1)\, f^*(x_2)\right) \\ &= \lim_{N\to\infty} \sum_{i=0}^{N-1} \frac{f_i(x_1)\, f_i^*(x_2)}{N} \end{aligned} \tag{8.136}$$

For $x_1 = x_2 = x$,

$$\begin{aligned} R_{f_x f_x} &= E\left(f(x)\, f^*(x)\right) \\ &= E\left(|f(x)|^2\right) \end{aligned} \tag{8.137}$$

$|f(x)|^2$ is measurable and presents the energy content of the signal.

8.6.6 Systems and Random Signal

This section examines the response of a deterministic system to a stochastic or random input signal. A system is deterministic when its impulse response function is deterministic. For linear time-invariant (LTI) systems (formally defined in Chap. 9), the impulse response $h(x)$ operates on a signal $f(x)$ through the process of convolution:

$$g(x) = \int_{-\infty}^{\infty} f(y)h(x-y)dy = \int_{-\infty}^{\infty} f(x-y)h(y)dy \tag{8.138}$$

$f(x)$ is the input signal to the system with expected value μ_{f_x}. The expected value of the output signal is expressed as

$$\mu_{g_x} = E\left(g(x)\right) = E\left(\int_{-\infty}^{\infty} f(y)h(x-y)dy\right) \tag{8.139}$$

The linearity property of expected value and integral operators allows the interchange of the operations:

$$\begin{aligned} \mu_{g_x} = E\left(g(x)\right) &= \int_{-\infty}^{\infty} E\left(f(y)h(x-y)\right) dy \\ &= \int_{-\infty}^{\infty} E\left(f(y)\right) h(x-y)dy \\ &= \int_{-\infty}^{\infty} \mu_{f_y} h(x-y)dy \\ &= \mu_{f_x} * h(x) \end{aligned} \tag{8.140}$$

The impulse response of a system $h(x)$ is deterministic and hence comes out of the expectation operator. The expected value of the output signal is simply the convolution of the expected value of the input signal with the system's impulse response function.

To obtain the correlation between the input and output of a system, let us multiply $g^*(x_2)$ on both sides of the Eq. (8.138):

$$g(x_1)g^*(x_2) = \int_{-\infty}^{\infty} f(x_1 - y)g^*(x_2)h(y)dy \tag{8.141}$$

Taking the expectation of both sides,

$$\begin{aligned} E\left(g(x_1)g^*(x_2)\right) &= \int_{-\infty}^{\infty} E\left(f(x_1 - y)g^*(x_2)\right)h(y)dy \\ R_{g_{x_1}g_{x_2}} &= \int_{-\infty}^{\infty} R_{f_{x_1-y}g_{x_2}}h(y)dy \\ R_{g_{x_1}g_{x_2}} &= R_{f_{x_1}g_{x_2}} * h(x_1) \end{aligned} \tag{8.142}$$

The convolution is carried out with respect to x_1, and x_2 is a parameter. An interesting outcome can be derived by examining the conjugate system. For a system with an impulse response $h(x)$, the conjugate system is defined as having an impulse response $h^*(x)$. A real system serves as its own conjugate. By applying logic similar to that used to arrive at Eq. (8.142), it can also be shown that if the input to this conjugate system is $R_{f_{x_1}g_{x_2}}$, the resulting output will be $R_{g_{x_1}g_{x_2}}$.

The results presented so far in this section apply to deterministic as well as random signals. For fully random or incoherent signals,

$$R_{f_{x_1}f_{x_2}} = 0 \qquad x_1 \neq x_2 \tag{8.143}$$

For $x_1 = x_2 = x$,

$$R_{f_x f_x} = E\left(f(x), f^*(x)\right) = E\left(|f(x)|^2\right) \tag{8.144}$$

Thus, for an incoherent signal,

$$R_{f_{x_1}f_{x_2}} = f(x_1)\,\delta(x_2 - x_1) \tag{8.145}$$

Completely random signals are also known as white noise. If impulse function $h(x)$ defines a system that maps $f(x)$ to $g(x)$, then the system defined by $|h(x)|^2$ map $E\left(|f(x)|^2\right)$ to $E\left(|g(x)|^2\right)$:

$$E\left(|g(x)|^2\right) = E\left(|f(x)|^2\right) * \left|h(x)^2\right| \tag{8.146}$$

This equation is fascinating because it reveals the behavior of a system when responding to incoherent signals. System response to incoherent signals can be predicted by knowing how it responds to deterministic signals (i.e., by knowing

$h(x)$, $|h(x)|^2$ can be determined). The system's transfer function is defined as the Fourier transform of its incoherent impulse response function:

$$\mathscr{F}\{|h(x)|^2\} = \mathscr{F}\{h(x)h^*(x)\} = \mathscr{F}\{h(x)\} * \mathscr{F}\{h^*(x)\} \tag{8.147}$$

Denoting $\mathscr{F}\{h(x)\}$ as $H(\omega)$,

$$\mathscr{F}\{|h(x)|^2\} = H(\omega) * H^*(-\omega) \tag{8.148}$$

A statistically stationary signal is a random process whose statistical properties do not change over time, i.e., the signal behaves consistently regardless of when it is observed. A statistically stationary signal has the following requirements:

$$E\,(f(x)) = \eta_{f(x)} = \mu \quad \forall x \tag{8.149}$$

The average value of the signal remains unaltered for the entire range of random variables:

$$E\left((f(x) - \mu)^2\right) = \sigma^2, \quad \forall x \tag{8.150}$$

The variance or the measure of the spread of the signal also remains constant over for the entire range. The autocorrelation function, which measures how the signal values at different times relate to each other, must depend only on the lag ($\tau x_2 - x_1$ the difference) between two points, not on their absolute values themselves. The autocorrelation function $R_{x_1x_2}$ is defined as

$$R_{f_{x_1}f_{x_2}} = E\left(f\,(x_1)\,f^*\,(x_2)\right) = E\left(f\,(x_1)\,f^*\,(x_1 + \tau)\right) = R_{f_\tau} \tag{8.151}$$

In simpler terms, for a signal to be stationary, its mean, variance, and autocorrelation must remain unaltered for the entire range of random variables.

A signal is strictly stationary if all of its statistical moments (mean, variance, higher-order moments) follow the invariant requirement stated above; however, if only the first two moments (mean and autocovariance) are invariant, then the signal is wide-sense stationary. Most practical applications focus on wide-sense stationery.

The ergodicity principle is a fundamental concept in statistics and signal processing that applies to random processes. It describes a condition under which the averaged behavior of a single realization of a process can represent the ensemble-averaged behavior across many realizations. In simpler terms, a process is ergodic if a single sample for the process over a long period has the same statistical properties (e.g., mean, variance) as obtained by averaging over an ensemble of many different realizations of the process at a single point in time. The time average of the mean value of a process should equal the ensemble average of the mean:

$$\lim_{T\to\infty}\frac{1}{T}\int_0^T f(x)\,dx = E\left(f(x)\right) \tag{8.152}$$

$E\left(f(x)\right)$ is the ensemble average (expected value) of the process for any x, and the left-hand side is the time average over a long period of time T. Similarly, for the autocorrelation function, the time average of the autocorrelation must equal the ensemble average:

$$\lim_{T\to\infty}\frac{1}{T}\int_0^T f(x)f^*(x+\tau)\,dx = E[f(x)f(x+\tau)] \tag{8.153}$$

τ is the lag between two points in time.

Stationarity is a necessary condition for ergodicity but not sufficient on its own. A process must be stationary (its statistical properties do not change over time) to be ergodic, but not all stationary processes are ergodic. However, the ergodicity implies that one can infer the long-term statistical behavior of a process from a single time series (realization), making it very useful in practical applications, like signal processing, where only one realization of a signal is often available.

8.7 Convolution and Probability Distribution

The Aristotelian dogma on the immutability of the heavens was first effectively challenged by Danish astronomer Tycho Brahe (1546–1601). Brahe observed a new star in 1572, which gradually diminished and disappeared over the next 18 months. As the astronomical measurements were not accurate, Brahe advocated incorporating repeated astronomical measurements but did not specify any specific way to utilize those measurements. In the seventeenth century, German astronomer Kepler (1571–1630) and Italian scientist Galileo (1564–1642) made many brilliant astronomical discoveries. Societies dominated by the Catholic church of Rome were not happy with these scientific discoveries as they used to believe understanding the unknown of the universe was against God's will. Catholic church society banned astronomical books by Copernicus, Galileo, Kepler, and Descartes. All experimental results are liable to some inevitable observational errors, regardless of whether measurements are made by different or even by the same observer at different times or places. For accuracy, astronomers used to take repeated measurements and used median, mean, or mode to express astronomical readings. The debate for the best alternative between mean and median has existed among astronomers for centuries. Anglo-Irish physicist Robert Boyle (1627–1691) argued for more careful measurement to determine precise astronomical observation rather than representing observation in terms of averaging. In representing repeated measurements through their central tendency, astronomers started error theory and fitting equations to data. Galileo used several sets of observations for the new star of 1572 to compare two hypotheses on the star's position. Galileo postulated in his work ("*Dialogue*

Concerning the Two Chief Systems of the World-Ptolemaic and Copernican, 1632) that true distance has a unique length, and all observations contain errors. These errors are symmetric around the true value, and small errors are more frequent than larger ones. However, Galileo did not address how to estimate the true distance (Stigler, 1986; Hald, 2003).

British mathematician Thomas Simpson (1710–1761) noted mean as the best possible value of a series of observations or measurements and introduced continuous error distribution. Simpson (1757a) introduced a quite strange probability distribution of an error theory. For n, independent observation is sorted in the bound like $-v$, $(-v+1), \ldots, -1, 0, 1, \ldots, (v-1)$, v, Simpson (1757a) first proposed that their probability is proportional to

$$r^{-v}, r^{-v+1}, \ldots, r^{-1}, r^{0}, r^{1}, \ldots, r^{1-v}, r^{v} \tag{8.154}$$

Later Simpson proposed the second method, where the probability is proportional to

$$r^{-v}, 2r^{-v+1}, 3r^{-v+2}, \ldots, (v+1)r^{0}, \ldots, 3r^{v-2}, 2r^{v-1}, r^{v} \tag{8.155}$$

In a simple case, putting $r = 1$, Eq. (8.154) gives a uniform error distribution (Fig. 8.24a), which is symmetric but inconsistent with Galileo's postulation (the frequency of the small error is higher than that of the large error). Simpson's second error curve (Fig. 8.24b) from Eq. (8.155) was consistent with Galileo's postulation. However, it was a linear curve having a sharp discontinuity at the mean. In 1770, Italian mathematician Joseph-Louis Lagrange (1736–1813) published his celebrated memoir on determining the best value from a set of experimental observations. Lagrange (1770) had shown the process of estimating the limits of the error bound for a measurement. Lagrange also argued about the finite-size error domain and placement of mean error within the limit. Using the least square criterion, he introduced an equation for the error curve of the following form:

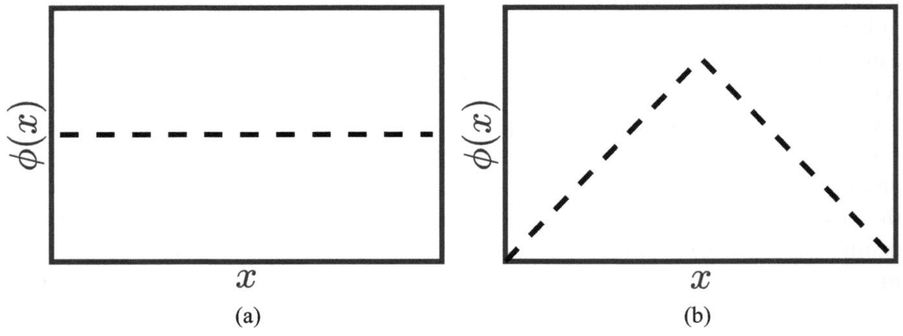

Fig. 8.24 Error distribution curve by Simpson (1757b). (**a**) First error curve. (**b**) Second error curve

$$\phi(x) = ke^{-h^2x^2} \tag{8.156}$$

Its sum equals one, which is the probability distribution curve:

$$k\int_{-\infty}^{\infty} e^{-h^2x^2}dx = \frac{k}{h}\sqrt{\pi} = 1 \tag{8.157}$$

1805 French mathematician Adrien-Marie Legendre (1753–1833) formulated the principle of least squares. Legendre (1806) is counted as one of the clearest and most elegant introductions of a new statistical method in the history of statistics. In the least square theory, the most probable value for the observed quantities is one, for which the sum of the squares of the individual errors is minimum. If $x_0, x_1, \ldots x_n$ are n observations and x is the best possible value, then as per the least square principle, the function sum of the square of deviation is minimum:

$$f(x) = \sum_{k=1}^{n}(x - x_k)^2 \tag{8.158}$$

When $f(x)$ is minimum, then $f'(x) = 0$

$$\sum_{k=1}^{n}(x - x_k) = 0 \tag{8.159}$$

$f(x)$ is minimum because $f''(x) = 2n > 0$. So, the arithmetic mean is the best possible value for a large set of observations:

$$\bar{x} = \frac{1}{n}\sum_{k=1}^{n}x_k \tag{8.160}$$

Pierre-Simon Laplace (1749–1827), a renowned French mathematician and astronomer, was a key figure in the early nineteenth-century development of probability theory. His groundbreaking work on the analytic theory of probability, as highlighted in Laplace (1820), became one of the most influential contributions to mathematics during that era. In addition to his focus on probability, Laplace advanced research on planetary motion, continuing the work initially pioneered by Isaac Newton. Newton's gravitational model aligned well with the available data, but later observations revealed small discrepancies, known as gravitational perturbations, between the predicted and actual planetary orbits. These perturbations arose due to gravitational interactions among the planets themselves. Laplace applied his mathematical prowess to model these perturbations, taking into account the various forces acting between planets and demonstrating that these interactions did not threaten the overall stability of the solar system.

Laplace's deterministic worldview shaped his belief that both the future and the past could be predicted with precision, provided the present state of the universe was fully understood. He argued that any uncertainty in predicting events stemmed purely from human ignorance. This belief was foundational to Laplace's approach to probability, which rejected the idea of randomness in the universe and instead embraced a deterministic framework. Beyond his significant contributions to celestial mechanics, Laplace modernized and formalized probability theory, particularly in the context of statistical inference and error analysis.

In his second major work, Laplace (1829), Laplace explored the theory of errors, applying probability to a wide range of fields, including theology, mechanics, public health, and actuarial science. He demonstrated how probability could be used to analyze measurements and calculate the most accurate or true values. Laplace proposed that positive and negative errors in a large set of observations tend to cancel each other out, causing the average to converge to the true value. He was also the first to introduce the error curve for the probability distribution, as shown in Fig. 8.25a, assuming the curve was symmetric about zero and decreased monotonically as the magnitude of the error increased. He modeled the error distribution with the decay rate proportional to the error, expressed as $\frac{d\phi(x)}{dx} = -m\phi(x)$, which led to the form of the error distribution:

$$\phi(x) = \frac{m}{2} e^{-m|x|} \tag{8.161}$$

Laplace's first error curve was not smooth at $x = 0$. Three years later, Laplace proposed a second curve. If a is the supremum of all possible error $-a \le x \le a$, considering symmetry and the sum of the entire probability equal to one, the second error curve takes the form

$$\phi(x) = \frac{1}{2a} \ln\left(\frac{a}{|x|}\right) \tag{8.162}$$

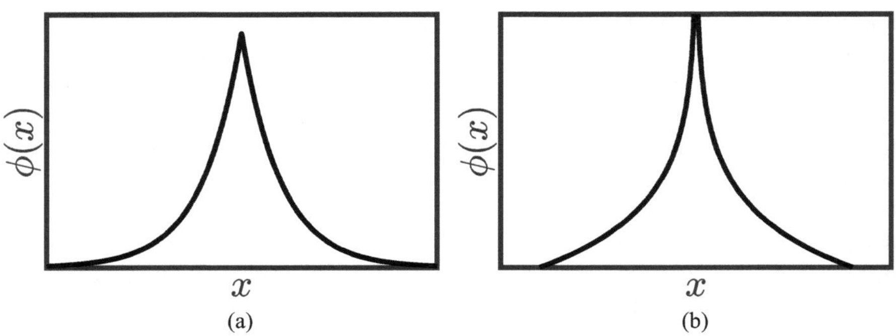

Fig. 8.25 Error distribution curve by Laplace. (**a**) First error curve. (**b**) Second error curve

Despite the finite error domain size, the second error curve had singularity at $x = 0$ (Fig. 8.25b). Laplace was also aware that his error curves might not represent the precise representation of the law of errors. Hence, Laplace suggested a limitation of his error analysis to very delicate observations, like the transit of Venus across the sun. In contemporary times, mass vaccination was used to fight against smallpox. Laplace discussed the probable effect of vaccination on the population growth rate after eliminating a deadly disease. This was an early attempt to forecast the problems of unrestricted population growth.

Italian astronomer Giuseppe Piazzi discovered a heavenly body in January 1801 and had mistaken it for a planet. Piazzi named it Ceres, but unfortunately, Ceres disappeared 6 weeks later (Teets & Whitehead, 1999). This exciting event fascinated many European astronomer communities for the accurate determination of Ceres and its orbital locus. Young German mathematician Carl Friedrich Gauss (1777–1858) suggested searching for different areas of the sky, which the earlier astronomers neglected. Moreover, Gauss turned out to be correct. Gauss (1809) proposed a new error curve (also known as Gaussian distribution) analogous to what Galileo thought centuries ago. Gauss showed that the (sample) mean maximizes the likelihood of errors only when the error curve is of the following form:

$$\phi(x) = \frac{h}{\sqrt{\pi}} e^{-h^2 x^2} \tag{8.163}$$

Gauss considered h to be precision in the measurement process and derived the probability integral between any two limits x_0 and x as

$$P(x) = \frac{h}{\sqrt{\pi}} \int_{x_0}^{x} e^{-h^2 x^2} dx \tag{8.164}$$

This is also known as the Gaussian integral equation. The Gaussian error curve is smooth throughout the domain and has a horizontal tangent at the inflation point. Gauss also agreed that average is the most likely measurement among the several measurements of the same quantity. Gauss used the least square criterion to locate the orbit that best fit the observation. The correct ascension of Mars was consistent with Gaussian distribution (Quetelet, 1846). Average became legitimated as the least squares estimator of the data. Gauss was the first to justify it by appealing to the probability theory. In 1810, Laplace read Gauss's work and came up with a revolutionary connection between the least square estimation of error and the primitive form of the central limit theorem. According to Laplace, the Gaussian curve distributes the aggregate of a large number of errors. Laplace quoted "a science, which began with the consideration of play, has risen to the most important object of human knowledge."

In 1860, Scottish celebrity physicist James Clerk Maxwell (1831–1879) used Gauss-Laplace synthesis to express the probability for velocity distributions of molecules in gases based on classical thermodynamics. British mathematician John

Venn (1834–1923) considered Gaussian distribution as the law of error in his book entitled "*Logic of Chance*" in 1867. In 1892, Henri Poincaré (1854-1912) wrote in the preface of his book "*Thermodynamique*" that "everybody firmly believe in the law of errors, because mathematicians imagine that it is a fact of observation, and observers that it is a theorem of mathematics." Gauss-Laplace error curve is the most compelling rationale finding and is regarded as one of the significant milestones in the history of science. The Gauss-Laplace synthesis was a staple in astronomy and other physical sciences till mid of the 1800s. Post 1900, it was fully diffused into other branches of science, engineering, and social sciences (Tabak, 2004).

8.7.1 Central Limit Theorem

The central limit theorem (CLT) is a significant contribution to modern mathematics by Laplace. Distribution of a sample mean tends toward Gauss-Laplace distribution as the sample size substantially increases, regardless of the original distribution of the population, from which random species are sampled. If $x_1, x_2, x_3, \ldots, x_n$ are i.i.d. random variables, with a finite population mean $|\mu| < \infty$ and standard deviation $0 < \sigma < \infty$, then a modified random variable

$$Z_n = \frac{\mu_{\bar{x}} - \mu}{\sigma/\sqrt{n}} = \frac{x_1 + x_2 + x_3 + \cdots + x_n - n\mu}{\sqrt{n}\sigma} \tag{8.165}$$

follows the Gauss-Laplace distribution under the limiting condition $n \to \infty$. Random variable Z_n is known as standard normal variable having expected value, $E(Z_n) = 0$, and variance, $\mathrm{Var}(Z_n) = 1$. The probability density function of the population in terms of random variable x is expressed as

$$f\left(x|\mu, \sigma^2\right) = \frac{1}{\sqrt{2\pi}\sigma} e^{-\frac{(x-\mu)^2}{2\sigma^2}} \tag{8.166}$$

μ being the mean and σ the standard deviation. Gauss-Laplace distribution is also known as the normal distribution. CLT is applicable for both continuous and discrete random variables. A sampling of statistics from a population with a distribution other than normal is also normally distributed when the sampling size is large. Six tokens are labeled with an index of 1–6 and placed in a box. The probability distribution of getting any value between index 1 and 6 by randomly picking one token from the box is uniform, having a probability of 0.1667. If a token from the box is randomly drawn twice with replacement, its average frequency takes a triangular form. When the token is randomly drawn from the box 31 times with replacement, then the frequency of their average gets a Gauss-Laplace distribution shape. Sampling of large-size samples from exponential and other distribution populations also results in the normal distribution. The sampling takes a universal form, even though the original population distribution is not normal. When the original population is normally distributed, its sampling mean is also normally distributed. Often, sampling from a skewed population follows normal distribution after taking the logarithm. Such distribution is known as log-normal distribution. Despite unlimited sampling, the sampling mean of a few populations never follows the Gauss-Laplace distribution.

The central limit theorem is a fundamental discovery in probability and statistics. Initially introduced by Laplace, the Gauss-Laplace distribution was first applied to error measurement. However, Belgian polymath Lambert Adolphe Jacques Quetelet (1796–1874), a mathematician, astronomer, statistician, and sociologist, expanded its use. Quetelet (1817) analyzed chest measurements of Scottish soldiers, and when he plotted the frequency distribution of the measurements, the resulting bar plot resembled de Moivre's discrete distribution. Quetelet realized that large sets of identical measurements tend to follow the Gauss-Laplace distribution. Known as the father of quantitative social science, Quetelet applied this distribution to human physiological and social traits in 1835, suggesting that deviations from the mean were errors of nature (Quetelet, 1849). His work illuminated the nature of Gaussian distribution and encouraged its adoption in various scientific and social science fields. Victorian-era English statistician Sir Francis Galton (1822–1911) applied the bell curve to physical traits such as height, weight, strength, exam scores, and even sweet peas in 1863. He initiated standardizing these traits for better alignment with the bell curve. At the same time, American philosopher Charles Sanders Peirce (1839–1914) and German statistician Wilhelm Lexis (1837–1914) also utilized the Gauss-Laplace distribution to analyze social and economic data. Following them, figures like American astronomer Benjamin Apthorp Gould (1824–1896), Italian economist Luigi Bodio (1840–1920), and Luigi Perozzo (1856–1916) further extended the use of the Gauss-Laplace distribution to a range of scientific and social science disciplines.

At this point, one might naturally wonder how and why the Gauss-Laplace distribution came to be known as the normal distribution. Gauss originally used the term "normal" to describe the nature of the equation, referring to its orthogonal (perpendicular) properties rather than its common usage. However, by the nineteenth century, the word "normal" had taken on a broader meaning, as Charles Peirce and other natural philosophers used it to describe something commonly found across various fields. English mathematician and biostatistician Karl Pearson (1857–1936), a pioneer of mathematical statistics, helped popularize the term "normal distribution" in 1893. Pearson intended to give equal recognition to both Laplace and Gauss for their roles in developing the bell curve, previously known solely as the Gaussian distribution. However, Pearson overlooked the contributions of de Moivre, who had first discovered the bell curve. By the time Pearson recognized this oversight, the term "normal distribution" had gained widespread usage, making it unlikely to be renamed to honor all three figures. Pearson later clarified that calling the Gauss-Laplace distribution "normal" did not imply that other distributions are "abnormal."

The Gaussian distribution described in Eq. (8.166) demonstrates how data tends to cluster around a central value, with the majority of points near the mean and fewer values symmetrically decreasing as one moves farther away. This distribution plays a key role in numerous fields, including natural and social sciences, where many phenomena, such as human heights, test scores, and measurement errors, follow this pattern. Its significance is highlighted by the central limit theorem, which asserts that the sum of many independent random variables will approximate a normal distribution, regardless of their initial distributions. The bell-shaped probability density function (PDF) continues to intrigue engineers and scientists alike. As French physicist Gabriel Jonas Lippmann (1845–1921) once remarked, "Everyone believes in the normal approximation, the experimenters because they think it is a mathematical theorem, the mathematicians because they think it is an experimental fact."

Let us try to understand the mystery behind the interesting shape of the Gaussian curve. Two random variables, u and v, are said to be statistically independent when the probabilities of their simultaneous occurrence are equal to the product of their individual probabilities of occurrence:

$$P\left(u \leq a \,\&\, v \leq b\right) = P\left(u \leq a\right) P\left(v \leq b\right) \tag{8.167}$$

u and v have individual PDF as $f_1(u)$ and $f_2(v)$, respectively. The combined probability of occurrence of $a_1 \leq u \leq b_1$ and $a_2 \leq v \leq b_2$ is expressed as

$$\begin{aligned} P\left(a_1 \leq u \leq b_1 \,\&\, a_2 \leq v \leq b_2\right) &= \left(\int_{a_1}^{b_1} f_1(u)\, du\right)\left(\int_{a_2}^{b_2} f_2(v)\, dv\right) \\ &= \int_{a_2}^{b_2}\int_{a_1}^{b_1} f_1(v)\, f_2(u)\, du dv \end{aligned} \tag{8.168}$$

The probability of $u + v \leq c$ is expressed as

$$P\left(u + v \leq c\right) = \iint_{u+v\leq c} f_1(u)\, f_2(v)\, du dv \tag{8.169}$$

The integral in Eq. (8.169) is computed taking the joint probability distribution $f_{uv}(u, v) = f_1(v)\, f_2(u)$ in $u - v$ plane where $u + v \leq c$. The original integral $u - v$ plane is shown in Fig. 8.26, where the computation of double integral is difficult. Substituting the variables,

$$\begin{aligned} u &= x \\ v &= y - x \\ u + v &= y \end{aligned} \tag{8.170}$$

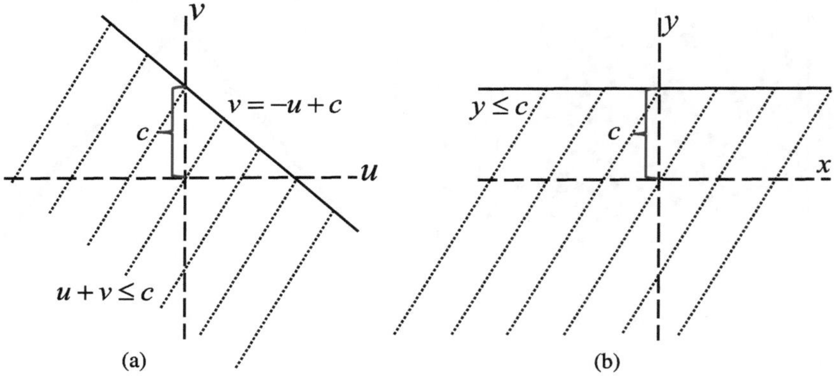

Fig. 8.26 Transformation of plane for integration. (**a**) $u - v$ plane. (**b**) $x - y$ plane

Equation (8.169) takes the following form:

$$\iint_{u+v\leq c} f_1(u)\, f_2(v)\, du dv = \int_{-\infty}^{c} \int_{-\infty}^{\infty} f_1(x)\, f_2(y-x)\, dx dy$$
$$= \int_{-\infty}^{c} \left(\int_{-\infty}^{\infty} f_1(x)\, f_2(y-x)\, dx \right) dy \tag{8.171}$$

The bracketed term on the right side is a function of y and the convolution of $f_1 * f_2$:

$$\iint_{u+v\leq c} f_1(u)\, f_2(v)\, du dv = \int_{-\infty}^{c} (f_1 * f_2)(y)\, dy \tag{8.172}$$

The joint probability density function $f_{uv}(u, v)$ equals to the convolution of their individual probability density function $(f_1 * f_2)$. This is true for any number of random variable $u_1, u_2, u_3, \cdot$ having PDF $f_1, f_2, f_3, \cdot$. Their joint probability becomes

$$f_{u_1 u_2 u_3 \ldots u_m} = (f_1 * f_2 * f_3 \ldots * f_m)(x) \tag{8.173}$$

If a random variable x is measured m times by m different operators, then their joint probability distribution becomes

$$(f_1 * f_2 * f_3 \ldots * f_m)(x) = f^{*m}(x) = \{F(\omega)\}^m \tag{8.174}$$

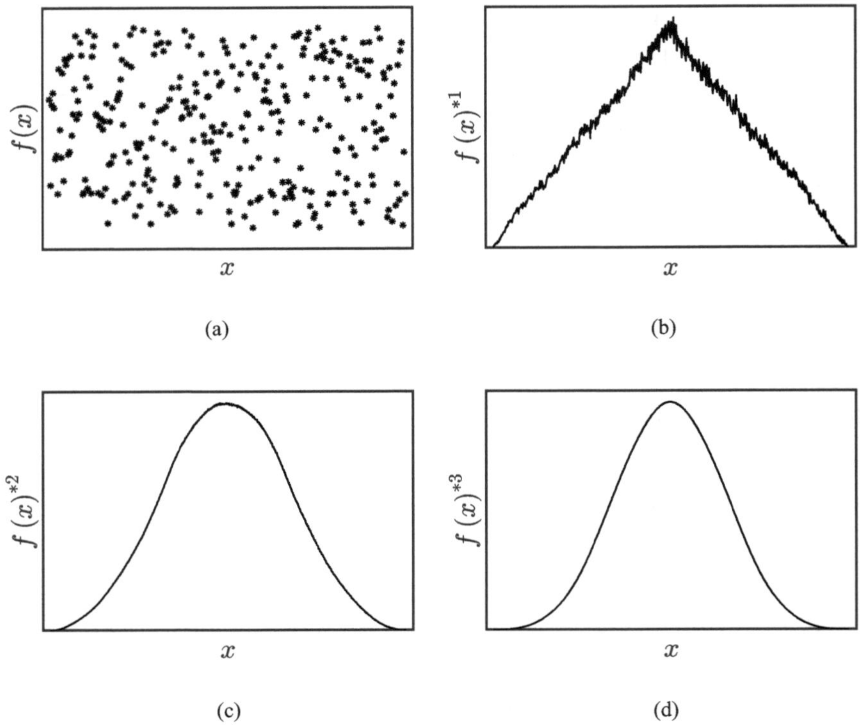

Fig. 8.27 Multiplication of signals in the frequency domain. (**a**) Random signal from measurement. (**b**) First convolution of random signal with itself. (**c**) Second self-convolution. (**d**) Third self-convolution

The convolution of a signal in the time domain corresponds to the multiplication of that signal in the frequency domain. When a random variable x is measured independently twice, and their results are convolved in the physical domain, this action scales the frequency of occurrence in the first trial by the frequency of occurrence in the second trial within the spectral domain. If m independent measurements are taken in the physical or time domain, the likelihood of the random variable x occurring in the spectral space is scaled to the power of m. Figure 8.27a shows a single random signal measurement, while Fig. 8.27d depicts the result of convolving that measurement three times. Convolution serves as a smoothing operator, combining the individual frequencies from different trials, effectively smoothing out the variability. Consequently, in experiments with very large sample sizes ($m \to \infty$), such as molecular energy distributions in kinetic gas theory or turbulence in fluid dynamics, the result tends to follow a bell curve probability density function (PDF), thanks to the infinite smoothing capacity of convolution. Convolution of the Π function with itself ($\Pi * \Pi$) yields the Λ function, whose Fourier transform produces the square of the $sinc$ function, while the Fourier transform of the Π function alone results in just the *sinc* function. This frequency multiplication smooths out the discontinuities in the Π function. Repeated convolution of Λ progressively leads to a perfectly smooth Gaussian distribution. The bell curve's smoothness is no longer an assumption but rather the outcome of rigorous mathematical reasoning.

8.8 Cross-correlation

Before diving into the mathematical definition, let us explore the concept of correlation more persuasively. Picture the enigmatic smile of Mona Lisa in Leonardo Da Vinci's iconic painting, as shown in Fig. 8.28a, and the brilliant smile of the genius Leonhard Euler in Fig. 8.28b. Now, imagine a fragment of a smiling lip in Fig. 8.28c, and the task is to determine whether this smile belongs to Mona Lisa or Euler. Correlation steps in like a delicate dance, sliding the test function (the smiling lips) across each image. As it moves, it measures the connection between the smile and each face. When the correlation reaches a perfect match, when the lips fit seamlessly, the mystery is unraveled, revealing whether the smile belongs to Euler or Mona Lisa (Fig. 8.29).

Fig. 8.28 Graphical representation of cross-correlation. (**a**) Mona Lisa. (**b**) Euler. (**c**) Smiling lips

(a) (b)

Fig. 8.29 Cross-correlation for template matching. (**a**) Detected template. (**b**) Cross-correlation result

At its core, correlation, often called the sliding dot product or inner product, is a powerful method for revealing the underlying similarities between two signals, aligning them in perfect synchrony. In the previous section, some mathematical operations on the signal through cross-correlation are performed; here, detailed insight is provided. Cross-correlation, a mathematical process, measures the degree of similarity between two signals or datasets by shifting one relative to the other. This technique evaluates how closely one signal mirrors another and finds widespread application in fields such as pattern recognition, time series analysis, and signal alignment across domains like communications, image processing, and geophysics.

The roots of cross-correlation are deeply embedded in the study of correlation and covariance within statistics. Prominent contributions came from English polymath Francis Galton (1822—1911) and British statistician Karl Pearson (1857–1936), who helped develop correlation coefficients, a metric for the strength and direction of a linear relationship between variables. In the 1890s, Pearson introduced the Pearson correlation coefficient. During World War II, cross-correlation was found to be practical in radar technology for detecting and locating objects by analyzing reflected signals. With the rise of digital computing, the efficient calculation of cross-correlation for large datasets became possible. Autocorrelation, a specific type of cross-correlation where a signal is compared with itself, became essential for identifying patterns and periodicities within a single time series. In contrast, cross-correlation extended this approach to multiple time series, allowing for the exploration of relationships between distinct signals.

In image processing, cross-correlation aids in pattern matching, feature detection, and image alignment. In neuroscience, it helps researchers explore the relationships between neural signals. In finance, cross-correlation is used to study temporal relationships between financial instruments. This technique is especially useful for locating a shorter, known feature within a longer signal. It finds applications in various fields, from pattern recognition and single particle analysis to electron tomography, cryptanalysis, and neurophysiology. Cross-correlation shares similarities with the convolution of two functions. In autocorrelation, where a signal is cross-correlated with itself, a peak will always appear at a zero lag, corresponding to the signal's energy. Python code for pattern detection from an image is provided below.

```
# Image template matching
import cv2
import numpy as np
import matplotlib.pyplot as plt

# Load the target image and the template image (both in
                                                grayscale)
image = cv2.imread('mona-lisa.jpg', cv2.IMREAD_GRAYSCALE)
template = cv2.imread('lips.jpg', cv2.IMREAD_GRAYSCALE)

# Get the width and height of the template
h, w = template.shape

# Perform template matching using cross-correlation
result = cv2.matchTemplate(image, template, cv2.
                                                TM_CCOEFF_NORMED)

# Find the location with the highest correlation value
min_val, max_val, min_loc, max_loc = cv2.minMaxLoc(result)

# Draw a rectangle around the matched region
top_left = max_loc
bottom_right = (top_left[0] + w, top_left[1] + h)
cv2.rectangle(image, top_left, bottom_right, 255, 2)

# Display the result
plt.figure(figsize=(5,5))
plt.subplot(1, 1, 1)
plt.title('Detected Template')
plt.imshow(image, cmap='gray')
plt.xticks(fontsize=0)
plt.yticks(fontsize=0)
plt.show()
plt.figure(figsize=(5,5))
plt.subplot(1, 1, 1)
plt.title('Cross-Correlation Result')
plt.imshow(result, cmap='gray')
plt.xticks(fontsize=0)
plt.yticks(fontsize=0)
plt.show()
```

8.8.1 Mathematical Properties of Cross-correlation

Mathematically the cross-correlation between two functions $f(x)$ and $g(x)$ is defined as

$$R_{f_x g_x}(x) = (f \star g)(x) = \int_{-\infty}^{\infty} f^*(\xi) g(\xi + x) d\xi \tag{8.175}$$

$f^*(x)$ is the complex conjugate of $f(x)$ and x is called the displacement of lag. Highly correlated $f(x)$ and $g(x)$ have a maximum cross-correlation at a particular x. A feature in $f(x)$ at ξ also occurs later in $g(x)$ at $\xi + x$. Hence, $g(x)$ could be described as a lag of $f(x)$ by x. In the convolution operation $f * g(x)$, one signal is rotated and displaced by an amount x and then slides over the other signal. But in the case of cross-correlation of two functions $f(x)$ and $g(x)$, the complex conjugate of the function $f^*(x)$ is multiplied with the test function $g(x)$ displaced by an amount $-x$ and then slides capture the inner product. The function $R_{f_x g_x}(x) = (f \star g)(x)$ is known as the cross-correlation function. It is important to note that this function is the result of an integral transform like the convolution integral. Cross-correlation operation is associative and distributive with respect to addition; however, unlike convolution, it is not commutative in general, i.e., $f(x) \star g(x) \neq g(x) \star f(x)$.

$$(f \star g)(x) = \int_{-\infty}^{\infty} f^*(\xi) g(\xi + x) d\xi = \int_{-\infty}^{\infty} f^*(\xi - x) g(\xi) d\xi \tag{8.176}$$

For a complex signal,

$$R_{f_x g_x}(x) = R^*_{g_x f_x}(-x) \tag{8.177}$$

$R^*_{g_x f_x}(-x)$ is the complex conjugate of the cross-correlation between $g(x)$ and $f(x)$. For real-valued signals $f(x)$ and $g(x)$, the cross-correlation is symmetric with respect to the order of the signal:

$$R_{f_x g_x}(x) = R_{g_x f_x}(-x) \tag{8.178}$$

Cross-correlation is a linear operation. For two signals $f(x)$ and $g(x)$, and constants a and b, the cross-correlation satisfies the property

$$R_{af_x + bh_x g_x}(x) = aR_{f_x g_x}(x) + bR_{h_x g_x}(x) \tag{8.179}$$

This means the cross-correlation of a linear combination of signals is the linear combination of their individual cross-correlations.

Taking the Fourier transform of Eq. (8.175),

$$\mathscr{F}\{(f \star g)(x)\} = \frac{1}{2\pi} \int_{-\infty}^{\infty} \left(\int_{-\infty}^{\infty} f^*(\xi) g(\xi + x) d\xi \right) e^{-i\omega x} \, dx \tag{8.180}$$

Introducing $u = \xi + x$ to simplify the inner integral,

$$\begin{aligned}
\mathscr{F}\{(f \star g)(x)\} &= \frac{1}{2\pi}\int_{-\infty}^{\infty}\left(\int_{-\infty}^{\infty} f^*(\xi)g(u)du\right)e^{-i\omega x}\,dx \\
&= \frac{1}{2\pi}\int_{-\infty}^{\infty} f^*(\xi)\left(\int_{-\infty}^{\infty} g(u)e^{-i\omega(u-\xi)}du\right)dx \\
&= \frac{1}{2\pi}\int_{-\infty}^{\infty} f^*(\xi)\left(\int_{-\infty}^{\infty} g(u)e^{-i\omega u}du\right)e^{-i\omega(-\xi)}dx \\
&= \frac{1}{2\pi}\int_{-\infty}^{\infty} f^*(\xi)e^{-i\omega(-\xi)} - d\xi\left(\int_{-\infty}^{\infty} g(u)e^{-i\omega u}du\right) \\
&= 2\pi F^*(\omega)G(\omega)
\end{aligned} \tag{8.181}$$

The Fourier transform of the cross-correlation between two complex signals $f(x)$ and $g(x)$ in the time domain is equal to the product of the complex conjugate of the Fourier transform of $f(x)$ and the Fourier transform of $g(x)$.

Cross-correlation is closely related to convolution. The cross-correlation of two signals $f(x)$ and $g(x)$ is equivalent to the convolution of one signal with the time-reversed version of the other:

$$(f \star g)(x) = (f^*(-y) * g(y))(x) \tag{8.182}$$

Thus, cross-correlation can be computed using the convolution operation by flipping one of the signals.

The cross-correlation in discrete form is expressed as

$$R_{f_x g_x}[k] = \sum_{n=0}^{N-1} f^*[n]g[n+k] \tag{8.183}$$

In a special case, when a function $f(x)$ is cross-correlated with itself, it is known as the autocorrelation:

$$R_{f_x f_x}(x) = f(x) \star f(x) = \int_{-\infty}^{\infty} f^*(\xi)f(\xi + x)d\xi \tag{8.184}$$

Taking the Fourier transform of both sides,

$$\mathscr{F}\{R_{f_x f_x}(x)\} = \mathscr{F}\{f(x) \star f(x)\} = F^*(\omega)F(\omega) = \|F(\omega)\|^2 \tag{8.185}$$

The autocorrelation gives the total energy content at a position x:

$$\mathscr{F}\{R_{f_x f_x}(x)\} = \mathscr{F}\{f(x) \star f(x)\} = F^*(\omega)F(\omega) = \|F(\omega)\|^2 \tag{8.186}$$

The cross-correlation function can be normalized to ensure that the output is between -1 and 1, which provides a clearer measure of similarity:

$$R_{f_x f_x}[k] = \frac{\sum_{n=0}^{N-1} f[n]g[n+k]}{\sqrt{\sum_{n=0}^{N-1} |x[n]|^2 \sum_{n=0}^{N-1} |y[n]|^2}} \tag{8.187}$$

This is called the normalized cross-correlation and is often used in applications like pattern matching, where comparing signals with different magnitudes is desirable.

Figure 8.30a and b show the normalized cross-correlations and autocorrelation between two random signals. Figure 8.30c shows the cross-correlations between deterministic sin and cos signal. Figure 8.30d shows autocorrelations between two sin signals. Python code for normalized correlation for the one-dimensional signal is given below.

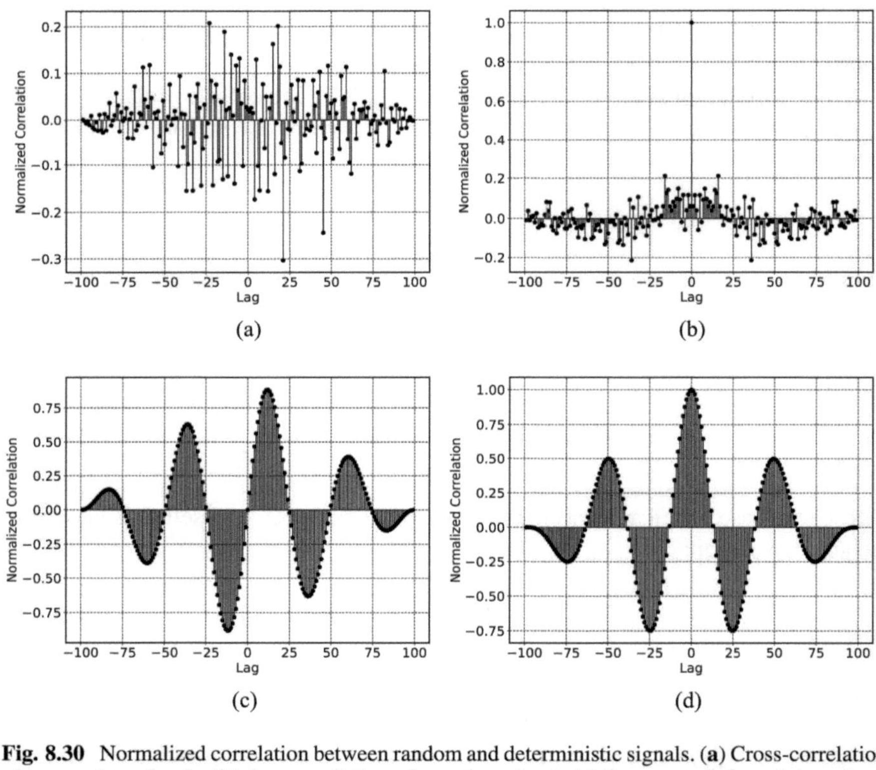

Fig. 8.30 Normalized correlation between random and deterministic signals. (**a**) Cross-correlation of two random signals. (**b**) Autocorrelation of a random signal. (**c**) Cross-correlation of sin and cos wave. (**d**) Autocorrelation of a sin wave

```
# Correlations between two one dimensional signals
import numpy as np
import matplotlib.pyplot as plt

# Step 1: Generate two random signals
signal_length = 100
randsignal1 = np.random.randn(signal_length)
randsignal2 = np.random.randn(signal_length)

# Step 2: Generate two deterministic signals
x = np.linspace(0, 4*np.pi, 100)
detsignal1 = np.sin(x)
detsignal2 = np.cos(x)

signal1 = randsignal1
signal2 = randsignal2
# Step 3: Compute cross-correlation between the two signals
correlation = np.correlate(signal1 - np.mean(signal1), signal2
                               - np.mean(signal2), mode='full'
                               )
# Step 3: Normalize the cross-correlation
# Normalize by the standard deviation and length of the signals
norm_correlation = correlation / (np.std(signal1) * np.std(
                               signal2) * signal_length)

# Step 4: Create a lag array
lags = np.arange(-signal_length + 1, signal_length)

# Plot cross-correlation between signal1 and signal2
fig = plt.figure()
fig.set_figheight(2.5)
fig.set_figwidth(4)
ax = fig.add_subplot(1, 1, 1)
ax.stem(lags, norm_correlation, 'k', markerfmt='ko', basefmt =
                               'k')
plt.xlabel('Lag', fontsize=20)
plt.ylabel('Normalized Correlation', fontsize=20)
plt.xticks(fontsize=20)
plt.yticks(fontsize=20)
plt.grid(True, which='both', color = 'black', linestyle = '--',
                                linewidth = 1.0)
ticklen = np.pi/2
ax = fig.gca()
for axis in ['top','bottom','left','right']:
    ax.spines[axis].set_linewidth(2.0)
ratio = 0.75
x_left, x_right = ax.get_xlim()
y_low, y_high = ax.get_ylim()
ax.set_aspect(abs((x_right-x_left)/(y_low-y_high))*ratio)
plt.show()
plt.close()
```

8.9 Closure

This chapter examines the fundamental concepts of convolution and their relationship with probability, establishing a foundation for understanding signal processing and the statistical analysis of random signals. The chapter began with a detailed exploration of convolution as a mathematical operation, where two functions are combined to produce a third function that reveals how one is modified by the other. This has been shown to be particularly significant in signal processing, where convolution describes the response of systems to external inputs. Various properties of convolution, such as commutativity, associativity, and distributivity, have been outlined to highlight how it simplifies signal analysis within linear systems. Attention was then given to circular convolution, a concept that emerges in the context of periodic signals and digital systems. Unlike linear convolution, circular convolution assumes the periodicity of signals, which makes it particularly useful in digital signal processing, where the signals are finite in length. The applications of convolution were also discussed, showing its importance in fields like image processing and system filtering, where it is utilized for tasks such as smoothing and edge detection.

A transition was made into probability theory, with a historical context provided to show how it evolved from studies on games of chance to a formal mathematical framework, largely attributed to figures like Blaise Pascal and Pierre de Fermat. Basic probability theory was then presented, with key concepts such as random variables, probability distributions, expectations, variance, and the law of large numbers being explained. These ideas were shown to be fundamental when analyzing random signals, where probabilistic methods are necessary for describing uncertainty and variability in real-world processes.

It was then demonstrated how convolution is used on random signals to determine the probability distribution of the sum of two independent random variables. By convolving their probability distributions, the distribution of the combined outcomes can be derived. This application of convolution has been recognized as a powerful tool for modeling cumulative effects in probabilistic systems, such as signal interference or queuing processes in communication systems. The normal distribution was also addressed, with emphasis placed on its central role in probability theory and statistics. Convolution derives the normal distribution from independent, identically distributed random variables through the central limit theorem, showcasing its importance in statistical theory.

Cross-correlation was explored as well, with its function of measuring the similarity between two signals as a function of time lag being highlighted. This technique has been emphasised for its utility in areas such as signal alignment, pattern recognition, and time-delay estimation in communication systems. The properties of cross-correlation, including its linearity and shift-invariance, were covered, along with normalisation techniques that allow cross-correlation to be compared across signals of different scales. Finally, applications of cross-correlation were considered, illustrating its wide use in fields such as signal processing (for detecting periodic patterns) and statistics (for analysing time-lagged relationships between variables). Cross-correlation has been widely applied to identify dependencies and relationships over time in time-series data in areas such as economics, finance, and biology.

Overall, the chapter has demonstrated the strong connections between convolution, probability, and correlation, showing how these mathematical tools are essential for analyzing, interpreting, and predicting behaviors in complex systems. Whether applied in engineering, data science, or physical sciences, these concepts form a critical basis for studying real-world signals, systems, and statistical relationships, and they serve as key tools for further research and application.

References

D'Alembert, J. L. R. (1756). *Recherches sur différens points importans du système du monde* (1st ed.). Chez David l'aîné.

David, F. N. (1962). *Games, gods and gambling* (1st ed.). Hafner.

Gauss, C. F. (1809). *Theoria motus corporum celestium* (Vol. 277). Perthes et Besser.

Goldberger, A., Amaral, L., Glass, L., Hausdorff, J., Ivanov, P., Mark, R., Mietus, J., Moody, G., Peng, C.-K., & Stanley, H. (2000). PhysioBank, PhysioToolkit, and PhysioNet: Components of a new research resource for complex physiologic signals. *Circulation, 101*, e2015–e220.

Hald, A. (2003). *A history of probability and statistics and their applications before 1750* (1st ed.). Wiley.

Lacroix, S., & Courcier, L. v. (1819). *Traité du calcul différentiel et du calcul intégral: S.F. Lacroix Tome troisieme contenant un traité des différences et des séries* (1st ed.). mme ve Courcier, imprimeur-libraire pour les sciences, rue du Jardinet-Saint-André-des-Arcs.

Lagrange, J. L. (1770). *Reflections on the algebraic resolution of éequations*. Prussian Academy.

Laplace, P. S. (1820). *Théorie analytique des probabilités*. Courcier.

Laplace, P.-S. (1829). *Essai philosophique sur les probabilités*. H. Remy.

Legendre, A. M. (1806). *Nouvelles méthodes pour la détermination des orbites des comètes: Sur laméthode des moindres quarrés*. Courier.

Quetelet, A. (1817). Statement of the sizes of men in different counties of scotland, taken from the local militia. *Edinburgh Medical and Surgical Journal, 13*, 260–264.

Quetelet, A. (1846). *Lettres à SAR le duc régnant de Saxe-Coburg et Gotha: sur la théorie des probabilités, appliquée aux sciences morales et politiques*. M. Hayez.

Quetelet, A. (1849). *Letters addressed to HRH the Grand Duke of Saxe Coburg and Gotha: On the theory of probabilities, as applied to the moral and political sciences*. C. & E. Layton.

Simpson, T. (1757a). A letter to the right honourable george macclesfield, president of the royal society, on the advantage of taking the mean, of a number of observations, in practical astronomy. *Philosophical Transactions, 49*, 82–93.

Simpson, T. (1757b). *Miscellaneous Tracts on Some Curious and Very Interesting Subjects in Mechanics, Physical-astonomy, and Speculative Mathematics: Wherein, the Precession of the Equinox, the Nutation of the Earth's Axis And, the Motion of the Moon in Her Orbit, are Determined*. J. Nourse over-against Katherine-Street in the Strand.

Stigler, S. M. (1986). *The history of statistics: The measurement of uncertainty before 1900* (1st ed.). The Belknap Press of Harvard University.

Tabak, J. (2004). *Probability and statistics: The science of uncertainty* (1st ed.). Facts on File.

Teets, D., & Whitehead, K. (1999). The discovery of ceres: How gauss became famous. *Mathematics Magazine, 72*(2), 83–93.

Chapter 9
Linear Systems

Nature is an infinite sphere of which the center is everywhere and the circumference nowhere.

— Blaise Pascal

Abstract This chapter thoroughly examines linear systems, with an emphasis on linear time-invariant (LTI) systems, essential in signal processing and systems analysis. It begins by introducing system fundamentals, including linearity, translation (shifting), and the effects of cascading systems. The concept of impulse response is then presented as a key method to predict system output based on given inputs. LTI systems are explored in detail, highlighting their importance due to their analytical simplicity and reliability. The chapter further discusses eigenfunctions and their role in defining system responses, along with the Fourier transform, which is used to analyze and design LTI systems. Essential system characteristics like causality and stability are examined, providing insights into system behavior over time. Finally, matched filters are introduced as a method to improve signal detection by aligning filters with specific signal characteristics.

Keywords Linear systems · Cascading · Impulse response · Linear time-invariant systems · LTI · Eigenfunctions · Causality · System stability · Poles · Zeros · Matched filters

9.1 Description of a System

In Chap. 2, a description of a system is already provided. A system can be considered a virtual machine that performs some mathematical task of uniquely mapping the cause $f(x)$ and the effect $g(x)$:

$$g(x) = L\{f(x)\} \tag{9.1}$$

or without the variable as

S. K. Jena, *Fourier, Laplace, and the Tangled Love Affair with Transforms*,
https://doi.org/10.1007/978-3-031-80165-5_9

$$g = L\{f\} \tag{9.2}$$

The system is completely characterized terminally if the nature of the dependence of the output on the input is known. For example, $g(x)$ could be the solution of an ordinary differential equation with forcing term $f(x)$, as is the case in lumped parameter systems; it could be the solution of a partial differential equation, as in transmission lines, heat transfer, or radiation; it could also be experimentally established if one has no information about the interior of the system. This chapter is concerned with the terminal properties of the system, i.e., the relationship between a cause $f(x)$ and the effect $g(x)$, not with the interior of the system. Signals are the functions of time or a spatial variable. The system operates on the signal in some way to produce another signal. This is the continuous case, where input and outputs are functions of a continuous variable. The discrete case will often arise for us by sampling continuous functions. Defining the domain L for a system is crucial, like some system that handles the works with the energy or power signal. The system is specified by a set of assumptions (linearity, time invariance, causality) that correspond to a "real, linear, passive, nondegenerate system with constant coefficients and zero initial conditions."

9.2 Linear System

A system is linear if

$$L\{a_1 f(x)\} = a_1 L\{f(x)\} \tag{9.3}$$

L is homogeneous and

$$L\{a_1 f_1(x) + a_2 f_2(x)\} = a_1 L\{f_1(x)\} + a_2 L\{f_2(x)\} \tag{9.4}$$

and additive for the signals $f_1(x)$, $f_1(x)$ and arbitrary scalars a_1 and a_2. A simple example of a linear system is the differentiation operation $L\{f(x)\} = \dfrac{df(x)}{dx}$ and the delay line $L\{f(x)\} = f(x - x_0)$. If a system can add and scale inputs and outputs but violates one or both Eqs. (9.3)–(9.3) is referred to as nonlinear. A linear system is often said to satisfy the principle of superposition, i.e., adding the inputs results in adding the outputs and scaling the input scales the output by the same amount. These properties can be directly extended to finite sums:

$$L\left\{\sum_{k=1}^{N} a_k f_k(x)\right\} = \sum_{k=1}^{N} a_k L\{f_k(x)\} \tag{9.5}$$

with proper assumptions of continuity of L and convergence of the sums to infinite sums. That is,

$$L\left\{\sum_{k=1}^{\infty} a_k f_k(x)\right\} = \sum_{k=1}^{\infty} a_k L\{f_k(x)\} \tag{9.6}$$

a is directly proportional to b, the most basic linear system example. From high school days, I usually think of the direct proportion in terms of a constant of proportionality, i.e., voltage is proportional to the current, $V = RI$, and the acceleration is proportional to the force, $a = (1/m)F$. However, the proportionality constant can be a function x, and the key property of linearity is still present:

$$L\{f(x)\} = h(x)f(x) \tag{9.7}$$

Then also L is linear because

$$\begin{aligned} L\{a_1 f_1(x) + a_2 f_2(x)\} &= h(x)\,(a_1 f_1(x) + a_2 f_2(x)) \\ &= a_1 h(x) f_1(x) + a_2 h(x) f_2(x) \\ &= a_1 L\{f_1(x)\} + a_2 L\{f_2(x)\} \end{aligned} \tag{9.8}$$

Suppose a system consists of a switch. When the switch is closed, the signal goes through unchanged, and when the switch is open, the signal does not go through at all (so by convention, what comes out the other end is the zero signal). Suppose that the switch is closed for $-\frac{1}{2} \le x \le \frac{1}{2}$. Is this a linear system? Sure; it is described precisely by multiplication with Π:

$$L\{f(x)\} = \Pi(x)f(x) \tag{9.9}$$

Sampling is a linear system. To sample a signal $f(x)$ with sample points spaced p apart is to form

$$L\{f(x)\} = \text{Ш}_p f(x) = \sum_{k=-\infty}^{\infty} f(kp)\delta(x - kp) \tag{9.10}$$

Sampling the sum of two signals is the sum of the sampled signals.

Analogous to simple direct proportion, matrix multiplication and integration represent the linear system with one level up in sophistication. Matrix multiplication is the most basic example of a discrete linear system:

$$F = Af \tag{9.11}$$

f and F are two vectors with n and m elements, respectively. A is an $m \times n$ matrix. This operation index form is expressed as

$$F_i = \sum_{j=1}^{n} A_{ij} f_j,\ i = 1, 2, 3, \ldots, m \tag{9.12}$$

In Einstein tensor notation (named after celebrity German-Swiss-American scientist Albert Einstein [1879–1955]), often the $\sum_{j=1}^{n}$ symbol is dropped, and the summation is performed over the repeated indices for the entire range:

$$F_i = A_{ij} f_j \tag{9.13}$$

The matrix is symmetric if $A_{ij} = A_{ji}$ and Hermitian (named after French mathematician Charles Hermite [1822–1901]) if $A_{ij} = A^*_{ji}$. A^*_{ji} is the complex conjugate. If $A_{ij} A^*_{ji} = \delta_{ij}$, then A_{ij} is a unitary or identity matrix having 1 along its main diagonal and zero for the rest of the element. $\delta_{ij} = 1$ if $i = j$; else 0 is the Kronecker delta function. Kronecker delta is often used in tensor algebra and is named after German mathematician and logician Leopold Kronecker(1823–1891). Kronecker delta should not be confused with Dirac's delta δ_0, which is a distribution. An nth-order linear dynamical system is expressed as

$$L\{f_i\} = e^{A_{ij}} f_i \tag{9.14}$$

where A is an $n \times n$ matrix.

$$\dot{x}_i = A_{ij} x_j \tag{9.15}$$

$\dot{x}$ represents the derivative of x. The system is associated with an initial value

$$x_i(0) = f_i(0) \tag{9.16}$$

(0) represents the initial state of the dynamical system, and the matrix $e^{A_{ij}}$ varies in time, and the system describes how the initial value f_i evolves over time:

$$F(x) = L\{f(x)\} = \int_{-\infty}^{\infty} k(x, y) f(y) dy \tag{9.17}$$

$k(x, y)$ is called the kernel and $f(y)$ is integrated against it:

$$\begin{aligned} L\{\alpha_1 f_1(x) + \alpha_2 f_2(x)\} &= \int_{-\infty}^{\infty} k(x, y)\,(\alpha_1 f_1(y) + \alpha_2 f_2(y))\,dy \\ &= \alpha_1 \int_{-\infty}^{\infty} k(x, y) f_1(y) dy + \alpha_2 \int_{-\infty}^{\infty} k(x, y) f_2(y) dy \\ &= \alpha_1 L\{f_1(x)\} + \alpha_2 L\{f_2(x)\} \end{aligned} \tag{9.18}$$

Equation (9.18) ensures that the integral transform process in Eq. (9.17) is a linear system. The Fourier integral discussed earlier is a linear system having the kernel $k(\omega, x) = e^{-i\omega x}$. Convolution is also an linear process:

$$L\{f(x)\} = (g * f)(x) = \int_{-\infty}^{\infty} g(x-y)f(y)dy \tag{9.19}$$

where the kernel is $k(x, y) = g(x - y)$. Analogous to continuous convolution, the discrete convolution discussed in last chapter is a linear system and determined through matrix multiplication:

$$\{g\} * \{f\} = h_i = L_{i,j} f_j \tag{9.20}$$

where the discrete convolution kernel $L_{i,j}$ is expressed as

$$L_{i,j} = g_{i-j} \tag{9.21}$$

$L_{i,j}$ is constant along the diagonals and is filled out, column by column, by the shifted versions of g_i. For example, $g_i = \{1, 2, 3, 4\}$, and then the $L_{i,j}$ is

$$L_{i,j} = \begin{bmatrix} 1 & 4 & 3 & 2 \\ 2 & 1 & 4 & 3 \\ 3 & 2 & 1 & 4 \\ 4 & 3 & 2 & 1 \end{bmatrix} \tag{9.22}$$

Square matrices that maintain constant values along each diagonal, though these constants may differ from one diagonal to another, are known as Toeplitz matrices, a concept named after the German mathematician Otto Toeplitz (1881–1940). These matrices are notable for their intriguing and diverse properties. When Toeplitz matrices are used in convolution operations, an additional layer of structure emerges due to the assumed periodicity in the columns. This subset of Toeplitz matrices is called circulant matrices (Golub & Van Loan, 2013).

9.3 Translation or Shifting

Both continuous and discrete signals are delayed or advanced to produce new signals, through a new linear operation:

$$L\{f(x)\} = f(x - \tau) \tag{9.23}$$

x is the current time, and the counting started at $x = 0$. $x - \tau$ is the delay in time by an amount τ for $\tau > 0$. $\tau < 0$ represents the advance in time. The linear translation in time and space is the convolution of the signal with Dirac's delta function:

$$f(x - \tau) = (\delta_\tau * f)(x) \tag{9.24}$$

In the discrete form, it is expressed as

$$f[m-n] = (\delta_n * f)[m] \tag{9.25}$$

The periodizing of the signal is a linear system by convoluting a signal with III_p:

$$L\{f(x)\} = g(x) = \left(\text{III}_p * f\right)(x) \tag{9.26}$$

In other words, the periodization of the sum of two signals is the sum of the periodizations.

9.4 Cascading Linear Systems

If L and M are two linear systems, then ML is also a linear system:

$$\begin{aligned}(ML)\left(\alpha_1 f_1(x) + \alpha_2 f_2(x)\right) &= M\left(L\left(\alpha_1 f_1(x) + \alpha_2 f_2(x)\right)\right)\\ &= M\left(\alpha_1 L f_1(x) + \alpha_2 L f_2(x)\right)\\ &= \alpha_1 ML f_1(x) + \alpha_2 ML f_2(x)\end{aligned} \tag{9.27}$$

ML is an important operation to compose or cascade two (or more) linear systems. The cascade operation is not always commutative $ML \neq LM$. If L is the linear system given by

$$L\{f(x)\} = \int_a^b k(x,y) f(y) dy \tag{9.28}$$

and M is another linear system (not necessarily given by an integral), then the composition ML is the linear system

$$M\{L\{f(x)\}\} = \int_a^b M(k(x,y)) f(y) dy \tag{9.29}$$

This is true when $k(x, y)$ is a signal and M can operate in writing $M(k(x, y))$ on $k(x, y)$ in its x-dependence. For example, if M is also given by integration against a kernel

$$M\{f(x)\} = \int_a^b l(x,y) f(y) dy \tag{9.30}$$

then

$$
\begin{aligned}
M\{L\{f(y)\}\} &= \int_a^b l(x,y)\{L\{f(y)\}dy \\
&= \int_a^b l(x,y)\left(\int_a^b k(y,z)f(z)dz\right)dy \\
&= \int_a^b \int_a^b l(x,y)k(y,z)f(z)dzdy
\end{aligned}
\tag{9.31}
$$

If all the necessary hypotheses are satisfied, this can be further written as

$$
\begin{aligned}
M\{L\{f(y)\}\} &= \int_a^b \int_a^b l(x,y)k(y,z)f(z)dzdy \\
&= \int_a^b \left(\int_a^b l(x,y)k(y,z)dy\right) f(z)dz \\
&= \int_a^b K(x,z)f(z)dz
\end{aligned}
\tag{9.32}
$$

The cascaded system $M\{L\{f(y)\}\}$ is also expressed as the integration against a new kernel:

$$
K(x,z) = \int_a^b l(x,y)k(y,z)dy \tag{9.33}
$$

This new kernel $K(x,z)$ is analogous to a matrix product.

9.5 The Impulse Response

Integral transform is not just the way, rather the only way to define a continuous linear system, and all linear systems follow this. The value of the signal is recovered by convoluting the signal with the δ function:

$$
f(x) = (\delta * f)(x) = \int_{-\infty}^{\infty} \delta(x-y)f(y)dy \tag{9.34}
$$

Applying a liner system L to $f(x)$ gives

$$
g(x) = L\{f(x)\} = \int_{-\infty}^{\infty} L\{\delta(x-y)\}f(y)dy \tag{9.35}
$$

$\delta(x-y)$ in the integrand only depends on x, so L operates on it. $L\{\delta(x-y)\}$ describes how the system responds to the impulse $\delta(x-y)$. Let

$$L\{\delta(x-y)\} = h(x,y) \tag{9.36}$$

In practice, this represents how the system responds to a very short, very peaked signal. The limit of such responses is the impulse response. Putting this into the earlier integral in Eq. (9.35) results in the superposition theorem:

$$g(x) = L\{f(x)\} = \int_{-\infty}^{\infty} h(x,y)f(y)dy \tag{9.37}$$

The linear system L integrates the input signal against an impulse response kernel. Can this be clarified further? Specifically, what does it mean for L to act on $\delta(x-y)$? Is this action related to a function or a distribution? The answer is indeed yes; all of these concepts can be precisely defined within the framework of distributions. The superposition theorem in this context is known as the Schwartz kernel theorem, and the impulse response, which is a distribution, is referred to as the Schwartz kernel. Additionally, a unique theorem applies here. It states that for each linear system L, there exists a unique kernel $h(x, y)$ such that

$$L\{f(x)\} = \int_{-\infty}^{\infty} h(x,y)f(y)dy \tag{9.38}$$

The uniqueness of the kernel is reassured. The impulse response has been identified when a linear system is expressed as an integral with a kernel. For example, the impulse response of the Fourier transform is given by $h(s, x) = e^{-2\pi isx}$:

$$\mathscr{F}\{f(x)\} = \int_{-\infty}^{\infty} e^{-2\pi isx} f(x)dx = \int_{-\infty}^{\infty} h(s,x)f(x)dx \tag{9.39}$$

This provides what was utilized in the Fourier transform before having a deep understanding of the impulse response. The Schwartz kernel theorem is widely regarded as one of the most challenging results in the theory of distributions.

Any linear system is represented through the integration against a kernel. Now, switching to the discrete version, the matrix A_{ij} works the same way the functional kernels worked for a continuous system. An n-dimensional vector f_i acts against a $m \times n$ matrix A_{ij} to produce a transform pair n-dimensional transform pair F_j:

$$F_j = A_{ij} f_i \tag{9.40}$$

9.6 Linear Time-Invariant (LTI) Systems

A linear time-invariant (LTI) system is one where the system's behavior remains consistent over time, i.e., the system produces the same output at any instant in the future. This concept is akin to running a computer program; if the program is successfully executed today, the program will generate the same results in the future. The only difference is the time at which the results are obtained. This time invariance is a fundamental assumption in analyzing many systems, particularly in mathematics and engineering. It simplifies the study and design of systems by allowing us to predict their behavior based on past performance. While we acknowledge that components can fail over time, making true time invariance impossible to maintain indefinitely, the assumption of time invariance is a useful and accurate approximation for most practical purposes. When a system is time-invariant, it greatly simplifies the analysis and the design processes, as the system's response to inputs can be reliably predicted over time.

The time-invariance property dictates that a shift in the input signal's timing should lead to an identical shift in the output signal's timing. This concept of time invariance is described in terms of shifts or differences in time, such as saying the system behaves the same tomorrow as it does today. This implies that the focus is on whether the system's behavior changes over a given time interval or between two specific moments rather than "absolute time" itself, which is not meaningful in this context; the differences between two points in time matter. Mathematically, if $g(x) = L\{f(x)\}$ represents the system's output at the current time, then for the system to be considered time-invariant or an LTI system, a delay in the input signal by a certain amount τ should result in a delay in the output signal by the same amount, with no other alterations. This relationship can be expressed as a formula:

$$g(x-\tau) = L\{f(x-\tau)\} \tag{9.41}$$

Delaying the input signal by 24 hours results in the output being delayed by 24 hours as well, with no other changes occurring. Sometimes, LTI is interpreted as a "linear translation-invariant" system, acknowledging that the variable might not always be time, but the operation involved is always a form of translation. Similarly, the term LSI, meaning "linear shift invariant," is also used to describe this concept.

How does the impulse response behave in an LTI system? For a general linear system, the impulse response $h(x, \tau) = L\{\delta(x-\tau)\}$ depends independently on both x and τ, meaning the response can vary for impulses occurring at different times. However, this is not true for an LTI system. For an LTI system, if the impulse response at $\tau = 0$ is given as

$$h(x) = L\{\delta(x)\} \tag{9.42}$$

then by the time invariance

$$h(x-\tau) = L\{\delta(x-\tau)\} \tag{9.43}$$

The impulse response does not depend independently on x and τ but rather only on their difference, $x-\tau$. The character of the impulse response means that the superposition integral assumes the form of a convolution:

$$g(x) = L\{f(x)\} = \int_{-\infty}^{\infty} h(x-\tau)f(\tau)d\tau = (h * f)(x) \tag{9.44}$$

Conversely, let us show that a linear system given by a convolution integral is time-invariant. Suppose

$$g(x) = L\{f(x)\} = (w * f)(x) = \int_{-\infty}^{\infty} w(x-\tau)f(\tau)d\tau \tag{9.45}$$

Then

$$L\{f(x-x_0)\} = \int_{-\infty}^{\infty} w(x-\tau)f(\tau - x_0)d\tau \tag{9.46}$$

Taking $s = \tau - x_0$, the above integral becomes

$$\begin{aligned} L\{f(x-x_0)\} &= \int_{-\infty}^{\infty} w(x-x_0-s)f(s)ds \\ &= (w * f)(x-x_0) = g(x-x_0) \end{aligned} \tag{9.47}$$

Therefore, L is indeed time-invariant, as intended. Additionally, when L is defined through convolution, $w(t-\tau)$ represents the impulse response. This is because, in a time-invariant system, the impulse response is characterized by

$$L\{\delta(x)\} = (w * \delta)(x) = w(x) \tag{9.48}$$

i.e.,

$$L\{\delta(x-\tau)\} = w(x-\tau) \tag{9.49}$$

This situation is highly satisfactory. To summarize the key points, if L is a linear system, then

$$L\{f(x)\} = \int_{-\infty}^{\infty} h(x,y)f(y)dy \tag{9.50}$$

where $h(x, y)$ is the impulse response

$$h(x,y) = L\{\delta(x-y)\} \tag{9.51}$$

The system is time-invariant if, and only if, it operates through convolution. In this case, the impulse response depends on the difference $x - y$, and the convolution is performed with this impulse response:

$$L\{f(x)\} = \int_{-\infty}^{\infty} h(x-y)f(y)dy = (h * f)(x) \tag{9.52}$$

Equation (9.52) represents the fundamental nature of convolution.

A discrete system given by a convolution operation is always linear time-invariant:

$$g[n] = (h * f)[n] = \sum_{m=-\infty}^{\infty} f[m]h[n-m] \tag{9.53}$$

However, a general discrete linear system is given in terms of matrix multiplication:

$$g[n] = \sum_{m=1}^{M} L[m][n]f[m] \tag{9.54}$$

for $n = 1, 2, \ldots, N$. But when is this matrix multiplication time-invariant? If the matrix L_{ij} associated with the system $g[n] = (h * f)[n]$ is a circulant matrix, filled out column by column by the shifted versions of $h[n]$, then it represents an LTI system. This is already shown in the previous section during the description of discrete convolution. This circulant matrix is given by $g[n] = (h * f)[n]$ where $h[n]$ is the first column of L_{ij}.

9.7 Eigenfunctions

An eigenfunction is a special type of function that, when passed through a linear operator, remains essentially unchanged except for a scaling factor known as the eigenvalue. Consider a linear operator L, which could be a differential operator, an integral operator, or any other linear transformation. A function $f(x)$ is said to be an eigenfunction of the operator L if applying L to $f(x)$ results in the same function $f(x)$ scaled by a constant λ, known as the eigenvalue. Mathematically, this relationship is expressed as

$$L\{f(x)\} = \lambda f(x) \tag{9.55}$$

Here, λ is the eigenvalue associated with the eigenfunction $f(x)$. A simple and classic example involves the derivative operator $\frac{d}{dx}$. The exponential function $f(x) = e^{\alpha x}$ is an eigenfunction of this operator. Applying the derivative operator to $f(x)$ gives

$$\frac{d}{dx}e^{\alpha x} = \alpha e^{\alpha x} \tag{9.56}$$

In this case, $L\{f(x)\} = \dfrac{d}{dx}e^{\alpha x}$ returns the original function $e^{\alpha x}$ multiplied by the constant α. Therefore, $e^{\alpha x}$ is an eigenfunction of the derivative operator, and the corresponding eigenvalue is α. The set of all possible eigenvalues of L is called its spectrum, which may be discrete, continuous, or a combination of both. Eigenfunctions are crucial because they simplify the analysis of linear systems. When a system can be described in terms of its eigenfunctions, its behavior becomes much easier to predict and understand.

Various examples of linear systems are covered, beginning with the simplest case of direct proportion. How do these examples hold up concerning time invariance? Does "direct proportion" meet the criteria? Unfortunately, it does not, except in the most straightforward scenario. Suppose that

$$L\{f(x)\} = h(x)f(x) \tag{9.57}$$

is time-invariant. Then for any τ, this invariant system becomes

$$L\{f(x-\tau)\} = h(x)f(x-\tau) \tag{9.58}$$

On the other hand,

$$L\{f(x-\tau)\} = (L\{f\})(x-\tau) = h(x-\tau)f(x-\tau) \tag{9.59}$$

From Eqs. (9.58) and (9.59)

$$h(x-\tau)f(x-\tau) = h(x)f(x-\tau) \tag{9.60}$$

For every input $f(x)$ and every τ, Eq. (9.59) holds good only if $h(x)$ is a constant. Hence, the relationship of direct proportion only defines a time-invariant linear system when the proportionality factor is constant. For example, consider a linear system for switching networks, where on/off a signal is performed by multiplying the signal with the $\Pi(x)$ function:

$$L\{f(x)\} = \Pi(x)f(x) \tag{9.61}$$

The impulse response for this liner operation becomes

$$h(x,\tau) = L\{\delta(x-\tau)\} = \Pi(x)\delta(x-\tau) = \Pi(\tau)\delta(x-\tau) \tag{9.62}$$

This shows that the impulse response is not only the function of $(x-\tau)$. The superposition integral against this kernel is

$$
\begin{aligned}
L\{f(x)\} &= \int_{-\infty}^{\infty} h(x,\tau) f(\tau) d\tau \\
&= \int_{-\infty}^{\infty} \Pi(\tau)\delta(x-\tau) f(\tau) d\tau \\
&= \Pi(x) f(x)
\end{aligned}
\tag{9.63}
$$

A linear operator L acts on any nonzero function $f(x)$ in the functional space and represents an LTI system if and only if $f(x)$ is the eigenfunction with eigenvalue λ:

$$
L\{f(x)\} = \lambda f(x) \tag{9.64}
$$

9.8 Translating in Time and Plugging into L

It has been observed that knowing when and how to incorporate time shifts into a formula is only sometimes straightforward. Consequently, a brief return to the definition of time invariance is recommended, along with a streamlined, mathematical representation and a perspective on how to approach it in terms of cascading systems. For those accustomed to this way of thinking, this method clarifies what should be "plugged in" to L. The first approach involves expressing the act of shifting by an amount b as an operation on a signal, bringing back the "translate by b" operator to define

$$
\tau_b\{f(x)\} = f(x-b) \tag{9.65}
$$

If $L\{f(x)\} = g(x)$, for the b shift

$$
L\{f(x-b)\} = g(x-b) = \tau_b\{g(x)\} \tag{9.66}
$$

The time-invariance property then says that

$$
L\{\tau_b f(x)\} = \tau_b L\{f(x)\} = L\{f(x-b)\} \tag{9.67}
$$

The placement of parentheses here is crucial, as it indicates that translating by b and then applying L (as on the left-hand side) has the same effect as applying L and then translating by b (as on the right-hand side). It is said that an LTI system L "commutes" with translation. Most succinctly expressed,

$$
L\{\tau_b\{\}\} = \tau_b\{L\{\}\} \tag{9.68}
$$

Commuting is another way to view time invariance from a different perspective. It has already been observed that translation by b is itself a linear system. When the operations of τ_b and L are combined in that order, the output $L\{f(x-b)\}$ is produced. To state that L is an LTI system is to indicate that the system τ_b followed by L yields the same result as the system L followed by τ_b. This concept can now be applied to convolution:

$$L\{f(x)\} = (f * g)(x) = \int_{-\infty}^{\infty} g(x-y)f(y)dy \tag{9.69}$$

Then

$$L\{f(x-x_0)\} = \int_{-\infty}^{\infty} g(x-y)f(y-x_0)dy \tag{9.70}$$

This can be written as

$$\begin{aligned} L\{\tau_{x0}\{f(x)\}\} &= \int_{-\infty}^{\infty} g(x-y)\tau_{x0}\{f(y)\}dy \\ &= \int_{-\infty}^{\infty} g(x-y)f(y-x_0)dy \end{aligned} \tag{9.71}$$

Substituting $z = y - x_0$

$$\begin{aligned} L\{\tau_{x0}\{f(x)\}\} &= \int_{-\infty}^{\infty} g(x-z-x_0)f(z)dz \\ &= (g * f)(x-x_0) = \tau_{x0}\{(g * f)(x)\} = \tau_{x0}\{L\{f(x)\}\} \end{aligned} \tag{9.72}$$

9.9 The Fourier Transform and LTI Systems

The LTI systems are defined by convolution. Given the LTI system

$$g(x) = (h * f)(x) \tag{9.73}$$

Taking the Fourier transforms,

$$G(s) = H(s)F(s) \tag{9.74}$$

where s is the spectral variable in Fourier space ($\omega = 2\pi s$). Or this can be written as

$$H(s) = \frac{F(s)}{G(s)} \tag{9.75}$$

$H(s)$ is called the transfer function of the system, which transfers the input signal $F(s)$ to the output signal $G(s)$. The transfer function $H(s)$ is typically expressed as a ratio of two polynomials in the complex variable s, representing the frequency. Poles are the set of values of s for which the denominator of $H(s)$ is zero. Zeros are the values of s for which the numerator of $H(s)$ is zero. The transfer function is invaluable for understanding how a system responds to different inputs. By analyzing the poles and zeros, engineers can determine the stability of the system, design controllers to modify the system's behavior, and predict how the system will respond to various inputs. Transfer functions for components are used to design and analyze systems assembled from components, particularly using the block diagram technique in electronics and control theory. The inverse of the transfer function $\mathscr{F}^{-1}\{H(s)\} = h(x)$ is the impulse response of the system, and $(h * f)(x)$ is called an LTI filter. For example, consider the filter

$$L\{f(x)\} = f(x)^2$$

If $f(x) = \cos 2\pi x$, then

$$L\{f(x)\} = \cos^2 2\pi x = \frac{1}{2} + \frac{1}{2}\cos 4\pi x$$

The input signal $f(x)$ has a single frequency of 1 Hz, while the nonlinear filter added a DC component of $1/2$ and a frequency component at 2 Hz. The nonlinear filters add extra frequency to the system. The LTI filter is a special filter that adds no new frequency to the system.

The impulse response of the LTI filter in Eq. (9.74) to the complex exponential of frequency ν is

$$\begin{aligned} G(s) &= H(s)\mathscr{F}^{-1}\{e^{2\pi i s x}\} \\ &= H(s)\delta(s-\nu) \\ &= H(\nu)\delta(s-\nu) \end{aligned} \tag{9.76}$$

$\delta(s-\nu) \rightleftharpoons e^{2\pi i \nu t}$ is the Fourier pair. The property of the δ function is used to obtain the third line in Eq. (9.76). If the inverse Fourier transform is the linear operation L, taking the inverse transform results in

$$L\{e^{2\pi i \nu x}\} = H(\nu)e^{2\pi i \nu x} \tag{9.77}$$

Equation (9.77) represents a striking result. $e^{2\pi i \nu x}$ is the eigenfunction of L and $H(\nu)$ represents the corresponding eigenvalues. Complex exponentials are fundamental to LTI systems. What about the sin and cos functions under an LTI system? When a cosine signal $\cos(2\pi\nu x)$ is fed to the system,

$$f(x) = \cos(2\pi\nu x) = \frac{1}{2}e^{2\pi i \nu x} + \frac{1}{2}e^{-2\pi i \nu x}$$

So the response is

$$
\begin{aligned}
L\{f(x)\} &= \frac{1}{2}H(\nu)e^{2\pi i\nu x} + \frac{1}{2}H(-\nu)e^{-2\pi i\nu x} \\
&= \frac{1}{2}H(\nu)e^{2\pi i\nu x} + \frac{1}{2}H^*(\nu)e^{-2\pi i\nu x} \\
&= \frac{1}{2}\left(H(\nu)e^{2\pi i\nu x} + \overline{H(\nu)e^{2\pi i\nu x}}\right) \\
&= \Re\{H(\nu)e^{2\pi i\nu x}\} \\
&= |H(\nu)|\cos\left(2\pi\nu x + \phi_H(\nu)\right)
\end{aligned}
$$

where

$$H(\nu) = |H(\nu)|e^{i\phi_H(\nu)}$$

$$(H(-\nu) = H^*(\nu) \text{ because } h(x) \text{ is real-valued})$$

The response is a cosine of the same frequency but with a changed amplitude and phase. A sine signal has a similar response. This shows that neither the cosine nor the sine are themselves eigenfunctions of L. It is only the complex exponential that is an eigenfunction.

9.10 Causality

The notion that "effects never precede their causes" (Dummett & Flew, 1954; Black, 1956) is a fundamental principle known as causality when applied to systems and their inputs and outputs. Causality implies that the past affects the present but not the reverse. Specifically, if L is a system and $L\{f(x)\} = g(x)$, the value of $g(x)$ at $x = x_0$ is determined solely by the values of $f(x)$ for $x < x_0$. More precisely, if $f_1(x) = f_2(x)$ for $x < x_0$, then $L\{f_1(x)\} = L\{f_2(x)\}$ for $x < x_0$, and this is true for any x_0.

At first glance, this might seem self-evident or trivial: shouldn't their outputs also be identical if two signals are the same? However, it is crucial to understand that $f_1(x)$ and $f_2(x)$ are assumed to exist for all time, and the system L might produce outputs based not only on input values up to a specific time x_0 but also on future values. Therefore, it is a critical requirement that a system be causal, meaning that its output values depend only on the input values up to the present.

The provided definition for a system L to be considered causal is quite general. However, different versions of this definition can arise based on additional assumptions about the system, and exploring these variations can be useful. For instance, the causality condition can be expressed more succinctly and conveniently when L

is linear. Specifically, if L is both linear and causal, then $f(x) = 0$ for $x < x_0$ implies that $g(x) = 0$ for $x < x_0$.

To understand why, consider the following reasoning: Let $u(x) \equiv 0$ represent the zero signal. Since L is linear, $L\{u(x)\} = 0$. Causality implies that if $f(x) = 0 = u(x)$ for $x < x_0$, then $g(x) = L\{f(x)\} = L\{u(x)\} = 0$ for $x < x_0$.

Conversely, assume that L is linear and that $f(x) = 0$ for $x < x_0$ implies $g(x) = 0$ for $x < x_0$, where $g(x) = L\{f(x)\}$. We assert that L is causal. If $f_1(x) = f_2(x)$ for $x < x_0$, then $f(x) = f_1(x) - f_2(x) = 0$ for $x < x_0$. According to the hypothesis about L, if $x < x_0$, then $L\{f(x)\} = L\{f_1(x) - f_2(x)\} = L\{f_1(x)\} - L\{f_2(x)\} = 0$, meaning $L\{f_1(x)\} = L\{f_2(x)\}$ for $x < x_0$.

These arguments together demonstrate that a linear system is causal if and only if $f(x) = 0$ for $x < x_0$ implies $g(x) = 0$ for $x < x_0$.

For an LTI system, the specific value of x_0 does not matter, allowing for a further simplification of the causality definition. If L is linear and causal, then $f(x) = 0$ for $x < 0$ implies $g(x) = 0$ for $x < 0$ (with $x < 0$ used instead of $x < x_0$). Conversely, if L is an LTI system where $f(x) = 0$ for $x < 0$ results in $L\{f(x)\} = 0$ for $x < 0$, we assert that L is causal. To prove this, consider a function $f(x)$ that is zero for $x < x_0$ and let $g(t) = L\{f(t)\}$. The signal $u(x) = f(x + x_0)$ is zero for $x < 0$, so $L\{u(x)\} = 0$ for $x < 0$. Due to time invariance, $L\{u(x)\} = L\{f(x + x_0)\} = g(x + x_0)$, meaning $g(x + x_0) = 0$ for $x < 0$, or $g(x) = 0$ for $x < x_0$. Therefore, an LTI system is causal if and only if $f(x) = 0$ for $x < 0$ implies $g(x) = 0$ for $x < 0$.

In many discussions of causal systems, only this final definition is often presented, with the assumptions of linearity and time invariance sometimes implied without being explicitly stated. In fact, it is common to encounter the following two definitions directly, typically provided without the preceding context. A function $f(x)$ is considered causal if it is zero for $x < 0$. An LTI system is deemed causal if causal inputs produce causal outputs. While this is an acceptable definition, it can seem somewhat stark.

Another way to define causality for LTI systems is through the impulse response. Since $\delta(x) = 0$ for $x < 0$, the impulse response $h(x) = L\{\delta(x)\}$ of a causal LTI system must satisfy $h(x) = 0$ for $x < 0$. Conversely, if the impulse response of an LTI system meets this condition and $f(x)$ is a signal where $f(x) = 0$ for $x < 0$, then the output $g(x) = (f * h)(x)$ will also be zero for $x < 0$, indicating that L is causal. This definition is the one found in Bracewell (1965), "an LTI system is causal if and only if its impulse response $h(x)$ is zero for $x < 0$".

Causality is sometimes referred to as the condition of "physical realizability" due to its "past determines the present" interpretation. Many systems governed by differential equations, such as current in RC circuits, are causal. The principle that a differential equation and initial conditions uniquely determine a solution is another common application of causality. While causality seems naturally suited to systems where "time is running" and concepts of "past" and "present" are meaningful, it is crucial to recognize that causality might not apply in situations where "everything is already present." For instance, optical systems may not adhere to causality because the variable involved is typically spatial rather than temporal, where the input could be an object and the output an image. In such cases, there is no past or present—

all the information exists simultaneously. Similarly, even when time is the relevant variable, a noncausal system might be desired, such as when filtering pre-recorded music. In this scenario, since the entire signal is known, it may be beneficial to consider the input's past, present, and future values to determine the desired output.

9.11 System Stability

A linear system is considered stable if it produces a bounded output in response to any bounded input. This condition is known as bounded-input, bounded-output (BIBO) stability. Specifically, if $|f(x)| < M$, then $|g(x)| < MI$, where I is a constant that does not depend on the input. This definition is equivalent to the requirement that the impulse response $h(x)$ of the system is absolutely integrable:

$$I = \int_{-\infty}^{\infty} |h(x)|dx < \infty \tag{9.78}$$

If Eq. (9.78) is true, then

$$|g(x)| = \left| \int_{-\infty}^{\infty} f(x - \tau)h(\tau)d\tau \right| \leq M \int_{-\infty}^{\infty} |h(\tau)|d\tau = MI \tag{9.79}$$

In the case of asymptotic stability of an LTI system, the output tends towards zero (equilibrium state of the system) in the absence of any input, irrespective of the initial conditions.

As noticed from the earlier discussion, the transfer function provides a basis for determining important system response characteristics without solving the complete differential equation. The transfer function is rational in the complex variable s:

$$H(s) = \frac{b_m s^m + b_{m-1} s^{m-1} + \cdots + b_1 s + b_0}{a_n s^n + a_{n-1} s^{n-1} + \cdots + a_1 s + a_0} \qquad m < n \tag{9.80}$$

It is often convenient to factor the polynomials in the numerator and denominator and to write the transfer function in terms of those factors:

$$H(s) = \frac{N(s)}{D(s)} = K \frac{(s - z_1)(s - z_2) \cdots (s - z_{m-1})(s - z_m)}{(s - p_1)(s - p_2) \cdots (s - p_{n-1})(s - p_n)} \tag{9.81}$$

The numerator and denominator polynomials $N(s)$ and $D(s)$ have real coefficients defined by the system's differential equation and $K = b_m/a_n$. The z_i are the roots of the equation $N(s) = 0$ and are known as the zeros of the system. The p_i are the roots of the equation $D(s) = 0$ and are known as the pole of the system. For $s = z_i$ the numerator $N(s) = 0$ and the transfer function vanishes:

$$\lim_{s \to z_i} H(s) = 0 \tag{9.82}$$

When $s = p_i$, the denominator polynomial $D(s) = 0$ and the value of the transfer function becomes unbounded:

$$\lim_{s \to p_i} H(s) = \infty \tag{9.83}$$

All coefficients of the polynomials $N(s)$ and $D(s)$ are real, which implies that the poles and zeros must either be purely real or occur in complex conjugate pairs. This means the poles are either in the form $p_i = \sigma_i$ or as a pair $p_i, p_{i+1} = \sigma_i \pm i\omega_i$. The presence of a single complex pole without a corresponding conjugate would result in complex coefficients in the polynomial $D(s)$. Similarly, the zeros of the system are either real or appear in complex conjugate pairs. The poles and zeros are intrinsic properties of the transfer function and, thus, the differential equation governing the input-output dynamics of the system. Along with the gain constant K, they fully define the differential equation and offer a complete description of the system's behavior.

Example 9.1 (Transfer function) A system is described by the differential equation

$$\frac{d^2g}{dx^2} + 5\frac{dg}{dx} + 6g = 2\frac{df}{dx} + 1$$

Taking the Laplace transform of this linear differential equation,

$$s^2G(s) + 5sG(s) + 6G(s) = 2sF(s) + F(s)$$

The transfer function in Laplace space becomes

$$\begin{aligned} H(s) &= \frac{G(s)}{F(s)} \\ &= \frac{2s+1}{s^2+5s+6} \\ &= \frac{1}{2}\frac{s+1/2}{(s+3)(s+2)} \\ &= \frac{1}{2}\frac{s-(-1/2)}{(s-(-3))(s-(-2))} \end{aligned}$$

Therefore, the system has a single real zero at $s = -1/2$ and a pair of real poles at $s = -3$ and $s = -2$.

Example 9.2 (Transfer Function) A system has a pair of complex conjugate poles $p_1, p_2 = -1 \pm 2i$, a single real zero $z_1 = -4$, and a gain factor $K = 3$. The transfer function is

$$\begin{aligned}H(s) &= K\frac{s-z}{(s-p_1)(s-p_2)}\\ &= 3\frac{s-(-4)}{(s-(-1+2i))(s-(-1-2i))}\\ &= 3\frac{(s+4)}{s^2+2s+5}\end{aligned}$$

$$F(s) = s+4$$

$$G(s) = s^2+2s+5$$

The differential equation is

$$\frac{d^2g}{dx^2} + 2\frac{dg}{dx} + 5g = 3\frac{df}{dx} + 12f$$

A system is characterized by its poles and zeros so that they allow reconstruction of the input/output differential equation. Generally, the poles and zeros of a transfer function may be complex, and the system dynamics may be represented graphically by plotting their locations on the complex s plane, whose axes represent the real and imaginary parts of the complex variable s. Such plots are known as pole-zero plots. Figure 9.1 shows a pole-zero plot for $H(s) = \frac{s^2+0.8s+1}{s^2+0.4s+0.4}$. It is usual to mark a zero location by a circle and a pole location a cross ($\times$). The location of the poles and zeros provides qualitative insights into the response characteristics of a system. The characteristic equation of a transfer function is obtained by making the transfer function's denominator equal to zero ($D(s) = 0$). The solution of these characteristic equations provides information about the stability of the system. If all the roots of the characteristic equation have negative real parts, the system is stable. The system is unstable if any root of the characteristic equation has a positive real part or a repeated root on the imaginary axis. A positive real part indicates the exponential growth of the system with time. Suppose all the roots of the characteristic equation have negative real parts except for one or more non-repeated roots on the imaginary axis. In that case, the system is marginally stable. Up to 2nd- or 3rd-order characteristic equation, the polynomial equations can be solved to obtain the solution and analyze the stability. However, computing the polynomial root to analyze the stability for the higher-order characteristic equation is computationally expensive. The nth-order characteristic equation is expressed as

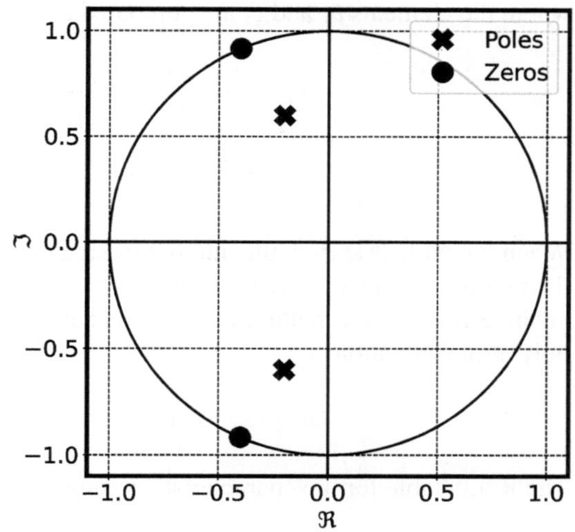

Fig. 9.1 Plot of poles and zeros

$$D(s) = a_n s^n + a_{n-1} s^{n-1} + \cdots + a_1 s + a_0 = 0 \tag{9.84}$$

This equation presents the presence of n number of close loops in the transfer function. German mathematician Adolf Hurwitz (1859–1919) independently noticed that all the coefficients of the polynomial must be of the same sign for a stable system. $H(s) = 4s^4 - 2s^3 + 8s^2 - 6s + 11$ represents an unstable system, while $H(s) = 4s^4 + 2s^3 + 8s^2 + 6s + 11$ and $H(s) = -6s^2 - 5s - 2$ represents a stable system. Further, Hurwitz noticed that no coefficients should vanish for a stable system. All power of s must be present for the stability of the system. $H(s) = 4s^4 + 8s^2 + 6s + 11$ represents a unstable system. Any complex polynomial obeying the above two conditions is known as a Hurwitz polynomial. Hurwitz polynomial is the necessary condition for stability but not the sufficient condition (Hurwitz, 1895). English mathematician Edward John Routh (1831–1907) derived an efficient recursive algorithm to ensure that the real part of the poles lies on the negative region by arranging the coefficient of the Hurwitz polynomial in a matrix form Routh (1877). This algorithm gives sufficient conditions for the stability of the system. Combined Routh-Hurwitz criterion ensures sufficient conditions for the stability of an LTI system. This method involves computing the coefficient of the characteristic equation in a particular pattern in the Routh table as follows. For the nth-order polynomial in Eq. (9.81), the table has $n + 1$ rows and the following structure:

a_n	a_{n-2}	a_{n-4}	$\cdots$
a_{n-1}	a_{n-3}	a_{n-5}	$\cdots$
b_1	b_2	b_3	$\cdots$
c_1	c_2	c_3	$\cdots$
$\vdots$	$\vdots$	$\vdots$	$\ddots$

where the elements b_i and c_i are computed as follows:

$$\begin{aligned} b_i &= \frac{a_{n-1} \times a_{n-2i} - a_n \times a_{n-(2i+1)}}{a_{n-1}} \\ c_i &= \frac{b_1 \times a_{n-(2i+1)} - a_{n-1} \times b_{i+1}}{b_1} \end{aligned} \tag{9.85}$$

When the table is sorted, the number of sign changes in the first column represents the number of nonnegative roots of the characteristic equation. The presence of any positive roots is the hallmark of the unstable system. For example, consider the polynomial equation:

$$D(s) = s^4 + 6s^3 + 11s^2 + 6s + 200 = 0$$

The Routh table for the characteristic system is

1	11	200	0
1	1	0	0
1	20	0	0
−19	0	0	0
20	0	0	0

The Routh table has two sign changes, so the system is unstable. Sometimes, the presence of poles on the imaginary axis creates a situation of marginal stability. For example, consider the system:

$$D(s) = s^6 + 2s^5 + 8s^4 + 12s^3 + 20s^2 + 16s + 16 = 0$$

The Routh table for this system is

1	8	20	16
2	12	16	0
2	12	16	0
0	0	0	0

In this case, the coefficients of the Routh array in a whole row become zero; thus, further solution of the polynomial for finding changes in sign is impossible. Then, another approach comes into play. The row of polynomials, which is just above the row containing the zeroes, is called the "auxiliary polynomial." For this case, the auxiliary polynomial is $A(s) = 2s^4 + 12s^2 + 16$, again equal to zero. The next step is to differentiate the above equation, which yields the polynomial $B(s) = 8s^3 + 24s^1$. The coefficients of the row containing zero now become "8" and "24." The process of the Routh array is proceeded using these values, which yield two points on the imaginary axis. These two points on the imaginary axis are the prime cause of marginal stability. A Python script for determining the Hurwitz-Routh matrix is given below.

```
import numpy as np

def routh_array(coeffs):
    n = len(coeffs)
    routh = np.zeros((n, int(np.ceil(n/2))))

    # Fill the first two rows of the Routh array
    routh[0, :len(coeffs[::2])] = coeffs[::2]
    routh[1, :len(coeffs[1::2])] = coeffs[1::2]

    # Calculate the remaining rows of the Routh array
    for i in range(2, n):
        for j in range(routh.shape[1] - 1):
            routh[i, j] = -(routh[i-1, 0]*routh[i-2, j+1] -
                                                routh[i-2, 0]*routh
                                                [i-1, j+1]) / routh
                                                [i-1, 0]

    return routh

def is_stable(routh):
    first_column = routh[:, 0]
    return np.all(first_column > 0)

# Example characteristic polynomial: s^4 + 2s^3 + 3s^2 + 4s + 5
coefficients = [1, 2, 3, 4, 5]

# Construct the Routh array
routh = routh_array(coefficients)
print("Routh Array:")
print(routh)

# Check the stability
if is_stable(routh):
    print("The system is stable.")
else:
    print("The system is unstable.")
```

9.12 Matched Filters

LTI systems are widely used in communication systems, where a key objective is to differentiate between a signal and noise and to design a filter capable of doing so. The filter should "respond strongly" to a specific signal and only to that signal. This is not about extracting or recovering a signal from noise; instead, it is about detecting the presence of the signal, similar to how radar works. If the filtered signal exceeds a certain threshold, an alert is triggered, indicating that the desired signal is likely present. This is a brief overview of this fundamental issue, but even in this condensed form, it aligns well with our focus. With $g(x) = (h * f)(x)$ and $G(s) = H(s)F(s)$, the goal is to design the transfer function $H(s)$ so that the system responds strongly (a concept to be defined) to a specific signal $f(x)$.

Suppose an incoming signal takes the form $f(x) + p(x)$, where $p(x)$ represents noise. The resulting output is given by $h(x) * (f(x) + p(x)) = g(x) + q(x)$, where $q(x)$, the noise's contribution to the output, possesses total energy.

$$\int_{-\infty}^{\infty} |q(x)|^2 dx \tag{9.86}$$

Using Parseval's theorem and the transfer function, the same can be written as

$$\int_{-\infty}^{\infty} |q(x)|^2 dx = \int_{-\infty}^{\infty} |Q(s)|^2 ds = \int_{-\infty}^{\infty} |H(s)|^2 |P(s)|^2 ds \tag{9.87}$$

A white noise comprises all possible frequencies occurring simultaneously, each with the same amplitude, similar to how white light contains all visible light frequencies combined. When considering random noise in the signal as white noise, the definition translates into a practical condition: $p(x)$ should have equal power at all frequencies. This implies that $|P(s)|$ remains constant, such as $|P(s)| = C$, and consequently, the output energy of the noise is determined by this constant power:

$$E_{\text{noise}} = C^2 \int_{-\infty}^{\infty} |H(s)|^2 ds \tag{9.88}$$

Using the Fourier inversion for $g(x) = (h * f)(x)$

$$g(x) = \int_{-\infty}^{\infty} G(s)e^{2\pi isx} ds = \int_{-\infty}^{\infty} H(s)F(s)e^{2\pi isx} ds \tag{9.89}$$

Using the Cauchy-Schwarz inequality

$$\begin{aligned} |g(x)|^2 &= \left| \int_{-\infty}^{\infty} H(s)F(s)e^{2\pi isx} ds \right|^2 \\ &\leq \int_{-\infty}^{\infty} |H(s)|^2 ds \int_{-\infty}^{\infty} |F(s)|^2 ds \end{aligned} \tag{9.90}$$

$\left| e^{2\pi isx} \right| = 1$. Comparing the energy of the noise output to the strength of the output signal,

$$\frac{|g(x)|^2}{E_{\text{noise}}} \leq \frac{1}{C^2} \int_{-\infty}^{\infty} |F(s)|^2 ds \tag{9.91}$$

The ratio $\frac{|g(x)|^2}{E_{\text{noise}}}$ defines the signal-to-noise ratio (SNR). The maximum SNR is achieved when there is equality in the Cauchy-Schwarz inequality. Therefore, the filter that yields the strongest response, meaning the highest SNR, when a given signal $f(x)$ is mixed with noise $f(t) + p(t)$ is one whose transfer function is proportional to $\overline{F(s)e^{2\pi i s t}}$, where $F(s) = \mathscr{F}\{f(x)\}$. This outcome is known as the matched filter theorem. The transfer function $H(s)$ should be shaped similarly to $F^*(s)$ to design a filter that responds most strongly to a specific signal $f(x)$ in terms of maximizing the SNR. When the filter is designed in this manner, the Parseval theorem applies.

$$\begin{aligned} |g(x)|^2 &= \left| \int_{-\infty}^{\infty} H(s)F(s)e^{2\pi i s x} ds \right|^2 \\ &= \frac{1}{C^2} E_{\text{noise}} \int_{-\infty}^{\infty} |F(s)|^2 \, ds = \frac{1}{C^2} E_{\text{noise}} \int_{-\infty}^{\infty} |f(x)|^2 \, dx \end{aligned} \tag{9.92}$$

Thus, the SNR is

$$\frac{|g(x)|^2}{E_{\text{noise}}} = \frac{1}{C^2} \int_{-\infty}^{\infty} |f(x)|^2 \, dx = \frac{1}{C^2} (\text{Energy of } f(x)) \tag{9.93}$$

9.13 Closure

This chapter explores the foundational concepts forming the backbone of linear system theory, which is crucial in signal processing. The discussion is initiated with the definition of linear systems, where their fundamental properties are emphasized, serving as the building blocks for more complex analyses. Translation or shifting is discussed to reveal how time shifting affects signals and their corresponding system outputs. The understanding was further advanced by examining the cascading of linear systems, which illustrated how interconnected systems influence each other's dynamics and how the overall system response can be derived from individual components. The concept of the impulse response is introduced as a critical tool for characterizing the behavior of linear systems, enabling the prediction of outputs for any given input using the principle of convolution.

The focus was then provided to the linear time-invariant (LTI) systems, representing a class of systems having time-invariant properties. The significance of LTI systems in simplifying the analysis of complex systems is highlighted, laying the groundwork for further exploration of signal processing techniques. Concepts of eigenfunctions are also explored, emphasizing their role in understanding system behavior, particularly how certain input signals retain their form when passed through a system, only being scaled by an eigenvalue. This discussion naturally led to the Fourier transform for analyzing LTI systems. Finally, the crucial concepts of causality and system stability are addressed.

Synthesis of these concepts has established a comprehensive understanding of linear systems and their profound implications in signal processing and control theory. The principles covered in this chapter are foundational and instrumental for advancing towards more complex topics, ensuring a solid grasp of the mechanics that govern modern engineering systems.

References

Black, M. (1956). Why cannot an effect precede its cause? *Analysis, 16*(3), 49–58.

Bracewell, R. (1965). *The Fourier transform and its applications* (1st ed.). McGraw Hill.

Dummett, A., & Flew, A. (1954). Symposium: Can an effect precede its cause? *Proceedings of the Aristotelian Society, Supplementary Volumes, 28*, 27–62.

Golub, G. H., & Van Loan, C. F. (2013). *Matrix computations* (4th ed). Johns Hopkins University Press.

Hurwitz, A. (1895). Ueber die Bedingungen, unter welchen eine Gleichung nur Wurzeln mit negativen reellen Theilen besitzt. *Mathematische Annalen, 46*(2), 273–284.

Routh, E. J. (1877). *A treatise on the stability of a given state of motion, particularly steady motion: being the essay to which the Adams prize was adjudged in 1877, in the University of Cambridge*. Macmillan and Company.

Chapter 10
Z Transform

The shortest path between two truths in the real domain passes through the complex domain.

— Jacques Hadamard

Abstract This chapter introduces the Z transform as a key tool for analyzing discrete-time signals and systems, with broad applications in digital signal processing and control systems. It begins with an overview of the Z transform's historical development and definition, followed by its core mathematical properties, including linearity, translation, multiplication, and convolution. Essential theorems such as Parseval's and the initial and final value theorems and techniques for handling partial derivatives in the Z-domain are covered. A detailed discussion of the inverse Z transform enables the reconstruction of original signals. Practical uses, like solving difference equations and defining transfer functions, are demonstrated, with a focus on system analysis through poles and zeros. The chapter also covers frequency response and the role of pole-zero configurations in assessing system stability and summing infinite series.

Keywords Z transform · Region of convergence · ROC · Initial value theorem · Final value theorem · Partial derivatives · Inverse Z transform · Difference equation · Transfer function

10.1 History of Z Transform

Unlike the Laplace and Fourier transform, the Z transform is named after a letter of the alphabet rather than a renowned mathematician. The origins of the Z transform can be traced back to the work of the French mathematician and statistician Abraham de Moivre (1667–1754) on generating functions in probability theory (de Moivre, 1730). Mathematically, the Z transform can be seen as a particular case of a Laurent series (named after the French mathematician Pierre Alphonse Laurent [1813–1854]), where the sequence of numbers under consideration serves as the

S. K. Jena, *Fourier, Laplace, and the Tangled Love Affair with Transforms*,
https://doi.org/10.1007/978-3-031-80165-5_10

coefficients in the Laurent expansion of an analytic function. This viewpoint underscores the Z transform's deep mathematical foundations and highlights its versatility and broad applicability across various fields of mathematics and engineering. The Z transform is regarded as the discrete counterpart of the Laplace transform. Gardner and Barnes (1942) introduced a method for solving linear, constant-coefficient difference equations using Laplace transforms based on jump functions, applied to ladder networks, transmission lines, and Bessel functions. However, Gardner's approach was quite complex. During World War II, Polish-American mathematician Witold Hurewicz (1904–1956) was inspired by the challenges of sampled-data control systems, which were increasingly relevant due to advancements in radar technology at the time. In 1947, Hurewicz sought to develop a simplified transform for sampled signals or sequences closely related to the Z transform. In hindsight, the Z transform could have been called the "Hurewicz transform," but the name has since been established. The Z transform was formally introduced by American electrical engineer John Ralph Ragazzini (1912–1988) and Azerbaijani-American mathematician Lotfi Asker Zadeh (1921–2017) (Ragazzini & Zadeh, 1952). It offered a systematic and effective method for solving linear difference equations with constant coefficients, which are prevalent in the analysis of discrete-time signals and systems. Additionally, Iraqi-born American engineer Eliahu Ibrahim Jury (1923–2020) made significant contributions by extending the Z transform, leading to what is now known as the modified or advanced Z transform. Jury's work enhanced the transform's applicability and robustness, particularly in dealing with initial conditions and providing a more comprehensive framework for analyzing digital control systems.

10.2 Definition of the Z Transform

A discrete-time Fourier transfer (DTFT) of a sequence $\{f[n]\}$ is expressed as

$$F\left(e^{i\omega}\right) = \sum_{n=-\infty}^{\infty} f[n]e^{-i\omega n} \tag{10.1}$$

DTFT exists if and only if the sequence $\{f[n]\}$ is absolutely summable, i.e., for finite energy signals. What if $\{f[n]\}$ is not an energy or power signal? Let us multiply r^{-n} with the sequence $\{f[n]\}$ for convergence, so that for $r \in \mathbb{R}$ $r > 0$, and $r^{-n}\{f[n]\}$ is absolutely summable. If a signal $f[n]$ is not Fourier transformable, then its exponentially weighted version, $\left(f[n]r^{-n}\right)$, maybe Fourier transformable for the positive real quantity $r > 1$. If $f[n]$ is Fourier transformable, $\left(f[n]r^{-n}\right)$ may still be transformable for some values of $r < 1$. The DTFT in Eq. (10.1) becomes

$$F\left(re^{i\omega}\right) = \sum_{n=-\infty}^{\infty} f[n]r^{-n}e^{-i\omega n} = \sum_{n=-\infty}^{\infty} f[n]\left(re^{-i\omega}\right)^{-n} \tag{10.2}$$

Fig. 10.1 Region of convergence (ROC)

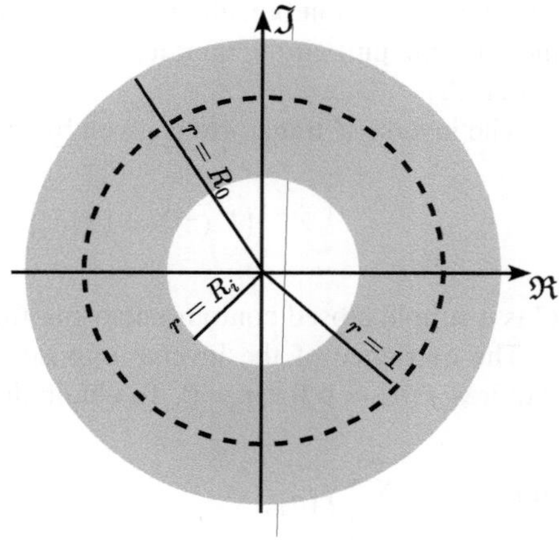

$z = re^{-i\omega}$ represents a complex number in the Argand plane. For a causal system $(f[n] = 0,\ n < 0)$, the above equation can be written as

$$Z\{f[n]\} = F(z) = \sum_{n=0}^{\infty} f[n]z^{-n} \tag{10.3}$$

Equation (10.3) is defined as the Z transfer of the discrete sequence $\{f[n]\}$. A causal assumption leads to the one-sided or unilateral version of the Z transform, which is mostly used for practical system analysis. Z transform can be regarded as the discrete version of the Laplace transform. Z is a linear transformation and can be considered as an operator mapping sequences of scalars into functions of the complex variable z. Figure 10.1 shows the z plane, where the Fourier transform is only valid for $r = 1$, that is, on the circumference of the unit circle. For $Z\{f[n]\}$ transfer, the complex variables change both magnitude and frequency on the Z plane, which can be thought of as a linear combination of variable amplitude sinusoids. The rate of change of the amplitude of the transform spectrum of a one-dimensional signal to two dimensions. The spectrum provides infinite spectral representations of the signal; that is, the spectral values along any appropriate closed contour of the two-dimensional spectrum could be used to reconstruct the signal. Therefore, a signal may be reconstructed using constant amplitude sinusoids, exponentially decaying sinusoids, exponentially growing sinusoids, or an infinite combination of these types of sinusoids.

The z transform, $F(z)$, represents a sequence only for the set of values of z for which it converges, that is, the magnitude of $F(z)$ is finite. The region that comprises this set of values in the z-plane (a complex plane used for displaying the z transform) is called the region of convergence (ROC). For a given positive number r, the equation $|z| = |a + ib| = r$ or $a^2 + b^2 = r^2$ describes a circle in the z-plane with center at the origin and radius r. Consequently, the condition $|z| > r$ for ROC specifies the region outside this circle. If the ROC of the z transform of a sequence includes the unit circle, then its DTFT can be obtained from $f\{[n]\}$ by replacing z with $e^{i\omega}$.

The inverse Z transform is given by the complex integral

$$Z^{-1}\{F(z)\} = \{f[n]\} = \frac{1}{2\pi i}\oint_C \{F[z]\}z^{n-1}dz \tag{10.4}$$

C is a simple closed contour enclosing the origin and lying outside the circle $|z| = r$. The existence of the inverse imposes restrictions on $\{f[n]\}$ for uniqueness. It requires $f[n] = 0$ for $n < 0$. To obtain the inversion integral, let us consider

$$\begin{aligned}\{[F(z)]\} &= \sum_{n=0}^{\infty} f[n]z^{-n} \\ &= f[0] + f[1]z^{-1} + f[2]z^{-2} + \cdots + f[n]z^{-n} + f[n+1]z^{-(n+1)} + \cdots\end{aligned} \tag{10.5}$$

Multiplying both sides by $(2\pi i)^{-1}\, z^{n-1}$ and integrating along the closed contour C, which usually encloses all singularities of $\{F[z]\}$,

$$\begin{aligned}\frac{1}{2\pi i}\oint_C F(z)z^{n-1}dz =& \frac{1}{2\pi i}\left[\oint_C f[0]z^{n-1}dz + \oint_C f[1]z^{n-2}dz\right.\\ &\left. + \cdots + \oint_C f[n]z^{-1}dz + \oint_C f[n+1]z^{-2}dz + \cdots\right]\end{aligned} \tag{10.6}$$

From Cauchy's fundamental theorem, all integrals on the right vanish except

$$\frac{1}{2\pi i}\oint_C f[n]\frac{dz}{z} = f[n] \tag{10.7}$$

This leads to the inversion integral for the Z transform in the form

$$Z^{-1}\{F(z)\} = \{f[n]\} = \frac{1}{2\pi i}\oint_C F(z)z^{n-1}dz \tag{10.8}$$

The bilateral Z transform is expressed as

$$Z\{f[n]\} = F(z) = \sum_{n=-\infty}^{\infty} f[n]z^{-n} \tag{10.9}$$

for all complex numbers z in the Argand plane, where the series converges. Equation (10.9) reduces to the unilateral Z transform in Eq. (10.3) if $\{f[n]\} = 0$ for $n < 0$.

The inverse Z transform is given by a complex integral similar to Eq. (10.4). For $r = 1$, the bilateral Z transform in Eq. (10.9) gives the DTFT of the sequence as given in Eq. (10.1).

Example 10.1 (*Z* **Transform)** $f(n) = \delta_0$

$$Z(\delta_0) = \sum_{n=0}^{\infty} \delta_0(n) z^{-n} = 1$$

ROC is for all z

Example 10.2 (*Z* **Transform)** $f(n) = \delta_m$

$$Z(\delta_m) = \sum_{n=0}^{\infty} \delta_m(n) z^{-n} = z^{-m}$$

ROC is for all $|z| > 0$

Example 10.3 (*Z* **Transform)** $f(n) = a^n \ , n \geq 0$

$$Z(a^n) = \sum_{n=0}^{\infty} \left(\frac{a}{z}\right)^n = \frac{1}{1 - \frac{a}{z}} = \frac{z}{z - a}$$

For $|z| = a$, the Z transform blows up to infinity and the series does not converge (Fig. 10.2). These locations are called pole in the complex plane. The series converges well for $|z| > a$. For $a = 1$, we obtain

$$Z(\mathcal{H}(n)) = Z(1) = \sum_{n=0}^{\infty} z^{-n} = \frac{z}{z - 1}$$

ROC for the unit step/Heaviside function is $|z| > 1$.

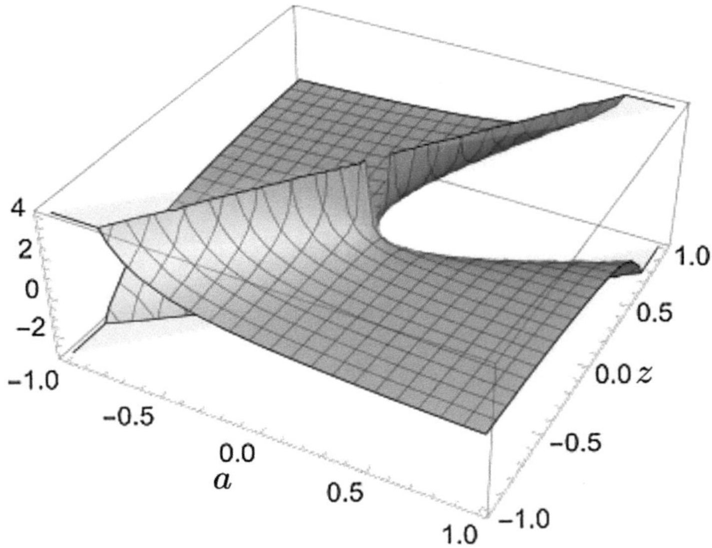

Fig. 10.2 The pole-zero plot of the Z transform of $f(n) = a^n$

Example 10.4 (*Z* **Transform**) $f(n) = n^p$ From the definition

$$Z\left(n^p\right) = \sum_{n=0}^{\infty} n^p z^{-n} = \sum_{n=0}^{\infty} n^{p-1} n z^{-(n+1)} z = z \sum_{n=0}^{\infty} n^{p-1} n z^{-(n+1)}$$

Z transform for $f(n) = n^{(p-1)}$

$$Z\left(n^{p-1}\right) = \sum_{n=0}^{\infty} n^{p-1} z^{-n}$$

Differentiating both sides with respect to z

$$\frac{d}{dz}\left(Z\left(n^{p-1}\right)\right) = -\sum_{n=0}^{\infty} n^{p-1} n z^{-(n+1)}$$

Substituting the above, $Z\left(n^p\right)$ becomes

$$Z\left(n^p\right) = -z \frac{d}{dz}\left(Z\left(n^{p-1}\right)\right)$$

(continued)

Example 10.4 (continued)
For $p = 1$

$$Z(n) = -z\frac{d}{dz}\left(Z\left(n^0\right)\right) = -z\frac{d}{dz}(Z(1)) = -z\frac{d}{dz}\left(\frac{z}{z-1}\right)$$
$$= -z\left(\frac{(z-1)-z}{(z-1)^2}\right) = \frac{z}{(z-1)^2}$$

ROC is $|z| > 1$ (Fig. 10.3).

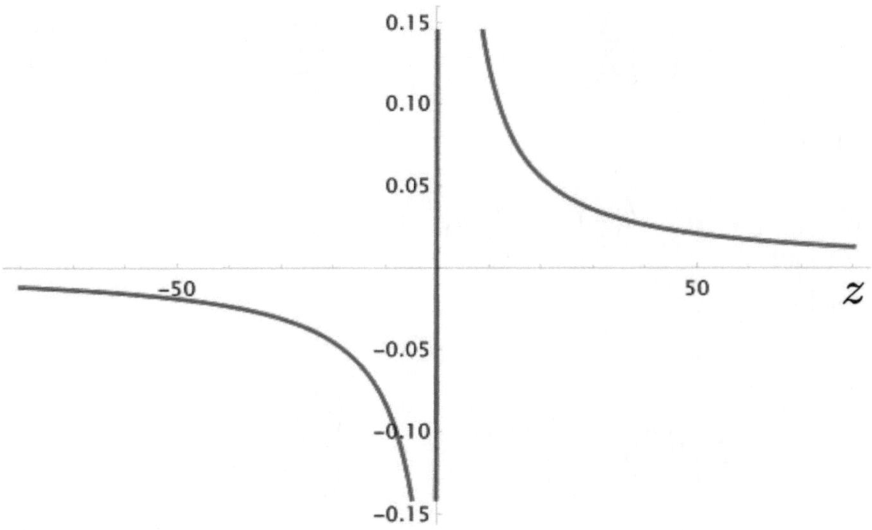

Fig. 10.3 The pole-zero plot of the Z transform of $f(n) = n$

Example 10.5 (Z **Transform**) $f(n) = na^n$ for $n \geq 0$

$$Z\left(na^n\right) = \sum_{n=0}^{\infty} n\left(\frac{a}{z}\right)^n$$

Simplifying the RHS

$$\sum_{n=0}^{\infty} n\left(\frac{a}{z}\right)^n = 0 + \left(\frac{a}{z}\right) + 2\left(\frac{a}{z}\right)^2 + 3\left(\frac{a}{z}\right)^3 + \cdots$$

$$= \left(\frac{a}{z}\right)\left(1 + 2\left(\frac{a}{z}\right) + 3\left(\frac{a}{z}\right)^2 + 4\left(\frac{a}{z}\right)^3 + \cdots\right)$$

$$= \left(\frac{a}{z}\right)\sum_{n=0}^{\infty}(n+1)\left(\frac{a}{z}\right)^n = \left(\frac{a}{z}\right)\left(\sum_{n=0}^{\infty} n\left(\frac{a}{z}\right)^n + \sum_{n=0}^{\infty}\left(\frac{a}{z}\right)^n\right)$$

Then

$$\left(\left(\frac{z}{a}\right) - 1\right)\sum_{n=0}^{\infty} n\left(\frac{a}{z}\right)^n = \sum_{n=0}^{\infty}\left(\frac{a}{z}\right)^n$$

and

$$\sum_{n=0}^{\infty} n\left(\frac{a}{z}\right)^n = \frac{\sum_{n=0}^{\infty}\left(\frac{a}{z}\right)^n}{\left(\left(\frac{z}{a}\right) - 1\right)} = \frac{\frac{1}{1-\left(\frac{a}{z}\right)}}{\left(1-\left(\frac{z}{a}\right)\right)} = \frac{za}{(z-a)^2}$$

For $|z| = a$, the Z transform blows up to infinity and the series does not converge (Fig. 10.4). The ROC is $|z| > a$.

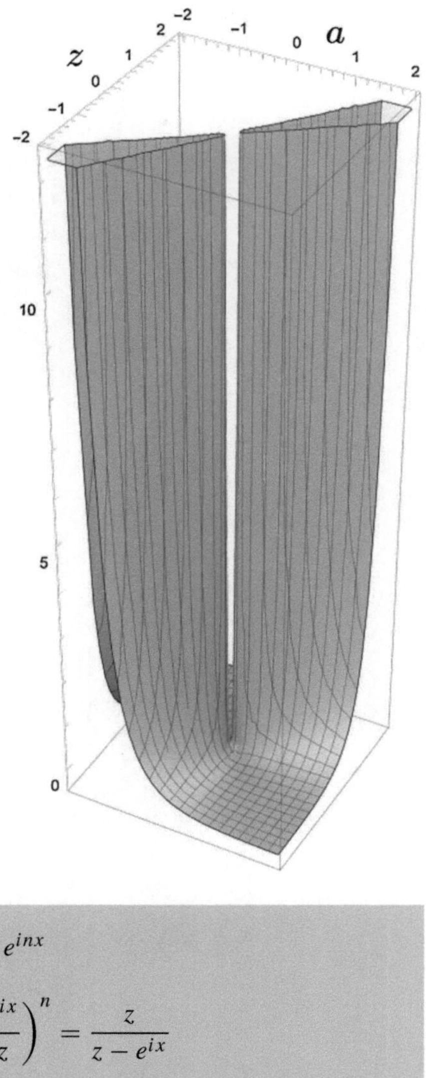

Fig. 10.4 The pole-zero plot of the Z transform of $f(n) = na^n$

Example 10.6 (Z **Transform**) $f(n) = e^{inx}$

$$Z\left(e^{inx}\right) = \sum_{n=0}^{\infty}\left(\frac{e^{ix}}{z}\right)^n = \frac{z}{z - e^{ix}}$$

Using the relationship $e^{inx} = \cos nx + i \sin nx$

$$Z(\cos nx) = \frac{z(z - \cos x)}{z^2 - 2z\cos x + 1}$$

$$Z(\sin nx) = \frac{z \sin x}{z^2 - 2z\cos x + 1}$$

The ROC is $|z| > 1$ (Fig. 10.5).

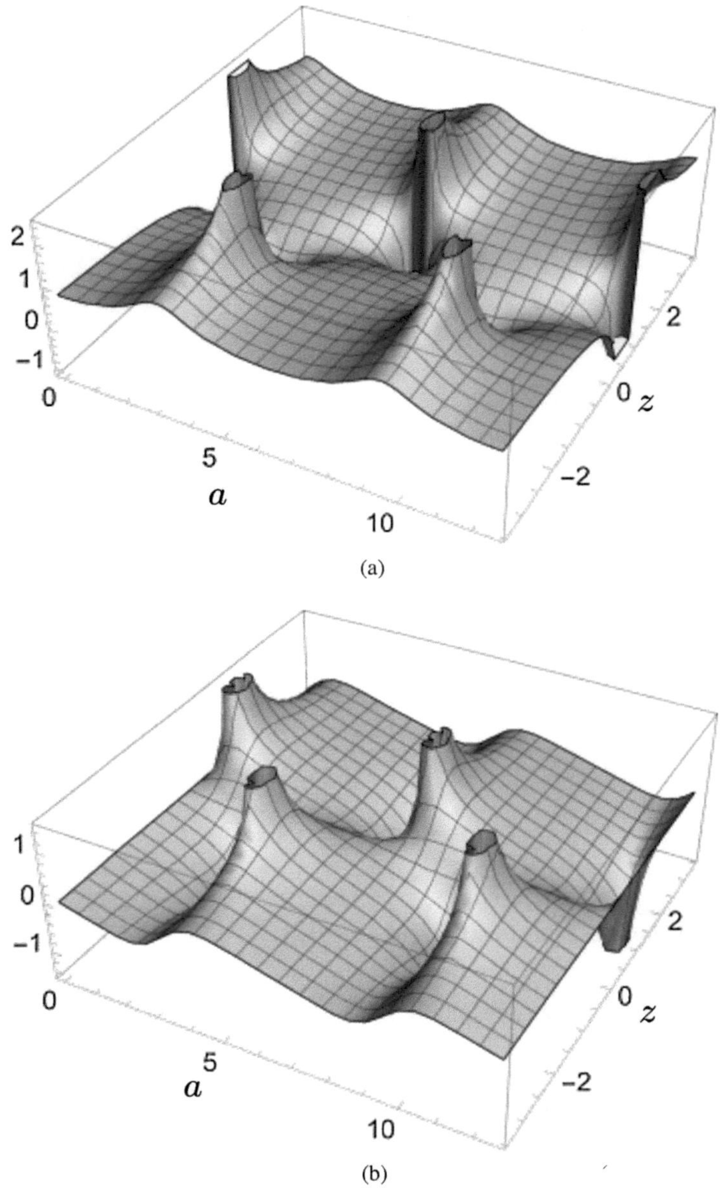

Fig. 10.5 The pole-zero plot of the Z transform of $f(n) = e^{inx}$. (**a**) Real. (**b**) Imaginary

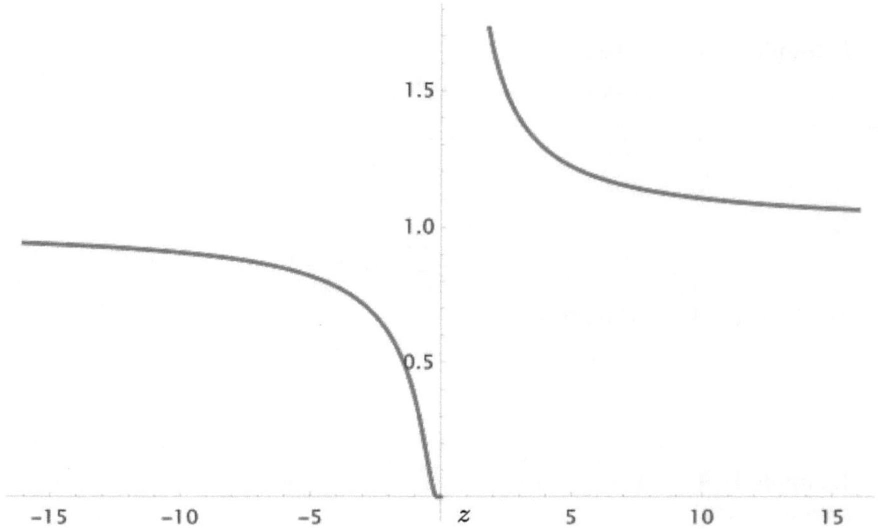

Fig. 10.6 The Z transform of $f(n) = \dfrac{1}{n!}$

Example 10.7 (*Z* **Transform**) $f(n) = \dfrac{1}{n!}$

$$Z\left(\frac{1}{n!}\right) = \sum_{n=0}^{\infty} \frac{1}{n!} z^{-n} = \sum_{n=0}^{\infty} \frac{\left(z^{-1}\right)^n}{n!} = e^{\frac{1}{z}}$$

ROC is for all z (Fig. 10.6).

Example 10.8 (*Z* **Transform**) $f(n) = \cosh nx$

$$\begin{aligned} Z(\cosh nx) &= \frac{1}{2} Z\left(e^{nx} + e^{-nx}\right) \\ &= \frac{1}{2}\left[\frac{z}{z - e^{x}} + \frac{z}{z - e^{-x}}\right] \\ &= \frac{z(z - \cosh x)}{z^2 - 2z\cosh x + 1} \end{aligned}$$

$f(n) = \sinh nx$

(continued)

Example 10.8 (continued)

$$Z(\sinh nx) = \frac{1}{2} Z\left(e^{nx} - e^{-nx}\right)$$
$$= \frac{1}{2}\left[\frac{z}{z - e^{x}} - \frac{z}{z - e^{-x}}\right]$$
$$= \frac{z \sinh x}{z^2 - 2z\cosh x + 1}$$

The ROC is $|z| > 1$ (Fig. 10.7).

Example 10.9 (*Z* **Transform**) If $f(n)$ is a periodic sequence of integral period N, and

$$F_1(z) = \sum_{k=0}^{N-1} f(k) z^{-k}$$

the z transform of $f(k)$ by definition is expressed as

$$F(z) = Z\{f(n)\} = \sum_{n=0}^{\infty} f(n) z^{-n} = z^N \sum_{n=0}^{\infty} f(n+N) z^{-(n+N)}$$
$$= z^N \sum_{k=N}^{\infty} f(k) z^{-k}, \quad (n + N = k)$$
$$= z^N \left[\sum_{k=0}^{\infty} f(k) z^{-k} - \sum_{k=0}^{N-1} f(k) z^{-k}\right]$$
$$= \left\{z^N F(z) - z^N F_1(z)\right\}$$

Thus,

$$F(z) = \frac{z^N}{\left(z^N - 1\right)} F_1(z)$$

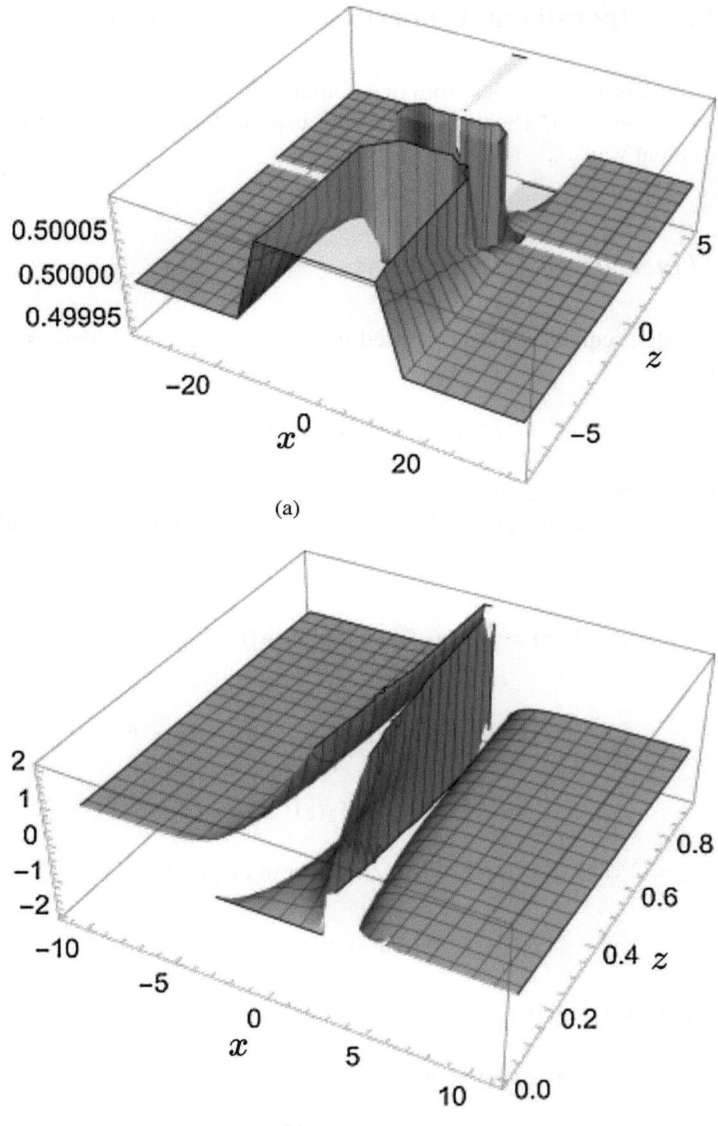

(a)

(b)

Fig. 10.7 The pole-zero plot of the Z transform of hyperbolic function. (**a**) $f(n) = \cosh nx$. (**b**) $f(n) = \sinh nx$

10.3 Basic Operational Properties of Z Transforms

Like other transform, properties from one domain affect the characteristics in other domains in Z transform. These properties are used to find new transform pairs in more convenient ways.

10.3.1 Linearity of Z Transforms

Often a complex sequence is decomposed into several simpler sequences, for ease of handling.

$$f[n] = a_1 f_1[n] + a_2 f_2[n] \tag{10.10}$$

a_1 and a_2 are arbitrary constants. The Z transform of a linear combination of sequences is the same linear combination of the Z transforms of the individual sequences:

$$\begin{aligned} Z\{f[n]\} = F(z) &= \sum_{n=0}^{\infty}(a_1 f_1[n] + a_2 f_2[n])z^{-n} \\ &= a_1 \sum_{n=0}^{\infty} f_1[n]z^{-n} + a_2 \sum_{n=0}^{\infty} f_2[n]z^{-n} \\ &= a_1 F_1(z) + a_2 F_2(z) \end{aligned} \tag{10.11}$$

Let $f_1[n] = 2^n$ and $f_2[n] = 3^n$. The Z transform of $f_1[n]$ is

$$Z\{f_1[n]\} = Z\{2^n\} = \frac{z}{z-2}, \quad |z| > 2$$

The Z transform of $f_2[n]$

$$Z\{f_2[n]\} = Z\{3^n\} = \frac{z}{z-3}, \quad |z| > 3$$

Their linear combination

$$f[n] = 5f_1[n] + 4f_2[n] = 5 \cdot 2^n + 4 \cdot 3^n$$

Applying the linearity property

$$\begin{aligned} Z\{f[n]\} &= Z\{5 \cdot 2^n + 4 \cdot 3^n\} \\ &= 5Z\{2^n\} + 4Z\{3^n\} \end{aligned}$$

$$= 5\left(\frac{z}{z-2}\right) + 4\left(\frac{z}{z-3}\right)$$

$$= \frac{(9z-23)}{(z-2)(z-3)}$$

The ROC is $|z| > 3$. The ROC for $Z\{f[n]\}$, which is derived from the linear combination of two sequences $f_1[n]$ and $f_2[n]$, is the intersection of the ROCs of $Z\{f_1[n]\}$ and $Z\{f_2[n]\}$.

The Z transform of $2n + 3\sin\frac{n\pi}{4} - 5a^4$ using the linearity principle

$$\begin{aligned} Z\left\{2n + 3\sin\frac{n\pi}{4} - 5a^4\right\} &= 2Z\{n\} + 3Z\left\{\sin\frac{n\pi}{4}\right\} - 5Z\left\{a^4\right\} \\ &= 2Z\{n\} + 3Z\left\{\sin\frac{n\pi}{4}\right\} - 5a^4 Z\{1\} \\ &= \frac{2z}{(z-1)^2} + \frac{3z\sin\frac{\pi}{4}}{z^2 - 2z\cos\frac{\pi}{4} + 1} - \frac{5a^4 z}{z-1} \end{aligned}$$

The Z transform of $(n+1)^2$ using the linearity principle

$$\begin{aligned} Z\left\{(n+1)^2\right\} &= Z\left\{n^2 + 2n + 1\right\} \\ &= Z\left\{n^2\right\} + 2Z\{n\} + Z\{1\} \\ &= \frac{z^2+z}{(z-1)^3} + \frac{2z}{(z-1)^2} + \frac{z}{z-1} \end{aligned}$$

The Z transform of $\sin\left(\frac{n\pi}{2} + \frac{\pi}{4}\right)$

$$\begin{aligned} Z\left\{\sin\left(\frac{n\pi}{2} + \frac{\pi}{4}\right)\right\} &= Z\left\{\sin\frac{n\pi}{2}\cos\frac{\pi}{4} + \cos\frac{n\pi}{2}\sin\frac{\pi}{4}\right\} \\ &= \cos\frac{\pi}{4} Z\left\{\sin\frac{n\pi}{2}\right\} + \sin\frac{\pi}{4} Z\left\{\cos\frac{n\pi}{2}\right\} \\ &= \frac{1}{\sqrt{2}}\left[\frac{z\sin\frac{\pi}{2}}{z^2 - 2z\cos\frac{\pi}{2} + 1} + \frac{z\left(z - \cos\frac{\pi}{2}\right)}{z^2 - 2z\cos\frac{\pi}{2} + 1}\right] \\ &= \frac{1}{\sqrt{2}}\left[\frac{z}{z^2+1} + \frac{z^2}{z^2+1}\right] = \frac{1}{\sqrt{2}}\left[\frac{z+z^2}{z^2+1}\right] \end{aligned}$$

10.3.2 Translation

The shift or translation property is used to express the transform of the shifted version $f[n \pm m]$, of a sequence $f[n]$ in terms of its transform $F(z)$. If $f[n] \Longleftrightarrow F(z)$ are transform pairs and for $m \geq 0$

$$
\begin{aligned}
Z\{f[n-m]\} &= \sum_{n=0}^{\infty} f[n-m]z^{-n}, \quad (n-m=r) \\
&= z^{-m} \sum_{r=-m}^{\infty} f[r]z^{-r} = z^{-m} \sum_{r=0}^{\infty} f[r]z^{-r} + z^{-m} \sum_{r=-m}^{-1} f[r]z^{-r} \\
&= z^{-m} \left[F(z) + \sum_{r=-m}^{-1} f[r]z^{-r} \right]
\end{aligned}
\tag{10.12}
$$

$$
\begin{aligned}
Z\{f[n+m]\} &= \sum_{n=0}^{\infty} f[n+m]z^{-n}, \quad (n+m=r) \\
&= z^{m} \sum_{r=m}^{\infty} f[r]z^{-r} = z^{m} \sum_{r=0}^{\infty} f[r]z^{-r} - z^{m} \sum_{r=0}^{m-1} f[r]z^{-r} \\
&= z^{m} \left[F(z) - \sum_{r=0}^{m-1} f[r]z^{-r} \right]
\end{aligned}
\tag{10.13}
$$

for $m = 1, 2, 3, \ldots$, then the right shift becomes

$$
\begin{aligned}
Z\{f[n-1]\} &= z^{-1}F(z) - f[-1]z \\
Z\{f[n-2]\} &= z^{-2} \left[F(z) + \sum_{r=-2}^{-1} f[r]z^{-r} \right]
\end{aligned}
$$

Similarly, the left shift becomes

$$
\begin{aligned}
Z\{f[n+1]\} &= z\{F(z) - f[0]\} \\
Z\{f[n+2]\} &= z^2\{F(z) - f[0]\} - zf[1] \\
Z\{f[n+3]\} &= z^3\{F(z) - f[0]\} - z^2 f[1] - zf[2]
\end{aligned}
$$

All these results are widely used for the solution of initial value problems .

10.3.3 Multiplication

If $f[n] \Longleftrightarrow F(z)$ are transform pairs,

$$\begin{aligned} Z\{nf[n]\} &= \sum_{n=0}^{\infty} nf[n]z^{-n} = z\sum_{n=0}^{\infty} nf[n]z^{-(n+1)} \\ &= z\sum_{n=0}^{\infty} f[n]\left\{-\frac{d}{dz}z^{-n}\right\} = -z\frac{d}{dz}\left\{\sum_{n=0}^{\infty} f[n]z^{-n}\right\} = -z\frac{d}{dz}F(z) \end{aligned} \tag{10.14}$$

For example,

$$\delta_0 \Longleftrightarrow 1 \quad \text{and} \quad n\delta_0 = 0$$

$$1 \Longleftrightarrow \frac{z}{z-1} \text{ and } n \Longleftrightarrow \frac{z}{(z-1)^2}$$

$$\cos nx \Longleftrightarrow \frac{z(z-\cos x)}{z^2 - 2z\cos x + 1}$$

$$\begin{aligned} Z\{n\cos nx\} &= -z\frac{d}{dz}\left[\frac{z(z-\cos x)}{z^2-2z\cos x+1}\right] \\ &= -z\left[\frac{-z^2\cos x + 2z - \cos x}{\left(z^2-2z\cos x+1\right)^2}\right] \\ &= \frac{z^3\cos x - 2z^2 + z\cos x}{\left(z^2-2z\cos x+1\right)^2} \end{aligned}$$

$$\begin{aligned} Z\{a^n f[n]\} &= \sum_{n=0}^{\infty} a^n f[n]z^{-n} = \sum_{n=0}^{\infty} f[n]\left(\frac{z}{a}\right)^{-n} \\ &= F\left(\frac{z}{a}\right), \quad |z| > |a| \end{aligned} \tag{10.15}$$

Multiplication of $f[n]$ by a^n corresponds to scaling the frequency variable z. For example,

$$1 \Longleftrightarrow \frac{z}{z-1} \quad \text{and} \quad (2)^n \Longleftrightarrow \frac{\frac{z}{2}}{\left(\frac{z}{2}-1\right)} = \frac{z}{z-2}$$

The pole at $z = 1$ is shifted to the point $z = 2$, by multiplying with $(2)^n$.

With $a = -1$ and $f[n] \Longleftrightarrow F(z)$, $(-1)^n f[n] \Longleftrightarrow F(-z)$.
For example, $1 \Longleftrightarrow \dfrac{z}{z-1}$ and $(-1)^n \Longleftrightarrow \dfrac{-z}{-z-1} = \dfrac{z}{z+1}$.
If $a = e^b$, then

$$Z\{\left(e^b\right)^n\} = \frac{ze^{-b}}{ze^{-b}-1} = \frac{z}{z-e^b}, \quad |z| > \left|e^b\right| \tag{10.16}$$

When $b = ix$,

$$Z\{e^{inx}\} = \frac{z}{z-e^{ix}}, \quad |z| > 1 \tag{10.17}$$

10.3.4 Division

$$\begin{aligned} Z\left\{\frac{f[n]}{n+m}\right\} &= \sum_{n=0}^{\infty} \frac{f[n]}{n+m} z^{-n}, \quad (m \geq 0) \\ &= -z^m \sum_{n=0}^{\infty} f[n] \left[-\int_0^z \xi^{-(n+m+1)} d\xi\right] \\ &= -z^m \int_0^z \xi^{-(m+1)} \left[\sum_{n=0}^{\infty} f(n)\xi^{-n}\right] d\xi \\ &= -z^m \int_0^z \xi^{-(m+1)} F(\xi) d\xi = -z^m \int_0^z \frac{F(\xi)d\xi}{\xi^{m+1}} \end{aligned} \tag{10.18}$$

For $m = 0, 1, 2, \ldots$

10.3.5 Convolution

The Z transform of the convolution of two discrete-time signals is equal to the product of their individual Z transforms. This is a key property that simplifies the analysis of linear time-invariant (LTI) systems in the Z domain. If $Z\{f[n]\} = F(z)$ and $Z\{g[n]\} = G(z)$, then the Z transform of the convolution $f[n] * g[n]$ is given by

$$
\begin{aligned}
Z\{f[n] * g[n]\} &= \sum_{n=-\infty}^{\infty} z^{-n} \sum_{m=-\infty}^{\infty} f[n-m]g[m] \\
&= \sum_{m=\infty}^{\infty} g[m]z^{-m} \sum_{n=\infty}^{\infty} f[n-m]z^{-(n-m)} \\
&= \sum_{m=\infty}^{\infty} g[m]z^{-m} \sum_{r=-\infty}^{\infty} f[r]z^{-r} \quad (r = n - m) \\
&= Z\{f[n]\}Z\{g[n]\} \\
&= F(z)G(z)
\end{aligned} \tag{10.19}
$$

Equation (10.19) is the convolution theorem for the bilateral Z transform. The Z transform of the product $f[n]g[n]$ is given by

$$
Z\{f[n]g[n]\} = \frac{1}{2\pi i} \oint_C F(w)G\left(\frac{z}{w}\right)\frac{dw}{w} \tag{10.20}
$$

C is a closed contour enclosing the origin in the domain of convergence of $F(w)$ and $G\left(\frac{z}{w}\right)$.

For two simple causal sequences

$$
f_1[n] = (0.5)^n
$$

(a decaying exponential) and its transform

$$
F_1(z) = Z\{(0.5)^n\} = \frac{z}{z - 0.5}, \quad |z| > 0.5
$$

$$
f_2[n] = 1
$$

(a unit step sequence) and its transform

$$
F_2(z) = Z\{1\} = \frac{z}{z-1}, \quad |z| > 1
$$

The convolution $f_1[n] * f_2[n]$ requires summing over all possible shifts of $f_1[n]$ and $f_2[n]$, which is time-consuming. However, using the Z transform property

$$
f_1[n] * f_2[n] = \frac{z}{(z-0.5)} \frac{z}{(z-1)} = \frac{z^2}{(z-1)(z-0.5)}
$$

The convolution of two sequences in the time domain corresponds to the multiplication of their Z transforms in the Z domain. This property is extremely useful for analyzing the output of LTI systems, where the system output is the convolution of the input signal and the system's impulse response.

10.3.6 Parseval's Formula

Parseval's formula in the context of Z transforms shows that the energy of a discrete-time signal can be equivalently computed in the Z domain by integrating the squared magnitude of its Z transform. If $F(z) = Z\{f[n]\}$ and $G(z) = Z\{g[n]\}$, then

$$\sum_{n=-\infty}^{\infty} f[n]g^*[n] = \frac{1}{2\pi}\int_{-\pi}^{\pi} F\left(e^{i\theta}\right) G^*\left(e^{i\theta}\right) d\theta \tag{10.21}$$

In particular,

$$\sum_{n=-\infty}^{\infty} |f[n]|^2 = \frac{1}{2\pi}\int_{-\pi}^{\pi} \left|F\left(e^{i\theta}\right)\right|^2 d\theta \tag{10.22}$$

For the Z transform, Parseval's theorem provides a way to compute the total energy of a discrete-time signal using its Z transform.

10.3.7 Initial Value Theorem

The initial value theorem (IVT) in the context of the Z transform provides a straightforward method to determine the initial value of a discrete-time signal directly from its Z transform without the need to perform an inverse Z transform. From the definition,

$$F(z) = \sum_{n=0}^{\infty} f[n]z^{-n} = f[0] + \frac{f[1]}{z} + \frac{f[2]}{z^2} + \cdots \tag{10.23}$$

If $f[n]$ is a causal discrete-time signal with Z transform $F(z)$, the initial value theorem states that the initial value $x[0]$ of the signal can be found from Eq. (10.23) by letting $z \to \infty$, and hence

$$f[0] = \lim_{z\to\infty} F(z) \tag{10.24}$$

$$f[1] = \lim_{z\to\infty} z\,(F(z) - f[0]) \tag{10.25}$$

If $f[0] = 0$, then

$$f[1] = \lim_{z\to\infty} zF(z) \tag{10.26}$$

The Z transform $F(z)$ must exist for the theorem to be applicable, which typically means the region of convergence (ROC) of $F(z)$ includes $|z| \to \infty$. This theorem is especially useful in analyzing discrete-time systems and signals in the Z domain, as it allows for the quick determination of the signal's behavior at the beginning of the sequence.

Consider a discrete-time signal $f[n] = 2 \cdot (0.5)^n$; then

$$\begin{aligned} F(z) &= \sum_{n=0}^{\infty} 2 \cdot (0.5)^n \cdot z^{-n} \\ &= 2 \cdot \sum_{n=0}^{\infty} \left(\frac{0.5}{z}\right)^n \\ &= \frac{2z}{z - 0.5}, \quad |z| > 0.5 \end{aligned}$$

From the initial value theorem

$$x[0] = \lim_{z\to\infty} \frac{2z}{z - 0.5}$$

As $z \to \infty$, the term -0.5 becomes negligible, so

$$x[0] = \lim_{z\to\infty} \frac{2z}{z} = 2$$

$x[0] = 2$, matches with the initial value of the signal in the time domain.

For another example

$$F(z) = \frac{z}{(z-a)(z-b)}$$

applying IVT

$$f[0] = \lim_{z\to\infty} \frac{z}{(z-a)(z-b)} = 0$$

$$f[1] = \lim_{z\to\infty} zF(z) = 1$$

10.3.8 Final Value Theorem

The final value theorem (FVT) in the Z transform provides a way to determine the steady-state (long-term) behavior of a discrete-time signal directly from its Z transform. If $Z\{f[n]\} = F(z)$, then using the linearity and translation properties of the Z transform

$$Z\{f[n+1] - f[n]\} = z\{F(z) - f[0]\} - F(z) \tag{10.27}$$

which can be expressed as

$$\sum_{n=0}^{\infty}[f[n+1] - f[n]]z^{-n} = (z-1)F(z) - zf[0] \tag{10.28}$$

Taking the limit $z \to 1$

$$\lim_{z\to 1}\sum_{n=0}^{\infty}[f[n+1] - f[n]]z^{-n} = \lim_{z\to 1}(z-1)F(z) - f[0] \tag{10.29}$$

$$\lim_{n\to\infty}[f[n+1] - f[0]] = f(\infty) - f(0) = \lim_{z\to 1}(z-1)F(z) - f(0) \tag{10.30}$$

$$\lim_{n\to\infty} f[n] = \lim_{z\to 1}\{(z-1)F(z)\} \tag{10.31}$$

provided the limits exist. Desoer and Zadeh (1963) derived the rigorous proof of this theorem. The final value theorem allows you to bypass the need to perform an inverse Z transform to find $x[n]$ and directly evaluate the steady-state value. This is especially useful in analyzing the behavior of signals in control systems, where the final output of a system as time approaches infinity is of interest.

For example, for a discrete-time signal $f[n]$ with its Z transform,

$$F(z) = \frac{3z}{(z-0.5)}$$

Applying the final value theorem

$$\begin{aligned}\lim_{n\to\infty} f[n] &= \lim_{z\to 1}(z-1)\cdot\frac{3z}{(z-0.5)} \\ &= \frac{3(1)(1-1)}{1-0.5} = \frac{0}{0.5} = 0\end{aligned}$$

The final value of $f[n] = 0$ as $n \to \infty$ shows a decaying system, where the response eventually dies out.

10.3.9 The Z Transform of Partial Derivatives

f is a sequence of both n and a:

$$\begin{aligned} Z\left\{\frac{\partial}{\partial a} f[n,a]\right\} &= \sum_{n=0}^{\infty}\left[\frac{\partial}{\partial a} f[n,a]\right] z^{-n} \\ &= \frac{\partial}{\partial a}\left[\sum_{n=0}^{\infty} f[n,a] z^{-n}\right] \\ &= \frac{\partial}{\partial a}[Z\{f[n,a]\}] \end{aligned} \tag{10.32}$$

For example,

$$\begin{aligned} Z\left\{ne^{an}\right\} &= Z\left\{\frac{\partial}{\partial a} e^{na}\right\} = \frac{\partial}{\partial a} Z\left\{e^{na}\right\} \\ &= \frac{\partial}{\partial a}\left(\frac{z}{z-e^a}\right) = \frac{ze^a}{(z-e^a)^2} \end{aligned}$$

10.4 The Inverse Z Transform

The inverse Z transform, expressed by the complex integral equation (10.8), can be calculated using the Cauchy residue theorem. However, executing this complex integral is often impractical. Instead, there are simpler methods to determine the inverse Z transform of a given $F(z)$ by using the definition provided in Eq. (10.3). One such method involves expanding $F(z)$ as a power series in z, as shown below:

$$F(z) = f[0] + f[1]z^{-1} + f[2]z^{-2} + \cdots + f[n]z^{-n} + \cdots \tag{10.33}$$

The coefficient of z^{-n} in this expansion is

$$f[n] = Z^{-1}\{F(z)\} \tag{10.34}$$

If $F(z)$ is given by a series

$$F(z) = \sum_{n=-\infty}^{\infty} a_n z^{-n}, \quad r_1 < |z| < r_2 \tag{10.35}$$

In this case, the inverse Z transform is unique and equals $\{f[n] = a_n\}$ for all n. If the domain where $F(z)$ is analytic includes the unit circle $|z| = 1$, and F is single-valued within this region, then $F\left(e^{i\theta}\right)$ becomes a periodic function with a period of 2π. As a result, it can be expressed as a Fourier series, where the coefficients of this series correspond to the inverse Z transform of $F(z)$ and are given by

$$Z^{-1}\{F(z)\} = f[n] = \frac{1}{2\pi}\int_{-\pi}^{\pi} F\left(e^{i\theta}\right) e^{in\theta} d\theta \tag{10.36}$$

Example 10.10 (Inverse Z Transform) $F(z) = \dfrac{z}{z-a}$

$$\begin{aligned} F(z) &= \frac{z}{z-a} = \left(1 - \frac{a}{z}\right)^{-1} \\ &= 1 + az^{-1} + a^2z^{-2} + \cdots + a^n z^{-n} + \cdots \end{aligned}$$

so that $f[0] = 1$, $f[1] = a$, $f[2] = a^2, \ldots, f[n] = a^n, \ldots$; then

$$f[n] = Z^{-1}\left\{\frac{z}{(z-a)}\right\} = a^n$$

Example 10.11 (Inverse Z Transform) $F(z) = Z^{-1}\left\{e^{\frac{1}{z}}\right\}$

$$e^{\frac{1}{z}} = 1 + 1 \cdot z^{-1} + \frac{1}{2!}z^{-2} + \cdots + \frac{1}{n!}z^{-n} + \cdots$$

$$f[n] = Z^{-1}\left\{e^{\frac{1}{z}}\right\} = \frac{1}{n!}$$

Partial fractions and the convolution are used to determine the inverse Z transform.

Example 10.12 (Inverse Z **Transform)** $F(z) = \dfrac{z}{z^2 - 6z + 8}$ This can be expressed as

$$F(z) = \frac{z}{(z-2)(z-4)} = \frac{1}{2}\left(\frac{z}{z-4} - \frac{z}{z-2}\right)$$

It follows from the table of Z transforms that

$$Z^{-1}\{F(z)\} = f[n] = \frac{1}{2}\left[Z^{-1}\left\{\frac{z}{z-4}\right\} - Z^{-1}\left\{\frac{z}{z-2}\right\}\right] = \frac{1}{2}\left(4^n - 2^n\right)$$

Example 10.13 (Inverse Z **Transform)** $H(z) = \dfrac{z^2}{(z-a)(z-b)}$

This can be split into two components:

$$F(z) = \frac{z}{z-a}$$
$$G(z) = \frac{z}{z-b}$$

The inverse transform of each component is

$$Z^{-1}\{F(z)\} = f[n] = a^n$$
$$Z^{-1}\{G(z)\} = g[n] = b^n$$

Using the convolution

$$Z^{-1}\{F(z)G(z)\} = \sum_{m=0}^{n} a^{n-m} b^m = a^n \sum_{m=0}^{n} \left(\frac{b}{a}\right)^m$$

$$= a^n \left\{ \frac{1 - \left(\frac{b}{a}\right)^{n+1}}{1 - \frac{b}{a}} \right\} = \frac{a^{n+1}}{(a-b)} \left\{ 1 - \left(\frac{b}{a}\right)^{n+1} \right\}$$

Example 10.14 (Inverse Z Transform) $F(z) = \dfrac{3z^2 - z}{(z-1)(z-2)^2}$

$F(z)$ is expressed in the form of partial fractions as

$$F(z) = \frac{3z^2 - z}{(z-1)(z-2)^2} = 2\frac{z}{(z-1)} - 2\frac{z}{(z-2)} + \frac{5}{2}\frac{2z}{(z-2)^2}$$

The inverse Z transform becomes

$$\begin{aligned} f[n] &= Z^{-1}\left\{\frac{2z}{(z-1)}\right\} - Z^{-1}\left\{2 \cdot \frac{z}{(z-2)}\right\} + \frac{5}{2}Z^{-1}\left\{\frac{2z}{(z-2)^2}\right\} \\ &= 2 - 2^{n+1} + \frac{5}{2} \cdot n2^n = 2 - 2^{n+1} + 5 \cdot n2^{n-1} \end{aligned}$$

Example 10.15 (Inverse Z Transform) The inverse transform of $H(z) = \dfrac{z(z+1)}{(z-1)^3}$ is

$$H(z) = \frac{z(z+1)}{(z-1)^3} = \frac{z}{(z-1)^2}\left(\frac{z+1}{z-1}\right) = \frac{z}{(z-1)^2}\left[\frac{z}{z-1} + \frac{1}{z-1}\right]$$

This can be split into two parts:

$$F(z) = \frac{z}{(z-1)^2}$$

$$G(z) = \frac{z}{z-1} + \frac{1}{z-1}$$

The inverse Z transform of these two functions is

$$f[n] = n$$

$$g[n] = g_1(n) + g_1(n-1)$$

Using the convolution

$$Z^{-1}\left\{\frac{z(z+1)}{(z-1)^3}\right\} = f[n] * g[n] = \sum_{m=0}^{n} m[g_1(n-m) + g_1(n-m-1)] = n^2$$

Example 10.16 (Inverse *Z* Transform)

$$F(z) = \frac{z}{z-1}, \quad |z| > 1 \quad \text{and} \quad G(z) = \frac{z}{z-1}, \quad |z| < 1$$

$$f[n] = Z^{-1}\{F(z)\} = \begin{cases} 1, & n \geq 0 \\ 0, & n < 0, \end{cases}$$

$$g[n] = Z^{-1}\{G(z)\} = \begin{cases} 1, & n \leq 0 \\ 0, & n \geq 0 \end{cases}$$

The inverse *Z* transform of $z(z-1)^{-1}$ is not unique. In general, the inverse *Z* transform is not unique, unless its region of convergence is specified.

10.5 Applications of *Z* Transforms

The *Z* transform is a powerful tool used in discrete-time signal processing and control systems for analyzing linear, time-invariant systems. It converts discrete-time signals into the complex frequency domain, making solving difference equations easier, analyzing system stability, and designing digital filters. The *Z* transform simplifies convolution operations, turning them into algebraic multiplication, and provides insights into the system's behavior by examining poles and zeros. Its inverse helps reconstruct the original time-domain signal, allowing for comprehensive analysis and design in digital signal processing.

10.5.1 Solution of Difference Equation

Just as the Laplace transform is utilized to solve linear differential equations, the *Z* transform can be employed to solve linear difference equations. Delayed sequences can be converted into algebraic functions of *Z* by applying one of the delay rules associated with *Z* transforms. For an *N*th-order linear difference equation, *N* pieces of information are required to determine the output for a given input. The properties of the *Z* transform previously discussed are used to convert the difference equations that describe the input-output relationship into a straightforward set of linear algebraic equations involving the *Z* transforms of the input and output sequences.

Example 10.17 (First-Order Difference Equation) A difference equation is expressed as

$$f[n+1] - f[n] = 1$$

With initial value

$$f[n] = 0$$

the Z transform of this difference equation becomes

$$z[F(z) - f[0]] - F(z) = \frac{z}{z-1}$$

or

$$F(z) = \frac{z}{(z-1)^2}$$

The inverse z transform gives the sequence

$$Z^{-1}\left\{\frac{z}{(z-1)^2}\right\} = f[n] = n$$

Example 10.18 (First-Order Difference Equation) A difference equation is expressed as

$$f[n+1] + 2f[n] = n$$

with initial condition

$$f[0] = 1$$

The Z transform to this difference equation becomes

$$z\{F(z) - f[0]\} + 2F(z) = \frac{z}{(z-1)^2}$$

or

(continued)

Example 10.18 (continued)

$$F(z) = \frac{z}{z+2} + \frac{z}{(z+2)(z-1)^2}$$

$$= \frac{z}{z+2} + \frac{1}{9}\cdot\frac{z}{z+2} + \frac{3}{9}\cdot\frac{z}{(z-1)^2} - \frac{1}{9}\cdot\frac{z}{z-1}$$

$$= \left(\frac{10}{9}\right)\frac{z}{(z+2)} + \frac{3}{9}\cdot\frac{z}{(z-1)^2} - \frac{1}{9}\frac{z}{(z-1)}$$

The inverse z transform gives the sequence

$$f[n] = \frac{1}{9}\left[10(-2)^n + 3n - 1\right]n$$

Example 10.19 (The Fibonacci Sequence) The Fibonacci sequence is defined as a sequence in which every term is the sum of the two proceeding terms. The Fibonacci sequence in the difference equation form is expressed as

$$f[n+1] = f[n] + f[n-1]$$

with the initial condition

$$f[1] = f[0] = 0$$

The Z transform of the above sequence becomes

$$F(z) = \frac{z^2}{z^2 - z - 1}$$

The inverse Z transform gives the solution

$$f[0] = Z^{-1}\left\{\frac{z^2}{z^2 - z - 1}\right\} = Z^{-1}\left\{\frac{z^2}{(z-a)(z-b)}\right\}$$

$a = \frac{1}{2}(1+\sqrt{5})$ and $b = \frac{1}{2}(1-\sqrt{5})$

(continued)

Example 10.19 (continued)
The Fibonacci sequence is

$$f[n] = \frac{a^{n+1} - b^{n+1}}{(a-b)}, \quad n = 0, 1, 2, \ldots$$

The Fibonacci sequence is given by $1, 1, 2, 3, 5, \ldots$.

Example 10.20 (Second-Order Difference Equation) The second-order difference equation is given as

$$f[n+2] - 3f[n+1] + 2f[n] = 0$$

with the initial condition

$$f[0] = 1, f[1] = 2$$

The Z transform of this difference equation becomes

$$z^2\{F(z) - f[0]\} - zf[1] - 3[z\{F(z) - f[0]\}] + 2F(z) = 0.$$

or

$$\left(z^2 - 3z + 2\right) F(z) = \left(z^2 - z\right)$$

$$F(z) = \frac{z}{(z-2)}$$

The inverse Z transform becomes

$$f[n] = Z^{-1}\left\{\frac{z}{(z-2)}\right\} = 2^n$$

Example 10.21 (Periodic Solution) The second-order difference equation is given as

$$f[n+2] - f[n+1] + f[n] = 0$$

(continued)

Example 10.21 (continued)
with initial conditions

$$f[0] = 1$$
$$f[1] = 2$$

The Z transform of the above sequence becomes

$$\left\{z^2 F(z) - z^2 - 2z\right\} - \{zF(z) - z\} + F(z) = 0$$

or

$$F(z) = \frac{z^2 + z}{(z^2 - z + 1)} = \frac{\left(z^2 - \frac{1}{2}z\right)}{z^2 - z + 1} + \frac{\sqrt{3}\left(\frac{\sqrt{3}}{2}z\right)}{z^2 - z + 1}$$

The inverse Z transform of the above function is

$$f[n] = \cos\left(\frac{n\pi}{3}\right) + \sqrt{3}\sin\left(\frac{n\pi}{3}\right)$$

Example 10.22 (Second-Order Nonhomogeneous Difference Equation)
The second-order difference equation is given as

$$f[n+2] - 5f[n+1] + 6f[n] = 2^n$$

with initial conditions

$$f[0] = 1$$
$$f[1] = 0$$

The Z transformation gives

$$\left(z^2 - 5z + 6\right) F(z) = z^2 - 5z + \frac{z}{z-2}$$

or

(continued)

Example 10.22 (continued)

$$F(z) = z\left[\frac{z-5}{(z-2)(z-3)} + \frac{1}{(z-2)^2(z-3)}\right]$$
$$= z\left[\left(\frac{3}{z-2} - \frac{2}{z-3}\right) + \left(\frac{1}{z-3} - \frac{1}{z-2} - \frac{1}{(z-2)^2}\right)\right]$$
$$= z\left[\frac{2}{z-2} - \frac{1}{z-3} - \frac{1}{(z-2)^2}\right]$$

The inverse Z transform gives the sequence

$$f[n] = 2^{n+1} - 3^n - n2^{n-1}$$

10.5.2 Transfer Function

The convolution operation in the time domain represents the input-output relationship of a linear time-invariant (LTI) system. Since convolution corresponds to multiplication in the frequency domain, it is expressed as

$$h[n] = \sum_{m=0}^{\infty} f[m]g[n-m] \Longleftrightarrow H(z) = F(z)G(z) \tag{10.37}$$

The system input is denoted by $f[m]$, and the output is represented by $h[n]$. The impulse response is denoted by $g[n]$, with $F(z)$, $H(z)$, and $G(z)$ being their respective transforms. Since multiplication by $G(z)$ transfers the input to the output, $G(z)$ is referred to as the transfer function of the system. The transfer function, which is the transform of the impulse response, characterizes the system in the frequency domain, similar to how the impulse response characterizes it in the time domain. For stable systems, the frequency response $G\left(e^{j\omega}\right)$ is obtained from $G(z)$ by substituting z with $e^{j\omega}$. Any input with a nonzero spectral amplitude for all values of z within the region of convergence (ROC) can be applied to the system, and the response can be determined. The ratio of the Z transforms, $H(z)$ of the output to $F(z)$ of the input, is given by $G(z) = \dfrac{H(z)}{F(z)}$.

Suppose a linear time-invariant (LTI) system governed by the following first-order difference equation:

$$h[n] - ah[n-1] = f[n] \tag{10.38}$$

$f[n]$ is the input and $h[n]$ is the output of the system with a constant coefficient a. Taking the Z transform of both sides of the equation, assuming zero initial conditions $f[0] = 0$,

$$H(z) - az^{-1}H(z) = F(z) \tag{10.39}$$

Simplification of Eq. (10.39) gives

$$H(z)\left(1 - az^{-1}\right) = F(z) \tag{10.40}$$

Now, the transfer function is determined as

$$G(z) = \frac{H(z)}{F(z)} = \frac{1}{1 - az^{-1}} = \frac{z}{z - a} \tag{10.41}$$

This transfer function $G(z) = \dfrac{z}{z-a}$ characterizes the behavior of the system in the frequency domain. It indicates how the system responds to any input $f[n]$ in terms of its output $h[n]$. This transfer function has a pole at $z = a$ and a zero at $z = 0$. The location of the pole determines the system's stability and frequency response.

Since the transform of a delayed signal is its transform multiplied by a factor, the transfer function can be obtained by taking the transform of the difference equation characterizing a system. Consider the difference equation of a causal LTI discrete system:

$$\begin{aligned} &h[n] + a_{K-1}h[n-1] + a_{K-2}h[n-2] + \cdots + a_0 h[n-K] \\ &= b_M f[n] + b_{M-1} f[n-1] + \cdots + b_0 f[n-M] \end{aligned} \tag{10.42}$$

Taking the Z transform of both sides with an assumption of zero initial conditions,

$$\begin{aligned} &H(z)\left(1 + a_{K-1}z^{-1} + a_{K-2}z^{-2} + \cdots + a_0 z^{-K}\right) \\ &= F(z)\left(b_M + b_{M-1}z^{-1} + \cdots + b_0 z^{-M}\right) \end{aligned} \tag{10.43}$$

The transfer function $G(z)$ is obtained as

$$G(z) = \frac{H(z)}{F(z)} = \frac{b_M + b_{M-1}z^{-1} + \cdots + b_0 z^{-M}}{1 + \left(a_{K-1}z^{-1} + a_{K-2}z^{-2} + \cdots + a_0 z^{-K}\right)} = \frac{\sum\limits_{l=0}^{M} b_{M-1}z^{-1}}{1 + \sum\limits_{i=1}^{K} a_{K-1}z^{-l}} \tag{10.44}$$

Expressing the transfer function as the positive powers of z is often convenient:

$$G(z) = \frac{z^{K-M}\left(b_M z^M + b_{M-1}z^{M-1} + \cdots + b_0\right)}{z^K + \left(a_{K-1}z^{K-1} + a_{K-2}z^{K-2} + \cdots + a_0\right)} \tag{10.45}$$

10.5.3 Characterization of a System by Its Poles and Zeros

Zeros are the values of z where the transfer function $G(z)$ becomes zero. They represent the frequencies at which the system output will be zero for a given input. Poles are the values of z where the transfer function $G(z)$ becomes infinite. They determine the stability and the natural response of the system. Characterizing a system by its poles and zeros using the Z transform is a powerful method in signal processing. Poles and zeros provide insight into the behavior, stability, and frequency response of a system. By using the pole-zero representation of the Z transform, the transfer function is written as

$$G(z) = B\frac{z^{K-M}(z-z_1)(z-z_2)\cdots(z-z_M)}{(z-p_1)(z-p_2)\cdots(z-p_K)} = Bz^{K-M}\frac{\prod_{i=1}^{M}(z-z_i)}{\prod_{i=1}^{K}(z-p_i)} \tag{10.46}$$

B is a constant. As the coefficients of the $G(z)$ polynomials are real for practical systems, the zeros and poles are real-valued, or they always occur as complex conjugate pairs. When equated to zero, the numerator and denominator polynomials of $G(z)$ have M and K roots, respectively:

$$\{z_1, z_2, \ldots, z_M\} \text{ and } \{p_1, p_2, \ldots, p_K\} \tag{10.47}$$

The roots represent complex frequencies. When $G(z)$ equals zero at $\{z_1, z_2, \ldots, z_M\}$, these frequencies are referred to as the zeros of $G(z)$. Conversely, when $G(z)$ becomes infinite at $\{p_1, p_2, \ldots, p_M\}$, these frequencies are known as the poles of $G(z)$. A linear system is fully characterized by its poles, zeros, and a constant factor. The poles dictate the time variation in the system's response, while the zeros affect the magnitude. If a pole or zero occurs m times, it is said to have a multiplicity of m and is called a second-order, third-order, etc., pole or zero. A pole or zero that does not repeat is known as a simple pole or zero.

The pole-zero plot of a system's transfer function, $G(z)$, visually represents key characteristics such as response speed, frequency selectivity, and stability. Poles with magnitudes much less than one correspond to a fast-responding system, where the transient response decays quickly. In contrast, poles with magnitudes closer to one result in a slower, more sluggish system. Complex conjugate poles within the unit circle lead to an oscillatory transient response that decays over time, with higher oscillation frequencies occurring for poles in the second and third quadrants of the unit circle. The system exhibits a steady oscillatory response if these complex conjugate poles lie directly on the unit circle. Poles on the positive real axis within the unit circle produce an exponentially decaying transient response, while poles on the negative real axis cause the transient response to alternate between positive and negative values.

The pole-zero plot also reveals how the system affects different frequency components of an input signal: frequencies near a zero are suppressed, whereas those near a pole are transmitted more effectively. Systems with poles symmetrically placed about the positive real axis inside the unit circle and near the unit circle in the passband exhibit low-pass characteristics, favoring the transmission of low-frequency signals. Zeros symmetrically positioned about the negative real axis in the stopband further enhance this low-pass behavior. Conversely, poles symmetrically located about the negative real axis inside the unit circle in the passband contribute to high-pass characteristics, where high-frequency signals are transmitted more effectively than low-frequency signals. The pole-zero plot also provides insight into the system's stability, which is discussed in further detail later. For example, for a transfer function

$$G(z) = \frac{(z - 0.5)(z + 1)}{(z - 0.8)(z + 0.3)} \tag{10.48}$$

The zeros are at $z = 0.5$ and $z = -1$, while the poles at $z = 0.8$ and $z = -0.3$. Figure 10.8 graphically represents the location of poles and zeros in the Z plane. The poles at $z = 0.8$ and $z = -0.3$ lie inside the unit circle ($|z| < 1$), so the system is stable. Zeros at $z = 0.5$ and $z = -1$ create notches in the frequency response, meaning the system output will be significantly reduced or nullified at certain frequencies. The pole at $z = 0.8$ causes the system to have a resonant frequency near this pole, where the system's response will be amplified.

10.5.4 Frequency Response and the Locations of the Poles and Zeros

The frequency response of a system is used to characterize its performance. For instance, a high-fidelity amplifier should maintain a distortion-free response up to 20 kHz. In system design, the required frequency response is often specified in advance. Mathematically, it is derived from $G(z)$ by substituting z with $e^{i\omega}$. Here, $z = re^{i\omega}$, and on the unit circle, $r = 1$. The frequency response represents the system's reaction to a continuous complex exponential signal with unit magnitude and zero phase. Since this type of signal begins at time $-\infty$, the response observed at any finite time reflects the system's steady state. In practice, the frequency response is approximated by applying a sinusoidal signal to the system at a finite starting

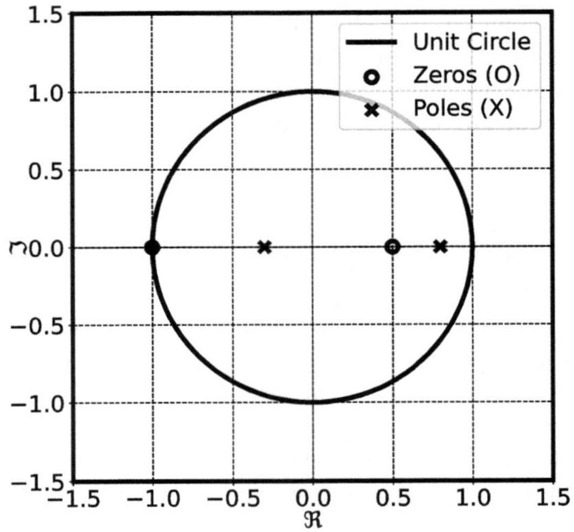

Fig. 10.8 The pole-zero plot of the Z transform of $G(z) = \dfrac{(z-0.5)(z+1)}{(z-0.8)(z+0.3)}$

time and measuring the response after the transient effects have decayed to an insignificant level. For the M-order numerator and N-order denominator, this is achieved by replacing z with $e^{i\omega}$ in $G(z)$:

$$G(z)|_{z=e^{i\omega}} = \frac{H\left(e^{i\omega}\right)}{F\left(e^{i\omega}\right)} = \frac{h[0] + h[1]e^{-i\omega} + h[2]e^{-i2\omega} + \cdots + h[M]e^{-iM\omega}}{f[0] + f[1]e^{-i\omega} + f[2]e^{-i\omega} + \cdots + f[N]e^{-iN\omega}} \tag{10.49}$$

The DFT is used to evaluate the frequency response. The DFT is equivalent to polynomial evaluation at the roots of unity. The DFTs of the numerator and denominator polynomials of $G(z)$ are computed separately, and the frequency response is obtained by performing term-by-term division. Although the frequency response is continuous, it is typically computed at a finite number of points.

For example, consider a transfer function:

$$G(z) = \frac{3}{4z + 1.0} \tag{10.50}$$

Replacing z by $e^{i\omega}$ in the transfer function,

$$G(e^{i\omega}) = \frac{3}{4e^{i\omega} + 1.0} \tag{10.51}$$

Figure 10.9 displays the magnitude and phase of the frequency response. These graphical expressions are helpful in understanding the frequency response and are known as Bode plots. Bode diagram is named after Dutch ancestry American engineer Hendrik Wade Bode (1905–1982). The Bode diagram graphically represents the complex transfer function of a linear time-invariant continuous system (LTC

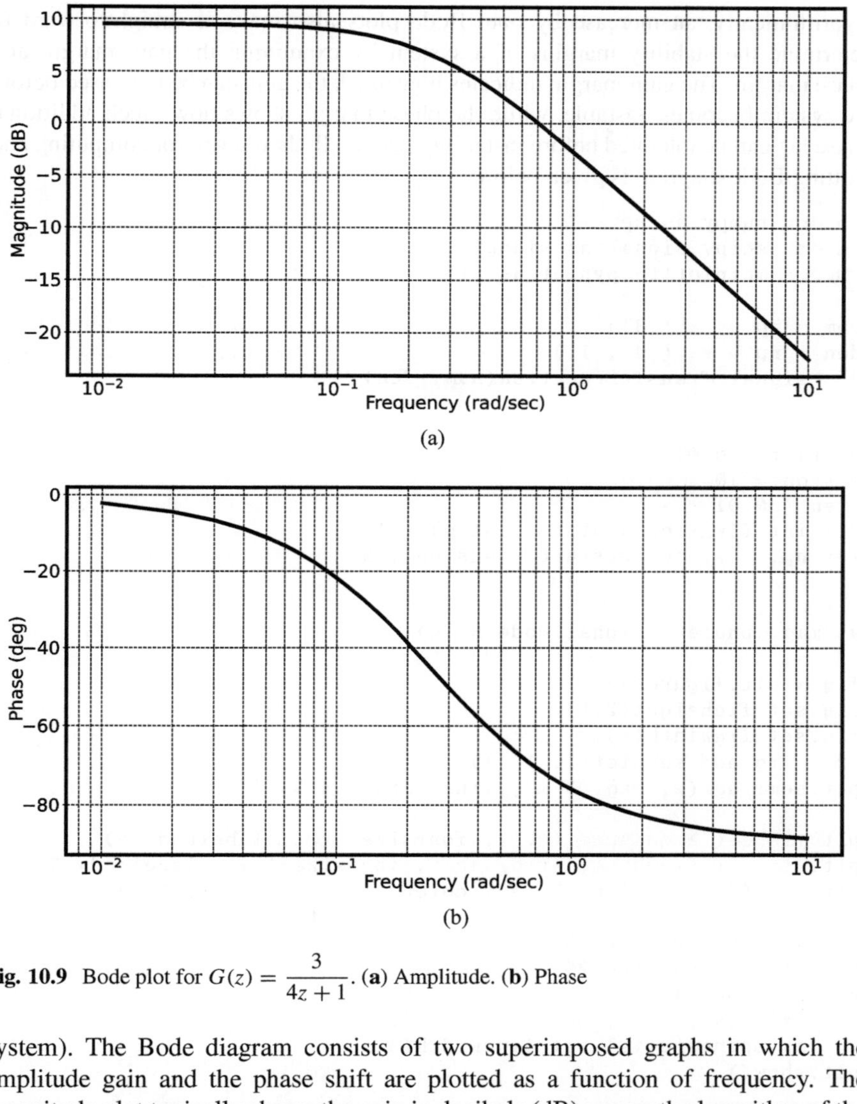

Fig. 10.9 Bode plot for $G(z) = \frac{3}{4z+1}$. (**a**) Amplitude. (**b**) Phase

system). The Bode diagram consists of two superimposed graphs in which the amplitude gain and the phase shift are plotted as a function of frequency. The magnitude plot typically shows the gain in decibels (dB) versus the logarithm of the frequency, while the phase plot shows the phase shift in degrees versus the logarithm of the frequency. Bode plots are particularly useful for analyzing the behavior of systems in the frequency domain, especially in control systems and filter design.

A key feature of Bode plots is their ability to illustrate the effects of poles and zeros on the system's frequency response. For example, a pole generally causes the magnitude to decrease by 20 dB per decade and introduces a phase shift of $-90°$. Conversely, a zero increases the magnitude by 20 dB per decade and introduces a phase shift of $+90°$. A decade is a frequency increase by a factor of ten, or logarithmically, an increase by one. Bode plots also make it straightforward to determine the stability margins of a system by examining the gain margin and phase margin. The gain margin indicates how much the gain can be increased before the system becomes unstable, while the phase margin shows how much additional phase lag can be tolerated before instability occurs. Python script for computing and plotting Bode diagram is given below.

```
import numpy as np
import scipy.signal as signal
import matplotlib.pyplot as plt

num = np.array([3])                  # Numerator coefficients
den = np.array([4 , 1])              # Denominator coefficients
H = signal.TransferFunction(num, den)

# Frequencies
w_start = 0.01
w_stop = 10
step = 0.01
N = int ((w_stop-w_start )/step) + 1
w = np.linspace (w_start , w_stop , N)

# Bode Plot
w, mag, phase = signal.bode(H, w)

fig = plt.figure()
fig.set_figheight(2.5)
fig.set_figwidth(4)
ax = fig.add_subplot(1, 1, 1)
plt.semilogx(w, mag, "-k", linewidth = 4.0) # Bode Magnitude
                                              Plot
plt.ylabel("Magnitude (dB)", fontsize='20', labelpad=-5)
plt.xlabel("Frequency (rad/sec)", fontsize='20', labelpad=-5)
plt.grid(True, which='both', color = 'black', linestyle = '--',
                                              linewidth = 1.0)
plt.xticks(fontsize=20)
plt.yticks(fontsize=20)
ax = fig.gca()
for axis in ['top','bottom','left','right']:
    ax.spines[axis].set_linewidth(3.5)
plt.show()

fig = plt.figure()
fig.set_figheight(2.5)
fig.set_figwidth(4)
ax = fig.add_subplot(1, 1, 1)
plt.semilogx(w,  phase, "-k", linewidth = 4.0) # Bode Phase
                                              plot
plt.ylabel("Phase (deg)", fontsize='20', labelpad=-5)
plt.xlabel("Frequency (rad/sec)", fontsize='20', labelpad=-5)
plt.grid(True, which='both', color = 'black', linestyle = '--',
                                              linewidth = 1.0)
plt.xticks(fontsize=20)
```

```
plt.yticks(fontsize=20)
ax = fig.gca()
for axis in ['top','bottom','left','right']:
    ax.spines[axis].set_linewidth(3.5)
plt.show()
```

10.5.5 System Stability

The zero-input response of a system is determined solely by the locations of its poles. A system is deemed stable if its zero-input response, resulting from finite initial conditions, converges. It is considered marginally stable if the zero-input response either settles at a constant value or oscillates with a constant amplitude, and it is classified as unstable if the zero-input response diverges. Marginally stable systems, such as oscillators, produce a bounded zero-input response and are commonly used. The response associated with each pole p of a system takes the form $r^n e^{in\theta}$, where r and θ represent the magnitude and phase of the pole, respectively. If $r < 1$, then r^n approaches zero as n tends to infinity. If $r > 1$, then r^n approaches infinity as n increases. For $r = 1$, $r^n = 1$ for all n. However, if poles of order greater than one lie on the unit circle, the response tends to infinity because the expression for the response includes a factor that depends on n. Poles of any order that lie within the unit circle do not lead to instability. Therefore, it can be concluded from the pole locations of a system that all the poles, of any order, of a stable system must lie inside the unit circle. That is, the ROC of $G(z)$ must include the unit circle. Any pole lying outside the unit circle or any pole of order more than one lying on the unit circle makes a system unstable. A system is marginally stable if it has no poles outside the unit circle and has poles of order one on the unit circle. Figure 10.8 shows pole locations of some transfer functions and the corresponding impulse responses. If all the poles of a system lie inside the unit circle, the bounded-input bounded-output stability condition is satisfied. However, the converse is not necessarily true since the impulse response is an external description of a system and may not include all its poles. The bounded-input bounded-output stability condition is not satisfied by a marginally stable system.

10.5.6 Summation of Infinite Series

The Z transform simplifies the process of summing infinite series by transforming the problem into the frequency domain, where algebraic methods can be applied. By evaluating the Z transform at specific points, the sum of an infinite series can be easily determined, making this method a valuable tool in signal processing and system analysis. Here are a few examples.

Example 10.23 (Summation of Infinite Series) The sequence

$$g[n] = \sum_{k=0}^{n} f[k]$$

can also be written as

$$g[n] = g[n-1] + f[n]$$

Taking the Z transform of both sides

$$G(z) = F(z) + z^{-1}G(z)$$

or

$$G(z) = \frac{z}{(z-1)}F(z)$$

so

$$Z\{g[n]\} = Z\left\{\sum_{k=0}^{n} f[k]\right\} = \frac{z}{(z-1)}F(z)$$

Applying the final value theorem, in the limit as $z \to 1$, gives

$$\lim_{n\to\infty} \sum_{k=0}^{n} f[k] = \lim_{z\to 1}(z-1)\cdot\frac{z}{z-1}F(z) = F(1)$$

Example 10.24 (Summation of Infinite Series) From the multiplicity operation in the Z transform,

$$Z\left\{x^n f[n]\right\} = F\left(\frac{z}{x}\right)$$

Setting $f[n] = \frac{1}{n!}$ so that $F(z) = e^{\frac{1}{z}}$,

$$Z\left\{\frac{x^n}{n!}\right\} = e^{\frac{x}{z}}$$

(continued)

Example 10.24 (continued)
Using the final value theorem

$$\sum_{n=0}^{\infty} \frac{x^n}{n!} = \lim_{z \to 1} e^{\frac{x}{z}} = e^x$$

Example 10.25 (Summation of Infinite Series)

$$Z\left\{x^{n+1}\right\} = \frac{zx}{z-x}$$

This can be processed as

$$\begin{aligned}
Z\left\{\frac{x^{n+1}}{n+1}\right\} &= z\int_z^{\infty} \frac{zx}{(z-x)} \cdot \frac{dz}{z^2} \\
&= xz\int_z^{\infty} \frac{dz}{z(z-x)} \\
&= xz\left[\frac{1}{x}\log\left(\frac{z-x}{z}\right)\right]_z^{\infty} \\
&= -z\log\left(\frac{z-x}{z}\right)
\end{aligned}$$

Replacing x by $(-x)$ in this result,

$$Z\left\{(-1)^n\frac{x^{n+1}}{n+1}\right\} = z\log\left(\frac{z+x}{z}\right)$$

Applying the final value theorem

$$\sum_{n=0}^{\infty}(-1)^n \cdot \frac{x^{n+1}}{n+1} = \lim_{z \to 1} z\log\left(\frac{z+x}{z}\right) = \log(1+x)$$

Example 10.26 (Summation of Infinite Series)

$$Z\{\sin nx\} = \frac{z \sin x}{z^2 - 2z \cos x + 1}$$

So

$$Z\left\{a^n \sin nx\right\} = F\left(\frac{z}{a}\right) = \frac{az \sin x}{a^2 - 2az \cos x + z^2}$$

Applying the final value theorem

$$\sum_{n=0}^{\infty} a^n \sin nx = \lim_{z \to 1} F\left(\frac{z}{a}\right) = \frac{a \sin x}{a^2 - 2a \cos x + 1}$$

10.6 Closure

This chapter explored the Z transform as a fundamental tool in analyzing and designing discrete-time systems. The Z transform extends the concept of the Fourier transform by providing a broader framework that is particularly useful for understanding the behavior of systems described by linear difference equations. The Z transform can transform complex time-domain operations into more manageable algebraic operations in the Z-domain, facilitating the analysis of system stability, frequency response, and system dynamics.

The chapter began with defining the Z transform and discussing its properties, such as linearity, time-shifting, and convolution, which are essential for manipulating and interpreting signals and systems. The importance of the region of convergence (ROC) has been emphasized, as it determines the conditions under which the Z transform exists and dictates the stability and causality of the system. The chapter also covered the inverse Z transform, which allows to revert back to the time domain, enabling a complete understanding of the system's behavior. Subsequently, the applications of the Z transform are explored, like solving linear difference equations. The relationship between poles, zeros, and the system's response was examined in detail, highlighting how the pole-zero configuration directly impacts the stability, frequency response, and overall behavior of a system.

The Z transform provides a powerful means to transition between the time and frequency domains, offering insights into the stability and performance of discrete-time systems. With this knowledge, readers are expected to be well-equipped to tackle more advanced topics in signal processing, control theory, and related fields.

References

de Moivre, A. (1730). *Miscellanew, analytica de seriebus et quatratoris*. J. Tonson & J. Watts.
Desoer, C. A., & Zadeh, L. A. (1963). *Linear system theory: The state space approach* (1st ed.). McGraw-Hill.
Gardner, M. F., & Barnes, J. L. (1942). *Transients in linear systems studied by the Laplace transformation* (Vol. 1). Wiley.
Ragazzini, J. R., & Zadeh, L. A. (1952). The analysis of sampled-data systems. *Transactions of the American Institute of Electrical Engineers, Part II: Applications and Industry, 71*(5), 225–234.

Chapter 11
Higher-Dimensional Fourier Analysis

Mathematics compares the most diverse phenomena and discovers the secret analogies that unite them.
— Joseph Fourier

Abstract This chapter covers higher-dimensional Fourier analysis, broadening the principles of Fourier transforms into multiple dimensions. It begins with a clear introduction to fundamental concepts and then delves into the distinctive properties and mathematical framework of the multidimensional Fourier transform. The discussion includes the Fourier transform of separable functions, simplifying the analysis of complex multivariable systems by breaking them into manageable components. Additionally, the chapter examines radial functions in two dimensions, highlighting the importance of symmetry in higher-dimensional spaces. It also discusses higher-dimensional impulse responses, illustrating their significance in signal processing and system analysis when exposed to multidimensional stimuli. Finally, the chapter covers the principles of higher-dimensional sampling, which is vital for accurately capturing and reconstructing signals across multiple dimensions. By integrating these concepts, this chapter offers a thorough understanding of how higher-dimensional Fourier analysis enhances the analysis and processing of multivariable data, making it invaluable for various applications in engineering, physics, and applied mathematics.

Keywords Separable functions · $circ$ function · $jinc$ function · Two dimensional Dirac comb · Band-limits · Higher dimensional sampling · Higher dimensional impulse response · Radial functions · Multi dimensional Fourier transform

11.1 Introduction

When considering the one-dimensional Fourier transform, the natural association is the transformation of signals between the time and frequency domains. However,

S. K. Jena, *Fourier, Laplace, and the Tangled Love Affair with Transforms*,
https://doi.org/10.1007/978-3-031-80165-5_11

multidimensional transforms become essential for meaningful analysis in certain situations, particularly in studying phenomena like turbulence. A prime example is the analysis of eddies in turbulent flow. Though eddies lack a precise definition, they are commonly understood as small, rotating structures within a turbulent flow field. It is often necessary to transform the data from physical space into spectral or Fourier space to characterize these eddies by their size and velocity. The transformation from physical to Fourier space is pivotal in this context. In a Cartesian framework, the position or velocity of an eddy is defined by three coordinates. Upon transforming into Fourier space, the spatial distance is represented by a wave number, which encapsulates information about the scale of these eddies. Chapter 1 delves into the specification of wave number, providing a detailed exploration of how this concept bridges physical and Fourier spaces. Thus, multidimensional Fourier transforms are far from abstract mathematical constructs; they have direct physical interpretations, particularly in fields like turbulence analysis.

Mathematically, the multidimensional Fourier transform can be generalized to n dimensions, revealing an expansive framework that opens up new possibilities for analysis. The beauty of Fourier analysis in higher dimensions lies in its ability to maintain the elegance of its one-dimensional counterpart while simultaneously offering richer insights. For instance, grouping eddies in the flow field by their wave number effectively sorts them based on their size, regardless of their specific location in the flow field. This offers a practical method for understanding complex flow fields in a more organized manner. Imaging is another significant application of Fourier transforms in higher dimensions, particularly in two and three dimensions. Although distinctions exist between the one-dimensional case and higher-dimensional scenarios, the core concepts extend naturally. The mathematical definitions and principles developed in one dimension can be seamlessly adapted to higher dimensions, often using vector notation to preserve the simplicity and structure of the one-dimensional formulation. While general definitions are provided for n-dimensional cases, much of the focus will remain on two-dimensional scenarios, as these are the most commonly encountered in practical applications, such as in imaging and turbulence analysis.

This exploration of higher-dimensional Fourier transforms reinforces their utility and highlights the conceptual continuity that links the one-dimensional case to its multidimensional extensions. A vector on $\mathbf{R}^n$ space is expressed as

$$\mathbf{x} = (x_1, x_2, \ldots, x_n) \tag{11.1}$$

For example, the position vector of a particle in Cartesian space is expressed as $ax_1 + bx_2 + cx_3$. Cartesian space is named after French mathematician and philosopher René Descartes (1596–1650). Each vector in physical space has a corresponding wave number vector in spectral space:

$$\boldsymbol{\xi} = (\xi_1, \xi_2, \ldots, \xi_n) \tag{11.2}$$

$\mathbf{x}$ and $\boldsymbol{\xi}$ are in a reciprocal relationship. Suppose $\mathbf{x}$ represents the size in m in the Cartesian space; the corresponding wave number vector $\boldsymbol{\xi}$ has a dimension m^{-1}. Often, it is convenient to express the $\mathbf{x}$ and $\boldsymbol{\xi}$ vector in component form x_i and ξ_i. The dot product between $\mathbf{x}$ and $\boldsymbol{\xi}$ in Einstein tensor notation is expressed as

$$\mathbf{x}\boldsymbol{\xi} = (x_1\xi_1, x_2\xi_2, \ldots, x_n\xi_n) = x_i\xi_i \tag{11.3}$$

The dot product governs the geometry of $\mathbf{R}^n$. The dot product of a physical vector and its corresponding wave number vector is a dimensionless quantity. The dot product $x_i\xi_i$ combines with the complex exponential to facilitate a n-dimensional dimensionless transform kernel $e^{-ix_i\xi_i}$. So, the n-dimensional Fourier transform of a real or complex-valued function $f(\mathbf{x})$ is expressed as

$$\mathscr{F}\{f(\mathbf{x})\} = F(\boldsymbol{\xi}) = \frac{1}{(2\pi)^n}\int_{\mathbf{R}^n} e^{-ix_i\xi_i} f(\mathbf{x})d\mathbf{x} \tag{11.4}$$

And the inverse Fourier transform is expressed as

$$\mathscr{F}^{-1}\{F(\boldsymbol{\xi})\} = f(\mathbf{x}) = \int_{\mathbf{R}^n} e^{ix_i\xi_i} F(\boldsymbol{\xi})\, d\boldsymbol{\xi} \tag{11.5}$$

The exponential now incorporates the dot product of the vectors $\mathbf{x}$ and $\boldsymbol{\xi}$, which serves as the key to generalizing the definitions from one dimension to higher dimensions while maintaining the appearance of a one-dimensional case. The integration takes place over the entire space $\mathbf{R}^n$, and as an n-fold multiple integral, each x_j (or ξ_j for $\mathcal{F}^{-1}$) ranges from $-\infty$ to ∞. Ultimately, the structure of the Fourier transform and its inverse retains the same form as before. For a two-dimensional case, the forward Fourier transform becomes

$$\mathscr{F}\{f(x_1, x_2)\} = F(\xi_1, \xi_2) = \frac{1}{(2\pi)^2}\int_{-\infty}^{\infty}\int_{-\infty}^{\infty} f(x_1, x_2)e^{-i(x_1\xi_1+x_2\xi_2)}dx_1dx_2 \tag{11.6}$$

The inverse Fourier transform becomes

$$\mathscr{F}^{-1}\{F(\xi_1, \xi_2)\} = f(x_1, x_2) = \int_{-\infty}^{\infty}\int_{-\infty}^{\infty} F(\xi_1, \xi_2)\, e^{i(x_1\xi_1+x_2\xi_2)}d\xi_1 d\xi_2 \tag{11.7}$$

Like in a one-dimensional case, the above integral exists if and only if the energy content of the function $f(x_1, x_2)$ is finite, i.e., it is a rapidly decreasing function (Schwartz functions, $L1$, $L2$, etc.).

The exponential transform kernel $e^{\mp ix_i\xi_i} = 1$, when

$$x_i\xi_i = n, \quad n \text{ an integer} \tag{11.8}$$

The complex exponential has zero phase for such conditions.

11.2 Properties of Multidimensional Fourier Transform

The one-dimensional Fourier transform's familiar algebraic properties are present in the higher-dimensional setting. $f(-\mathbf{x})$ is the signal reversal for a signal $f(\mathbf{x})$. For convenience let us write $f(-\mathbf{x}) = f^-(\mathbf{x})$. $F(-\boldsymbol{\xi})$ is the reverse transfer signal for the original Fourier transfer signal $F(\boldsymbol{\xi})$. For convenience let us use the notation $F(-\boldsymbol{\xi}) = F^-(\boldsymbol{\xi})$. The Fourier integral of reversal signal $f(-\mathbf{x})$ is expressed as

$$\mathscr{F}\{f(-\mathbf{x})\} = \frac{1}{(2\pi)^n}\int_{\mathbf{R}^n} e^{-ix_i\xi_i} f(-\mathbf{x})d\mathbf{x} \tag{11.9}$$

Putting $\mathbf{y} = -\mathbf{x}$ in Eq. (11.9)

$$\begin{aligned}\mathscr{F}\{f(-\mathbf{x})\} = \mathscr{F}\{f^-(\mathbf{x})\} &= -\frac{1}{(2\pi)^n}\int_{\mathbf{R}^{n-}} e^{iy_i\xi_i} f(\mathbf{y})d\mathbf{y}\\ &= \frac{1}{(2\pi)^n}\int_{\mathbf{R}^n} e^{-iy_i(-\xi_i)} f(\mathbf{y})d\mathbf{y} = F(-\boldsymbol{\xi}) = F^-(\boldsymbol{\xi})\end{aligned} \tag{11.10}$$

$\mathbf{R}^{n-}$ represents the inverted limits of the integration. This definition makes the duality results read exactly as in the one-dimensional case, i.e., the Fourier transform of a reverse signal is the reverse of the Fourier transform of the original signal.

$$\mathscr{F}\{f(-\mathbf{x})\} = F^-(\boldsymbol{\xi}) = \frac{1}{(2\pi)^n}\int_{\mathbf{R}^n} e^{ix_i\xi_i} f(\mathbf{x})d\mathbf{x} = \frac{1}{(2\pi)^n}\mathscr{F}^{-1}\{f(\mathbf{x})\} \tag{11.11}$$

The Fourier transform of a reversed signal is $\frac{1}{(2\pi)^n}$ times its inverse Fourier transform. The symmetric behavior of the Fourier transform kernel results in this duality in higher dimensions. From Eq. (11.11)

$$\mathscr{F}\{f(\mathbf{x})\} = \frac{1}{(2\pi)^n}\mathscr{F}^{-1}\{f(-\mathbf{x})\} \tag{11.12}$$

Taking the Fourier transform of Eq. (11.12) again

$$\mathscr{F}\{\mathscr{F}\{f(\mathbf{x})\}\} = \frac{1}{(2\pi)^n} f(-\mathbf{x}) = \frac{1}{(2\pi)^n} f(-\mathbf{x}) \tag{11.13}$$

Taking the Fourier transform twice for a function equals to $\frac{1}{(2\pi)^n}$ multiplication of the reverse signal. For even functions

$$f^{-}(\mathbf{x}) = f(\mathbf{x}) \tag{11.14}$$

And for odd functions

$$f^{-}(\mathbf{x}) = -f(\mathbf{x}) \tag{11.15}$$

Geometric interpretations of odd and even functions are graphically not easy for higher dimensions.

The higher-dimension Fourier transform is also a linear operator. This means that if $f(\mathbf{x})$ and $g(\mathbf{x})$ are two functions, and a and b are constants, then,

$$\begin{aligned}\mathscr{F}\{af(\mathbf{x}) + bg(\mathbf{x})\} &= a\mathscr{F}\{f(\mathbf{x})\} + b\mathscr{F}\{g(\mathbf{x})\} \\ &= aF(\boldsymbol{\xi}) + bF(\boldsymbol{\xi})\end{aligned} \tag{11.16}$$

If $f(\mathbf{x}) \rightleftharpoons F(\boldsymbol{\xi})$ are higher-dimensional Fourier pairs, and $f(\mathbf{x})$ is translated by a vector $\mathbf{x}_0$, i.e., $f(\mathbf{x} - \mathbf{x}_0)$, then the Fourier transform of the shifted function is

$$\mathscr{F}\{f(\mathbf{x} - \mathbf{x}_0)\} = e^{-i\mathbf{x}_0\cdot\boldsymbol{\xi}}F(\boldsymbol{\xi}) \tag{11.17}$$

The Fourier transform is multiplied by a phase factor $e^{-i\mathbf{x}_0\cdot\boldsymbol{\xi}}$ by translating the original function an amount $\mathbf{x}_0$. A shift in the spatial domain introduces a phase shift in the frequency domain. When the original function is multiplied with a phase shift $e^{-i\mathbf{x}_0}$, then the Fourier transform becomes

$$\mathscr{F}\{e^{-i\mathbf{x}_0}f(\mathbf{x})\} = F(\boldsymbol{\xi} - \mathbf{x}_0) \tag{11.18}$$

If $f(\mathbf{x})$ is scaled by a factor a, i.e., $f(a\mathbf{x})$, then its Fourier transform $\mathscr{F}\{\boldsymbol{\xi}\}$ is scaled by a factor $1/a$ in the frequency domain, along with a normalization factor:

$$\mathscr{F}\{f(a\mathbf{x})\} = \frac{1}{|a|^n}F\left(\frac{\boldsymbol{\xi}}{a}\right) \tag{11.19}$$

If different components of the vector $\mathbf{x}$ are scaled with different factors, for example, in two-dimensional scaling of $\mathbf{x}$ as $a_1x_1 + a_2x_2$, it has a Fourier transform

$$\mathscr{F}\{f(a_1x_1, a_2x_2)\} = \frac{1}{|a_1||a_2|}F\left(\frac{\xi_1}{a_1}, \frac{\xi_2}{a_2}\right) \tag{11.20}$$

A second version of the stretching properties is noticed in the higher dimension, which is absent in one dimension. The stretching property of the two-dimensional $(2D)$ Fourier transform can be understood in terms of a linear change of variables in the spatial domain, which transforms the spatial coordinates in a way that scales or skews the signal or image. This linear transformation can be represented mathematically by a transformation matrix applied to the spatial coordinates, which predictably affects the frequency domain.

In two dimensions, a linear change of variables is expressed as a transformation of the spatial coordinates (x_1, x_2) to new coordinates (x_1', x_2') via a linear transformation matrix A:

$$\begin{bmatrix} x_1' \\ x_2' \end{bmatrix} = A \begin{bmatrix} x_1 \\ x_2 \end{bmatrix} = \begin{bmatrix} a_{11} & a_{12} \\ a_{21} & a_{22} \end{bmatrix} \begin{bmatrix} x_1 \\ x_2 \end{bmatrix} \tag{11.21}$$

Matrix operations $A\mathbf{x} = \mathbf{x}'$ are commonly used to solve systems of linear equations. However, there is not a single, universal way to represent every mathematical operation, as the same operation can take on different meanings depending on the context. For example, when a matrix is multiplied by a vector, it can represent a variety of transformations, not just numerical solutions. Geometrically, this operation often transforms the vector, such as rotating, stretching, or compressing it in space. The matrix defines how this transformation occurs, giving the operation a deeper significance beyond just solving equations. The matrix A typically contains important geometric information, and the specific meaning of A depends on the context of the problem. If A is performing a rotation or change of basis, the elements of A might correspond to the direction cosines between the old and new coordinate axes. In such a case, A would describe how the orientation of the vector $\mathbf{x}$ changes relative to the axes, essentially encoding the angles between directions before and after the transformation. A could store other linear transformations like scaling, reflection, or shear, in which case the matrix's entries would encode the extent of these transformations rather than just direction cosines.

Let $f(x_1, x_2)$ be a 2D function (such as an image) and applying a linear change of variables to get a new function $f(A^{-1}(x_1', x_2'))$, which represents the image stretched or transformed in the spatial domain. The Fourier transform of $f(x_1, x_2)$ is denoted by $F(\xi_1, \xi_2)$, where ξ_1 and ξ_2 are the frequency variables corresponding to x_1 and x_2. When a linear transformation is applied to the spatial coordinates, the Fourier transform is affected as follows:

$$\mathscr{F}\{f(A^{-1}(x_1', x_2'))\} = \frac{1}{|\det(A)|} F(A^{-T}(\xi_1, \xi_2)) \tag{11.22}$$

The matrix A^{-T} is the inverse transpose (equals to transpose of the inverse too) of the transformation matrix A, which defines how the frequency domain is transformed. $\det(A)$ is the determinant of the matrix A and serves as a scaling factor in the frequency domain to preserve the total energy of the signal or image.

For simple or independent stretching,

$$A = \begin{bmatrix} a_1 & 0 \\ 0 & a_2 \end{bmatrix} \tag{11.23}$$

The effect on the Fourier transform is

$$\mathscr{F}\{f(a_1x_1, a_2x_2)\} = \frac{1}{|a_1a_2|}F\left(\frac{\xi_1}{a_1}, \frac{\xi_2}{a_2}\right) \tag{11.24}$$

This represents independent scaling of the spatial dimensions, where a_1 stretches (or shrinks) along the x_1-axis and a_2 along the x_2-axis. The factor $\frac{1}{|a_1a_2|}$ ensures that the overall energy of the signal is preserved.

Stretching in the spatial domain causes shrinking in the frequency domain. Specifically, when the signal is stretched along the x_1-axis by a_1, the frequency components along the ξ_1-axis are compressed by $\dfrac{1}{a}$. The same applies to the x_2-axis and ξ_2-axis with factor a_2. Shearing or rotation in the spatial domain (when A is not diagonal) results in a more complex frequency domain transformation involving both scaling and rotation of the frequency components. The frequency components (ξ_1, ξ_2) are transformed by A^{-T}, which means that any stretching, rotation, or shearing in the spatial domain results in a corresponding inverse transformation in the frequency domain. The amplitude of the Fourier transform is scaled by $\dfrac{1}{|\det(A)|}$, which adjusts the energy to account for the area change in the spatial domain due to the transformation.

When the transformation matrix A performs a rotation, let us see how this affects the Fourier transform of a function in 2D space. The 2D rotation matrix A for an angle θ is given by

$$A = R(\theta) = \begin{bmatrix} \cos\theta & -\sin\theta \\ \sin\theta & \cos\theta \end{bmatrix} \tag{11.25}$$

This matrix rotates any vector $\mathbf{x} = [x_1, x_2]^T$ counterclockwise by an angle θ. A point (x_1', x_2') in the rotated frame corresponds to a point (x_1, x_2) in the original frame according to

$$\begin{bmatrix} x_1' \\ x_2' \end{bmatrix} = R(\theta)\begin{bmatrix} x_1 \\ x_2 \end{bmatrix} = \begin{bmatrix} \cos\theta & -\sin\theta \\ \sin\theta & \cos\theta \end{bmatrix}\begin{bmatrix} x_1 \\ x_2 \end{bmatrix} \tag{11.26}$$

Thus, the rotated coordinates (x_1', x_2') in terms of (x_1, x_2) are

$$x_1' = x_1\cos\theta + x_2\sin\theta \tag{11.27}$$

$$x_2' = -x_1\sin\theta + x_2\cos\theta \tag{11.28}$$

Therefore, the rotated function $f(R(\theta)\mathbf{x})$ becomes $f(x_1', x_2')$. The Fourier transform of the rotated function $f(R(\theta)\mathbf{x_1})$ is

$$\mathscr{F}\{f(R(\theta)\mathbf{x})\} = \frac{1}{(2\pi)^2}\int_{-\infty}^{\infty}\int_{-\infty}^{\infty} f(R(\theta)\mathbf{x})e^{-i(\xi_1x_1+\xi_2x_2)}\,dx_1\,dx_2 \tag{11.29}$$

Substituting the rotation matrix into the coordinates results

$$f(R(\theta)\mathbf{x}) = f(x_1', x_2') = f(x_1\cos\theta + x_2\sin\theta, -x_1\sin\theta + x_2\cos\theta) \tag{11.30}$$

Now substituting this to the Fourier integral in Eq. (11.29)

$$F_{\text{rot}}(\xi_1, \xi_2) = \frac{1}{(2\pi)^2}\int_{-\infty}^{\infty}\int_{-\infty}^{\infty} \Big[f(x_1\cos\theta + x_2\sin\theta, -x_1\sin\theta + x_2\cos\theta) \\ e^{-i(\xi_1 x_1 + \xi_2 x_2)}\, dx_1\, dx_2 \Big] \tag{11.31}$$

The inverse rotation gives

$$x_1 = x_1'\cos\theta - x_2'\sin\theta \tag{11.32}$$

$$x_2 = x_1'\sin\theta + x_2'\cos\theta \tag{11.33}$$

The Fourier integral after these substituting becomes

$$F_{\text{rot}}(\xi_1, \xi_2) = \frac{1}{(2\pi)^2}\int_{-\infty}^{\infty}\int_{-\infty}^{\infty} f(x_1', x_2') e^{-i(\xi_1(x_1'\cos\theta - x_2'\sin\theta) + \xi_2(x_1'\sin\theta + x_2'\cos\theta))} \\ \times dx_1'\, dx_2' \tag{11.34}$$

Simplifying the terms in the exponential

$$\begin{aligned} &\xi_1(x_1'\cos\theta - x_2'\sin\theta) + \xi_2(x_1'\sin\theta + x_2'\cos\theta) \\ &\qquad = \xi_1 x_1'\cos\theta - \xi_1 x_2'\sin\theta + \xi_2 x_1'\sin\theta + \xi_2 x_2'\cos\theta \\ &\qquad = x_1'(\xi_1\cos\theta + \xi_2\sin\theta) + x_2'(-\xi_1\sin\theta + \xi_2\cos\theta) \end{aligned} \tag{11.35}$$

After substituting this into the Fourier integral

$$\begin{aligned} F_{\text{rot}}(\xi_1, \xi_2) =& \frac{1}{(2\pi)^2}\int_{-\infty}^{\infty}\int_{-\infty}^{\infty} f(x_1', x_2') e^{-i(x_1'(\xi_1\cos\theta + \xi_2\sin\theta) + x_2'(-\xi_1\sin\theta + \xi_2\cos\theta))} \\ &\times dx_1'\, dx_2' \\ &\frac{1}{(2\pi)^2}\int_{-\infty}^{\infty}\int_{-\infty}^{\infty} f(x_1', x_2') e^{-i(x_1'\xi_1' + x_2'\xi_2')}\, dx_1'\, dx_2' \end{aligned} \tag{11.36}$$

where

$$\xi_1' = \xi_1\cos\theta + \xi_2\sin\theta \tag{11.37}$$

$$\xi_2' = -\xi_1 \sin\theta + \xi_2 \cos\theta \tag{11.38}$$

The Fourier transform of the rotated function $f(R(\theta)\mathbf{x})$ is

$$\mathscr{F}\{f(R(\theta)\mathbf{x})\} = F(\xi_1', \xi_2') = F(\xi_1 \cos\theta + \xi_2 \sin\theta, -\xi_1 \sin\theta + \xi_2 \cos\theta) \tag{11.39}$$

If a function $f(x, y)$ is rotated by an angle θ in the spatial domain, its Fourier transform is rotated by $-\theta$ in the frequency domain. There is no change in the magnitude of the Fourier transform, only in the orientation of the frequency components. This follows from the fact that the transpose and inverse of a rotation matrix are also rotations but in the opposite direction.

11.3 Multidimensional Fourier Transform of Separable Functions

Often a function $f(x_1, \ldots, x_n)$ of n variables can be written as a product of n functions of one variable, as in

$$f(x_1, x_2 \ldots, x_n) = f_1(x_1) f_2(x_2) \cdots f_n(x_n) \tag{11.40}$$

This method is called the variable separable method and is often used to solve partial differential equations (details provided in Chap. 11). When a function is factored in this way, its higher-dimensional Fourier transform is calculated as the product of the lower-dimensional Fourier transform of the factors. For a two-dimensional case,

$$\begin{aligned}
\mathscr{F}\{f(x_1, x_2)\} &= \frac{1}{(2\pi)^2} \int_{\mathbf{R}^2} e^{-ix_i\xi_i} f(\mathbf{x}) d\mathbf{x} \\
&= \frac{1}{(2\pi)^2} \int_{-\infty}^{\infty} \int_{-\infty}^{\infty} e^{-i(x_1\xi_1 + x_2\xi_2)} f(x_1, x_2)\, dx_1 dx_2 \\
&= \int_{-\infty}^{\infty} \int_{-\infty}^{\infty} e^{-i\xi_1 x_1} e^{-i\xi_2 x_2} f_1(x_1) f_2(x_2)\, dx_1 dx_2 \\
&= \int_{-\infty}^{\infty} \left(\int_{-\infty}^{\infty} e^{-i\xi_1 x_1} f_1(x) dx_1 \right) e^{-i\xi_2 x_2} f_2(x_2)\, dx_2 \\
&= \mathscr{F}\{f_1\}(x_1) \int_{-\infty}^{\infty} e^{-i\xi_2 x_2} f_2(x_2)\, dx_2 \\
&= \mathscr{F}\{f_1(x_1)\} \mathscr{F}\{f_2(x_2)\} \\
&= F(\xi_1) F(\xi_2)
\end{aligned} \tag{11.41}$$

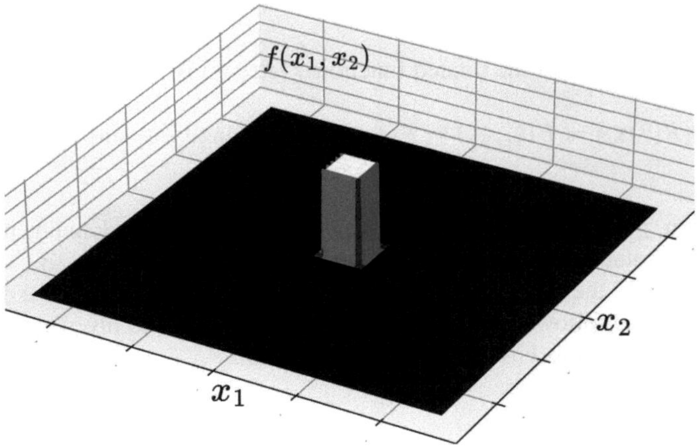

Fig. 11.1 Two-dimensional rectangular pulse function

In general,

$$\mathscr{F}\{f(\xi_1, x_2, \ldots \xi_n)\} = \mathscr{F}\{f_1(x_1)\}\mathscr{F}\{f_2(x_2)\}\cdots\mathscr{F}\{f_n(x_n)\} \\ = F(\xi_1)F(\xi_2)\cdots F(\xi_n) \tag{11.42}$$

Not all functions fit into the variable separable method. The higher-dimensional rectangular pulse function is the simplest function that fits into Eq. (11.40)'s description. Figure 11.1 shows a two-dimensional rectangular pulse centered at the origin with a square of unitary side length and value zero outside this square.

$$\Pi(x_1, x_2) = \begin{cases} 1 & -\frac{1}{2} < x_1 < \frac{1}{2}, -\frac{1}{2} < x_2 < \frac{1}{2} \\ 0 & \text{otherwise} \end{cases} \tag{11.43}$$

The two-dimensional rectangular pulse can be reduced to a variable separable form as

$$\Pi(x_1, x_2) = \Pi(x_1)\,\Pi(x_2) \tag{11.44}$$

The Fourier transform of the two-dimensional rectangular pulse can be calculated from the product of two one-dimensional pulses:

$$\mathscr{F}\{\Pi(\xi_1, \xi_2)\} = \mathscr{F}\{\Pi(\xi_1)\}\mathscr{F}\{\Pi(\xi_2)\} \\ \frac{1}{(2\pi)^2} sinc\left(\frac{\xi_1}{2}\right) sinc\left(\frac{\xi_2}{2}\right) \tag{11.45}$$

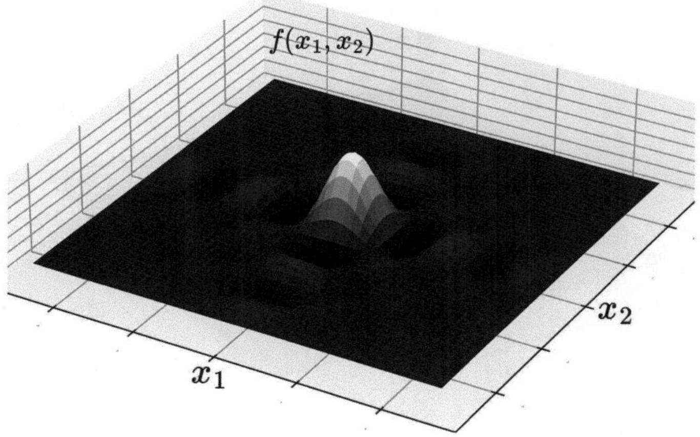

Fig. 11.2 Two-dimensional *sinc* function

Figure 11.2 shows the graphical representation of the dimensional *sinc* function. The rectangular pulse can be stretched or shrunk in either direction, having value one in $-a_1/2 < x_1 < a_1/2$, $-a_2/2 < x_2 < a_2/2$ and zero outside that rectangle.

$$\Pi_{a_1 a_2}(x_1, x_2) = \Pi_{a_1}(x_1)\,\Pi_{a_2}(x_2) \tag{11.46}$$

The Fourier transform of this is

$$\mathscr{F}\{\Pi_{a_1 a_2}(x_1, x_2)\} = \frac{1}{4a_1 a_2 \pi^2} sinc\left(\frac{\xi_1}{2a_1}\right) sinc\left(\frac{\xi_1}{2a_2}\right) \tag{11.47}$$

Figure 11.3 shows the two-dimensional *sinc* functions for $a_2 = 2a_1$. For n-dimensional Π

$$\Pi(x_1, x_2, \ldots, x_n) = \begin{cases} 1 & -\frac{1}{2} < x_k < \frac{1}{2}, \quad k = 1, \ldots, n \\ 0 & \text{otherwise} \end{cases} \tag{11.48}$$

which factors as

$$\Pi(x_1, x_2, \ldots, x_n) = \Pi(x_1)\,\Pi(x_2) \cdots \Pi(x_n) \tag{11.49}$$

The Fourier transform of the n-dimensional Π is

$$\mathscr{F}\{\Pi(x_1, x_2, \ldots, x_n)\} = \frac{1}{(2\pi)^n} sinc\left(\frac{\xi_1}{2}\right) sinc\left(\frac{\xi_2}{2}\right) \cdots sinc\left(\frac{\xi_n}{2}\right) \tag{11.50}$$

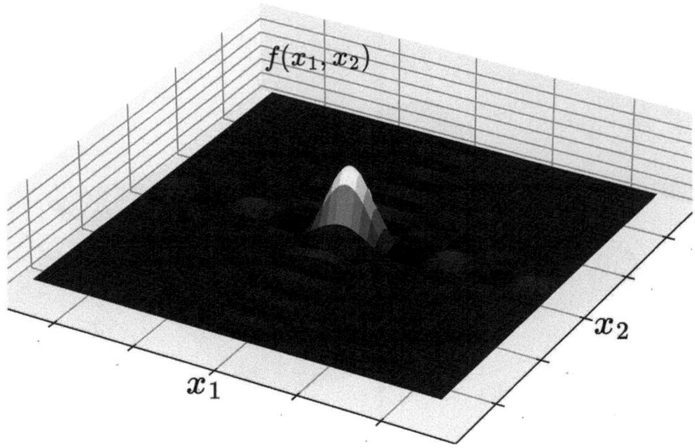

Fig. 11.3 Two-dimensional *sinc* function for $a_2 = 2a_1$

Gaussian is another class of separable function. By analogy to the one-dimensional case, the most natural Gaussian to use in connection with Fourier transforms is

$$g(\mathbf{x}) = e^{-\pi|\mathbf{x}|^2} = e^{-\pi\left(x_1^2+x_2^2+\cdots+x_n^2\right)} \tag{11.51}$$

This factors as a product of n one-variable Gaussians:

$$g\left(x_1, \ldots, x_n\right) = e^{-\pi\left(x_1^2+x_2^2+\cdots+x_n^2\right)} = e^{-\pi x_1^2} e^{-\pi x_2^2} \ldots e^{-\pi x_n^2} \tag{11.52}$$

Taking the Fourier transform of Eq. (11.52)

$$\begin{aligned}
\mathscr{F}\{g(\mathbf{x})\} &= \frac{1}{(2\pi)^n} e^{-\frac{\xi_1^2}{4\pi}} e^{-\frac{\xi_2^2}{4\pi}} \ldots e^{-\frac{\xi_n^2}{4\pi}} \\
&= \frac{1}{(2\pi)^n} e^{-\frac{\xi_1^2+\xi_2^2+\cdots+\xi_n^2}{4\pi}} \\
&= \frac{1}{(2\pi)^n} e^{-\xi^2/4\pi}
\end{aligned} \tag{11.53}$$

Figure 11.4 shows the two-dimensional Gaussian function.

11.4 Radial Functions in 2D and Their Fourier Transforms

One additional aspect of the two-dimensional case is considered for use in various applications. A function on $\mathbf{R}^2$ is referred to as radial (or radially symmetric/-

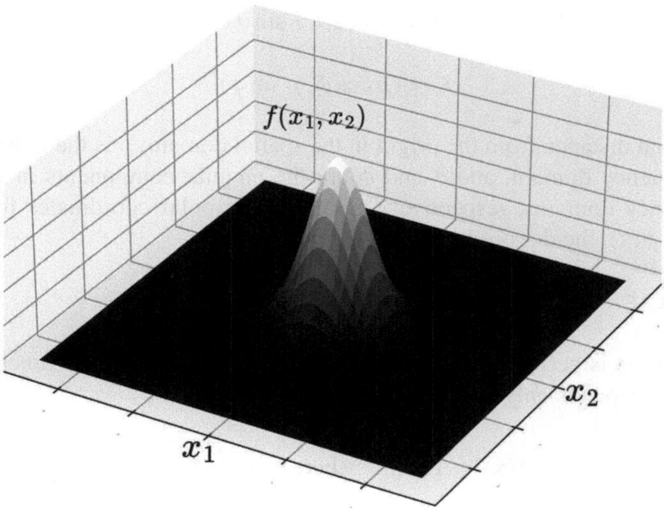

Fig. 11.4 Two-dimensional Gaussian function

circularly symmetric) if it depends solely on the distance from the origin. In polar coordinates, this distance is denoted by r, meaning that a radial function depends only on r and not on θ (with the usual polar coordinates written as (r, θ)). Radial functions are often written in the form $f(\mathbf{x}) = f(r)$, where $r = \|\mathbf{x}\| = \sqrt{x_1^2 + x_2^2}$ is the radial distance from the origin. Since the Fourier transform is defined in Cartesian coordinates, expressing it in polar coordinates is more effective when dealing with radial functions. However, doing so requires introducing some special functions to achieve a compact expression form, making the process less straightforward. The Fourier transform of a radial function $f(r)$ in two dimensions also exhibits radial symmetry. The goal is to derive and describe the form of the Fourier transform of such a function.

For a general function $f(\mathbf{x}) = f(x_1, x_2)$, the two-dimensional Fourier transform is given by

$$\begin{aligned} F(\boldsymbol{\xi}) &= \int_{\mathbb{R}^2} f(\mathbf{x}) e^{-2\pi i \mathbf{x}\cdot\boldsymbol{\xi}}\, d\mathbf{x} \\ &= \int_{-\infty}^{\infty}\int_{-\infty}^{\infty} f(x_1, x_2) e^{-2\pi i(x_1\xi_1 + x_2\xi_2)} dx_1 dx_2 \end{aligned} \tag{11.54}$$

Here, $\boldsymbol{\xi}$ denotes the frequency rather than the angular frequency, which explains the disappearance of the factor $\dfrac{1}{(2\pi)^2}$ in the forward transform and the presence of 2π in the exponential kernel. In polar coordinates, $\mathbf{x}$ and $\boldsymbol{\xi}$ can be expressed as

$$\mathbf{x} = (r\cos\theta, r\sin\theta) \tag{11.55}$$

$$\boldsymbol{\xi} = (\rho\cos\phi, \rho\sin\phi) \tag{11.56}$$

r is the radial distance from the origin in the spatial domain, ρ is the radial distance in the frequency domain, and θ and ϕ are the angular components in the spatial and frequency domains, respectively. Switching to polar coordinates for radially symmetric $f(\mathbf{x})$, the dot product $\mathbf{x} \cdot \boldsymbol{\xi}$ becomes

$$\mathbf{x} \cdot \boldsymbol{\xi} = r\rho(\cos\theta\cos\phi + \sin\theta\sin\phi) = r\rho\cos(\theta - \phi) \tag{11.57}$$

The integration is performed over r and θ for a radially symmetric function. Since the function depends only on r,

$$f(\mathbf{x}) = f(r), \quad \text{where } r = \sqrt{x_1^2 + x_2^2} \tag{11.58}$$

Thus, the $2D$ Fourier transform becomes

$$\int_{-\infty}^{\infty}\int_{-\infty}^{\infty} f(x_1, x_2)e^{-2\pi i(x_1\xi_1 + x_2\xi_2)}dx_1dx_2 = \int_{\mathbb{R}^2} f(r)e^{-2\pi i r\rho\cos(\theta-\phi)}\, r\, dr\, d\theta \tag{11.59}$$

Due to the radial symmetry, the Fourier transform only depends on the radial frequency ρ, and the angle ϕ can be set to zero (since the function is invariant under rotations). This simplifies the integral

$$F(\rho) = \int_0^{\infty}\int_0^{2\pi} f(r)e^{-2\pi i r\rho\cos\theta} r\, d\theta\, dr \tag{11.60}$$

The integral over θ is the standard integral of a complex exponential over a circle. This integral does not have an antiderivative and is expressed as the zero-order Bessel function of the first kind $J_0(2\pi r\rho)$:

$$\int_0^{2\pi} e^{-2\pi i r\rho\cos\theta}\, d\theta = 2\pi J_0(2\pi r\rho) \tag{11.61}$$

The Bessel function is named after German astronomer and mathematician Friedrich Wilhelm Bessel (1784–1846). The $2D$ Fourier transform becomes

$$F(\rho) = 2\pi\int_0^{\infty} f(r) r J_0(2\pi r\rho)\, dr \tag{11.62}$$

This is the general formula for the Fourier transform of a radially symmetric function in two dimensions. The inverse Fourier transform for a radial function is similarly given by

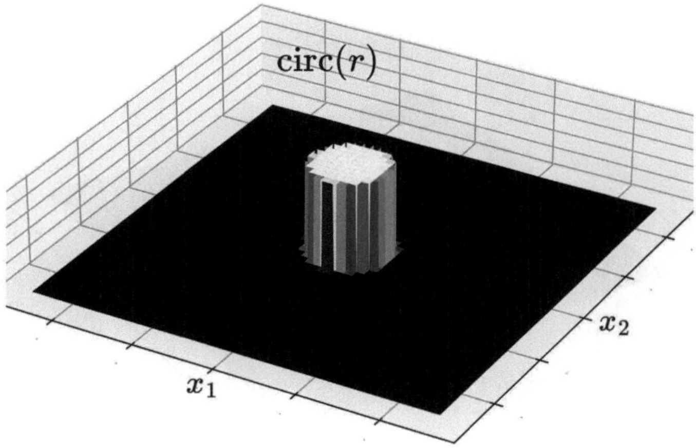

Fig. 11.5 Two-dimensional circ function

$$f(r) = 2\pi \int_0^\infty F(\rho)\rho J_0(2\pi r\rho)\, d\rho \tag{11.63}$$

If the Fourier transform $F(\rho)$ is known, the original radial function can be retrieved through a similar integral that includes the Bessel function J_0. The expression for $F(\rho)$ in terms of $f(r)$ is referred to as the zero-order Hankel transform of $f(r)$, but it essentially represents the Fourier transform of a radial function. The Hankel transform is named after German mathematician Hermann Hankel (1839–1873).

The circ is a useful radial function analogue to the rectangle pulse function and is defined as

$$\text{circ}(r) = \begin{cases} 1 & r < 1 \\ 0 & r \geq 1 \end{cases} \tag{11.64}$$

Figure 11.5 shows the graph of circ(r). The Fourier transform integration limits for r go only from 0 to 1:

$$\mathscr{F}\{\text{circ}(r)\} = 2\pi \int_0^1 J_0(2\pi r\rho) r dr \tag{11.65}$$

With the change of variable, $u = 2\pi r\rho$, and $du = 2\pi\rho dr$. The limits of integration go from $u = 0$ to $u = 2\pi\rho$. The integral becomes

$$\mathscr{F}\{\text{circ}(r)\} = \frac{1}{2\pi\rho^2} \int_0^{2\pi\rho} u J_0(u) du \tag{11.66}$$

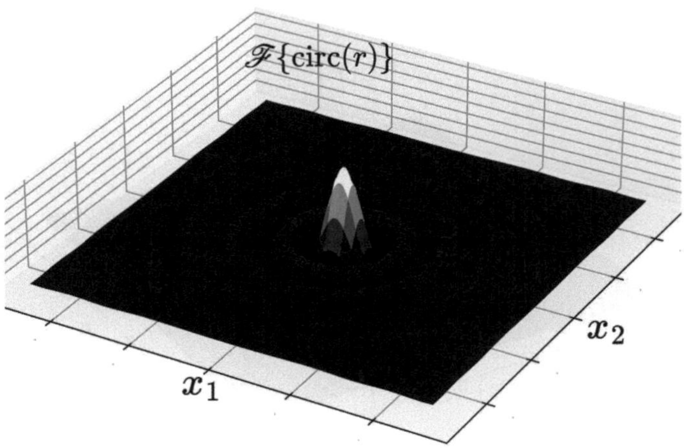

Fig. 11.6 Two-dimensional $\mathscr{F}\{\mathrm{circ}(r)\}$ function

The integral has been written in this way due to the introduction of an identity involving the first-order Bessel function of the first kind, which is expressed as follows:

$$\int_0^x u J_0(u)du = x J_1(x) \tag{11.67}$$

In terms of J_1, the Fourier integral is written as

$$\mathscr{F}\{\mathrm{circ}(r)\} = \frac{J_1(2\pi\rho)}{\rho} \tag{11.68}$$

The $\mathrm{jinc}(\rho)$ function is introduced as

$$\mathrm{jinc}(\rho) = \frac{J_1(\pi\rho)}{2\rho} \tag{11.69}$$

The Fourier integral in terms of $\mathrm{jinc}(\rho)$ function is expressed as

$$\mathscr{F}\{\mathrm{circ}(r)\} = 4\,\mathrm{jinc}(2\rho) \tag{11.70}$$

Figure 11.6 presents the graph of $\mathscr{F}\{\mathrm{circ}(r)\}$. The $\mathrm{jinc}(\rho)$ function looks like a radially symmetric version of the *sinc* function. There is a symmetrization process at work involving repeated convolutions (Fig. 11.6).

A $2D$ Gaussian function $f(r) = e^{-\pi r^2}$ shows the radial symmetry, having the Fourier transform as another Gaussian function $F(\rho) = \frac{1}{2\pi} e^{-\frac{\rho^2}{4\pi}}$.

11.5 Higher-Dimensional Impulse Response

A two-dimensional linear system maps an input $\{f(x_1, x_2); x_1, x_2 \in \mathbf{R}\}$ into an output $\{g(x_1, x_2); x_1, x_2 \in \mathbf{R}\}$ and characterized by its impulse response $h(x_1, x_2; \zeta_1, \zeta_2)$ to an input δ function at coordinates (ζ_1, ζ_2). The unit impulse in two dimensions is represented by the $\delta(x_1, x_2)$ symbol:

$$\delta(x_1, x_2) = \delta(x_1)\delta(x_2) \tag{11.71}$$

$\delta(x_1)$ and $\delta(x_2)$ are the standard one-dimensional Dirac delta functions in the x_1 and x_2 directions, respectively (separable form). The two-dimensional delta function has the following key properties:

$$\int_{-\infty}^{\infty}\int_{-\infty}^{\infty} \delta(x_1, x_2)\, dx_1\, dx_2 = 1 \tag{11.72}$$

A two-dimensional Dirac delta function infinitely peaked at the origin $(x_1, x_2) = (0, 0)$ and zero everywhere else, but its total integral over the entire plane is 1. Analogous to the 1D case, the sifting property of the Dirac delta function allows one to pick out the value of a function at a particular point. For a continuous function $f(x, y)$, the two-dimensional Dirac delta function satisfies

$$\int_{-\infty}^{\infty}\int_{-\infty}^{\infty} f(\zeta_1, \zeta_2)\delta(\zeta_1 - x_1, \zeta_2 - x_2) d\zeta d\eta = f(x_1, x_2) \tag{11.73}$$

A 2D Dirac delta function samples the value of $f(\zeta_1, \zeta_2)$ at the point (x_1, x_2). A 2D unit impulse can be defined through a circularly symmetric form and could be defined through a limiting sequence of $circ$ functions:

$$\delta(x_1, x_2) = \lim_{n \to \infty} \frac{n}{\pi} \operatorname{circ}\left(\sqrt{(nx_1)^2 + (nx_2)^2}\right) \tag{11.74}$$

The factor $\dfrac{1}{\pi}$ is used to ensure unit volume. Naturally, the equation above should not be interpreted literally, as the limit does not exist when $x_1 = x_2 = 0$, but equality between the left and right sides will hold if the limit is applied to integrals of the right-hand side.

In two dimensions, the unit impulse response of a linear system is often called the point-spread function. By direct analogy with the 11D case

$$g(x_1, x_2) = \int_{-\infty}^{\infty}\int_{-\infty}^{\infty} h(x_1, x_2; \zeta_1, \zeta_2) f(\zeta_1, \zeta_2)\, d\zeta_1 d\zeta_2 \tag{11.75}$$

A $2D$ superposition integral. The impulse response shifts with the movements of the impulse and does not change its shape. It is space invariant, and

$$h(x_1, x_2; \zeta_1, \zeta_2) = h(x_1 - \zeta_1, x_2 - \zeta_2) \tag{11.76}$$

and

$$g(x_1, x_2) = \int_{-\infty}^{\infty} \int_{-\infty}^{\infty} h(x_1 - \zeta_1, x_2 - \zeta_2) f(\zeta_1, \zeta_2)\, d\zeta_1 d\zeta_2 \tag{11.77}$$

Equation (11.77) is nothing but a 2D convolution integral. A space-invariant 2D filter is a shift-invariant system where the shift is in space instead of time. Taking the Fourier transform of Eq. (11.77),

$$G(\xi_1, \xi_2) = H(\xi_1, \xi_2) F(\xi_1, \xi_2) \tag{11.78}$$

$H(\xi_1, \xi_2)$ is the $2D$ transfer function.

A multidimensional Schwartz function can be described as follows: In the theory of tempered distributions, the concept of Schwartz functions serves as the foundation, functioning as the class of test functions. This framework operates similarly in higher dimensions as in the one-dimensional case. The function and all of its partial derivatives (including mixed partial derivatives) decrease more rapidly than any power of the coordinates as the coordinates tend to infinity to define a Schwartz function in several variables. This ensures that the function exhibits rapid decay. The condition of the rapid decrease can be expressed in multiple equivalent forms, including stipulating that certain criteria related to the function's behavior and its derivatives be satisfied.

$$|\mathbf{x}|^p \left|\partial^q \varphi(\mathbf{x})\right| \to 0 \quad \text{as} \quad |\mathbf{x}| \to \infty \tag{11.79}$$

p is a positive integer that just gives a power of $|\mathbf{x}|$, and q is a multi-index. This means that $q = (q_1, \ldots, q_n)$, each q_i a positive integer, so that ∂^q is supposed to mean

$$\frac{\partial^{q_1+\cdots+q_n}}{(\partial x_1)^{q_1} (\partial x_2)^{q_2} \cdots (\partial x_n)^{q_n}} \tag{11.80}$$

The higher-dimension δ function is the distribution defined by the pairing

$$\langle \delta, \varphi \rangle = \varphi(0, \ldots, 0) \quad \text{or} \quad \langle \delta, \varphi \rangle = \varphi(\mathbf{0}) \quad \text{in vector notation} \tag{11.81}$$

$\varphi(x_1, , \ldots, x_n)$ is a Schwartz function. The pairing in terms of integration is expressed as

$$\int_{\mathbf{R}^n} \varphi(\mathbf{x}) \delta(\mathbf{x}) d\mathbf{x} = \varphi(\mathbf{0}) \tag{11.82}$$

As before,

$$f(\mathbf{x})\delta(\mathbf{x}) = f(\mathbf{0})\delta(\mathbf{x}) \tag{11.83}$$

If f is a smooth function, then from convolution

$$(f * \delta)(\mathbf{x}) = f(\mathbf{x}) \tag{11.84}$$

The shifted delta function $\delta(\mathbf{x}-\mathbf{b}) = \delta(x_1 - b_1, x_2 - b_2, \ldots, x_n - b_n)$ or $\delta_{\mathbf{b}} = \tau_{\mathbf{b}}\delta$ has the corresponding properties:

$$f(\mathbf{x})\delta(\mathbf{x}-\mathbf{b}) = f(\mathbf{b})\delta(\mathbf{x}-\mathbf{b}) \quad \text{and} \quad f * \delta(\mathbf{x}-\mathbf{b}) = f(\mathbf{x}-\mathbf{b}) \tag{11.85}$$

In some cases, it is useful to know that the delta function can be factorized into one-dimensional deltas, as in

$$\delta(x_1, x_2, \ldots, x_n) = \delta_1(x_1)\delta_2(x_2)\cdots\delta_n(x_n) \tag{11.86}$$

Subscripts have been added to the δs on the right-hand side to label them according to the individual coordinates, offering certain advantages. Although, as a general principle, the multiplication of distributions is neither defined nor possible, this is an instance where such an operation is meaningful. The formula is valid because each side operates on a Schwartz function. This can be verified by considering the two-dimensional case, where the pairing is loosely represented as an integral. Then, on one side, the following result will be obtained:

$$\int_{\mathbf{R}^2} \varphi(\mathbf{x})\delta(\mathbf{x})d\mathbf{x} = \varphi(0, 0) \tag{11.87}$$

by definition of the 2-dimensional delta function. On the other hand,

$$\begin{aligned} &\int_{\mathbf{R}^2} \varphi(x_1, x_2)\delta_1(x_1)\delta_2(x_2)\,dx_1dx_2 \\ &= \int_{-\infty}^{\infty}\left(\int_{-\infty}^{\infty} \varphi(x_1, x_2)\delta_1(x_1)\,dx_1\right)\delta_2(x_2)\,dx_2 \\ &= \int_{-\infty}^{\infty} \varphi(0, x_2)\delta_2(x_2)\,dx_2 = \varphi(0, 0) \end{aligned} \tag{11.88}$$

So $\delta(x_1, x_2)$ and $\delta_1(x_1)\delta_2(x_2)$ have the same effect when integrated against a test function. The Fourier transform of a n-dimensional delta function is $\dfrac{1}{(2\pi)^n}$. The two-dimensional scaling of the delta function is expressed as

$$\begin{aligned}\delta\left(a_1x_1, a_2x_2\right) &= \delta_1\left(a_1x_1\right)\delta_2\left(a_2x_2\right)\\ &= \frac{1}{|a_1|}\delta_1\left(x_1\right)\frac{1}{|a_2|}\delta_2\left(x_2\right)\\ &= \frac{1}{|a_1|\,|a_2|}\delta_1\left(x_1\right)\delta_2\left(x_2\right) = \frac{1}{|a_1a_2|}\delta\left(x_1, x_2\right)\end{aligned} \tag{11.89}$$

and for n-dimensions

$$\delta\left(a_1x_1, \ldots, a_nx_n\right) = \frac{1}{|a_1\cdots a_n|}\delta\left(x_1, \ldots, x_n\right) \tag{11.90}$$

11.6 Higher-Dimensional Sampling

The two-dimensional Dirac comb, also known as the Sha function in $2D$, is an extension of the $1D$ Dirac comb into two dimensions (Fig. 11.6). It consists of an infinite sum of $2D$ Dirac delta functions regularly spaced on a grid. Mathematically, it is represented as

$$\text{III}_2(\mathbf{x}) = \sum_{m_1=-\infty}^{\infty}\sum_{m_2=-\infty}^{\infty}\delta(x_1 - m_1p_1, x_2 - m_2p_2), \tag{11.91}$$

$\mathbf{x} = (x_1, x_2)$ is the position vector in $2D$ space. p_1 and p_2 are the periodic spacing in the x_1- and x_2-directions, respectively. $\delta(x_1 - m_1p_1, x_2 - m_2p_2)$ is the $2D$ Dirac delta function centered at (m_1p_1, m_2p_2). This function creates a grid of delta spikes at regular intervals of p_1 and p_2, effectively serving as a $2D$ sampling function as shown in Fig. 11.7.

The $2D$ Fourier transform of the Dirac comb is another $2D$ Dirac comb in the frequency domain.

$$\mathscr{F}\{\text{III}_2(\mathbf{x})\} = \frac{1}{(2\pi)^2}\frac{1}{p_1p_2}\text{III}_2\left(\frac{n_1}{p_1}, \frac{n_2}{p_2}\right) \tag{11.92}$$

The resulting Dirac comb in the frequency domain has a period of $1/p_1$ and $1/p_2$ in the respective directions. n_1 and n_2 are the indices of the frequency summation.

A bandlimited $1D$ function defined over a single independent variable can be accurately reconstructed from samples taken at a uniform rate exceeding a critical minimum. Similarly, a two-dimensional (2D) function that is bandlimited in the (x_1, x_2) plane can also be perfectly reconstructed from uniformly spaced samples as long as the sampling density surpasses a certain lower threshold. However, the $2D$ sampling theory is notably more intricate than its $1D$ counterpart. This increased complexity arises from the additional degree of freedom introduced by the second

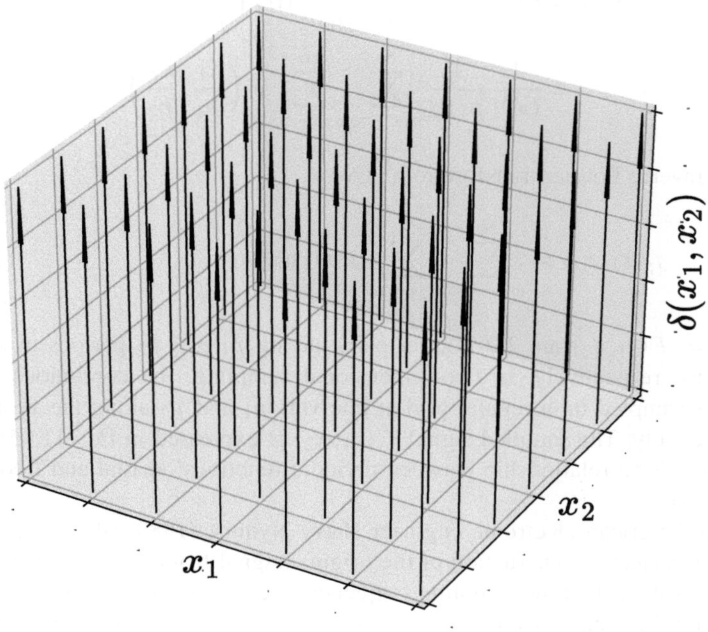

Fig. 11.7 Two-dimensional Dirac comb

dimension, allowing for more decadent behavior and new possibilities in signal reconstruction.

In $2D$, sampling refers to extracting continuous signal values at discrete intervals in both dimensions. This is effectively done by multiplying the continuous function $f(x_1, x_2)$ by the $2D$ Dirac comb $\text{III}_2(\mathbf{x})$. If $f(x_1, x_2)$ is the continuous function, the sampled function $f_s(x_1, x_2)$ is given by

$$\begin{aligned} f_s(x_1, x_2) &= f(x_1, x_2) \cdot \text{III}_2(x_1, x_2) \\ &= f(x_1, x_2) \sum_{m_1=-\infty}^{\infty} \sum_{n_1=-\infty}^{\infty} \delta(x_1 - m_1 p_1, x_2 - n_1 p_2) \\ &= \sum_{m_1=-\infty}^{\infty} \sum_{m_2=-\infty}^{\infty} f(m_1 p_1, m_2 p_2) \delta(x_1 - m_1 p_1, x_2 - m_2 p_2) \end{aligned} \tag{11.93}$$

This equation shows that the original continuous function $f(x_1, x_2)$ is sampled at the points $(m_1 p_1, m_2 p_2)$, and only the sampled values $f(m_1 p_1, m_2 p_2)$ are retained. Taking the Fourier transform of both sides in Eq. (11.93)

$$
\begin{aligned}
\mathscr{F}\{f_s(x_1, x_2)\} &= F(n_1, n_2) * \frac{1}{(\pi^2)} \frac{1}{p_1 p_2} \text{III}_2 \left(\frac{n_1}{p_1}, \frac{n_2}{p_2} \right) \\
&= \frac{1}{(\pi^2)} \frac{1}{p_1 p_2} \sum_{n_1=-\infty}^{\infty} \sum_{n_2=-\infty}^{\infty} F\left(\frac{k_1}{p_1}, \frac{k_2}{p_2} \right)
\end{aligned}
\tag{11.94}
$$

Taking the inverse Fourier transform of (11.94)

$$
f_s(x_1, x_2) = \sum_{m_1=-\infty}^{\infty} \sum_{m_2=-\infty}^{\infty} f(m_1 p_1, m_2 p_2) \tag{11.95}
$$

$f(x_1, x_2) \rightleftharpoons F(\xi_1, \xi_2)$ are $2D$ Fourier pairs having p_1 and p_2 periods in x_1 and x_2 directions, respectively. In two-dimensional sampling, the continuous signal $f(x_1, x_2)$ is sampled on a regular grid at intervals p_1 and p_2 along the x_1 and x_2 axes, respectively. The sampled signal $f_s(x_1, x_2)$ is expressed in Eq. (11.95). This formula provides a relationship between periodic functions' spatial and frequency representations.

Swedish-American electronic engineer Harry Nyquist (1889–1976) specified a criterion for perfect reconstruction of the original signal from its discrete samples. This criterion states that the sampling frequency in each spatial dimension must be at least twice the highest frequency present in the signal along that dimension. If $f_{\max,x_1}$ and $f_{\max,x_2}$ represent the maximum frequencies (band limits) of the signal $f(x_1, x_2)$ in the x_1 and x_2 directions. The Nyquist criterion requires the sampling intervals p_1 and p_2 to satisfy

$$
p_1 \leq \frac{1}{2 f_{\max,x_1}}, \quad p_2 \leq \frac{1}{2 f_{\max,x_2}} \tag{11.96}
$$

Alternatively, the sampling frequencies $f_s^{x_1}$ and $f_s^{x_2}$, which are the reciprocals of the sampling intervals, must satisfy

$$
f_s^{x_1} \geq 2 f_{\max,x_1}, \quad f_s^{x_2} \geq 2 f_{\max,x_2} \tag{11.97}
$$

Aliasing occurs when the sampling frequency is lower than twice the highest-frequency component in the signal. In $2D$, aliasing happens when the signal is under-sampled in either spatial dimension. In the $2D$ Fourier domain, the sampled signal's Fourier transform becomes periodic due to the repetition of the frequency content, as shown in Eq. (11.94). When under-sampling occurs, frequency components of the original signal overlap in the frequency domain, making it impossible to distinguish the original frequencies from the aliased ones. The result is distortion in the reconstructed signal. Nyquist criterion ensures that the original signal can be perfectly reconstructed without aliasing. Often, low-pass filtering operations are done with the sampled signal to remove high-frequency components; this helps in the accurate reconstruction of the continuous signal.

11.7 Closure

The higher-dimensional Fourier transform is a robust framework for analyzing functions and signals in two or more dimensions, facilitating the transition between the spatial and frequency domains. This transformation is crucial for understanding how complex multidimensional data can be represented in terms of its frequency content. In this chapter, the theoretical foundations of the higher-dimensional Fourier transform were explored, with familiar concepts from one-dimensional Fourier analysis being extended to multiple dimensions. The similarities and differences between the properties of the higher-dimensional Fourier transform and its $1D$ counterpart were examined, with particular attention given to how variable stretching in multiple dimensions requires specialized treatment.

When separable functions are addressed, a significant simplification in multidimensional Fourier analysis is noted. These functions, expressed as products of single-variable functions, allow complex multidimensional transforms to be reduced to more straightforward, one-dimensional transforms. This simplification enhances mathematical clarity and makes computations more efficient, enabling practical applications across various fields.

The Fourier analysis of radial functions is also covered, as these functions are frequently encountered in problems with circular or spherical symmetry. These functions exhibit unique behavior in the frequency domain, with their Fourier transforms simplifying into forms involving special functions such as Bessel functions, reflecting the intrinsic symmetry of the underlying problem. This is particularly useful in areas such as wave propagation and optics.

The multidimensional Dirac delta function is examined, as it generalizes the concept of an impulse or point source to multiple dimensions. This extension is deemed crucial for understanding system responses to localized inputs and singular events. It provides a mathematical foundation for analyzing how systems react to impulses or concentrated forces in space. It is considered a building block for more complex system analysis in physics and engineering. Another key aspect discussed is higher-dimensional sampling, which is essential for working with discrete data such as images, volumes, or multichannel signals. Proper sampling in higher dimensions ensures that signals can be accurately reconstructed without errors such as aliasing, as is also the case in $1D$.

The higher-dimensional Fourier transform is the backbone of advanced medical imaging techniques such as MRI and CT scans. Billion-dollar medical imaging industries depend on the Fourier transform's mathematical principles, including specialized methods like the Radon transform, to reconstruct internal body images from projection data. While these applications are both fascinating and impressive, a detailed exploration of such specialized topics is considered beyond the scope of this textbook, which is aimed at beginners. Readers interested in a more in-depth examination of the use of Fourier transforms in medical imaging are encouraged to refer to Kevles (1997) for a comprehensive discussion. The transformative power of Fourier analysis in higher dimensions extends far beyond what can be covered in a single chapter, influencing cutting-edge research and technology in fields as diverse as medical imaging, signal processing, and physics.

Reference

Kevles, B. (1997). *Naked to the bone: Medical imaging in the twentieth century*. Rutgers University Press.

Chapter 12
Fourier Analysis for the Solution of Differential and Integral Equations

We are servants rather than masters in mathematics.
—- Charles Hermite

Abstract This chapter covers the application of Fourier analysis in solving differential and integral equations, emphasizing its vital role in mathematical physics and engineering. The chapter systematically demonstrates how Fourier transforms can effectively solve ordinary differential equations (ODEs) through various methods and examples. Building on this foundation, the discussion transitions to integral equations, where Fourier transforms offer powerful techniques for tackling complex integral problems. The chapter further addresses partial differential equations (PDEs), introducing essential concepts such as the principle of superposition and the variable separation method, which are crucial for effective PDE solutions. Specific examples are presented, including detailed analyses of the one-dimensional and two-dimensional heat equations and the one-dimensional wave equation. The exploration of the two-dimensional Laplace equation with boundary conditions highlights the practical applications of Fourier analysis in addressing real-world issues. Finally, the chapter discusses the classification of partial differential equations, providing insights into their diverse nature and solutions. Through this thorough exploration, the chapter reinforces the significance of Fourier analysis in enhancing the understanding and resolution of differential and integral equations.

Keywords Ordinary differential equations · ODE · Partial differential equations · PDE · Integral equations · Bessel function · Fredholm integral equation · Principle of superposition · Heat equation · Wave equation · Elliptic PDE · Parabolic PDE · Hyperbolic PDE · Characteristic directions

12.1 Introduction

The Fourier transform has profoundly influenced the analysis and solution of ordinary differential equations (ODEs), partial differential equations (PDEs), and

S. K. Jena, *Fourier, Laplace, and the Tangled Love Affair with Transforms*,
https://doi.org/10.1007/978-3-031-80165-5_12

integral equations. In the nineteenth century, Jean-Baptiste Joseph Fourier introduced this concept to solve the heat conduction problem. Fourier aimed to solve the heat equation, a PDE describing heat distribution in a given region over time. His insight was that complex, periodic functions could be decomposed into simpler, sinusoidal components, making the heat equation more tractable. This outstanding concept leads to the development of the Fourier series and, later, the Fourier transform. The forward Fourier transform is defined as

$$\mathscr{F}\{f(x)\} = \frac{1}{2\pi}\int_{-\infty}^{\infty} e^{-i\omega x} f(x)dx = F(\omega) \tag{12.1}$$

and the inverse Fourier transform is defined as

$$\mathscr{F}^{-1}\{F(\omega)\} = \int_{-\infty}^{\infty} e^{i\omega x} F(\omega)d\omega = f(x) \tag{12.2}$$

The derivative of the Fourier transform is a concept that plays a crucial role in signal processing, physics, and various branches of engineering. Understanding how differentiation interacts with the Fourier transform allows for analyzing how signals behave in the frequency domain, particularly how different operations in the time domain affect their spectral characteristics.

If $f(x)$ is continuously differentiable and $f(x) \to 0$ as $|x| \to \infty$ (rapidly decreasing smooth function), then the Fourier transform of its derivative is expressed as

$$\mathscr{F}\left\{f'(x)\right\} = \frac{1}{2\pi}\int_{-\infty}^{\infty} e^{-i\omega x} f'(x)dx \tag{12.3}$$

Integrating equation (12.3), by parts, results in

$$\begin{aligned}\mathscr{F}\left\{f'(x)\right\} &= \frac{1}{2\pi}\int_{-\infty}^{\infty} e^{-i\omega x} f'(x)dx \\ &= \frac{1}{2\pi}\left[f(x)e^{-i\omega x}\right]_{-\infty}^{\infty} + \frac{i\omega}{2\pi}\int_{-\infty}^{\infty} e^{-i\omega x} f(x)dx \\ &= (i\omega)F(\omega)\end{aligned} \tag{12.4}$$

If $f(x)$ is continuously n-times differentiable and $f^{(m)}(x) \to 0$ as $|x| \to \infty$ for $m = 1, 2, \ldots, (n-1)$, then the Fourier transform of the nth derivative is

$$\mathscr{F}\left\{f^{(n)}(x)\right\} = (i\omega)^n \mathscr{F}\{f(x)\} = (i\omega)^n F(\omega) \tag{12.5}$$

Let $g(x, \omega)$ and $\dfrac{\partial g}{\partial \omega}$ be continuous functions with $-\infty < x < \infty$ and $-\infty < \omega < \infty$. Furthermore, let $\int_{-\infty}^{\infty} |g(x, \omega)|dx$ be finite and $|\dfrac{\partial f}{\partial \omega}| \leq h(x)$ where $h(x)$ is piecewise continuous and such that $\int_{-\infty}^{\infty} h(x)dx$ is finite. Then the Leibniz' rule allows the interchange of integral and differentiation operation as follows:

$$\frac{d}{d\omega}\int_{-\infty}^{\infty} g(x, \omega)dx = \int_{-\infty}^{\infty} \frac{\partial g(x, \omega)}{\partial \omega} dx \tag{12.6}$$

Using Leibniz' rule to differentiate the Fourier transform of $f(x)$, we obtain

$$\frac{dF(\omega)}{d\omega} = \frac{1}{2\pi}\frac{d}{d\omega}\int_{-\infty}^{\infty} f(x)e^{-i\omega x}dx = \frac{-i}{2\pi}\int_{-\infty}^{\infty} xf(x)e^{-i\omega x}dx = -i\mathscr{F}\{xf(x)\} \tag{12.7}$$

Differentiating $F(\omega)$ m times with ω results in

$$\frac{d^m F(\omega)}{d\omega^m} = \frac{(-i)^m}{2\pi}\int_{-\infty}^{\infty} x^m f(x)e^{-i\omega x}dx = (-i)^m \mathscr{F}\left\{x^m f(x)\right\} \tag{12.8}$$

if $f(x)$ is a continuous n times differentiable function. Furthermore, let $x^m f^{(r)}(x)$ for $r = 1, 2, \ldots, n$ satisfy Dirichlet conditions and absolutely integrable over $(-\infty, \infty)$. And $\omega^n F(\omega)$ possess an m times differentiable inverse Fourier transform. Then, provided $\lim_{|x|\to\infty} f^{(n-1)}(x) = 0$ (rapidly decreasing),

$$\mathscr{F}\left\{x^m f^{(n)}(x)\right\} = (i)^m \frac{d^m}{d\omega^m}\mathscr{F}\left\{f^{(n)}(x)\right\} = (i)^{m+n}\frac{d^m}{d\omega^m}\left[\omega^n F(\omega)\right] \tag{12.9}$$

The conditions imposed on $x^m f^{(n)}(x)$ and $\omega^n F(\omega)$ are necessary to ensure the existence of the Fourier transform.

The same operation can be extended for higher-dimension Fourier transform. For example, if $u(x, t)$ is a function of space variable x and time variable t, and its Fourier transform of the partial derivatives is expressed as

$$\mathscr{F}\{u(x, t)\} = \frac{1}{2\pi}\int_{-\infty}^{\infty} u(x, t)e^{-i\omega x}dx = U(\omega, t) \tag{12.10}$$

then

$$\mathscr{F}\left\{\frac{\partial u(x, t)}{\partial t}\right\} = \frac{dU(\omega, t)}{dt} \tag{12.11}$$

$$\mathcal{F}\left\{\frac{\partial^2 u(x,t)}{\partial t^2}\right\} = \frac{d^2 U(\omega,t)}{dt^2} \tag{12.12}$$

$$\mathcal{F}\left\{\frac{\partial u(x,t)}{\partial x}\right\} = i\omega U(\omega,t) \tag{12.13}$$

$$\mathcal{F}\left\{\frac{\partial^2 u(x,t)}{\partial x^2}\right\} = -\omega^2 U(\omega,t) \tag{12.14}$$

12.2 Fourier Transforms for the Solution of Ordinary Differential Equations

As shown in the last section, the Fourier transform simplifies the process of solving linear ODEs by converting differential equations in the time domain into algebraic equations in the frequency domain. For example, consider a linear ODE with constant coefficients. Applying the Fourier transform to both sides of the equation turns differentiation into multiplication by a frequency variable. This transformation reduces the problem to solving a simpler algebraic equation, which can then be inverted back into the time domain to obtain the solution.

Consider the nth-order (nonhomogeneous) linear differential equation with constant coefficients, which is expressed as

$$L\{h(x)\} = f(x) \tag{12.15}$$

$L\{\} = f(x)$ is an nth-order differential operator with constant coefficient that operates on function $h(x)$.

$$L\{\} \equiv a_n D^n + a_{n-1} D^{n-1} + \cdots + a_1 D + a_0 \tag{12.16}$$

$D \equiv \dfrac{d}{dx}$ and $a_n, a_{n-1}, \ldots, a_1, a_0$ are constants. Applying Fourier transform to both sides of the linear ODE in Eq. (12.15) results in

$$\mathcal{F}\{L\{h(x)\}\} = \mathcal{F}\{f(x)\} \tag{12.17}$$

which is simplified as

$$\left[a_n(i\omega)^n + a_{n-1}(i\omega)^{n-1} + \cdots + a_1(i\omega) + a_0\right] H(\omega) = F(\omega) \tag{12.18}$$

$\mathcal{F}\{h(x)\} = H(\omega)$ and $\mathcal{F}\{f(x)\} = F(\omega)$. Complex polynomial with constant coefficients in the left-hand side can be expressed as

$$P(i\omega) = a_n(i\omega)^n + a_{n-1}(i\omega)^{n-1} + \cdots + a_1(i\omega) + a_0$$
$$P(z) = a_n z^n + a_{n-1} z^{n-1} + \cdots + a_1 z + a_0 = \sum_{r=0}^{n} a_r (z)^r \tag{12.19}$$

So Eq. (12.18) is written as

$$P(i\omega)H(\omega) = F(\omega) \tag{12.20}$$

or

$$H(\omega) = \frac{F(\omega)}{P(i\omega)} = F(\omega)Q(\omega) \tag{12.21}$$

where $Q(\omega) = \dfrac{1}{P(i\omega)}$. Taking the inverse Fourier transform of Eq. (12.21),

$$h(x) = \mathscr{F}^{-1}\{F(\omega)Q(\omega)\} = \frac{1}{2\pi}\int_{-\infty}^{\infty} f(y)q(x-y)dy \tag{12.22}$$

(multiplication in Fourier space maps to convolution in the time domain). $q(x) = \mathscr{F}^{-1}\{Q(\omega)\} = \mathscr{F}^{-1}\{1/P(i\omega)\}$ is known explicitly. Equation (12.22) gives a particular solution to the nonhomogeneous ODE $L\{h(x)\} = f(x)$. The general solution can be found using the general solution to the homogeneous ODE with constant coefficients $L\{h(x)\} = 0$; this ODE can be solved using matrix theory.

Green's function, named after British mathematical physicist George Green (1793–1841), is the impulse response of an inhomogeneous linear differential operator defined on a domain with specified initial conditions or boundary conditions. When a given linear differential operator L acts on a function $h(x)$, the differential equation can be written as

$$L\{h(x)\} = f(x) \tag{12.23}$$

But when $u(x)$ is Dirac's delta function, the solution to this equation is Green's function $G(x, x_0)$, which satisfies

$$L\{G(x, x_0)\} = \delta(x - x_0) \tag{12.24}$$

Green's function $G(x, x_0)$ represents the response of the system at point x due to a unit impulse applied at point x_0. Once Green's function is known for a particular operator and boundary conditions, the solution to the original equation with any general source term $f(x)$ can be constructed as

$$u(x) = \int G(x, x_0) f(x_0)\, dx_0 \tag{12.25}$$

Physical interpretation of Eq. (12.22) can be done by allowing $f(x) = \delta(x)$, i.e., a suddenly applied impulse function. Then the solution of this differential equation is expressed as

$$L\{G(x)\} = \delta(x) \tag{12.26}$$

Putting $f(x) = \delta(x)$ in Eq. (12.22)

$$h(x) = \mathscr{F}^{-1}\{Q(\omega)\} = \frac{1}{2\pi}q(x) = \frac{1}{2\pi}G(x) \tag{12.27}$$

(convolution against delta function, gives the functional value at the point). So the general solution of Eq. (12.22) becomes

$$h(x) = \frac{1}{2\pi}\int_{-\infty}^{\infty} f(y)G(x-y)dy \tag{12.28}$$

In any physical system, $f(x)$ usually represents the input function, while $h(x)$ is referred to as the output obtained by the superposition principle. The Fourier transform of $G(x) = q(x)$ is called the admittance. In order to find the response to a given input, the Fourier transform of the input function is multiplied with the admittance, and then inverse Fourier transform is applied to the product. This idea is illustrated by solving some examples in this section.

Example 12.1 (First-Order ODE) The Fourier transform of $f(x) = e^{(-a^2x^2)}$ $(a > 0)$. The function $f(x)$ is continuous and differentiable for all x:

$$\int_{-\infty}^{\infty} \left|e^{(-a^2x^2)}\right| dx = \int_{-\infty}^{\infty} e^{(-a^2x^2)}dx = \frac{1}{a}\int_{-\infty}^{\infty} e^{(-u^2)}du = \frac{\sqrt{\pi}}{a}$$

The standard integral $\int_{-\infty}^{\infty} e^{(-u^2)}du = \sqrt{\pi}$. This shows that $f(x)$ is absolutely integrable over the interval $(-\infty, \infty)$, and so $f(x)$ has a Fourier transform. A straightforward calculation establishes that $f(x)$ satisfies the differential equation:

$$f' + 2a^2xf = 0$$

Taking the Fourier transform of both sides,

$$\mathscr{F}\left\{f'(x)\right\} + 2a^2\mathscr{F}\{xf(x)\} = 0$$

(continued)

Example 12.1 (continued)
Simplifying both sides using derivative properties of the Fourier transfer

$$i\omega F(\omega) + 2a^2 i F'(\omega) = 0$$

or

$$2a^2 F' + \omega F = 0$$

When variables are separated and both sides are integrated against ω,

$$\int \frac{F'}{F} d\omega = -\frac{1}{2a^2} \int \omega d\omega$$

This leads to the solution

$$\ln F(\omega) = -\frac{\omega^2}{4a^2} + \ln A$$

$$F(\omega) = A \exp\left[-\frac{\omega^2}{4a^2}\right]$$

The arbitrary integration constant has been written in the form $\ln A$. Using the initial condition $A = F(0)$. But from the definition of Fourier transform

$$F(0) = \frac{1}{2\pi} \int_{-\infty}^{\infty} e^{(-a^2x^2)} dx = \frac{1}{2\pi} \frac{\sqrt{\pi}}{a} = \frac{1}{2\sqrt{\pi}a}$$

So

$$\mathscr{F}\left\{e^{(-a^2x^2)}\right\} = F(\omega) = \frac{1}{2a\sqrt{\pi}} \exp\left[-\frac{\omega^2}{4a^2}\right] \quad (a > 0)$$

Example 12.2 (Bessel Function) The Bessel function $J_0(x)$ is named after German astronomer Friedrich Wilhelm Bessel (1784–1846). Bessel function $f(x) = J_0(x)$ is an even function that is defined for all x and satisfies Bessel's differential equation of order zero:

$$xf'' + f' + xf = 0$$

(continued)

Example 12.2 (continued)
Solution The Bessel function $J_0(x)$ does not satisfy the absolute integrability condition found. Absolute integrability is merely a sufficient condition that ensures the existence of the Fourier transform of a function $f(x)$, though not a necessary one. Functions exist that possess a Fourier transform even though this condition is violated, and $J_0(x)$ is such a function. Taking the Fourier transform of the differential equation,

$$\left(1-\omega^2\right)F' - \omega F = 0$$

where

$$F(\omega) = \frac{1}{2\pi}\int_{-\infty}^{\infty} J_0(x)e^{-i\omega x}\,dx$$

This linear first-order variable separable differential equation can be written as

$$\int \frac{F'}{F}d\omega = \int \frac{\omega}{1-\omega^2}d\omega$$

The integration gives

$$\ln F(\omega) = -\frac{1}{2}\ln\left(1-\omega^2\right) + \ln A$$

or

$$F(\omega) = \frac{A}{\left(1-\omega^2\right)^{1/2}} \quad \text{with } 0 < \omega^2 < 1$$

In this equation, the arbitrary integration constant has again been written in the form $\ln A$, and the restriction on ω^2 is necessary because the real logarithmic function is not defined for negative arguments. $A = F(0)$ with the standard result $\int_0^\infty J_0(x)dx = 1$ is used to determine A. $J_0(x)$ is an even function.

$$A = F(0) = \frac{1}{2\pi}\int_{-\infty}^{\infty} J_0(x)dx = \frac{2}{2\pi}\int_0^{\infty} J_0(x)dx = \sqrt{\frac{1}{\pi}}$$

Substituting A into $F(\omega)$ gives

$$\mathcal{F}\{J_0(x)\} = F(\omega) = \frac{1}{\pi\left(1-\omega^2\right)^{1/2}}\mathcal{H}(1-|\omega|)$$

(continued)

Example 12.2 (continued)
Multiplication with the Heaviside unit step function $\mathcal{H}(1 - |\omega|)$ is necessary because of the restriction imposed by the real logarithmic function that requires ω to be such that $0 < \omega^2 < 1$.

Example 12.3 (First-Order ODE) The given differential equation

$$f' + f = \frac{1}{2}e^{-|x|}, \quad -\infty < x < \infty$$

Taking the Fourier transform of both sides,

$$i\omega F(\omega) + F(\omega) = \frac{1}{\pi(\omega^2 + 1)}$$

Therefore,

$$F(\omega) = \frac{1}{\pi\left(\omega^2 + 1\right)(1 + \omega i)}$$

Applying the Fourier inversion integral

$$f(x) = \int_{-\infty}^{\infty} \frac{e^{ix\omega}}{\pi\left(\omega^2 + 1\right)(1 + \omega i)} d\omega$$

This complex integral can be evaluated using the residue theorem. For $x > 0$ the line integral is closed with an infinite semicircle in the upper half of the ω-plane. The integration along this arc equals zero by Jordan's formula. Within this closed contour is a second-order pole at $z = i$. Therefore,

$$\text{Res}\left[\frac{e^{ixz}}{\pi\left(z^2 + 1\right)(1 + zi)}\right] = \lim_{z \to i} \frac{d}{dz}\left[(z - i)^2 \frac{e^{ixz}}{i\pi(z - i)^2(z + i)}\right]$$

$$= \frac{xe^{-x}}{2\pi i} + \frac{e^{-x}}{2\pi i}$$

so

$$f(x) = (2\pi i)\left[\frac{xe^{-x}}{2\pi i} + \frac{e^{-x}}{4\pi i}\right] = \frac{e^{-x}}{2}(2x + 1)$$

(continued)

Example 12.3 (continued)

For $x < 0$, the line integral with an infinite semicircle is closed but this time it is in the lower half of the ω-plane. The contribution from the line integral along the arc vanishes by Jordan's lemma. Within the contour, we have a simple pole at $z = -i$. Therefore,

$$\text{Res}\left[\frac{e^{itz}}{\pi\left(z^2+1\right)(1+zi)}\right] = \lim_{z\to -i}(z+i)\frac{e^{itz}}{i(z+i)(z-i)^2} = -\frac{e^x}{4\pi i}$$

and

$$f(x) = (-2\pi i)\left(-\frac{e^x}{4\pi i}\right) = \frac{e^x}{2}$$

The minus sign in front of the $2\pi i$ results from the contour being taken in the clockwise direction or negative sense. A step function can be used to combine results from both half into a single expression:

$$f(x) = \frac{1}{2}e^{-|x|} + xe^{-x}H(x)$$

This is the particular or forced solution to the differential equation. The most general solution therefore requires that we add the complementary solution Ae^{-x}, yielding

$$y(x) = Ae^{-x} + \frac{1}{2}e^{-|x|} + xe^{-x}H(x)$$

And the arbitrary constant A is determined from the prescribed initial condition.

Example 12.4 (Second-Order Linear ODE) The second-order linear ODE is given as

$$-\frac{d^2u}{dx^2} + \alpha^2 u = f(x)$$

With $f(x) \in L^2(\mathbb{R})$ taking the Fourier transform of both sides,

$$-(i\omega)^2 U(\omega) + \alpha^2 U(\omega) = F(\omega)$$

(continued)

Example 12.4 (continued)
or

$$U(\omega) = \frac{F(\omega)}{\left(\omega^2 + \alpha^2\right)}$$

But

$$\mathscr{F}\left\{\frac{\pi}{\alpha} e^{-\alpha|x|}\right\} = \frac{1}{\omega^2 + \alpha^2}$$

So

$$\mathscr{F}\{u\} = \frac{1}{\omega^2 + \alpha^2} \mathscr{F}\{f\} = \mathscr{F}\left\{\frac{\pi}{\alpha} e^{-\alpha|x|}\right\} \mathscr{F}\{f\}$$

So the general solution is expressed through the following convolution operation:

$$u(x) == \frac{1}{2\pi} \frac{\pi}{\alpha} e^{-\alpha|x|} * f = \frac{1}{2\alpha} \int_{-\infty}^{\infty} e^{-\alpha|x-y|} f(y) dy$$

Example 12.5 (LR Circuit) The current $I(x)$ in a simple circuit containing the resistance R and inductance L satisfies the equation

$$L \frac{dI}{dx} + RI = E(x)$$

$E(x)$ is the applied electromagnetic force and R and L are constants. For a given $E(x) = E_0 e^{(-a|x|)}$, using the Fourier transform with respect to x,

$$(i\omega L + R)\hat{I}(\omega) = E_0 \frac{1}{\pi} \frac{a}{\left(a^2 + \omega^2\right)}$$

Or

$$\hat{I}(\omega) = \frac{aE_0}{iL\pi} \frac{1}{\left(\omega - \frac{iR}{L}\right)\left(a^2 + \omega^2\right)}$$

where $\hat{I}(\omega) = \mathscr{F}\{I\}$. Taking the inverse Fourier transform,

(continued)

Example 12.5 (continued)

$$I(x) = \frac{aE_0}{iL\pi}\int_{-\infty}^{\infty}\frac{e^{i\omega x}\,d\omega}{\left(\omega - \frac{iR}{L}\right)\left(a^2+\omega^2\right)}$$

This complex integral can be evaluated using the Cauchy residue theorem. For $x > 0$

$$I(x) = \frac{aE_0}{iL\pi}\cdot 2\pi i\left[\text{Residue at } \omega = \frac{iR}{L} + \text{Residue at } \omega = ia\right]$$

$$= \frac{2aE_0}{L}\left[\frac{e^{-\frac{R}{L}x}}{\left(a^2 - \frac{R^2}{L^2}\right)} - \frac{e^{-ax}}{2a\left(a - \frac{R}{L}\right)}\right]$$

$$= E_0\left[\frac{e^{-ax}}{R-aL} - \frac{2aLe^{-\frac{R}{L}x}}{R^2 - a^2L^2}\right]$$

Similarly, for $x < 0$, the residue theorem gives

$$I(t) = -\frac{aE_0}{i\pi L}\cdot 2\pi i[\text{ Residue at } \omega = -ia]$$

$$= -\frac{2aE_0}{L}\left[\frac{-Le^{ax}}{(aL+R)2a}\right] = \frac{E_0e^{at}}{(aL+R)}$$

At $x = 0$, the current is continuous, and therefore,

$$I(0) = \lim_{x\to 0} I(x) = \frac{E_0}{(R+aL)}$$

If $E(x) = \delta(x)$, then $\hat{E}(\omega) = \frac{1}{2\pi}$ and the solution is obtained by using the inverse Fourier transform:

$$I(x) = \frac{1}{2\pi iL}\int_{-\infty}^{\infty}\frac{e^{i\omega x}}{\left(\omega - \frac{iR}{L}\right)}d\omega$$

by the theorem of residues,

(continued)

Example 12.5 (continued)

$$= \frac{1}{L}[\text{ Residue at } \omega = iR/L]$$

$$= \frac{1}{L} e^{\left(-\frac{Rx}{L}\right)}$$

Thus, the current tends to zero as $x \to \infty$ as expected.

Example 12.6 (The Bernoulli-Euler Beam Equation) The vertical deflection $u(x)$ of an infinite beam on an elastic foundation under the action of a prescribed vertical load $W(x)$. The deflection $u(x)$ satisfies the ordinary differential equation:

$$EI \frac{d^4u}{dx^4} + \kappa u = v(x), \quad \infty < x < \infty$$

EI is the flexural rigidity and κ is the foundation modulus of the beam. With the assumption that $v(x)$ has a compact support and u, u', u'', u''' all tend to zero as $|x| \to \infty$, the ODE can be written as

$$\frac{d^4u}{dx^4} + a^4 u = w(x)$$

$a^4 = \kappa/EI$ and $w(x) = v(x)/EI$. Taking the Fourier transform of both sides,

$$(i\omega)^4 U(\omega) + a^4 U(\omega) = W(\omega)$$

or

$$U(\omega) = \frac{W(\omega)}{\omega^4 + a^4}$$

The inverse Fourier transform gives the solution

$$u(x) = \int_{-\infty}^{\infty} \frac{W(\omega)}{\omega^4 + a^4} e^{i\omega x} d\omega$$

$$= \frac{1}{2\pi} \int_{-\infty}^{\infty} \frac{e^{i\omega x}}{\omega^4 + a^4} d\omega \int_{-\infty}^{\infty} w(\xi) e^{-i\omega\xi} d\xi$$

$$= \int_{-\infty}^{\infty} w(\xi) G(\xi, x) d\xi$$

(continued)

Example 12.6 (continued)
where

$$G(\xi, x) = \frac{1}{2\pi}\int_{-\infty}^{\infty}\frac{e^{i\omega(x-\xi)}}{\omega^4 + a^4}d\omega = \frac{1}{\pi}\int_0^{\infty}\frac{\cos\omega(x-\xi)d\omega}{\omega^4 + a^4}$$

The last equality results from the Euler equation $e^{i\theta} = \cos\theta + i\sin\theta$, and the sin *theta* is an odd function. The last integral is evaluated using complex contour integration (Sheehan & Debnath, 1972). Using the residue theorem,

$$G(\xi, x) = \frac{1}{2a^3}\exp\left(-\frac{a}{\sqrt{2}}|x-\xi|\right)\sin\left[\frac{a(x-\xi)}{\sqrt{2}} + \frac{\pi}{4}\right]$$

In particular, the explicit solution is due to a concentrated load of unit strength acting at some point x_0, i.e., $w(x) = \delta(x - x_0)$. Then the solution for this case becomes

$$u(x) = \int_{-\infty}^{\infty}\delta(\xi - x_0)\,G(\xi, x)d\xi = G(x, x_0)$$

Therefore, the kernel $G(x, \xi)$ in the solution represents the deflection as a function of x caused by a unit point load applied at ξ. Consequently, the deflection resulting from a point load of magnitude $w(\xi)d\xi$ at ξ is given by $w(\xi)d\xi \cdot G(x, \xi)$. This expression thus signifies the cumulative effect of all such incremental deflections.

12.3 Fourier Transforms for the Solution of Integral Equations

In the context of integral equations, the Fourier transform is employed to convert convolution-type integral equations into algebraic equations in the frequency domain. This is particularly useful because convolution in the time or spatial domain corresponds to multiplication in the frequency domain. Once the transformed equation is solved, the inverse Fourier transform is applied to recover the solution in the original domain.

Example 12.7 (Fredholm Integral Equation) Fredholm integral equations are named after Swedish mathematician Erik Ivar Fredholm (1866–1927). Fredholm (1903) first introduced and studied these equations. These integral equations can be written in the general form

$$f(x) = g(x) + \lambda \int_a^b K(x, y)g(y)\,dy$$

$g(x)$ is the unknown function, $f(x)$ is a given function, $K(x, y)$ is the kernel, and λ is a parameter. Fredholm integral equations marked a significant milestone in mathematical analysis, particularly in the fields of functional analysis and operator theory. A Fredholm integral against a convolution kernel is given as

$$f(x) = g(x) + \lambda \int_{-\infty}^{\infty} g(y)h(x - y)dy$$

Taking the Fourier transform,

$$F(\omega) = G(\omega) + 2\pi\lambda G(\omega)H(\omega)$$

Or

$$G(\omega) = \frac{F(\omega)}{2\pi\lambda H(k) + 1}.$$

The inverse Fourier transform leads to a formal solution:

$$g(x) = \int_{-\infty}^{\infty} \frac{F(\omega)e^{i\omega x}d\omega}{2\pi\lambda H(\omega) + 1}$$

Choosing $h(x) = \dfrac{x}{2\lambda\,|x|} = \dfrac{1}{2\lambda\,\text{sgn}(x)}$

$$H(\omega) = \frac{1}{2\pi i\lambda\omega}$$

Using this value of $H(\omega)$ in the integral

(continued)

Example 12.7 (continued)

$$
\begin{aligned}
g(x) &= \int_{-\infty}^{\infty} (i\omega) \frac{F(\omega) e^{i\omega x} d\omega}{1 + i\omega} \\
&= \int_{-\infty}^{\infty} \mathscr{F}\{f'(x)\} \mathscr{F}\{2\pi e^{-x}\} e^{i\omega x} d\omega \\
&= f'(x) * e^{-x} \\
&= \int_{-\infty}^{\infty} f'(y) * e^{(y-x)} dy
\end{aligned}
$$

Example 12.8 (Integral Equation) The given integral equation is

$$
\int_{-\infty}^{\infty} f(x-y) f(y) dy = \frac{1}{x^2 + a^2}
$$

so

$$
f(x) * f(x) = \frac{1}{x^2 + a^2}
$$

Taking the Fourier transform of both sides,

$$
2\pi F(\omega) F(\omega) = \frac{e^{-a|\omega|}}{2a}
$$

$$
F(\omega) = \frac{1}{2\sqrt{\pi a}} e^{-\frac{a}{2}|\omega|}
$$

The inverse Fourier transform gives the solution

$$
\begin{aligned}
f(x) &= \frac{1}{2\sqrt{\pi a}} \int_{-\infty}^{\infty} e^{-\frac{a}{2}|\omega|} e^{ix\omega} d\omega \\
&= \frac{1}{2\sqrt{\pi a}} \left[\int_0^{\infty} e^{\{-\omega(\frac{a}{2}+ix)\}} d\omega + \int_0^{\infty} e^{\{-\omega(\frac{a}{2}-ix)\}} d\omega \right] \\
&= \frac{1}{2\sqrt{\pi a}} \left[\frac{4a}{(4x^2 + a^2)} \right] = \sqrt{\frac{a}{\pi}} \cdot \frac{2}{(4x^2 + a^2)}
\end{aligned}
$$

Example 12.9 (Integral Equation) The given integral equation is

$$\int_{-\infty}^{\infty} \frac{f(y)dy}{(x-y)^2 + a^2} = \frac{1}{(x^2 + b^2)}, \quad b > a > 0$$

This can be written as

$$f(x) * \frac{1}{x^2 + a^2} = \frac{1}{(x^2 + b^2)}, \quad b > a > 0$$

Taking the Fourier transform of both sides,

$$2\pi F(\omega).\mathscr{F}\left\{\frac{1}{x^2 + a^2}\right\} = \frac{e^{-b|\omega|}}{2b}$$

$$2\pi F(\omega)\frac{e^{-a|\omega|}}{2a} = \frac{e^{-b|\omega|}}{2b}$$

$$F(\omega) = \frac{1}{2\pi}\left(\frac{a}{b}\right) e^{\{-|\omega|(b-a)\}}$$

The inverse Fourier transform leads to the solution

$$\begin{aligned} f(x) &= \frac{a}{2\pi b}\int_{-\infty}^{\infty} e^{\{i\omega x - |\omega|(b-a)\}} d\omega \\ &= \frac{a}{2\pi b}\left[\int_0^{\infty} e^{[-\omega\{(b-a)+ix\}]} d\omega + \int_0^{\infty} e^{[-\omega\{(b-a)-ix\}]}\right] d\omega \\ &= \frac{a}{2\pi b}\left[\frac{1}{(b-a)+ix} + \frac{1}{(b-a)-ix}\right] \\ &= \left(\frac{a}{\pi b}\right)\frac{(b-a)}{(b-a)^2 + x^2} \end{aligned}$$

Example 12.10 (Integral Equation) The given integral equation is

$$f(x) + 4\int_{-\infty}^{\infty} e^{-a|x-y|} f(y)dy = g(x)$$

This can be written as

(continued)

Example 12.10 (continued)

$$f(x) + \{4e^{-a|x|} * f(x)\} = g(x)$$

Taking the Fourier transform of both sides,

$$F(\omega) + \frac{8a}{\left(a^2 + \omega^2\right)} F(\omega) = G(\omega)$$

So

$$F(\omega) = \frac{\left(a^2 + \omega^2\right)}{\left(a^2 + \omega^2 + 8a\right)} G(\omega)$$

Taking the inverse Fourier transform gives

$$f(x) = \int_{-\infty}^{\infty} \frac{\left(a^2 + \omega^2\right)}{\left(a^2 + \omega^2 + 8a\right)} G(\omega) e^{i\omega x} d\omega$$

If $a = 1$ and $g(x) = e^{-|x|}$ so that $G(\omega) = \dfrac{1}{\pi\left(1 + \omega^2\right)}$, then the solution becomes

$$f(x) = \frac{1}{\pi} \int_{-\infty}^{\infty} \frac{e^{i\omega x}}{\left(\omega^2 + 3^2\right)} d\omega$$

It represents a complex integral, and for $x > 0$, a semicircular closed contour in the lower half of the complex plane gives

$$f(x) = \frac{1}{3} e^{-3x}$$

Similarly, for $x < 0$, a semicircular closed contour in the upper half of the complex plane is used:

$$f(x) = \frac{1}{3} e^{3x}$$

Thus, the final solution is

$$f(x) = \frac{1}{3} e^{(-3|x|)}$$

12.4 Fourier Transforms for the Solution of Partial Differential Equations

A partial differential equation (PDE) is an equation that involves one or more partial derivatives of an unknown function, which depends on at least two variables. Usually, one of these variables represents time t, while the others correspond to spatial coordinates x. PDEs are crucial in modelling dynamic systems and phenomena in fluid dynamics, elasticity, heat transfer, electromagnetism, and quantum mechanics. Their applications are much broader than those of ordinary differential equations (ODEs), which typically describe simpler physical systems. The Fourier transform is an effective tool for solving PDEs, especially over infinite or semi-infinite domains. For example, in the case of the heat equation, the Fourier transform allows for the separation of variables, converting the PDE into an ODE in the frequency domain, which is generally easier to solve. The solution can then be inverted back to the original space-time domain. The Fourier transform's ability to handle boundary conditions and initial value problems makes it invaluable in physics and engineering, where PDEs are commonly used. This section will illustrate how the Fourier transform can be applied to solve boundary and initial value problems for various linear PDEs.

12.4.1 The Principle of Superposition of Solutions

Before solving the PDEs, a few concepts help to clarify why the Fourier analysis is so special for the solution of PDEs. In Chap. 9, details about the linear system are discussed. An eigenfunction of a linear system is defined as

$$L\{f(x)\} = \lambda f(x) \tag{12.29}$$

$L\{\}$ is an operator just like the differential operator $\left(\dfrac{d}{dx}\right)$ that works on a function, and the output is a constant multiplication of the function. Such functions of the operator are called the eigenfunction, and the value of λ is called the eigenvalues. If $f_1(x)$ and $f_2(x)$ are solutions of the differential equation, then any linear combination $c_1 f_1(x) + c_2 f_2(x)$ is also a solution. For a given differential equation, multiple linearly independent solutions can correspond to the same eigenvalue. The general solution is then expressed as a linear combination of these independent solutions:

$$f(x) = c_1 f_1(x) + c_2 f_2(x) + \cdots + c_n f_n(x) \tag{12.30}$$

where $f_1(x), f_2(x), \ldots, f_n(x)$ are linearly independent solutions, and $c_1, c_2, \ldots, c_n$ are constants determined by initial or boundary conditions. Now, the question is obvious: Why have these solutions been added? The superposition principle of the linear operator $L\{\}$ is the reason why solutions are added together. From the discussion on linear system in Chap. 9, it is clear that the operator $L\{\}$ is linear when it satisfies the additivity $(L\{f_1(x) + f_2(x)\} = L\{f_1(x)\} + L\{f_2(x)\})$ and homogeneity $(L\{cf(x)\} = cL\{f(x)\})$ for some constant c. The linearity of the differential equation allows for the superposition (addition) of solutions. The general solution to a linear differential equation must account for all possible behaviors of the system. Different solutions (e.g., sine and cosine functions in trigonometric solutions or exponential functions in solutions involving complex roots) capture different aspects of the system's behavior. The specific form of the solution depends on initial or boundary conditions. These conditions are used to determine the constants $c_1, c_2, \ldots, c_n$. By adding solutions together, we ensure that the general solution is flexible enough to satisfy any set of conditions, i.e., completeness.

Before moving to the solution of PDEs, understanding the first-order and second-order eigenfunction ordinary differential equations is helpful. A first-order eigenvalue differential equation can be written as

$$\frac{df(x)}{dx} + \lambda f(x) = 0 \tag{12.31}$$

where λ is the eigenvalue and $f(x)$ is the eigenfunction. It can be rearranged as

$$\frac{df(x)}{f(x)} = -\lambda dx \tag{12.32}$$

The general solution of the first-order linear homogeneous differential equation can be obtained by integrating both sides:

$$\ln|f(x)| = -\lambda x + c \tag{12.33}$$

c is the constant of integration. Exponentiating both sides gives

$$f(x) = Ce^{-\lambda x} \tag{12.34}$$

So the exponential functions are the eigenfunctions of the differential operator. C is an arbitrary constant that depends on initial or boundary conditions. The second-order eigenvalue differential equation is

$$\frac{d^2 f(x)}{dx^2} + \lambda f(x) = 0 \tag{12.35}$$

λ is the eigenvalue and $f(x)$ is the eigenfunction. A close inspection can reveal that the same exponential function is also the eigenfunction of the second-order differential equation. A second-order homogeneous linear differential equation has the general form

$$a\frac{d^2 f(x)}{dx^2} + b\frac{df(x)}{dx} + cf(x) = 0 \tag{12.36}$$

a, b, and c are constants, and $f(x)$ is the unknown function. The goal is to find the general solution for $f(x)$. The solution process for such an equation begins by converting it into an auxiliary equation (also called the characteristic equation). This is done by assuming a solution of the form

$$f(x) = e^{rx} \tag{12.37}$$

r is a constant that needs to be determined. Substituting this into the original equation (12.36):

$$a\left(r^2 e^{rx}\right) + b\left(re^{rx}\right) + ce^{rx} = 0 \tag{12.38}$$

Since $e^{rx} \neq 0$, both sides can be divided by e^{rx}, leading to the characteristic equation:

$$ar^2 + br + c = 0 \tag{12.39}$$

This is a quadratic equation in r, and its solutions can be found using the quadratic formula:

$$r = \frac{-b \pm \sqrt{b^2 - 4ac}}{2a} \tag{12.40}$$

The nature of the solutions for r depends on the discriminant $\Delta = b^2 - 4ac$. There are three cases to consider

12.4.1.1 Case 1: Distinct Real Roots ($\Delta > 0$)

When the discriminant is positive ($\Delta > 0$), the characteristic equation has two distinct real roots: r_1 and r_2. The general solution to the differential equation is given by

$$f(x) = c_1 e^{r_1 x} + c_2 e^{r_2 x} \tag{12.41}$$

For example, consider the equation

$$\frac{d^2 f}{dx^2} - 3\frac{df}{dx} + 2f = 0$$

The characteristic equation is

$$r^2 - 3r + 2 = 0$$

Factoring

$$(r-1)(r-2) = 0$$

Thus, $r_1 = 1$ and $r_2 = 2$. The general solution is

$$f(x) = c_1 e^x + c_2 e^{2x}$$

12.4.1.2 Case 2: Repeated Real Roots ($\Delta = 0$)

When the discriminant is zero ($\Delta = 0$), the characteristic equation has a repeated real root $r_1 = r_2 = r$. In this case, the general solution takes the form

$$f(x) = (c_1 + c_2 x)e^{rx} \tag{12.42}$$

The factor $c_2 x$ is included to ensure that the solution set is linearly independent. For example, consider the equation

$$\frac{d^2 f}{dx^2} - 4\frac{df}{dx} + 4f = 0$$

The characteristic equation is

$$r^2 - 4r + 4 = 0$$

This can be factored as

$$(r-2)^2 = 0$$

giving a repeated root $r = 2$. The general solution is

$$f(x) = (c_1 + c_2 x)e^{2x}$$

12.4.1.3 Case 3: Complex Conjugate Roots ($\Delta < 0$)

When the discriminant is negative ($\Delta < 0$), the characteristic equation has complex conjugate roots $r_1 = \alpha + i\beta$ and $r_2 = \alpha - i\beta$. In this case, the general solution is written as

$$f(x) = e^{\alpha x} \left(c_1 \cos(\beta x) + c_2 \sin(\beta x)\right) \tag{12.43}$$

where α is the real part and β is the imaginary part of the roots, and c_1 and c_2 are constants. Consider the equation

$$\frac{d^2 f}{dx^2} + 2\frac{df}{dx} + 5f = 0$$

The characteristic equation is

$$r^2 + 2r + 5 = 0$$

Using the quadratic formula,

$$r = \frac{-2 \pm \sqrt{2^2 - 4 \cdot 1 \cdot 5}}{2 \cdot 1} = \frac{-2 \pm \sqrt{4 - 20}}{2} = \frac{-2 \pm \sqrt{-16}}{2} = \frac{-2 \pm 4i}{2}$$

Thus, the roots are $r_1 = -1 + 2i$ and $r_2 = -1 - 2i$. The general solution is

$$f(x) = e^{-x} \left(c_1 \cos(2x) + c_2 \sin(2x)\right)$$

The specific constants c_1 and c_2 are determined from the general solution by applying the initial conditions (such as $f(0) = f_0$ and $f'(0) = f_1$) or boundary conditions are applied. These conditions are substituted into the general solution and its derivative, resulting in a system of equations that can be solved for the unknown constants.

Consider the second-order differential equation with the Dirichlet boundary conditions:

$$\frac{d^2 f(x)}{dx^2} + \lambda f(x) = 0$$

with boundary conditions $f(0) = 0$ and $f(L) = 0$. The general solution for $\lambda > 0$ is

$$f(x) = C_1 \cos(\sqrt{\lambda} x) + C_2 \sin(\sqrt{\lambda} x)$$

Applying the boundary condition $f(0) = 0$,

$$f(0) = C_1 \cos(0) + C_2 \sin(0) = C_1 = 0$$

So the solution reduces to

$$f(x) = C_2 \sin(\sqrt{\lambda} x)$$

Applying the second boundary condition $f(L) = 0$,

$$y(L) = C_2 \sin(\sqrt{\lambda}L) = 0$$

For nontrivial solutions ($C_2 \neq 0$),

$$\sin(\sqrt{\lambda}L) = 0$$

This gives

$$\sqrt{\lambda}L = n\pi, \quad \text{for } n = 1, 2, 3, \ldots$$

Therefore, the eigenvalues are

$$\lambda_n = \left(\frac{n\pi}{L}\right)^2$$

And the corresponding eigenfunctions are

$$f_n(x) = C_2 \sin\left(\frac{n\pi x}{L}\right)$$

An important observation from the solution of the ODE is that the function's solution is represented as a sum of several sinusoids. This aligns with the concept of the Fourier series and Fourier transforms, where any complex function can be broken down into a sum of sinusoids, which serve as a basis for that function. This is why Fourier analysis plays a key role in solving differential equations. By expressing complex functions through simple sinusoidal components, Fourier series simplifies the process. This approach becomes even more useful when applying Fourier analysis to solve partial differential equations (PDEs), as it will follow the same principle.

12.4.2 The Concept of Variable Separation Method in PDE Solutions

Partial differential equations are the function of more than one independent variable. The method of variable separation is one of the most fundamental techniques used to solve partial differential equations (PDEs). This approach is based on the idea that the solution to a PDE can be expressed as the product of functions, each depending on only one of the independent variables. It is particularly effective for solving linear PDEs with specific boundary and initial conditions, such as the heat equation, wave equation, and Laplace's equation.

Consider a partial differential equation involving two or more independent variables, for example, x and t. The basic assumption in the method of separation of variables is that the solution $u(x, t)$ can be written as a product of two functions, where one function depends solely on x and the other depends solely on t:

$$u(x, t) = X(x)T(t) \tag{12.44}$$

$X(x)$ is a function of the spatial variable x, and $T(t)$ is a function of the temporal variable t. A simple example is the one-dimensional heat equation, which models heat conduction along a rod. The heat equation is given by

$$\frac{\partial u(x,t)}{\partial t} = \alpha \frac{\partial^2 u(x,t)}{\partial x^2} \tag{12.45}$$

where α is a constant and is known as the thermal diffusivity. Using the assumption that $u(x, t) = X(x)T(t)$, and substituting this into the heat equation, gives

$$X(x)\frac{dT(t)}{dt} = \alpha T(t)\frac{d^2X(x)}{dx^2} \tag{12.46}$$

Now, dividing both sides of the equation by $X(x)T(t)$,

$$\frac{1}{T(t)}\frac{dT(t)}{dt} = k\frac{1}{X(x)}\frac{d^2X(x)}{dx^2} \tag{12.47}$$

At this stage, observe that the left-hand side depends only on t, while the right-hand side depends solely on x. For the equality to be valid for all values of x and t, both sides must be equal to a constant, denoted as $-\lambda$. This leads to two separate ordinary differential equations (ODEs):

$$\frac{1}{T(t)}\frac{dT(t)}{dt} = -\lambda \quad \Rightarrow \quad \frac{dT(t)}{dt} = -\lambda T(t) \tag{12.48}$$

$$k\frac{1}{X(x)}\frac{d^2X(x)}{dx^2} = -\lambda \quad \Rightarrow \quad \frac{d^2X(x)}{dx^2} = -\frac{\lambda}{k}X(x) \tag{12.49}$$

Each of these ODEs can now be solved independently.

12.4.3 One-Dimensional Heat Equation

While working in Grenoble, Joseph Fourier developed the theory of heat conduction through his groundbreaking mathematical concepts, experiments, and observations. He completed a significant memoir titled "On the propagation of heat in solid bodies" and submitted it to the Academy of Sciences in Paris in 1807 for a research award. Later, in 1822, he published his landmark work, "La Théorie Analytique de la Chaleur (The Analytical Theory of Heat)." In this theory, Fourier proposed that the heat flow between two neighboring molecules is proportional to the small temperature difference between them. His approach to heat conduction was highly mathematical, involving a general equation for heat propagation with varying initial and boundary conditions. Fourier's interest in heat diffusion stemmed from his desire to determine the ideal depth for his wine cellar to maintain a consistent temperature throughout the year. He aimed to understand how heat travels across the surface and into the ground. An ideal cellar would keep a steady temperature year-round, and because heat spreads through the Earth's crust in waves, the hottest surface day would not match the hottest day in the cellar. His fundamental heat conduction equation in one dimension is given as

$$\rho c_p \frac{\partial u(x,t)}{\partial t} = k \frac{\partial^2 u(x,t)}{\partial x^2} \tag{12.50}$$

ρc_p represents the heat capacity of the body, and k is the thermal conductivity of material. The left-hand side of the heat equation presents the rate of the heat flux through the system, and the right-hand side expresses the temperature gradient across the system. Equation (12.50) is rewritten as

$$\frac{\partial u(x,t)}{\partial t} = \alpha \frac{\partial^2 u(x,t)}{\partial x^2} \tag{12.51}$$

$u(x,t)$ is the temperature at position x at time t. α is the thermal diffusivity (proportionality constant), which is the property of the material. In simple words, the heat flux rate through a slab or rod is directly proportional to the temperate difference at both ends of the rod. This equation typically comes with initial and boundary conditions. Consider a rod of length L having a length larger than its diameter $L \gg d$. The system's domain is $x \in [0, L]$. Both ends of the rod are kept at constant temperature (Dirichlet boundary conditions), say $u(0,t) = u(L,t) = 0$ for all $t \geq 0$. The domain's initial temperature distribution at $t = 0$ is expressed as $u(x,0) = f(x)$. The temperature in the rod varies with time and along the length, so the temperature distribution in the rod can be assumed as follows:

$$u(x,t) = X(x)T(t) \tag{12.52}$$

$X(x)$ is a function of x only, and $T(t)$ is a function of t only. Substituting this into the heat equation,

$$X(x)\frac{dT(t)}{dt} = \alpha T(t)\frac{d^2 X(x)}{dx^2} \tag{12.53}$$

Dividing both sides by $\alpha X(x)T(t)$ gives

$$\frac{1}{\alpha T(t)}\frac{dT(t)}{dt} = \frac{1}{X(x)}\frac{d^2X(x)}{dx^2} = \lambda \tag{12.54}$$

The left side depends only on t and the right side only on x, so that both sides must equal a constant λ. For $\lambda = 0$ or $\lambda > 0$, the only solution $u(x,t) = 0$ satisfies the equation $u = X(x)T(t)$ and the boundary conditions. So, for the trivial solution, $\lambda < 0$ is expressed as $\lambda = -p^2$. This leads to two ordinary differential equations (ODEs):

$$\frac{d^2X(x)}{dx^2} + p^2X(x) = 0 \tag{12.55}$$

$$\frac{dT(t)}{dt} + \alpha p^2T(t) \tag{12.56}$$

For $\lambda < 0$, the general solution of the spatial ODE in Eq. (12.55) becomes

$$X(x) = A\cos(px) + B\sin(px) \tag{12.57}$$

From the boundary conditions at $x = 0$ and $x = L$,

$$u(0,t) = X(0)T(t) = 0 \tag{12.58}$$

and

$$u(L,t) = X(L)T(t) = 0 \tag{12.59}$$

From this boundary conditions $A = 0$ and $X(L) = B\sin(pL) = 0$ with $B \neq 0$. This condition is satisfied when

$$pL = n\pi \tag{12.60}$$

Taking $B = 1$, the general equation (12.57) is expressed as

$$X_n(x) = \sin\left(\frac{n\pi x}{L}\right) \tag{12.61}$$

The solution of the temporal ODE equation (12.56) becomes

$$\frac{dT_n(t)}{dt} + \alpha\left(\frac{n\pi x}{L}\right)^2\left(\frac{n\pi}{L}\right)^2 T_n(t) = 0 \tag{12.62}$$

The solution of this first-order differential equation is

$$T_n(t) = C_n e^{-\alpha\left(\frac{n\pi}{L}\right)^2 t} \tag{12.63}$$

The general solution is a superposition of all possible solutions:

$$u(x,t) = X(x)T(t) = \sum_{n=1}^{\infty} C_n \sin\left(\frac{n\pi x}{L}\right) e^{-\alpha\left(\frac{n\pi}{L}\right)^2 t} \tag{12.64}$$

The coefficients can be determined using the initial conditions:

$$u(x,0) = \sum_{n=1}^{\infty} C_n \sin\left(\frac{n\pi x}{L}\right) = f(x) \tag{12.65}$$

Equation (12.63) is the Fourier series expansion of the initial value function $f(x)$, and the coefficients C_n can be determined from

$$C_n = \frac{2}{L}\int_0^L f(x)\sin\left(\frac{n\pi x}{L}\right) dx \tag{12.66}$$

The solution of the heat equation depends upon the convergence of the Fourier series of initial conditions in Eq. (12.66). The solution can be established by assuming $f(x)$ is continuous or piecewise continuous. Because of the exponential factor, all the terms in Eq. (12.64) approach zero as t approaches infinity. Figure 12.1 presents the solution of the heat equation for $f(x) = 10\sin\pi x$. The temperature in the domain decays with time, and the rate of decay increases with n.

Often, instead of the value of temperature value at the boundary, the value of heat flux, i.e., $\frac{\partial u}{\partial x}$ is known, and this kind of boundary condition is known as the Neumann condition (named after the German mathematician Carl Gottfried Neumann [1832–1925]).

Consider a condition when the rod is insulated at one end $x = 0$; this modifies the boundary condition as

$$\frac{\partial u}{\partial x}(0,t) = 0, \quad \text{for } t > 0 \tag{12.67}$$

On the other end of the rod at $x = L$, a Dirichlet boundary condition is prescribed as

$$u(L,t) = 0, \quad \text{for } t > 0 \tag{12.68}$$

The boundary condition affects the solution, and Eq. (12.57) needs to be differentiated for applying the Neumann boundary condition in Eq. (12.67).

$$\frac{dX}{dx}(x) = -Ap\sin(px) + Bp\cos(px) \tag{12.69}$$

At $x = 0$, applying Eq. (12.67),

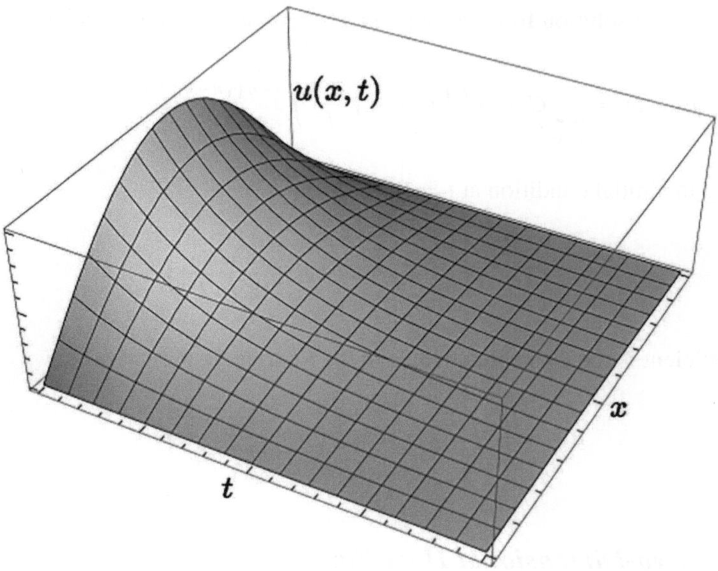

Fig. 12.1 Temperature distribution pattern with time

$$\frac{dX}{dx}(0) = -Ap\sin(px) + Bp\cos(px) = 0 \tag{12.70}$$

which gives $B = 0$ as $p \neq 0$. So the spatial solution becomes

$$X(x) = A\cos(px) \tag{12.71}$$

Applying the Dirichlet boundary condition at $x = L$,

$$0 = A\cos(pL) \tag{12.72}$$

Therefore, $pL = \left(n + \frac{1}{2}\right)\pi$ for $n = 0, 1, 2, \ldots$. This gives the eigenvalues

$$p_n = \frac{\left(n + \frac{1}{2}\right)\pi}{L} \tag{12.73}$$

The spatial part of the solution becomes

$$X_n(x) = A_n \cos\left(\left(n + \frac{1}{2}\right)\frac{\pi x}{L}\right) \tag{12.74}$$

And the general solution from the superposition of all the modes become

$$u(x,t) = \sum_{n=0}^{\infty} C_n \cos\left(\left(n+\frac{1}{2}\right)\frac{\pi x}{L}\right) e^{-\alpha\left(\left(n+\frac{1}{2}\right)\frac{\pi}{L}\right)^2 t} \tag{12.75}$$

Applying the initial condition at $t = 0$ results in

$$f(x) = \sum_{n=0}^{\infty} C_n \cos\left(\left(n+\frac{1}{2}\right)\frac{\pi x}{L}\right) \tag{12.76}$$

The coefficient is the C_n is obtained from the Fourier cosine series as

$$C_n = \frac{2}{L}\int_0^L f(x)\cos\left(\left(n+\frac{1}{2}\right)\frac{\pi x}{L}\right) dx \tag{12.77}$$

12.4.4 Two-Dimensional Heat Equation

In a two-dimensional space, the Fourier law of heat conduction is expressed as

$$\frac{\partial u(x,y,t)}{\partial t} = \alpha\left(\frac{\partial^2 u(x,y,t)}{\partial x^2} + \frac{\partial^2 u(x,y,t)}{\partial y^2}\right) \tag{12.78}$$

The temperature $u(x, y, t)$ at a point is the function of a two-dimensional space (x, y) and time t. α is the thermal diffusivity. The domain is assumed to be infinite in both x and y without any internal heat generation. The initial temperature distribution in the domain is $u(x, y, 0) = f(x, y)$. A two-dimensional Fourier transform is defined as

$$\hat{u}(k_x, k_y, t) = \frac{1}{(2\pi)^2}\int_{-\infty}^{\infty}\int_{-\infty}^{\infty} u(x,y,t)e^{-i(k_x x + k_y y)}\,dx\,dy \tag{12.79}$$

k_x and k_y are the wave numbers corresponding to the x- and y-directions, respectively. Applying the Fourier transform to the heat conduction equation (12.78),

$$\mathscr{F}\left\{\frac{\partial u(x,y,t)}{\partial t}\right\} = \mathscr{F}\left\{\alpha\left(\frac{\partial^2 u(x,y,t)}{\partial x^2} + \frac{\partial^2 u(x,y,t)}{\partial y^2}\right)\right\} \tag{12.80}$$

The Fourier transform of the time derivative becomes

$$\frac{\partial \hat{u}(k_x, k_y, t)}{\partial t} \tag{12.81}$$

The Fourier transform of the second derivative with respect to x and y yields

$$\mathscr{F}\left\{\frac{\partial^2 u(x, y, t)}{\partial x^2}\right\} = -k_x^2 \hat{u}(k_x, k_y, t) \tag{12.82}$$

$$\mathscr{F}\left\{\frac{\partial^2 u(x, y, t)}{\partial y^2}\right\} = -k_y^2 \hat{u}(k_x, k_y, t) \tag{12.83}$$

The heat conduction equation in Fourier space is a first-order ordinary differential equation in time for $\hat{u}(k_x, k_y, t)$:

$$\frac{\partial \hat{u}(k_x, k_y, t)}{\partial t} = -\alpha(k_x^2 + k_y^2)\hat{u}(k_x, k_y, t) \tag{12.84}$$

For direct integration, the variables needs to be separated as

$$\frac{1}{\hat{u}(k_x, k_y, t)} \frac{\partial \hat{u}(k_x, k_y, t)}{\partial t} = -\alpha(k_x^2 + k_y^2) \tag{12.85}$$

Integrating both sides with respect to t,

$$\ln \hat{u}(k_x, k_y, t) = -\alpha(k_x^2 + k_y^2)t + C(k_x, k_y) \tag{12.86}$$

Exponentiating both sides gives the general solution:

$$\hat{u}(k_x, k_y, t) = Ce^{-\alpha(k_x^2+k_y^2)t} \tag{12.87}$$

The Fourier transform of the initial temperature distribution $f(x, y)$ is

$$\hat{u}(k_x, k_y, 0) = \mathscr{F}\{f(x, y)\} \tag{12.88}$$

Thus, the solution in the Fourier space is

$$\hat{u}(k_x, k_y, t) = \mathscr{F}\{f(x, y)\}e^{-\alpha(k_x^2+k_y^2)t} \tag{12.89}$$

To obtain the solution in the physical domain $u(x, y, t)$, the inverse Fourier transform of Eq. (12.89) is taken:

$$u(x, y, t) = \int_{-\infty}^{\infty} \int_{-\infty}^{\infty} \hat{u}(k_x, k_y, t)e^{i(k_x x + k_y y)}\, dk_x\, dk_y \tag{12.90}$$

Now, substituting the expression for $\hat{u}(k_x, k_y, t)$ results is

$$u(x, y, t) = \int_{-\infty}^{\infty} \int_{-\infty}^{\infty} \mathscr{F}\{f(x, y)\} e^{-\alpha(k_x^2+k_y^2)t} e^{i(k_x x+k_y y)} \, dk_x \, dk_y \tag{12.91}$$

Equation (12.91) gives the temperature distribution $u(x, y, t)$ at any time t for a given the initial temperature distribution $f(x, y)$.

For example, the initial temperature distribution at the center of an infinite domain is given by the delta function:

$$f(x, y) = \delta(x)\delta(y)$$

The Fourier transform of the initial condition becomes

$$\mathscr{F}\{f(x, y)\} = \mathscr{F}\{\delta(x)\delta(y)\} = 1$$

Thus, the solution in the Fourier domain is

$$\hat{u}(k_x, k_y, t) = e^{-\alpha(k_x^2+k_y^2)t}$$

Taking the inverse Fourier transform,

$$u(x, y, t) = \int_{-\infty}^{\infty} \int_{-\infty}^{\infty} e^{-\alpha(k_x^2+k_y^2)t} e^{i(k_x x+k_y y)} \, dk_x \, dk_y = \frac{\pi}{\alpha t} e^{-\frac{x^2+y^2}{4\alpha t}}$$

The solution for the diffusion of heat from an instantaneous point source at (0, 0) is a Gaussian function (Fig. 12.2), which continuously damps out with an increase in time.

12.4.5 Solutions for the 2D Laplace Equations with Boundary Conditions

The steady-state Fourier law of heat conduction is expressed as

$$\left(\frac{\partial^2 u(x, y)}{\partial x^2} + \frac{\partial^2 u(x, y)}{\partial y^2} \right) = 0 \tag{12.92}$$

Equation (12.92) is also known as the Laplace equation. Marquis de Laplace (1825) first introduced this equation in his influential five-volume work, Trait 'e de m 'ecanique c 'eleste (Celestial Mechanics). Its standard form, $\nabla^2 u = 0$, arises in a wide range of physical phenomena, including electrostatics, fluid dynamics, and gravitational potential. Laplace's study of the equation began during his work on celestial mechanics and potential theory in the late eighteenth century. Laplace was interested in describing the behavior of gravitational fields in space and extended

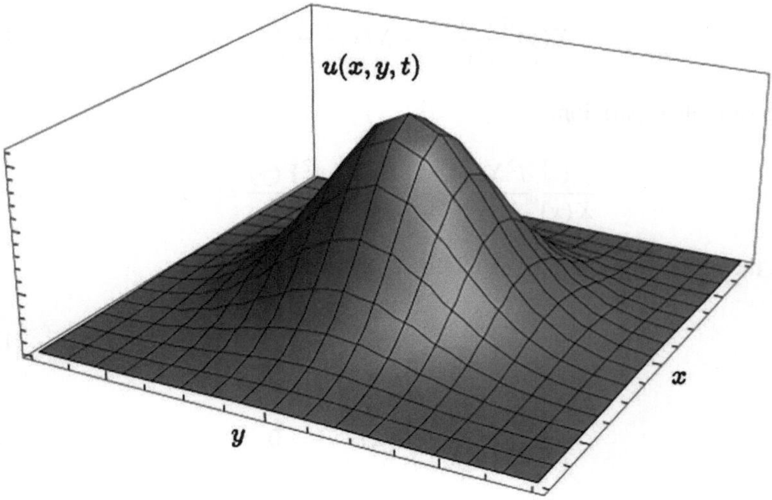

Fig. 12.2 Heat diffusion for an instantaneous point source at (0, 0) in the infinite domain

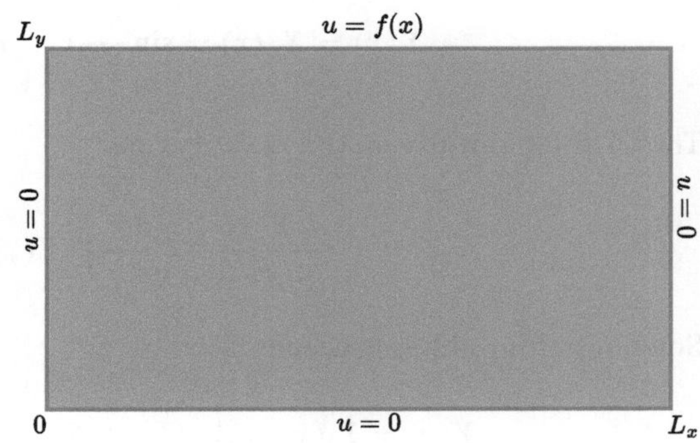

Fig. 12.3 Solution of two-dimensional steady-state heat transfer on a rectangular plate

Isaac Newton's theories of gravity. His work culminated in the Laplace equation, which describes a field in equilibrium where there is no source or sink of the quantity being modelled. Laplace equation continues to be a core concept in modern mathematics and physics, with applications ranging from solving problems in electrostatics to describing steady-state heat distribution.

Let us solve the steady-state heat conduction in a rectangular plate (Fig. 12.3) having $x \in [0, L_x]$ and $y \in [0, L_y]$, with the Dirichlet boundary conditions $u(0, y) = u(L_x, y) = u(x, 0) = 0$ and $u(x, L_y) = f(x)$. The solution can be assumed to be of the form

$$u(x, y) = X(x)Y(y) \tag{12.93}$$

Substituting this to the heat conduction equation,

$$Y(y)\frac{d^2X(x)}{dx^2}+X(x)\frac{d^2Y(y)}{dy^2}=0 \tag{12.94}$$

Doing variable separation,

$$\frac{1}{X(x)}\frac{d^2X(x)}{dx^2}=-\frac{1}{Y(y)}\frac{d^2Y(y)}{dy^2}=-\lambda \tag{12.95}$$

Then, the PDE is split into three ODEs.

$$\frac{d^2X(x)}{dx^2}+\lambda X(x)=0 \tag{12.96}$$

$$\frac{d^2Y(y)}{dy^2}-\lambda Y(y)=0 \tag{12.97}$$

Solving equation (12.96), with the boundary conditions $X(0)=0$ and $X(L_x)=0$, gives $\lambda=(n\pi/L_x)^2$ and corresponding nonzero solutions:

$$X(x)=X_n(x)=\sin\frac{n\pi}{L_x}x, \quad n=1,2,\ldots \tag{12.98}$$

The ODE for Y with $\lambda=(n\pi/L_x)^2$ becomes

$$\frac{d^2Y(y)}{dy^2}-\left(\frac{n\pi}{L_x}\right)^2 Y(y)=0 \tag{12.99}$$

Solutions of Eq. (12.99) become

$$Y(y)=Y_n(y)=A_n e^{\frac{n\pi y}{L_x}}+B_n e^{-n\pi y L_x} \tag{12.100}$$

The boundary condition $Y(0)=0$ on the lower side of the rectangle implies that $Y_n(0)=A_n+B_n=0$ or $B_n=-A_n$. This gives

$$Y_n(y)=A_n\left(e^{\frac{n\pi y}{L_x}}-e^{\frac{-n\pi y}{L_x}}\right)=2A_n\sinh\frac{n\pi y}{L_x} \tag{12.101}$$

Writing $2A_n=A_n^*$, the eigenfunctions are obtained as

$$u_n(x,y)=X_n(x)Y_n(y)=A_n^*\sin\frac{n\pi x}{L_x}\sinh\frac{n\pi y}{L_x} \tag{12.102}$$

Equation (12.102) satisfies the boundary condition of $u=0$ on the rectangle's left, right, and lower sides. Let us consider the infinite series to get a solution that satisfies the boundary condition $u(x,L_y)=f(x)$ on the upper side:

$$u(x, y) = \sum_{n=1}^{\infty} u_n(x, y) \tag{12.103}$$

So

$$u(x, b) = f(x) = \sum_{n=1}^{\infty} A_n^* \sin \frac{n\pi x}{L_x} \sinh \frac{n\pi L_y}{L_x} \tag{12.104}$$

This can be written as

$$u(x, b) = f(x) = \sum_{n=1}^{\infty} \left(A_n^* \sinh \frac{n\pi L_y}{L_x} \right) \sin \frac{n\pi x}{L_x} \tag{12.105}$$

From Eq. (12.105), the expressions in the parentheses must be the Fourier coefficients of $f(x)$:

$$A_n^* \sinh \frac{n\pi L_y}{L_x} = \frac{2}{L_x} \int_0^{L_x} f(x) \sin \frac{n\pi x}{L_x} dx \tag{12.106}$$

So

$$A_n^* = \frac{2}{L_x \sinh \dfrac{n\pi L_y}{L_x}} \int_0^{L_x} f(x) \sin \frac{n\pi x}{L_x} dx \tag{12.107}$$

So the solution for temperature distribution inside the rectangular domain is expressed as

$$u(x, y) = \sum_{n=1}^{\infty} u_n(x, y) = \sum_{n=1}^{\infty} A_n^* \sin \frac{n\pi x}{L_x} \sinh \frac{n\pi y}{L_x} \tag{12.108}$$

The Laplace equation (12.92) also governs the electrostatic potential of electrical charges in any region that is free of these charges. Thus, the steady-state heat problem can also be interpreted as an electrostatic potential problem. Equation 12.108 expresses the potential in the rectangle when the upper side of the rectangle is at potential $f(x)$, and the other three sides are grounded. This illustrates that entirely different physical systems may have the same mathematical model and can thus be treated by the same mathematical methods.

What happens to Solution (12.108), the domain is unbounded in both sides of the x direction $-\infty < x < \infty$ and bounded in one side of $y = 0$ and also unbounded for $0 < y < \infty$? On this unbounded domain, when the Dirichlet-type boundary condition is applied at the lower plane at $u(x, 0) = f(x)$, then the problem is called Dirichlet's problem in the half plane. The Fourier transform of the Laplace equation

(12.92) helps to analyze the problem for an unbounded domain. Taking the Fourier transform of $u(x, y)$ with respect to x results in

$$\hat{u}(k, y) = \frac{1}{2\pi} \int_{-\infty}^{\infty} u(x, y) e^{-ikx} \, dx \tag{12.109}$$

$\hat{u}(k, y)$ is the unknown function representing the potential, temperature, etc. in Fourier space and k represents the wave number vector in the y direction. Now substituting the results from Eq. (12.109) into Laplace equation (12.92),

$$\frac{d^2\hat{u}(k, y)}{dy^2} - k^2\hat{u}(k, y) = 0 \tag{12.110}$$

The ordinary differential equation (12.109) is the Fourier-transformed Laplace equation, having the general solution as

$$\hat{u}(k, y) = A(k)e^{-|k|y} + B(k)e^{|k|y} \tag{12.111}$$

$A(k)$ and $B(k)$ are constants (which may depend on k) that are determined from the boundary conditions:

$$u(x, 0) = f(x), \quad -\infty < x < \infty \tag{12.112}$$

and

$$u(x, y) \to 0 \text{ as } |x| \to \infty, \quad y \to \infty \tag{12.113}$$

$B(k) = 0$, to ensure the condition in Eq. (12.112). This prevents the solution to grow exponentially with y. Thus, the solution becomes

$$\hat{u}(k, y) = A(k)e^{-|k|y} \tag{12.114}$$

Taking Fourier transform of the boundary condition in the lower plane (Eq. (12.112)),

$$\hat{u}(k, 0) = \frac{1}{2\pi} \int_{-\infty}^{\infty} f(x) e^{-ikx} \, dx = \hat{f}(k) \tag{12.115}$$

At $y = 0$, the solution becomes

$$\hat{u}(k, 0) = A(k) \tag{12.116}$$

The solution in the Fourier domain becomes

$$\hat{u}(k, y) = \hat{f}(k)e^{-|k|y} \tag{12.117}$$

The inverse Fourier transform of Eq. (12.117) gives the solution in half plane:

$$u(x, y) = \int_{-\infty}^{\infty} \hat{f}(k) e^{-|k|y} e^{ikx}\, dk \tag{12.118}$$

In an alternative approach, from the convolution theorem,

$$u(x, y) = \frac{1}{2\pi} \int_{-\infty}^{\infty} f(\xi) g(x - \xi) d\xi \tag{12.119}$$

where

$$g(x) = \mathscr{F}^{-1}\left\{e^{-|k|y}\right\} = \frac{2y}{\left(x^2 + y^2\right)} \tag{12.120}$$

Consequently, the solution becomes

$$u(x, y) = \frac{y}{\pi} \int_{-\infty}^{\infty} \frac{f(\xi) d\xi}{(x - \xi)^2 + y^2}, \quad y > 0 \tag{12.121}$$

This is the well-known Poisson integral formula in the half plane. It is noted that

$$\lim_{y \to 0+} u(x, y) = \int_{-\infty}^{\infty} f(\xi) \left[\lim_{y \to 0+} \frac{y}{\pi} \cdot \frac{1}{(x - \xi)^2 + y^2}\right] d\xi = \int_{-\infty}^{\infty} f(\xi) \delta(x - \xi) d\xi \tag{12.122}$$

Cauchy's definition of the delta function is used, that is,

$$\delta(x - \xi) = \lim_{y \to 0+} \frac{y}{\pi} \cdot \frac{1}{(x - \xi)^2 + y^2} \tag{12.123}$$

Instead of the functional value $u(x, 0) = f(x)$, when the normal derivative is prescribed at the lower bound of y, the problem is known as Neumann's problem in the half plane:

$$\frac{\partial u}{\partial y}(x, 0) = g(x) \tag{12.124}$$

Taking the Fourier transform of this condition with respect to x gives

$$\int_{-\infty}^{\infty} \frac{\partial u}{\partial y}(x, 0) e^{-ikx}\, dx = \hat{g}(k) \tag{12.125}$$

where $\hat{g}(k)$ is the Fourier transform of $g(x)$. Now, differentiating equation (12.114) with respect to y and evaluating it at $y = 0$ gives

$$\frac{\partial \hat{u}}{\partial y}(k, 0) = -|k|A(k) \tag{12.126}$$

Therefore, the transformed boundary condition becomes

$$\hat{g}(k) = -|k|A(k) \tag{12.127}$$

$A(k)$ becomes

$$A(k) = -\frac{\hat{g}(k)}{|k|} \tag{12.128}$$

Thus, the solution in the Fourier domain is

$$\hat{u}(k, y) = -\frac{\hat{g}(k)}{|k|}e^{-|k|y} \tag{12.129}$$

The solution in the spatial domain can be obtained by taking the inverse Fourier transform of Eq. (12.129):

$$\begin{aligned} u(x, y) &= \int_{-\infty}^{\infty} \hat{u}(k, y)e^{ikx}\, dk \\ &= -\int_{-\infty}^{\infty} \frac{\hat{g}(k)}{|k|}e^{-|k|y}e^{ikx}\, dk \end{aligned} \tag{12.130}$$

This is the solution to the Neumann problem for Laplace's equation in the half plane.

French mathematician Siméon-Denis Poisson (1781–1840) extended Laplace's work by introducing a nonzero source term in the equation to account for the effects of a distributed source:

$$\frac{\partial^2 u(x, y)}{\partial x^2} + \frac{\partial^2 u(x, y)}{\partial y^2} = f(x, y) \tag{12.131}$$

u is the potential function, and f represents a source term like charge density in electrostatics or mass density in gravity. The Poisson equation is a generalization of the Laplace equation and is central to mathematical physics, particularly in describing fields influenced by sources or sinks.

12.4.6 The One-Dimensional Wave Equation

The wave equation and its solution for a vibrating string have had profound implications across many fields. In physics, the equation models strings, sound, and water waves. In engineering, it forms the basis for understanding vibrations in structures and materials. French mathematician and physicist Jean le Rond d'Alembert (1717–1783) in 1747 sought to model the vibration of a stretched string. The equation describes how the shape of the vibrating string evolves over time. D'Alembert's solution was notable for introducing two travelling waves moving in opposite directions. This insight was revolutionary, as it showed that the motion of a vibrating string could be understood in terms of simple wave propagation.

Around the same time, the Swiss mathematician Leonard Euler was also investigating vibrating strings. Euler made significant contributions by generalizing d'Alembert's solution. While d'Alembert's work focused on smooth initial conditions (without discontinuities), Euler extended the analysis to more general shapes of the string, recognizing that the solution could accommodate a wider variety of initial configurations. Euler's approach laid the groundwork for the later development of the Fourier series, as he recognized that the motion of the string could be represented as the sum of simpler oscillatory motions. In 1753, Swiss mathematician and physicist Daniel Bernoulli advanced the discussion by proposing that the motion of a vibrating string could be decomposed into a sum of sine and cosine functions. He argued that the solution to the wave equation could be expressed as a superposition of simple harmonic motions. This idea was an early form of what is now called the Fourier series, though the rigorous development of Fourier analysis would come later. Bernoulli's insight helped explain how complex vibrations of a string could be understood as combinations of simpler modes, each corresponding to a different frequency or harmonic.

Joseph Fourier's later work on heat conduction was built on the foundations laid by Bernoulli, Euler, and d'Alembert. In his 1807 memoir on heat, Fourier formalized the theory of the Fourier series, providing a mathematical framework for representing any periodic function as an infinite sum of sine and cosine terms. While Fourier's work was focused on the heat equation, his mathematical methods were soon recognized as invaluable for solving the wave equation. The use of the Fourier series allowed for the systematic solution of the wave equation with arbitrary initial conditions, offering a more powerful approach to understanding vibrating strings and other wave phenomena.

The one-dimensional wave equation is given by

$$\frac{\partial^2 u(x,t)}{\partial t^2} = c^2 \frac{\partial^2 u(x,t)}{\partial x^2} \tag{12.132}$$

$u(x,t)$ is the displacement at position x at time t. c is the speed of the wave. Boundary and initial conditions typically accompany this equation. For simplicity, let us consider a string of length L is tied at both end as shown in Fig. 12.4. This can be expressed with Dirichlet boundary conditions:

$$u(0,t) = 0 \quad t \geq 0 \tag{12.133}$$

$$u(L,t) = 0 \quad t \geq 0 \tag{12.134}$$

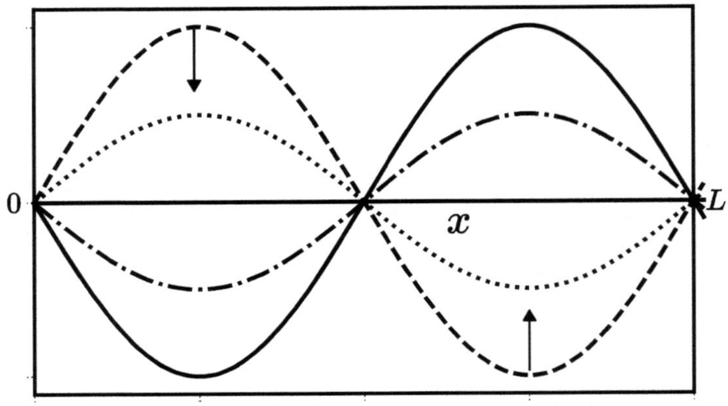

Fig. 12.4 Second normal mode of vibration for various values of t

These conditions imply that the displacement at both ends of the string is zero at all times. Also the initial displacement and velocity is provided as

$$u(x, 0) = f(x) \quad \text{for} \quad 0 \leq x \leq L \tag{12.135}$$

$$\frac{\partial u}{\partial t}(x, 0) = g(x) \quad \text{for} \quad 0 \leq x \leq L \tag{12.136}$$

To solve this problem, let us assume that the solution $u(x, t)$ can be written as a product of two functions:

$$u(x, t) = X(x)T(t) \tag{12.137}$$

where $X(x)$ is a function of position x, and $T(t)$ is a function of time t. Substituting this into the wave equation (12.132),

$$X(x)\frac{d^2T(t)}{dt^2} = c^2T(t)\frac{d^2X(x)}{dx^2} \tag{12.138}$$

Dividing both sides by $X(x)T(t)$,

$$\frac{1}{T(t)}\frac{d^2T(t)}{dt^2} = c^2\frac{1}{X(x)}\frac{d^2X(x)}{dx^2} = \lambda \tag{12.139}$$

Since the left-hand side depends only on t and the right-hand side depends only on x, both sides must be equal to a constant (λ). This leads to two ordinary differential equations:

$$\frac{d^2X(x)}{dx^2} - \lambda X(x) = 0 \tag{12.140}$$

$$\frac{d^2T(t)}{dt^2} - \lambda c^2 T(t) = 0 \tag{12.141}$$

Let us first solve the spatial ODE for $X(x)$. The nontrivial solution for the problem exists, when $\lambda < 0$. Taking $\lambda = -p^2$, the general solution of the ODE in Eq. (12.140) becomes

$$X(x) = A \cos px + B \sin px \tag{12.142}$$

Applying the boundary condition in Eq. (12.133),

$$X(0) = A = 0 \tag{12.143}$$

Applying the boundary condition in Eq. (12.134),

$$X(L) = B \sin pL = 0 = \sin n\pi \tag{12.144}$$

For $n = 1, 2, \ldots$. So

$$p = \frac{n\pi}{L} \tag{12.145}$$

Infinitely many solutions for $X(x)$ becomes

$$X_n(x) = B_n \sin\left(\frac{n\pi}{L}x\right) \tag{12.146}$$

Now the temporal ODE becomes

$$\frac{d^2T(t)}{dt^2} + \frac{n^2\pi^2c^2}{L^2}T(t) = 0 \tag{12.147}$$

This is a second-order homogeneous differential equation with the general solution:

$$T_n(t) = C_n \cos\left(\frac{n\pi c}{L}t\right) + D_n \sin\left(\frac{n\pi c}{L}t\right) \tag{12.148}$$

The general solution to the wave equation is the sum of all modes n:

$$u(x,t) = \sum_{n=1}^{\infty}\left(C_n \cos\left(\frac{n\pi c}{L}t\right) + D_n \sin\left(\frac{n\pi c}{L}t\right)\right)\sin\left(\frac{n\pi}{L}x\right) \tag{12.149}$$

Then the initial conditions can now be used to determine the coefficients C_n and D_n. At $t = 0$,

$$u(x, 0) = \sum_{n=1}^{\infty} C_n \sin\left(\frac{n\pi}{L}x\right) = f(x) \tag{12.150}$$

Thus, the coefficients C_n are the Fourier sine series coefficients of the initial displacement $f(x)$:

$$C_n = \frac{2}{L}\int_0^L f(x)\sin\left(\frac{n\pi}{L}x\right) dx \tag{12.151}$$

Differentiating $u(x, t)$ with respect to t and evaluating at $t = 0$,

$$\frac{\partial u}{\partial t}(x, 0) = \sum_{n=1}^{\infty} \frac{n\pi c}{L} D_n \sin\left(\frac{n\pi}{L}x\right) = g(x) \tag{12.152}$$

Thus, the coefficients D_n are the Fourier sine series coefficients of the initial velocity $g(x)$:

$$D_n = \frac{2}{n\pi c}\int_0^L g(x)\sin\left(\frac{n\pi}{L}x\right) dx \tag{12.153}$$

The final solution to the one-dimensional wave equation is

$$\begin{aligned} u(x, t) = &\sum_{n=1}^{\infty}\left(\frac{2}{L}\int_0^L f(x)\sin\left(\frac{n\pi}{L}d\right) dx\right)\cos\left(\frac{n\pi c}{L}t\right)\sin\left(\frac{n\pi}{L}x\right) \\ &+\sum_{n=1}^{\infty}\left(\frac{2}{n\pi c}\int_0^L g(x)\sin\left(\frac{n\pi}{L}x\right) dx\right)\sin\left(\frac{n\pi c}{L}t\right)\sin\left(\frac{n\pi}{L}x\right) \end{aligned} \tag{12.154}$$

The solution is a superposition of harmonic waves, each with a frequency determined by the wave number n and the wave speed c (Fig. 12.5). The coefficients C_n represent the contribution of the initial displacement to each harmonic mode, while D_n represents the contribution of the initial velocity. The boundary conditions ensure that the wave remains fixed at both ends of the string at $x = 0$ and $x = L$ for all times.

For simplicity let us consider the zero initial velocity condition $g(x) = 0$. Then the solution in Eq. (12.154) becomes

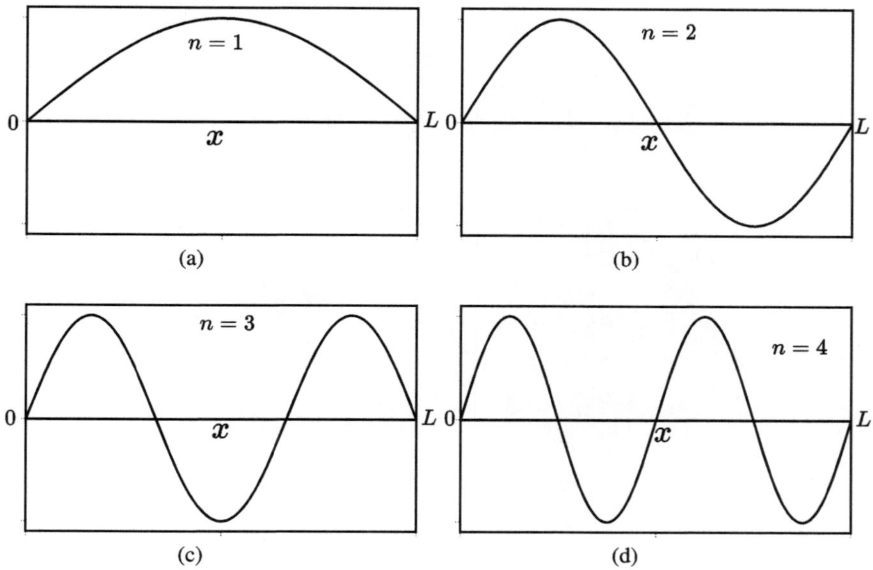

Fig. 12.5 Normal modes of the vibrating string. (**a**) $n = 1$. (**b**) $n = 2$. (**c**) $n = 3$. (**d**) $n = 4$

$$
\begin{aligned}
u(x,t) &= \sum_{n=1}^{\infty} \left(\frac{2}{L} \int_0^L f(x) \sin\left(\frac{n\pi}{L} d\right) dx \right) \cos\left(\frac{n\pi c}{L} t\right) \sin\left(\frac{n\pi}{L} x\right) \\
&= \sum_{n=1}^{\infty} \left(\frac{1}{L} \int_0^L f(x) \sin\left(\frac{n\pi}{L} d\right) dx \right) \\
&\quad \times \left[\sin\left\{\frac{n\pi}{L}(x - ct)\right\} + \sin\left\{\frac{n\pi}{L}(x + ct)\right\} \right]
\end{aligned}
\tag{12.155}
$$

These two series are those obtained by substituting $x - ct$ and $x + ct$, respectively, for the variable x in the Fourier sine series for $f(x)$. Since the initial deflection $f(x)$ is continuous on the interval $0 \leq x \leq L$ and zero at the endpoints, $u(x,t)$ is a continuous function of both variables x and t for all values of the variables.

D'Alembert (1743, 1750) introduced two travelling waves moving in opposite directions to model the evolved shape of the vibrating string. d'Alembert combined the space and time variable to form two new set of variables ξ and η as

$$\xi = x - ct \tag{12.156}$$

$$\eta = x + ct \tag{12.157}$$

These new variables represent waves travelling to the right (with speed c) and to the left (with speed $-c$), respectively (Fig. 12.6). In terms of these new variables, the

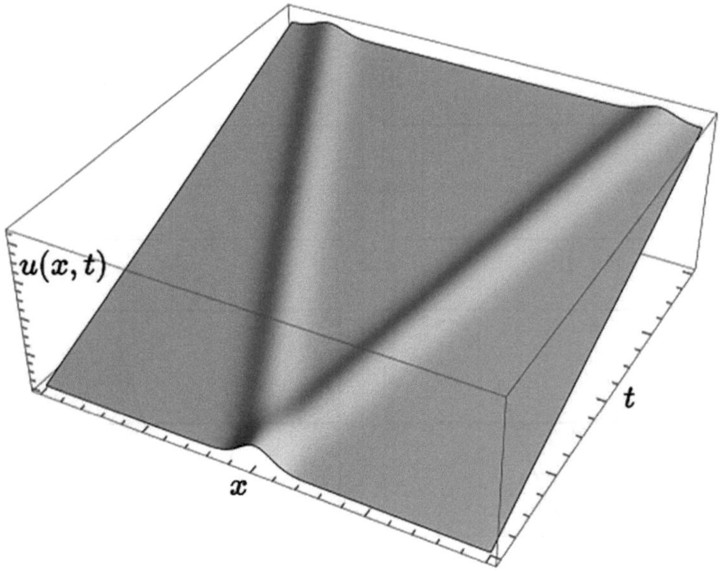

Fig. 12.6 Two characteristics direction of wave propagation

partial derivatives transform as follows:

$$\frac{\partial}{\partial x} = \frac{\partial}{\partial \xi} + \frac{\partial}{\partial \eta} \tag{12.158}$$

$$\frac{\partial}{\partial t} = -c\frac{\partial}{\partial \xi} + c\frac{\partial}{\partial \eta} \tag{12.159}$$

Substituting this to the wave equation (12.132),

$$\left(-c\frac{\partial}{\partial \xi} + c\frac{\partial}{\partial \eta}\right)^2 u = c^2\left(\frac{\partial}{\partial \xi} + \frac{\partial}{\partial \eta}\right)^2 u \tag{12.160}$$

Simplifying this expression gives

$$\frac{\partial^2 u}{\partial \xi \partial \eta} = 0 \tag{12.161}$$

This is a much simpler equation than the original wave equation. It states that the second mixed partial derivative of u with respect to ξ and η is zero. The general solution to Eq. (12.161) is

$$u(\xi, \eta) = F(\xi) + G(\eta) \tag{12.162}$$

where $F(\xi)$ and $G(\eta)$ are arbitrary functions determined by the initial conditions. Transforming back to the original variables $\xi = x - ct$ and $\eta = x + ct$,

$$u(x,t) = F(x - ct) + G(x + ct) \tag{12.163}$$

Equation (12.162) represents the superposition of two travelling waves, $F(x - ct)$ moving to the right and $G(x + ct)$ moving to the left.

Initial conditions are used to determine the specific form of $F(\xi)$ and $G(\eta)$. At $t = 0$, the displacement is $u(x, 0) = f(x)$. Substituting $t = 0$ into the general solution,

$$u(x,0) = F(x - c \cdot 0) + G(x + c \cdot 0) = F(x) + G(x) = f(x) \tag{12.164}$$

Differentiating $u(x,t)$ with respect to t gives

$$\frac{\partial u}{\partial t}(x,t) = -cF'(x - ct) + cG'(x + ct) \tag{12.165}$$

At $t = 0$, $\frac{\partial u}{\partial t}(x, 0) = g(x)$

$$\frac{\partial u}{\partial t}(x,0) = -cF'(x) + cG'(x) = g(x) \tag{12.166}$$

Simplifying equation (12.166),

$$F'(x) - G'(x) = \frac{g(x)}{c} \tag{12.167}$$

From Eq. (12.164),

$$G(x) = f(x) - F(x) \tag{12.168}$$

Substituting equations (12.168) to (12.167),

$$F'(x) - (f'(x) - F'(x)) = \frac{g(x)}{c} \tag{12.169}$$

Simplifying,

$$2F'(x) = f'(x) + \frac{g(x)}{c} \tag{12.170}$$

Therefore,

$$F'(x) = \frac{1}{2}\left(f'(x) + \frac{g(x)}{c}\right) \tag{12.171}$$

Integrating both sides with respect to x,

$$F(x) = \frac{1}{2}\left(f(x) + \frac{1}{c}\int_0^x g(\xi)\,d\xi\right) \tag{12.172}$$

Using $G(x) = f(x) - F(x)$,

$$G(x) = \frac{1}{2}\left(f(x) - \frac{1}{c}\int_0^x g(\xi)\,d\xi\right) \tag{12.173}$$

The final d'Alembert solution is obtained by substituting $F(x - ct)$ and $G(x + ct)$ into the general solution in Eq. (12.162):

$$u(x, t) = \frac{1}{2}\left[f(x - ct) + f(x + ct)\right] + \frac{1}{2c}\int_{x-ct}^{x+ct} g(\xi)\,d\xi \tag{12.174}$$

d'Alembert's solution provides a clear physical interpretation of the wave equation as a superposition of two travelling waves, one moving to the right and the other to the left. In Eq. (12.174), the first term $\frac{1}{2}[f(x - ct) + f(x + ct)]$ represents two waves travelling in opposite directions with speed $-c$ and c, respectively. It also shows how both the initial displacement and the initial velocity influence the wave's evolution over time. The second term $\frac{1}{2c}\int_{x-ct}^{x+ct} g(\xi)\,d\xi$ accounts for the initial velocity distribution, spreading over the interval $[x - ct, x + ct]$ as the wave propagates.

12.4.7 Classification of Partial Differential Equations

Partial differential equations (PDEs) involve multiple independent variables and the partial derivatives of an unknown function. They are crucial in describing various physical systems, such as heat conduction, wave propagation, fluid dynamics, and electromagnetism. Classifying PDEs is essential in understanding the behavior of these physical systems and choosing appropriate methods to solve the equations. PDEs can be classified based on the nature of their solutions, the characteristics of the differential operator, and the types of boundary and initial conditions.

A general second-order PDE in two variables x and y can be written as

$$A(x, y)\frac{\partial^2 u}{\partial x^2} + 2B(x, y)\frac{\partial^2 u}{\partial x \partial y} + C(x, y)\frac{\partial^2 u}{\partial y^2} + D(x, y, u, u_x, u_y) = 0 \tag{12.175}$$

where A, B, and C are coefficients that may depend on x, y, and sometimes the unknown function u. The term $D(x, y, u, u_x, u_y)$ includes first-order derivatives or lower-order terms, but the classification depends primarily on the second-order derivatives.

The classification of second-order PDEs is based on the nature of the quadratic form associated with the second-order partial derivatives. By analogy to conic sections, PDEs can be classified as elliptic, parabolic, or hyperbolic using the discriminant of the quadratic form:

$$\Delta = B^2 - AC \tag{12.176}$$

If $\Delta < 0$, the PDE is elliptic. These equations typically describe equilibrium states, where time is not an explicit variable, and the solution tends to be smooth. A classic example is the Laplace and the Poisson equation. Elliptic PDEs arise in physical problems like steady-state heat conduction, electrostatics, and potential theory. Elliptic PDEs typically require boundary conditions specified on a closed curve or surface, like the Dirichlet or Neumann boundary conditions.

If $\Delta = 0$, the PDE is parabolic. These equations describe processes that evolve over time towards a steady state. The most notable example is the transient heat equation discussed earlier in this chapter. Parabolic equations require an initial condition (to describe the state at $t = 0$) and boundary conditions to describe how the solution behaves on the edges of the domain. The solutions of the parabolic PDEs march forward in time.

If $\Delta > 0$, the PDE is hyperbolic. These equations typically describe wave propagation. Hyperbolic equations need both initial conditions (specifying the state of the system and its rate of change) and boundary conditions, usually on the edges of the domain.

Another critical way to classify second-order linear PDEs is based on their characteristic directions and disturbance propagation speeds. Characteristic curves are paths along which the PDE reduces to an ordinary differential equation (ODE). These curves are critical because they describe how disturbances propagate through the medium characterized by the PDE. For a second-order PDE in Eq. (12.175), the characteristic curves are determined by solving the characteristic equation:

$$A\left(\frac{dy}{dx}\right)^2 - 2B\left(\frac{dy}{dx}\right) + C = 0 \tag{12.177}$$

The solution of this equation gives the directions along which information propagates.

The characteristic equation has two real solutions for hyperbolic PDEs, indicating two distinct characteristic curves at each point. For example, the wave equation (12.132) has the characteristic equations:

$$\frac{dx}{dt} = \pm c \tag{12.178}$$

It has real characteristics, represented by straight lines in the $x-t$ plane, along which disturbances propagate with speed c. Disturbances at a point (x_0, t_0) affect only the points along these characteristic lines $x \pm ct = \text{constant}$, preserving causality. Hyperbolic PDE can be reduced to ODEs along these characteristic curves.

The characteristic equation for parabolic PDEs, like heat equation (12.51), has a repeated real root, and there is only one characteristic direction at each point:

$$\frac{dx}{dt} = \infty \tag{12.179}$$

This implies that the characteristic curves are vertical lines (parallel to the time axis), indicating that the solution at a point x depends on the initial conditions at that point and all other points due to diffusion. Disturbances spread infinitely fast along the characteristic direction. Solutions exhibit diffusive or smoothing behavior, where disturbances at a point affect the entire domain instantaneously. A disturbance at any point in space affects the solution at all future times but does not remain localized. Transformations or similarity solutions are often used due to the degeneracy in the characteristic equation.

Characteristic equations for elliptic PDEs, like the Laplace equation (12.92), have complex conjugate roots, indicating no real characteristic curves:

$$\frac{dy}{dx} \text{has no real solutions} \tag{12.180}$$

Disturbances do not propagate in the classical sense. The solution at a point depends on the values throughout the domain, leading to boundary value problems where the boundary conditions influence the entire solution. Solutions are harmonic functions representing steady-state situations without time dependence. Since there are no real characteristics, methods like separation of variables or conformal mapping are employed.

The first-order PDEs take the form

$$A(x, y, u)\frac{\partial u}{\partial x} + B(x, y, u)\frac{\partial u}{\partial y} = C(x, y, u) \tag{12.181}$$

These can describe more general, nonlinear systems. A well-known method for solving first-order PDEs is the method of characteristics, which reduces the PDE to a system of ordinary differential equations (ODEs) along curves called characteristics.

In the study of PDEs, "characteristic directions" have a significant physical meaning. These directions represent the lines or paths within the domain with discontinuous highest-order derivatives. This discontinuity suggests that, along these lines, the solution to the PDE does not exist, or there could be infinitely many possible solutions. The behavior of the solution in these directions plays a crucial role in how the system evolves.

The elliptic PDEs have no real characteristic directions. This absence has important physical implications. Elliptic PDEs govern phenomena that exhibit

Fig. 12.7 There are no specific characteristics or directions for parabolic PDEs. (**a**) A water drop hitting the surface of a still body of water. (**b**) Disturbance propagation in an infinite direction

smooth and continuous behavior throughout the domain. An example of this can be visualized by considering a water drop hitting the surface of a still body of water (Fig. 12.7a). When the drop impacts the water, disturbances are generated at the point of contact and spread through the fluid. These disturbances, or ripples, travel in all directions at infinite speed and aim to nullify the gradient (differences in the water level) to bring the system back to a state of equilibrium (Fig. 12.7b). In essence, the disturbances try to homogenize the domain, smoothing out any changes caused by the drop. After a short period, the water surface returns to its original undisturbed state if no further drops impact it. At distances far away from the point of impact, the velocity of the disturbances diminishes to zero, meaning the effect of the drop's impact becomes negligible. This behavior is a hallmark of elliptic PDEs, which implies that the solution is continuous and smooth throughout the entire domain. Even though the water surface was disturbed locally, the influence extends uniformly and continuously across the domain.

In problems governed by elliptic PDEs, any arbitrary point in the domain is influenced by all of its spatial neighbors. This interconnectedness of the points in the domain means that elliptic PDEs require well-defined boundary conditions for their solution. Despite any sharp discontinuities that may exist at the boundaries, the solution remains smooth and continuous within the domain because disturbances propagate at infinite speed in all directions. Therefore, the solution is always well-behaved, even in the presence of boundary value discontinuities.

Elliptic PDEs are commonly used to model physical processes where diffusion plays a role, such as heat conduction and mass diffusion. In such processes, the goal is often to reach equilibrium, where differences in temperature or concentration are levelled out over time, resulting in a smooth distribution throughout the domain. The absence of real characteristics in elliptic PDEs reflects the continuous and interconnected nature of these diffusion processes, ensuring that the solution is smooth everywhere in the domain.

Parabolic PDEs, such as the heat equation, describe processes that evolve over time, where one distinct real root reflects a certain type of behavior in the system.

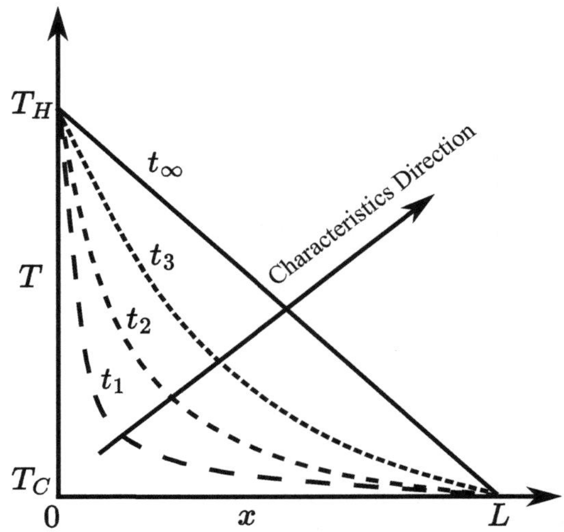

Fig. 12.8 One distinct characteristic direction of parabolic PDEs

This single real root implies that the process has a specific direction of time evolution, meaning that the solution progresses forward in time and is often referred to as "time-marching." In other words, the system starts from an initial state and evolves towards a steady state as time passes. To understand this concept more clearly, consider the example of transient heat conduction in a metal bar. Imagine a metal bar of length L is initially at a uniform temperature T_C. This uniform temperature represents the initial steady state. Suddenly, one end of the rod is heated to a higher temperature, T_H. This change disturbs the initial steady state, and heat begins to diffuse gradually along the length of the rod. Over time, the temperature distribution along the rod changes, with the heat travelling from the hotter end to the colder regions, eventually reaching a new steady state where the temperature is linearly distributed along the length of the rod. The temperature profile will be stable at this final steady state, with no further change over time (Fig. 12.8).

At the very beginning of the process, when the sudden change in temperature is applied at $t = t_0$, the parabolic PDE shows a discontinuity. This discontinuity reflects the fact that, before the heating event (for $t < t_0$), there is no solution to the PDE because the system has not yet begun evolving. Essentially, no information exists about the temperature distribution for times before the change was applied. Once the system starts evolving, the time evolution of the temperature distribution begins. The heat diffusion process progresses forward in time, meaning that the state of the system influences the temperature at any point along the rod at earlier moments. This forward progression is what we refer to as the characteristic direction for parabolic PDEs, indicating that the solution moves in a specific direction with respect to time. Over time, the system gradually approaches a new steady state, with the changes in temperature smoothing out. For parabolic PDEs, the solution depends heavily on the initial condition (in this case, the initial temperature distribution, T_C) and the boundary condition (the sudden application of T_H).

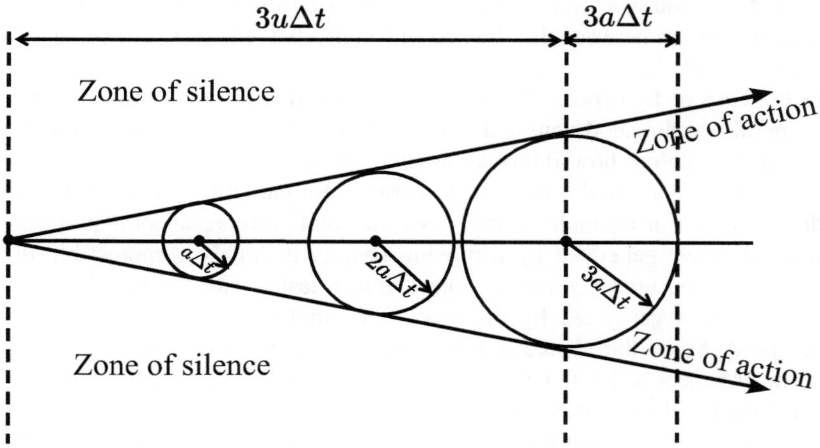

Fig. 12.9 Two characteristics direction of hyperbolic PDEs

Parabolic PDEs describe many time-dependent diffusion processes, such as heat conduction (as in this example), mass diffusion, and fluid flow. They capture the way systems evolve over time, starting from an initial state and moving towards equilibrium. In such processes, the solution always depends on the initial condition, and the characteristic direction represents a one-way progression towards a stable solution in time. As time progresses, the temperature distribution changes smoothly, but always with a clear reference to its initial state.

In contrast to parabolic or elliptic PDEs, hyperbolic PDEs are characterized by having two distinct real characteristics, which means that the solution propagates along two different directions in space-time. These characteristic directions represent the paths along which disturbances (or information) travel.

To understand this more clearly, let us consider the physical example of the wavefront generated by a source (such as an aircraft) moving at supersonic speed (Fig. 12.9). The source is moving at a velocity u greater than the speed of sound a. The Mach index is a measure of the speed of the source compared to the speed of sound $Ma = \dfrac{u}{a}$. The source is moving at speed u, and it generates sound waves that travel at the speed of sound a. Over a short time interval Δt, the sound wave emitted from the source moves a distance $a\Delta t$. Meanwhile, the source itself continues moving at its supersonic speed u and travels a distance $u\Delta t$. Suppose the source continues moving at this supersonic speed. In that case, it creates a shock wave pattern where disturbances are confined to two distinct lines that form the characteristic directions of the hyperbolic PDE. These lines divide the space into the zone of action (the region affected by the disturbances created by the moving source) and the zone of silence (where no disturbances are felt because the source has moved too fast for the sound waves to catch up). The disturbances are confined within these characteristic directions, which means that outside of this cone-shaped area (the Mach cone), no information or disturbance from the source can propagate. This sharp division between the affected and unaffected regions is a defining feature of hyperbolic PDEs.

In the case of hyperbolic PDEs, the entire solution must be framed in a spatiotemporal coordinate system that aligns with the characteristic directions. These characteristics define how information (like sound waves or other disturbances) moves through space and time, which is critical for capturing the wavelike nature of the solution. For example, in this case, the solution represents the propagation of a shock wave generated by a moving source. In highly compressible flows, such as those occurring when an object moves faster than the speed of sound, the disturbances generated by the source accumulate because they travel at a finite speed. Since the source moves faster than the sound waves it generates, these disturbances get left behind the moving source, creating a buildup. This accumulation of disturbances forms a shock wave or shock front. The shock front is the surface where the disturbances are suddenly relieved, and it represents a sharp discontinuity in the flow. Mathematically, this is a type of boundary condition called a "jump condition," where there is an abrupt change in variables like pressure, temperature, or velocity across the shock wave.

Because hyperbolic PDEs can produce such sharp discontinuities (like shock waves), special numerical methods are required to solve them accurately. Standard methods might fail to capture the sudden jump in values across the shock front, leading to inaccuracies in the solution. Efficient numerical schemes are specifically designed to handle these discontinuities, ensuring that the solution correctly captures the shock wave's behavior.

12.5 Closure

In this chapter, we explored the powerful role that Fourier analysis plays in solving ordinary differential equations (ODEs), integral equations, and partial differential equations (PDEs). Fourier analysis provides a systematic method to decompose complex functions into simpler sinusoidal components, offering a pathway to convert differential equations into algebraic forms or simpler differential equations in the frequency domain. This process greatly simplifies the solution of various equations, especially when boundary or initial conditions are involved.

For ODEs, Fourier transforms allow for efficient solutions, particularly in systems with periodic or oscillatory behavior. Integral equations often encountered in mathematical physics also benefit from Fourier methods, converting convolutions into products and simplifying the solution process. When it comes to PDEs, Fourier transforms are essential in solving equations such as the heat, wave, and Laplace equations, especially in infinite or semi-infinite domains. These transforms can transform PDEs into more manageable ODEs in the frequency domain, where standard solution techniques can be applied. Once the frequency domain solution is obtained, the inverse transform yields the solution in the original domain, providing clear insight into the behavior of the physical system.

In summary, Fourier analysis stands as an indispensable tool in the mathematical treatment of a wide range of physical problems. Its versatility in transforming complex problems into simpler forms has made it one of the cornerstone techniques in applied mathematics, physics, and engineering.

The solution of these ODEs, specifically PDEs, depends on the boundary and initial conditions. The boundary and initial conditions are often so complex that an analytical solution for the problem is impossible. So, despite a century of research on the topic, there are many PDEs like Navier-Stokes equations, whose solution still needs to be analytically possible in higher dimensions. As an alternative approach in the past few decades, numerical solutions using the finite difference method (FDM), finite volume method (FVM), and finite element method (FEM) are developed. These numerical methods give solutions for complex physical problems. However, the analytical solution of the PDEs and ODEs holds a special role in understanding and characterizing the system. For more details information on Fourier analysis and PDEs, interested readers may refer to Gonzalez-Velasco (1996), Hanna and Rowland (2008), Agarwal and O'Regan (2008), Strauss (2007), Asmar (2016), Borthwick (2017), Evans (2022).

References

Agarwal, R. P., & O'Regan, D. (2008). *Ordinary and partial differential equations: With special functions, Fourier series, and boundary value problems*. Springer Science & Business Media.

Asmar, N. H. (2016). *Partial differential equations with Fourier series and boundary value problems*. Courier Dover Publications.

Borthwick, D. (2017). *Introduction to partial differential equations*. Springer.

D'Alembert, J. L. R. (1743). *Traité de dynamique* (Vol. 1). Gauthier-Villars.

D'Alembert, J. L. R. (1750). Recherches sur la courbe que forme une courbe tendue, miseen vibration. *Histoire de l' Académie royale des sciences et belles-lettres de Berlin (HAB) pour l'année, 3*, 214–229.

Evans, L. C. (2022). *Partial differential equations* (Vol. 19). American Mathematical Society.

Fredholm, I. (1903). Sur une classe d'équations fonctionnelles. *Acta Mathematica, 27*, 365–390.

Gonzalez-Velasco, E. A. (1996). *Fourier analysis and boundary value problems*. Elsevier.

Hanna, J. R., & Rowland, J. H. (2008). *Fourier series, transforms, and boundary value problems*. Courier Corporation.

Marquis de Laplace, P. S. (1825). *Traité de mécanique céleste* (Vol. 5). Chez JBM Duprat, libraire pour les mathématiques, quai des Augustins.

Sheehan, J. P., & Debnath, L. (1972). On the dynamic response of an infinite Bernoulli-Euler beam. *Pure and Applied Geophysics, 97*, 100–110.

Strauss, W. A. (2007). *Partial differential equations: An introduction*. Wiley.

Index

S. K. Jena, *Fourier, Laplace, and the Tangled Love Affair with Transforms*,
https://doi.org/10.1007/978-3-031-80165-5

O

P

Q

R

S

MIX
Papier aus verantwortungsvollen Quellen
Paper from responsible sources
FSC® C105338
FSC
www.fsc.org

If you have any concerns about our products,
you can contact us on
ProductSafety@springernature.com

In case Publisher is established outside the EU,
the EU authorized representative is:
Springer Nature Customer Service Center GmbH
Europaplatz 3, 69115 Heidelberg, Germany

Printed by Libri Plureos GmbH
in Hamburg, Germany